Karin Hilck

Lady Astronauts, Lady Engineers, and Naked Ladies

Family Values and Social Change

Edited by
Isabel Heinemann

Volume 5

Karin Hilck

Lady Astronauts, Lady Engineers, and Naked Ladies

Women and the American space community during the Cold War, 1960s–1980s

DE GRUYTER
OLDENBOURG

Published with the kind support of the Walther Blohm foundation and the FAZIT foundation.

ISBN 978-3-11-062615-5
e-ISBN (PDF) 978-3-11-062982-8
e-ISBN (EPUB) 978-3-11-062618-6

Library of Congress Control Number: 2019931200

Bibliographic information published by the Deutsche Nationalbibliothek
The Deutsche Nationalbibliothek lists this publication in the Deutsche Nationalbibliografie; detailed bibliographic data are available on the Internet at http://dnb.dnb.de.

Typesetting: Integra Software Services Pvt. Ltd.
Printing and binding: CPI books GmbH, Leck
Cover image: Ed Walsh Studio, Pittsburgh, in: Jerrie Cobb Collection, Schlesinger Library, Radcliffe Institute, Harvard University.

MIX
Papier aus verantwortungsvollen Quellen
FSC® C083411

www.degruyter.com

Für Papa
Ich vermisse dich.

Martin Hagen
1956–2016

Contents

1 Introduction

> First, I believe that this nation should commit itself to achieving the goal, before this decade is out, of landing a man on the moon and returning him safely to the earth. No single space project in this period will be more impressive to mankind, or more important for the long-range exploration of space; and none will be so difficult or expensive to accomplish. We propose to accelerate the development of the appropriate lunar space craft. We propose to develop alternate liquid and solid fuel boosters, much larger than any now being developed, until certain which is superior. We propose additional funds for other engine development and for unmanned explorations – explorations which are particularly important for one purpose which this nation will never overlook: the survival of the man who first makes this daring flight. But in a very real sense, it will not be one man going to the moon – if we make this judgment affirmatively, it will be an entire nation. For all of us must work to put him there.[1]

This quote from President John F. Kennedy's famous address to Congress in 1961 captured the hearts and minds of the entire American nation and formalized an already existing competition: the space race between the Soviet Union and the United States. The decision to press forward with space exploration in only nine years can be considered as a major game changer for space exploration. Ralph G. Martin even calls Kennedy's speech about landing a man on the moon the 'real birth' of America's space program, because it gave the National Aeronautics and Space Administration (NASA) its clearest mandate to this day.[2] However, the reinvigoration of the space program did not only advance space technology itself, but also the space community as well as the Cold War and humankind across the globe. Nowadays, space technologies, like satellites, are indispensable in our daily life, but already in the 1960s the consequences of this decision could be felt on Earth.[3] New York Times columnist Arthur Krock regarded the American space program for this reason as "the greatest public works and employment operation ever instituted in the US," and historian

1 John F. Kennedy, 'Address before a Joint Session of Congress, May 25, 1961' < http://www.jfklibrary.org/Asset-Viewer/xzw1gaeeTES6khED14P1Iw.aspx> [accessed 1 August 2015].

2 Ralph G. Martin, *A Hero for Our Time. An Intimate Story of the Kennedy Years* (New York: Basic, 1985), p. 305; Mark E. Byrnes, *Politics and Space. Image Making by NASA* (Westport: Praeger Publishers, 1994), p. 40.

3 For more information about NASA spin-offs see: *Johnson Space Center*, 1978, Box GH 28, General Reference Series, General History VI, Johnson Space Center History Collection, University of Houston-Clear Lake (hereafter JSC); *Presentation to Officer's Wives' Club*, October 19, 1971, Box 746, Skylab Series, JSC; *The Family of Mann – Prospects for Progress*, January 1, 1971, Box 079-51, Apollo Series, JSC.

https://doi.org/10.1515/9783110629828-001

Bruce Mazlish compared the Apollo program to the construction of the railroad to illustrate the possible potential consequences for American society.[4] NASA Deputy Administrator Hugh Dryden even called the American space program "an instrument of social change."[5]

Those comparisons already indicate that the American space program was more than just a few daring astronauts flying to outer space. The fierce competitive drive to beat the Soviets in roaming extra-terrestrial bodies had a ripple effect that changed the path and landscape of American society and history in many ways. As Kennedy said, "it will not be one man going to the moon [. . .] it will be an entire nation. For all of us must work to put him there."[6] And the Americans – in various different ways – worked to accomplish the goal their president had set for them. Alongside the military, a new industrial sector, the aerospace industry, was augmented, and provided millions of Americans with jobs in various fields like engineering, mathematics, physics, computing, biology, life science, geology, astronomy, management and many more. New NASA installations and companies were founded and all the scientists, engineers and space workers flocked to their new sites of creation in the 1960s. Most of them brought their families with them. Their arrival created an upswing in housing development. It was not only the housing branch that got a boost from the space program, as the emerging communities also needed health care, education, municipal workers and of course retail providers and workers for the fast growing consumer base.[7]

So-called astrofuturists – writers and consumers of popular science and science fiction – politicians and opinion leaders, as well as the press, also aided the space program by setting the stage for the new American heroes – the astronauts. While the astrofuturists had been preparing the American imagination since the late nineteenth century for the extension of manifest destiny into space, quite a few press outlets were not that concerned with ethical standards of objective reporting and were completely absorbed in the national duty of supporting and defending the United States of America against its archenemy – the Soviet Union. They not only praised the American space program to the skies (or rather to space) and built the astronauts and their wives and families up as celebrity model American citizens, but also assisted in the concealment

4 Amitai Etzioni, *The Moondoggle: Domestic and International Implications of the Space Race* (Garden City: Doubleday, 1964), p. 96; Bruce Mazlish, *The Railroad and the Space Program: An Exploration in Historical Analogy* (Cambridge: MIT Press, 1965).

5 Edwin Diamond, *The Rise and Fall of the Space Age* (Garden City: Doubleday, 1964).

6 Kennedy.

7 Marina Benjamin, *Rocket Dreams* (London: Chatto & Windus, 2003), pp. 1–3.

of misconducts like adultery on the part of America's new "ordinary supermen" – the astronauts.[8]

Apparently, Americans of different occupational, social, ethnic and educational backgrounds and from various locations in the country pooled their know-how and passion and paved the way for the American victory in the space race – landing a man on the moon and returning him safely to the earth in 1969. They even built new communities surrounding this new goal in space exploration and augmented, dissolved, merged and transformed already existing communities which had aspired to explore and conquer space long before the Cold War had dictated the pace and course of not just American space exploration, but also of all space endeavours worldwide. These consortia of people who consecrated their lives to the exploration of space, hereafter called the space community in this study, can be considered as a sort of magnifying glass on American society during the Cold War. Considering that the American space program, and therewith the lion's share of the space community, was directly fashioned in response to the Sputnik Shock of October 4, 1957, the space community demonstrates how Americans lived and worked within changing social spaces in the second half of the twentieth century, navigating through, contributing and reacting to the falling and rising tides of the Cold War and the social denouements within American society.

The history of space exploration has been extensively written about. Especially the achievements and triumph of America's daring and brave astronauts have been celebrated, but credit has also been given to the creativity of the pioneers of space exploration and to the victories of robotic space exploration. One thing all the histories have in common: they mainly record the history of white men. Considering how many people within the space community worked toward the goal of triumphing over the Soviet Union in the space race and advancing the United States in space research, it seems unlikely that this achievement was solely accomplished by the few celebrated men. Of course,

8 The term "ordinary supermen" was used by an Air Force general who described what they were looking for in the United States' first astronauts. See: Scott M. Carpenter and others, *We Seven: By the Astronauts Themselves* (New York: Simon and Schuster, 1962), p. 7. However, before the introduction of the astronauts to the public, the press was not too friendly with NASA, complaining about the loss of rockets and payloads as well as the Soviet Union's lead over NASA in the space race. Harlen Makemson, *Media, NASA, and America's Quest for the Moon* (New York: Peter Lang Publishing, 2009), pp. 46–49; James L. Kaufman, *Selling Outer Space. Kennedy, the Media, and Funding for Project Apollo, 1961–1963* (Tuscaloosa: The University of Alabama Press, 1994), pp. 49–70; Dario Llinares, *The Astronaut: Cultural Mythology and Idealised Masculinity* (Newcastle: Cambridge Scholars Publishing, 2011), pp. 26–59; Lily Koppel, *The Astronaut Wives Club* (New York: Gran Central Publishing, 2013), pp. 52–67.

there was a great number of people supporting the celebrated few or trying to become one of them. These people rarely show up in the history books due to the nature of the traditional and most popular writing form within the field of history – political history. Furthermore, women and minorities oftentimes are not featured in works of historical writing because they had been dismissed as historically important figures due to their lack of power and/or membership in an undesirable ethnic or race group or, in the case of women, due to their assignment to the private sphere. Although women were encouraged to leave the private sphere of the home and join the workforce during time of war in the first half of the twentieth century, women went back to homemaking after the end of the war and enjoyed their status as housewives and mothers according to the prevalent postwar discourse. However, the happy housewife image of the 1950s and early 1960s has to be understood as a "cultural fantasy," not as a historical fact, because the postwar discourse about women was much more complex than women returning home to house and kids like the stereotype of postwar women might suggest.[9]

So far, historical narratives mainly emphasize the conservatism that shaped women and their identities. Women are portrayed as enthusiastic homemakers who were happy to go back to their domestic roles, which were infused with a national purpose and advertised as careers on the part of the state via guidance from experts who offered advice in every area of life. In this postwar narrative women wanted to stay in the private sphere because they regarded "the home as the bulwark against the dangers of the atomic age."[10] True enough, a lot of women did want to go back to the life they had before the war. However, women in the Cold War have so far mainly been discussed as passive symbols, not as active agents. Yet, there were spaces for women with a different idea of life and work. Although the Cold War induced social order often restricted women from pursuing their goals outside the home and organizing more efficiently because they had to be on the defense all the time, the postwar narrative about women and the sociocultural coding of femininity as well as masculinity has to be expanded and differentiated. The situation was much more complex than women being forced back to the private sphere by the likes of the

9 Joanne Meyerowitz, *Not June Cleaver: Women and Gender in Postwar America, 1945–1960* (Philadelphia: Temple University Press, 1994), p. 2; Helen Laville, *Cold War Women. The International Activities of American Women's Organisations* (Manchester: Manchester University Press, 2002), p. 2.

10 Elaine Tyler May, 'Explosive Issues: Sex, Women, and the Bomb', in *Recasting America: Culture and Politics in the Age of Cold War*, ed. by Lary May (Chicago: University of Chicago Press, 1989), pp. 154–70 (pp. 161–162).

seemingly undisputed leader of the household, like the stereotype of postwar women and men might suggest. The Cold War competition demanded a great deal of dedication on the part of all American citizens – including the women.

Therefore, the Cold War rhetoric and lifestyle provided areas of active agency for women and promoted women's participation in the public sphere to keep up with the Soviet Union, while at the same time promoting the necessity for American women to stay in the private sphere and remain a blissful homemaker due to the correlation drawn by the U.S. government between the domestic fulfillment of women and the success of the American economic and cultural way of life. Evidently, women had the opportunity to be active agents in various social spaces before the forming of the women's (liberation) movement changed the emphasis of the Cold War debate about women and officially created more agency for women. Thus, American society did not transition from approaching women as passive symbols to approaching them as active agents, but used parallel approaches according to the needs of the nation during the Cold War. Nevertheless, the postwar narrative tends to downplay women's agency and does not particularly challenge the gendered configuration of the public/private division, favoring instead the established discourse about the public/private social order in the area of American history as well as in the field of Cold War relations.[11]

It is the goal of this gender history to challenge the established postwar narrative of women, or rather gender relations, in the history of American society and add more social spaces, where women could exercise agency, to the discourse. Using the American spaceflight program as a unique entry point into the field, the book wants to find out what the sociocultural coding of masculinity and femininity in the space community can tell us about the mindset, cultural structures and social spaces of American society during the Cold War. Thus, the space community, a place where technology, military, science, culture and politics in the United States intersect, serves as a lens into American society during the Cold War, thereby building upon Glen Asner's application of social history to the field of space history, expanding it to gender history. According to Asner,

11 Suzanne Le-May Sheffield, 'Women's Advancement in Science since WWII', in *Gender in Science Workbook*, ed. by Jennifer Kasi Jackson (Dubuque: Kendall/Hunt Publishing, 2006), pp. 3–20; Ruth Rosen, *The World Split Open. How the Modern Women's Movement Changed America* (New York: Penguin Books, 2000); Meyerowitz; Laville, *Cold War Women. The International Activities of American Women's Organisations*; Helen Laville, 'Gender and Women's Rights in the Cold War', in *The Oxford Handbook of the Cold War*, ed. by Richard H. Immerman and Petra Goedde (Oxford: Oxford University Press, 2013), pp. 523–39.

> [r]ather than a side note in space history, the social history of NASA can be understood as a central element of U.S. history if we discuss it in the context of the federal government's expanded role in science, education, and technological development in the twentieth century, and with references to the individuals, communities, and social groups that defined their own existence and their ambitions in the shadow of the space program.[12]

The book aims to bring less famous people and social groups into the analysis of space history. This means not only recognizing their existence in the space community, but acknowledging them as active forces instead of casting them aside as passive subjects. The analysis will mainly focus on women in space. Like Marie Lathers in her book *Space Oddities* about fictional representations of women in space, this book understands women who did not travel to space but were linked to space – e.g. through their husbands or as precursors – as "representations of women in outer space."[13] Lather's approach can reveal more active female agency not only in popular culture representations of women in outer space, but also within the space community and divert the emphasis of discourse from the astronauts – the most visible members of the space community. It is the ambition of this project to write about women in space, or rather to make space for women in space. However, the history of women in space will be contextualized, and not just added to the political canon of space history.[14]

Although the primary focus of analysis in the research will be on the women of the space community, the project is interested in the (changing) sociocultural coding of femininity as well as masculinity because "[f]emale bodies are constructed against the backdrop in which male bodies are accepted as the norm, an inscription process shaped by the masculine context of space travel."[15] Thus, the cultural coding of femininity is always assigned and performed in interplay with the sociocultural coding of masculinity due to the prevalent gender belief system in American society. It is

> a set of beliefs and opinions about males and females and about the purported qualities of masculinity and femininity. This system includes stereotypes of women and men,

12 Glen Asner, 'Space History from the Bottom Up: Using Social History to Interpret the Societal Impact of Spaceflight', in *Societal Impact of Spaceflight*, ed. by Steven J. Dick and Roger D. Launius (Washington, D.C.: National Aeronautics and Space Administration Office of External Relations History Division, 2007), pp. 387–406 (p. 392).

13 Marie Lathers, *Space Oddities Women and Outer Space in Popular Film and Culture, 1960–2000* (London: Continuum International Publishing, 2010), p. 13.

14 Asner, p. 397.

15 Monica J. Casper and Lisa Jean Moore, 'Inscribing Bodies, Inscribing the Future: Gender, Sex, and Reproduction in Outer Space', *Sociological Perspectives*, 38 (1995), 311–33 (p. 316).

> attitudes toward the appropriate roles and behaviors of women and men, and attitudes toward individuals who are believed to differ in some significant way from the modal pattern.[16]

This book wants to analyze the gender belief system in the space community and examine the gender performances by NASA employees and other people involved in the space sector in order to fathom and broaden the history of the male connotated institution that is NASA. In this regard, it is important to note that this project follows the notion of gender history, assuming that gender is socially and politically constructed and mapped onto the binary, constructed biological sex.[17] However, men and women are not just completely defined and stereotyped as man or woman through the gender belief system arisen from public discourse – masculinity, femininity, sex and gender are not static. Men and women are actively involved in performing their respective gender identities, consolidating but also often challenging and re-negotiating them.[18] This book wants to explore how the sexes have interacted with each other and performed their roles in the space community in different decades during the Cold War, using gender as a useful tool for analysis, therewith opening space history up to re-interpretation.

Because, "[o]nce women begin to demonstrate their independent existence, males too become gendered. Men are forced to re-examine themselves, and the comforting image of a stronger, more creative, more rational sex breaks down."[19] It is exactly this process of reexamination, and the response of the men to the women acting as active and passive agents in the space community, that will be of interest in this book. This is relevant with regard to the fact that the sociocultural coding of masculinity in the space community was very much influenced by the code of conduct of the armed forces because a lot of the early members of the space community were drafted directly from the different

---

16 Kay Deaux and Mary E. Kite, 'Thinking about Gender', in *Analyzing Gender. A Handbook of Social Science Research*, ed. by Myra Marx Ferree and Beth B Hess (Newbury Park: Sage Publications, 1987), pp. 92–117 (p. 97).

17 Anne Cranny-Francis and others, *Gender Studies. Terms and Debates* (London: Palgrave Macmillan, 2003), pp. 1–7; Mary M. Talbot, *Language and Gender* (Cambridge: Polity, 2010), pp. 7–14; Angelika Wetterer, 'Konstruktion von Geschlecht: Reproduktionsweisen Der Zweigeschlechtlichkeit', in *Handbuch Frauen- Und Geschlechterforschung*, ed. by Ruth Becker and Beate Kortendiek (Wiesbaden: VS Verlag für Sozialwissenschaften, 2008), II, 126–36 (pp. 126–136).

18 Judith Butler, *Gender Trouble: Feminism and the Subversion of Identity* (New York: Routledge, 1999), p. xv; Sonya Rose, *What Is Gender History?* (Cambridge: Polity Press, 2010); Joan W. Scott, 'Gender: A Useful Category of Historical Analysis', *The American Historical Review*, 91 (1986), 1053–75.

19 Brian Attebery, *Decoding Gender in Science Fiction* (New York: Routledge, 2002), p. 7.

military branches. These men usually displayed a particular set of gendered attributes, such as aggression, rationality, and a sense of invulnerability.[20] Although the gender practice in the military differs by rank, it stands to reason that these military masculinities contributed to the challenge with which women were faced when exercising agency in the various social spaces within the space community, because the reproduction of military norms of conduct not only promotes a particular type of masculinity "but a masculinity linked to a specific family structure, a specific home, a specific femininity and sexuality."[21]

In order to discover what the (changing) sociocultural coding of masculinity and femininity can tell us about the mindset, cultural structures as well as social spaces of the space community and ultimately American society during the Cold War, this book intends to look at several different groups of women interacting in different social spaces within the space community and the responses and reactions they were faced with from fellow women and the men of the community. Instead of presenting an exhaustive account of all the women in the space community, I chose a selection of representative examples from different social spaces in the space community. Considering this book's definition of the space community, which will be discussed in the next chapter, a comprehensive account of all female members of the community would be an impossible task to tackle. However, the selected examples will give insight into the several layers of female participation in the community and will function as representations. Before these selected examples are analyzed, it is necessary to focus on the concept of the space community for a bit in order to put the examples into proper context.

The following chapter, *Mapping the space community*, will build the conceptual framework for this book. It intends to identify and locate the different groups who might belong to the space community, such as astronauts, NASA and contractors' employees, scientists, engineers, secretaries, politicians, space enthusiasts, astrofuturists, space families, the press, the military, academia,

---

20 Paul R. Higate, '"Soft Clerks" and "Hard Civvies". Pluralizing Military Masculinities', in *Military Masculinities: Identity and the State*, ed. by Paul R Higate (Westport: Praeger, 2003), pp. 27–42 (p. 29).

21 Jodi Dean, *Aliens in America. Conspiracy Cultures from Outerspace to Cyperspace* (Ithaca: Cornell University Press, 1998), p. 86; Higate, p. 30; Edith A Disler, *Language and Gender in the Military. Honorifics, Narrative, and Ideology in Air Force Talk* (Amherst: Cambria Press, 2008); Deborah Harrison and Lucie Lalibeté, 'Gender, the Military, and Military Family Support', in *Wives and Warriors. Women and the Military in the United States and Canada*, ed. by Laurie Weinstein and Christie C. White (Santa Barbara: Praeger, 1997), pp. 35–53 (pp. 36–37).

independent researchers, industry and connections to related communities such as the international space community, the science community or American society. The chapter will put the groups in relation to one another and detect their inherent commonalities and differences. Doing so will provide this project with a definition and understanding of 'space community.' Furthermore, it will study the space community's stance regarding the Cold War and, more importantly, its role in shaping and contributing to it. However, the main objective of the chapter is to ascertain if women had access to, and room for agency, in the space community. This allows for a discussion in the following chapters about how far women were a recognizable presence in the community and what their jobs, roles, and contributions say about gender relations as well as contemporary social norms and values in the space community and American society.

Following the conceptual section, the selected examples of different groups of women who participated in the space community in one way or another will be introduced. The different cases will not be presented in a chronological order, but rather in a topical one. Each chapter will look into a different social space within the space community. A social space is a physical or virtual space produced through social practices, patterns, and relationships, as well as material conditions, and shapes and is shaped by cultural values and economic prospects. Socially produced space is perpetuated and reproduced on a daily basis through the performance and execution of administrative policies, technological systems, and social conventions. Because social spaces can be considered the cornerstones to daily life, it is important to know that they are neither ubiquitous nor static. They are socially produced and subject to manipulation and change.[22] Researching the women of the space community within the social spaces in which they perform and come under scrutiny or admiration instead of undertaking a chronological analysis of their work, public, and private life, gives us greater insight into the practices and patterns that create the sociocultural coding of masculinity and femininity in the space community. Ultimately, the analysis of the women in varying social spaces of the space community will provide us with a profound insight into the mindset, and cultural as well as social structures of American society during the Cold War.

Thus, each chapter will discuss a specific social space. However, each chapter will still address chronology and outline the women's position within

22 Henri Lefebvre, *The Production of Space* (Oxford: Blackwell, 1991); Michel Foucault, 'Andere Räume', in *Stadt-Räume*, ed. by Martin Wentz (Frankfurt a. M.: Campus, 1991), pp. 65–72; *The People, Place & Space Reader*, ed. by Jen Jack Gieseking and others (New York: Routledge, 2014).

the social space and the community over time as social practices and relationships develop. Furthermore, each chapter will consider the women's relationship with other groups of women as well as the men with whom they shared the social spaces in the community. The use of language will be taken into consideration as well because language can function as a gatekeeper for the community. All the chapters will also look into the concept of race and into how far the question of race was relevant to the (prospective) women of the community.

The first chapter to discuss a selected group of women will introduce the most visible women in the space community to the readers. It will present the displaying of space community women and their supporters and adversaries in public. The chapter *Media space* will thus explore the stories and the discourse surrounding all the women in the space community who were regularly featured in the national and international media, such as the first female Shuttle astronauts in the late 1970s and 1980s and the astronauts' forerunners, namely Valentina Tereshkova and the so-called Mercury 13 in the early 1960s. Although the women of the Mercury 13 never made it to space, they put the idea of women travelling to space into people's heads and on the national agenda. It is important to look at these American trailblazers in order to put the first female astronauts into context within the tradition of American thinking about a female presence in space. Moreover, it is necessary to highlight the Soviet cosmonaut Valentina Tereshkova in this regard. As the first woman worldwide to fly to space, she not only stirred the gender norms in the space community but fueled a transnational Cold War discourse about gender norms. Furthermore, the chapter will analyze the role of Christa McAuliffe, a teacher set to fly to space but tragically died in the explosion of the Space Shuttle Challenger, in the public discourse of the 1980s in order to find out if the death of a woman in a space accident changed the public perception of female participation in spaceflight and contributed to a change of the gender norms in the United States. To retrace a possible change in the discourse from the 1960s to the 1980s, it is important to also devote attention to other groups of women in the public space, like some female NASA scientists who were regularly featured in the media or the wives of NASA's first astronauts. Although these women never wanted or planned to pierce the space frontier or become space celebrities, they were very much in the spotlight and added another quality to the discourse about women and the space community. The chapter will also look into female social innovators of the 1970s, who publicly tried to change administrative and social structures within the space community.

The next chapters intend to shed light on the women who were not as visible as those who were regularly featured in the media. They show the individual perspectives of women who worked in the space community's offices, labs, and other workspaces. Although rarely featured in the public space, these women are just as important to the space community and the decoding of the gender norms in the space community. Thus the chapters *Office space*, *Lab space*, and *Work space* will take a closer look at overlapping groups of women in different workspaces and their interactions with each other and with male colleagues in the 1960s and 1970s. They will look into the situation of women who worked in the space industry or for NASA as engineers, scientists, artists, nurses, clerks, secretaries, writers or administrators. The chapters will analyze the role of these women in the space community and their contribution to the stagnation or transformation of the sociocultural coding of gender relations, especially with regard to the Cold War, herewith testing the hypothesis that women who worked for the space effort on the ground were regarded (and treated) differently by their male and female colleagues in the community than the women who wanted to fly to space.

Another group who worked hard to give women agency in the space community will be analyzed in the next chapter – the staff of the Equal Employment Opportunity Office at NASA. The chapter *Equal space* is interesting regarding the role of women and gender norms in the space community because it looks at the official federal and NASA guidelines in matters of gender equality in the 1970s and the people in the community who tried to implement them. After public and individual perspectives, the chapter adds an institutional level of analysis to the book. The chapter helps to determine if the space community was an early or late adopter of federal guidelines, or even a trailblazer or denier of equality.

Finally, the conceptual part of the second chapter will be brought together with all the different examples of social spaces in the space community where women and men work, perform, and interact with each other. This analysis might lead to a profound insight into the social and cultural structures and gender relations in the space community, potentially illuminating different times and social spaces within the community and possibly back up the hypothesis that American society did not transition from approaching women as beauty queens to approaching them as active agents, as astronauts, but used parallel approaches according to the needs of the nation during the Cold War. Thus, this systematic analysis of the connection between gender, space, and the Cold War could add a new dimension to space history as well as expand the discourse in American history about gender relations and the opportunities of women in the twentieth century.

2 Mapping the space community

According to the online Oxford Dictionaries, space is “the physical universe beyond the earth’s atmosphere,” “the near-vacuum extending between the planets and stars, containing small amounts of gas and dust.”[1] Although space is an actual place, space is also much more to a variety of people. Space is not just an outer earthly place but also a canvas for dreamers, visionaries, and idealists who ascribe metaphorical meanings such as opportunity, freedom or infinite possibilities to this canvas. Space is an area of national and personal pride, a (potential) military battleground, a rhetorical battleground and a field of work and business for many trades, as well as a part of political strategies.[2] Therefore,

> space is not just a place, but an umbrella concept for a whole set of ideas, technologies, activities, and aspirations having to do with the exploration and utilization of that larger environment in which our planet floats, and with the often unstated hope of journeying outward to have new experiences, open up new possibilities, and escape the limitations and frustrations of life on Earth. It arches over an amazing variety of interests, political views, and personalities.[3]

Included under the umbrella concept of space are, for example, idealists who use the concept of space with the intention to improve conditions on Earth. They feel humankind has evolved into one single vulnerable community on the ‘spaceship Earth’ which should work together in order to preserve their spaceship, their planet, from steering off course towards destruction.[4] Contrary to environmental space idealists, other space dreamers are drawn to the idea of destruction, or rather a military presence, in space because achievements in space symbolize “the technological power and organizing capacities

---

1 ‘Space’ <http://www.oxforddictionaries.com/de/definition/englisch/space> [accessed 31 August 2015].

2 Michael Sheehan, *The International Politics of Space* (New York: Routledge, 2007), pp. 1–7.

3 Michael A. G. Michaud, *Reaching for the High Frontier. The American Pro-Space Movement 1972–84* (New York: Praeger, 1986), p. xx. Furthermore, the meaning of space can also change at different times and in different geographical and sociocultural places. Linda Billings, ‘Cultural Functions of Space Exploration’, in *Remembering the Space Age*, ed. by Steven J. Dick (Washington, D.C.: National Aeronautics and Space Administration, 2008), pp. 409–12 (p. 409).

4 Barbara Ward, *Spaceship Earth* (New York: Columbia University Press, 1966); R. Buckminster Fuller, *Operating Manual for Spaceship Earth*, ed. by Jaime Snyder (Baden: Lars Müller Publishers, 1968).

https://doi.org/10.1515/9783110629828-002

of a nation."[5] They like to imagine and talk about space to "symbolically invoke military might and threats without explicitly naming them."[6] Another group of space enthusiasts comes from a more economical way of thinking. They allude to the business of space, emphasizing that not one single (tax) dollar of the money used for space endeavors has ever been spent in space but solely on earth.[7] However, most space dreamers and visionaries imagine space to be a new, grand playground with infinite possibilities for exploration, discovery, and conquest. Outer space is a site to envision meetings with other species, a potential place for a new home, a space to seek and find glory, or see visions of the future come to life with cutting edge new technology.[8]

All of the visionaries have one thing in common, regardless of how and why they project their wildest dreams onto the extraterrestrial canvas of space – imagination. It does not matter which part of the macro-concept of space appeals to those who do the imagining, they all have to do just that – imagine space. Along with funding, imagination is the most important necessity for spaceflight.[9] Humankind's leap into the heavens has been imagined for as long as there have been humans on this Earth. Their visions and fantasies about a future where humans venture forth into the cosmos helped to create and shape all the space programs worldwide, along with their technologies. However, the people doing the imagining are all creatures of their own cultures, brimming with stereotypes, molded by their history and constricted by societal rules about what is possible and proper. Hence, there can be no imagining the great unknown that is space without the dreamer's culture and era exerting an influence; imagination and

---

5 James E. Webb and Robert McNamara to the Vice President, May 8, 1961 in: John M. Logsdon, *Exploring the Unknown. Selected Documents in the History of the U.S. Civil Space Program. Vol. I: Organizing for Exploration* (Washington, D.C.: NASA History Office, 1995), pp. 443–444.

6 Alexei Kojevnikov, 'The Cultural Spaces of the Soviet Cosmos', ed. by James T. Andrews and Asif A. Siddiqi (Pittsburgh: University of Pittsburgh Press, 2011), pp. 15–27 (p. 22).

7 Howard E. McCurdy, *Space and the American Imagination* (Washington: Smithsonian Institution Press, 1997), p. 207.

8 W. Patrick McCray, *The Visioneers: How a Group of Elite Scientists Pursued Space Colonies, Nanotechnologies, and a Limitless Future* (Princeton: Princeton University Press, 2013); De Witt Douglas Kilgore, *Astrofuturism. Science, Race, and Visions of Utopia in Space* (Philadelphia: University of Pennsylvania Press, 2003); Tom D. Crouch, *Aiming for the Stars. The Dreamers and Doers of the Space Age* (Washington, D.C.: Smithsonian institution Press, 1999); Barbara Marx Hubbard, *The Hunger of Eve. One Woman's Odyssey toward the Future* (Eastsound: Island Pacific NW); Michaud; Frank H. Winter, *Prelude to the Space Age. The Rocket Societies 1924–1940* (Washington, D.C.: Smithsonian Institution Press, 1983); Jodi Dean.

9 Walter A. McDougall, *The Heavens and the Earth. A Political History of the Space Age* (New York: Basic Books Publishers, 1985), p. 20.

culture feed on each other symbiotically, thus complementing one another. Therefore, it is important to note that space, and thereby the implementation of space, the space programs and spaceflight, "is a manifestation of culture, a product of culture, but it is also embedded in culture. The influences travel both ways, and it is well to recognize this at the outset" in order to understand that there are limitations even to imagination and that it is not easy to overcome these restrictions.[10] Accordingly, the two-way circuit of culture also applies to the people and composition of the space community, especially in regard to gender performance.

Considering that space is an umbrella concept arching over many areas and forms of imagination and activity, reaching from imperial conquest to peaceful exploration, from national ventures to private enterprises or international endeavors, it is quite difficult to map the community, or rather communities, included under it. Whereas it is considerably simpler to map the actual traffic between Earth and space and the technology and people needed for these space ventures, it is admittedly harder to map flights of the imagination, where dreams, aspirations and hopes pave the way for future endeavors. However, according to Marina Benjamin, "their course seems at least as logical and has evolved faster than that of technology."[11]

For the mapping of the space community it is helpful to emphasize the conceptual distinction between space as a mission, as a calling, or a vision to be achieved in some grand futuristic adventure, and space as a program in pursuit of defined goals. The difference is a subtle one, but should be recognized because the different conceptualizations of space determines, at least in part, the public's as well as the community members' attitudes and behaviors toward space and membership in the space community. MacDaniel thus emphasizes that when

> we think of space as a mission, we can be comfortably vague and general in terms of what we do about it or want others to do. [...] If we regard space as a place in which specific activities are to be carried out, however, we are forced into more concrete lines of thinking. We are forced to develop a plan for achieving the space mission.[12]

---

10 Steven J. Dick, 'Space, Time, and Aliens. The Role of Imagination in Outer Space', in *Imagining Outer Space. European Astroculture in the Twentieth Century*, ed. by Alexander C. T. Geppert (New York: Palgrave Macmillan, 2012), pp. 27–44 (p. 29).

11 Benjamin, p. 4.

12 William E. MacDaniel, *Space is a Place, not a Mission*, Newsletter of the Space Settlement Studies Project Niagara University, Fall 1981, GH 58, Speeches & Articles, General Reference Series, JSC.

Thus, vaguely defined visions of the utilization of space are so general and inclusive in their way of envisioning that they can stay vague about the membership of the space community, unlike the space program which cannot stay as vague but has to draw clear lines while trying to be as inclusive with its membership as possible in order to maintain support and funding, which in return influences the opportunities and possibilities of women in the space community.

Keeping in mind that space is an umbrella concept, there are a lot of possible space community compositions, and thus aspiring space community members, due to the different ideas and concepts of space. In order to research to what extent women had or gained access to space related work and visions, it would be essential to know the build-up and membership of the community. The literature, and even the community members themselves, quite often refer to equivocal terminology like "the fraternity," "the inner circle," "spaceworld," "space science and exploration community," "the extended family," "community of space boosters," "space culture," "NASA family," or "space community" without further explanations.[13] Mary Wiegers, writing for the Washington Post, for example, once stated:

> 'The space program is very hungry. It devours people.' Astronaut Wally Shirra once remarked. Besides devouring between \$20 billion and \$40 billion, the goal of putting a man on the moon by 1970 has devoured the time, energy, emotional and family life of thousands who work in the space community.[14]

Her concept of the space community in this article is mainly build around NASA's human spaceflight program including people who worked for NASA or other space companies which were NASA contractors, as well as their families. While she apparently includes family members of NASA and its contractors'

---

13 Harriett G. Jenkins, interview by Jennifer Ross-Nazzal, August 5, 2011, p. 44 and 63, edited Oral History Transcript, NASA Headquarters Oral History Project, JSC; Kathy Sawyer, '73 Seconds of Silence for Challenger's Crew', *Washington Post*, January 29, 1987; William E. Burrows, *This New Ocean. The Story of the First Space Age* (New York: Random House, 1998), pp. 434, 438, 439, 463, 479, 506; Howard E. McCurdy, 'Has Spaceflight Had an Impact on Society? An Interpretative Framework', in *Societal Impact of Spaceflight*, ed. by Steven J. Dick and Roger D. Launius (Washington, D.C.: National Aeronautics and Space Administration Office of External Relations History Division, 2007), pp. 3–17 (p. 221); Cathleen Lewis, 'From the Kitchen into Orbit. The Convergence of Human Spaceflight and Krushchev's Nascent Consumerism', in *Into the Cosmos Space Exploration and Soviet Culture*, ed. by James T. Andrews and Asif A. Siddiqi (Pittsburgh: University of Pittsburgh Press, 2011), pp. 213–39 (p. 215).

14 Mary Wiegers, 'Space Community: Emphasis on Excellence', *Washington Post*, September 13, 1969. See also: John D. Odegard, *Letter to Eilene Galloway*, 19.09.1986, Galloway, Eilene, 000747, NASA Historical Reference Collection, Washington, D.C.

employees, she omits the general visionaries and all other people who are involved with space but not in the field of human spaceflight. In a speech addressing all NASA employees, Ruth Bates Harris, Deputy Director of the NASA Equal Employment Opportunity Office, once stated: "All of our employees must be made to feel a part of the time – of the NASA family, no matter where they work or for how much."[15] She used the popular term 'NASA family' to describe the space community, or at least a part of it. According to her, all people working for NASA were part of the community – including administrators, secretarial and menial workers, and scientists who worked in other areas of the space program besides human spaceflight. Barbara Morgan, an educator astronaut, issued a statement after the Challenger accident: "To the NASA Family, All of us in the Teacher in Space Project share your grief over the loss of the Challenger crew. Christa McAuliffe was our friend, but she had also become part of the NASA family. Like a family we have shared a common sadness."[16] She also referred to the NASA family. However, her understanding of the 'NASA family' does not conform to Harris' definition. She also included the families of the NASA employees as well as people who had applied to NASA but were rejected, like the Teacher in Space Project Candidates, in her definition of the NASA family, of the space community. What is interesting, though, is that both definitions of 'family' did not refer to the American Cold War nuclear family concept of father, mother and kids but to a culturally superceded extended family concept of previous centuries.

The same term, like NASA family or space community, can therefore be used to describe differing ideas about the space community, while two different terms might refer to the same idea about space community membership. All the terms used to describe the space community may refer to the same people and activities or overlapping groups of people, but they all refer somehow to people involved with space in one way or another. There are apparently different concepts of community at play when it comes to the ambiguous 'space community.' In a 1955 review of ancient and modern literature about community, George A. Hillery identified 94 meanings for the word 'community.' They all differed in meaning due to different concepts,

15 Ruth Bates Harris, 'Mrs. Harris calls on NASA employees to work together', *Newsletter: NASA Activities*, January 15, 1972, GR 1701, NASA Information Summaries, NASA Activities, General Reference Series, JSC.

16 'Teacher program continues', *Space News Roundup*, February 28, 1986. See also: 'In Memoriam', *Space News Roundup*, February 7, 1986; Michael Collins, 'Riding the Beast', *Washington Post*, January 30, 1986. The use of the term 'NASA family' in a space community context was particularly popular in times of grief arising from space related accidents.

paradigms, and measures of competing disciplines. He also observed that around 70 percent of the definitions drew on geography, shared interests and identities. However, all of them included a social relationship between the community members in one way or the other.[17] Thus, community "is generally understood as a very particular kind of social network that is made up of tightly bound solidarities and densely knit ties of reciprocity."[18] In the past, community was mainly understood as a tight-knit, face-to-face primary group. Location, such as a neighborhood, was an important factor for the community. In this sense, Gerard Delanty defined community as "a sense of belonging based on shared experiences, a common language, kinship ties, and above all, inhabiting a common spatial life world."[19] Accordingly, Larry Lyon composed the following definition of community: "people living within a specific area, sharing common ties, and interacting with one another."[20] These geography based definitions would definitely cover some parts of the space community, but would leave out other groups who felt they were part of the space community as well.

In today's world, community has become a pretty vague concept. People constantly refer to community, in this case to the space community, without reflecting on the concept, like, for example, John D. Odegard, the dean of the University of North Dakota, in 1986. In a letter to the 'Grand Dame of Space' Eilene Galloway, Odegard thanked her for her "insight into the inner workings of the space community."[21] The reason for the constant use of the community concept without further reflection is that it is not a concept to us anymore. Community is a word that feels good and is full of promise for the people 'in the community' and makes people 'outside of the community' curious. Community is an idea manifest in our doxa – the collective knowledge people think with but do not think about.[22] However, the implementation of the meaning of the term can still vary from person to person, despite the term's manifestation in the doxa.

---

17 George A. Hillery, 'Definitions of Community, Areas of Agreement', *Rural Sociology*, 20 (1955), 111–23.

18 Tony Blackhawk, *Key Concepts in Community Studies* (London: Sage, 2010), p. 75.

19 Gerard Delanty, *Community* (London: Routledge, 2003), p. 55.

20 Larry Lyon, *The Community in Urban Society* (Long Grove: Waveland Press, 1999), p. 5.

21 John D. Odegard, *Letter to Eilene Galloway*, September 19, 1986, 000747, NASA Historical Reference Collection, Washington, D.C. See also: Linda Billings, 'Eilene Galloway, the Woman Who Helped Create NASA, Dies at Age 102', 2009 <http://www.nasa.gov/topics/history/features/galloway_obit.html> [accessed 17 March 2016].

22 Zygmunt Baumann, *Gemeinschaften: Auf Der Suche Nach Sicherheit in Einer Bedrohlichen Welt*, (Frankfurt a. M.: Suhrkamp, 2009).

2.1 A short history of the international space community

This chapter now tries to identify and locate the different groups who might belong to the space community in order to put them in relation to one another and determine what kind of relationship they stand in with regard to each other. The purpose of the mapping is to provide this project with an understanding of 'space community' in order to assess the position of women in the space community and analyze their access and agency in the community.

2.1.1 The bifurcated birth of astronautics – science fiction

Considering that the golden age of space travel occurred long before anyone or anything actually travelled to space in the first half of the twentieth century, it is quite enlightening to look into the history of astronautics and the formation of the space community. Unlike most sciences, astronautics originated from an art form or at the very least has two closely intertwined beginnings. One beginning is the theoretical treatises and engineering work of Konstantin Tsiolkovsky, Robert Goddard, Herman Oberth, Max Valier, Robert Esnault-Pelterie, Walter Hohmann and later Werner von Braun, who, to some extent, were inspired by the output of the other beginning – the art of space inspired fiction by Edward Everett Hale, Jules Verne, Kurt Lasswitz, and H.G. Wells.[23] Their imagination and the ideas of their successors basically paved the way for space travel, in preparing society of the wonder, glory and fear of the approaching space age.[24] The bifurcated birth of astronautics makes the community surrounding space unique but therefore also harder to map, especially that part of it consisting of writers, artists, and their audiences producing and consuming science fiction. While both beginnings "convey a certain amount of innocence, if not naiveté," the start of science fiction was especially carefree.[25]

23 Ron Miller, 'Spaceflight and Popular Culture', in *Societal Impact of Spaceflight*, ed. by Steven J. Dick and Roger D. Launius (Washington, D.C.: National Aeronautics and Space Administration Office of External Relations History Division, 2007), pp. 501–12; W. D. Kay, *Defining NASA. The Historical Debate over the Agency's Mission* (Albany: State University of New York Press, 2005), p. 27.

24 Emily Rosenberg, 'Far Out: The Space Age in American Culture', in *Remembering the Space Age: Proceedings of the 50th Anniversary Conference*, ed. by Steven J. Dick (Washington, D.C.: National Aeronautics and Space Administration. Office of External Relations, 2008), pp. 157–84 (p. 159); Robert Heinlein, 'Science Fiction: Its Nature, Faults, and Virtues', in *The Science Fiction Novel. Imagination and Social Criticism*, ed. by Basil Davenport (Chicago: Advent, 1969), pp. 14–48 (p. 46).

25 Kay, p. 27.

The prolific writer Jules Gabriel Verne wrote over one hundred stories about extraordinary journeys. Five of them featured his favorite theme – space flight. His novels *From the Earth to the Moon* and *Around the Moon*, published in 1865 and 1870, are deemed to be two of the most significant novels for the genre of science fiction as well as astronautics. In his research for his novels Verne consulted more than 500 reference books in order to create a scientifically plausible spaceflight story. Although a lot of his ideas could not be proven to be scientifically sound, like the flight with a giant cannon shell, they were inspiring to his readers who were drawn to the fictional part of the story as well as to those readers who were interested in the scientific part of his novels. In 1869 Edward Everett Hale imagined, in his novel *the Brick Moon*, two big flywheels to fling the first Earth satellite into space. Hale and Verne, as well as numerous other science fiction writers, such as H. G. Wells, inspired a lot of idealists and aspiring engineers all over the world to figure out the problems humankind was faced with when travelling to the stars.[26]

Science fiction writers dealt with quite a few problems of spacecraft engineering long before the engineers and they also solved a lot for them.[27] They were astrofuturists – writers and visionaries who imagine and write about space exploration based on scientific facts and not just mere fantasy.[28] However, accuracy in science fiction writing, including screen plays for movies and TV-shows, considerably lessened over time.[29] Philip Francis Nowlan's science fiction pulp story *Buck Rogers* from 1928, for example, was pure fantasy.[30]

At the beginning of the twentieth century, especially after World War I, all forms of modern science became synonymous with progress and people from all over the world and all walks of life were interested in the sciences that would take them to the future. A popular way to express one's curiosity and visions about the future in the United States, especially during the interwar period, was to consume and contribute to, for example, Hugo Gernsback's science-based pulp magazines *Amazing Stories* and *Science Wonder Stories*. During this time, science fiction took shape as a distinct genre at the intersection of popular

26 Herbert R. Lottmann, *Jules Verne: An Exploratory Biography* (New York: St. Martin's Press, 1996); Adam Mann, 'Strange Forgotten Space Station Concepts That Never Flew', 2012 <http://www.wired.com/2012/01/space-station-concepts/> [accessed 27 April 2016]; Kay, p. 27.

27 Ron Miller, p. 507; Tom D. Crouch, pp. 19–38; Winter, *Prelude to the Space Age. The Rocket Societies 1924–1940*, pp. 19–21.

28 Emily Rosenberg, pp. 158–59.

29 However, cinematic science fiction normally lags about 30 years behind the visions of written science fiction. See: Attebery, p. 174.

30 Ron Miller, p. 508.

science and popular culture. The astrofuturists gladly embraced science fiction as one of the few ways to circulate their ideas about space flight and the conquest of the stars before the rise of the space program.[31]

Science fiction was (and up to this day, still is) a very social and participatory genre. Science fiction readers, writers, editors, and fans formed personal but not necessarily direct relationships and bonded in social networks via local clubs and letters to the magazine or to each other in order to exchange their ideas about stories, progress and the future. Some people read the pulp magazines for science and others for adventure, but the majority of them believed that the stories held imaginative potential and a progressive disposition. Through the pulps, they were able to bond across countries and continents in order to imagine science together. Thus, at the beginning of the space age, imagining space was done within an international community.[32] Whoever wanted to join and share his or her vision was welcome to do so. However, John Cheng emphasizes that the "variety of their communications and interactions was both more and less personal than community in a conventional sense, especially in their mediated withdrawal into smaller, private social circles."[33] Even though the science fiction community of the interwar period cannot be classified as a community in a traditional sense, it can be regarded as networks of various future oriented interest groups of whom quite a few were especially interested in space and the technologies to bring humankind to space. Going through the pulp magazines of that period, it therefore cannot come as a surprise to find various announcements of the founding of space and rocket clubs. The American Interplanetary Society (later the American Rocket Society), the British Interplanetary Society as well as the German Verein für Raumschiffahrt (Society for Space Navigation) announced their existence in science fiction magazines.[34]

Although such societies did not do lobby work according to today's understanding of lobbying, these groups can be regarded as a pre-space age space lobby and pillars of the American Space program because they raised money for their cause like today's NGOs and spread their gospel of space using sensational science fiction terminology. They were all very inspired by and thus interested in science fiction stories and used them as a

31 John Cheng, *Astounding Wonder. Imagining Science and Science Fiction in Interwar America* (Philadelphia: University of Pennsylvania Press, 2012), pp. 1–7; Kilgore, p. 81.

32 Cheng.

33 Cheng, p. 77.

34 Cheng, p. 216.

means to connect to each other and potential supporters. However, they were very serious in their ambition to conquer space, because for them science fiction was the first step toward space science.[35] The American Interplanetary Society's secretary Nathan Schachner, for example, courted the readers of *Amazing Stories, Astounding Stories of Super Science* and *Wonder Stories* and offered active and associate membership for "lovers of science fiction" who wanted "a chance to assist in the bringing to realization of the dream of all interplanetary travel" in the summer of 1931, one year after the formation of the American Interplanetary Society (AIS).[36]

Especially AIS can be regarded as an offspring of science fiction. All AIS members were science fiction buffs and nine of the twelve founding members were writers or editors of *Science Wonder Stories*.[37] Even Hugo Gernsback joined AIS. Although they took their dream of flying to space very seriously, the emphasis of AIS was more propagandistic and less technological in the beginning. They wanted to spread their space vision while embracing and welcoming all people who thought alike in order to promote the space effort and make it a reality in the near future.[38] In the beginning of AIS, the members did not even need a scientific background to join the society but just had to be enthusiastic about spaceflight and the future. However, AIS wanted to gain recognition and endorsement from academics and luminaries of the field such as Robert H. Goddard, an American engineer, professor, physicist, and inventor who is often credited as the founding father of American rocketry. The man who had launched the world's first liquid-fuelled rocket in 1926 was very private and neither interested in sharing his ideas publicly nor becoming the posterchild of AIS, but although he did not want to get involved with the society directly, he did become a member.[39] Other academics in the early 1930s were also apprehensive about joining and endorsing AIS publicly because they did not want to risk their professional reputations by being involved

35 Cheng, pp. 252–253; Winter, *Prelude to the Space Age. The Rocket Societies 1924–1940*, p. 15; Kay, pp. 28–29.

36 Cheng, p. 251.

37 G. Edward Pendray, David Lasser, Charles P. Manson, Charles W. VanDevander, Fletcher Pratt, Nathan Schachner, Laurence E. Manning, William Lemkin, Adolph L. Fierst, Everett Long (or Roy A. Giles – the record is not clear) as well as Leatrice M. Pendray and Warren Fitzgerald – a woman and an African American – founded AIS in April 1930. See: Winter, *Prelude to the Space Age. The Rocket Societies 1924–1940*, p. 73.

38 Kilgore, p. 35.

39 Eventually AIS and Goddard parted ways due to disagreements. Tom D. Crouch, p. 61.

with the seemingly improbable and odd cause of space travel advocated principally by science fiction fantasisers.[40]

Besides academic attention, AIS was also seeking international bonds with fellow space enthusiasts. AIS was always monitoring what the other societies across the Atlantic were up to, especially the Germans. The German, British, and Soviet societies also kept an eye on the activities and publications of the American group in return. According to a 1932 AIS Bulletin, the society's membership was not restricted to the United States but was widespread across 21 states and nine foreign countries including Russia. There was an active exchange between certain members of the societies, too. They kept each other up to date by mail and even visited each other. Gawain Edward and Leatrice Pendray from AIS, for example, visited Willy Ley, vice-president of the Verein für Raumschiffahrt (VfR), in Berlin and were able to watch their rocket experiments first hand. Although neither the Pendrays nor Ley were fluent in the language of their counterparts, they were very excited and satisfied with the exchange of vision and knowledge.[41] Even though national pride was most definitely already evident, some American and European society members hoped for a universal space effort, representing all humankind. The president of AIS, Davis Lasser, for instance, proposed the formation of an International Interplanetary Commission and a Committee on Experiments in 1932 because he could "foresee the building of the first space ship only as a joint effort of a united earth."[42] The desire to launch an international space effort shows that space exploration, astrofuturism and thus the space community, was international and open-minded in its roots and outlook and willing to include everyone who was interested in the endeavour and willing to invest, emotionally and/or financially, in space exploration. The necessary astronomical capital needed for a space expedition was one of the main reasons to imagine space exploration as a joint international effort.[43] During the interwar years, in which governments did not show interest in the conquest of space at all, the astrofuturists of no country could imagine that space was to become an area of national pride. Thus, the roots of astrofuturism can be considered within the framework of interwar internationalism; however, through the

40 Tom D. Crouch, pp. 59–71; Winter, *Prelude to the Space Age. The Rocket Societies 1924–1940*, pp. 74–76; J.D. Hunley, 'The Enigma of Robert H. Goddard', *Technology and Culture*, 36 (1995), 327–50; David A. Clary, *Rocket Man: Robert H. Goddard and the Birth of the Space Age* (New York: Hyperion, 2003); Burrows, pp. 44–54; Chris Gainor, *To a Distant Day. The Rocket Pioneers* (Lincoln: University of Nebraska Press, 2008), pp. 36–52.

41 Winter, *Prelude to the Space Age. The Rocket Societies 1924–1940*, pp. 73–85.

42 Winter, *Prelude to the Space Age. The Rocket Societies 1924–1940*, p. 77.

43 Winter, *Prelude to the Space Age. The Rocket Societies 1924–1940*, p. 75.

commitment to the space race it became an American phenomenon.[44] But before the American government committed to the space project after World War II, the space community seemed to be open to members of all classes and backgrounds, even if they were predominantly Caucasian.

Were AIS, and the other societies for that matter, open to women, too? There was no official rule against women as equal society members; however, women were not specifically targeted in the societies' lobby work and not a lot of women participated in the societies. Out of the ten founding members of the German VfR only one was female, Hedwig Bernhard.[45] AIS's composition was very similar. Leatrice M. Pendray was the only woman among the twelve founders of the American space society.[46] While the German society was mainly comprised of engineers, which was an occupation mostly closed to women in the early twentieth century, the foundation of the American society was science fiction. This was an art form open to everyone – at least in principle. Yet, not only engineering and science, but also science fiction, were dominated by white men – within and outside of the texts. Because science fiction was devised, or at least made famous, by white men, science fiction is a white male product, too. It extents a white male elitist vision into space. As creatures of their own culture, white male Americans used their contemporary cultural forms and habits, also in regard to gender practices, as a baseline for their otherwise mind bending stories with complex elements. While science fiction would be *the* genre to break through known gender patterns because it allows, or rather aspires to, alternative visions in a different world, until the 1960s habits regarding gender were often transcribed unthinkingly or were knowingly used as a familiar pattern in an otherwise alien world.[47] Like the male heroes of the American space program, the heroes of science fiction also needed an audience to witness and to attest to their acts of heroism.[48] The function of most female characters in science fiction stories was to counterbalance the masculine ideals of the male heroes with a female domestic sensibility. In order to affirm the white male protagonist as the hero of the story, this particular gender dynamic was needed; the affirmation of the male as the hero required the presence of a

---

44 Kilgore, p. 3.

45 Winter, *Prelude to the Space Age. The Rocket Societies 1924–1940*, p. 35.

46 Winter, *Prelude to the Space Age. The Rocket Societies 1924–1940*, p. 73.

47 Kasi Jackson, 'Gender, Sex, and Sexuality in Speculative Fiction', in *Gender in Science Workbook*, ed. by Kasi Jackson (Dubuque: Kendall/Hunt Publishing, 2006), pp. 135–36 (p. 135); Kilgore, p. 28; Attebery, p. 5.

48 A more detailed explanation about the role of women as the witnesses of the American space effort can be found in the third chapter of this book. See also: Jodi Dean, pp. 69–81.

woman to witness his heroic acts.[49] As Brian Attebery remarked, part of the trouble with this particular narrative gender dynamic is that it

> generates the notion of a super male [but] offers no extrapolative path toward superwoman. Exaggerate the traits that the megatext associates with masculinity and you get the stronger, smarter, faster, more aggressive, more inventive superman of S[cience] F[iction] tradition. Exaggerate the feminine traits and you get someone who erases herself from the story. The female function, according to Darwin and Spencer, is to follow, to accommodate, to receive, to deny self."[50]

Considering that the women's function in science fiction texts was not to add their own flavor of personality and ideas to the mix but to deny self in order to accommodate all requests and desires of the male protagonist and to attest to his lion-hearted acts of heroism, it is quite interesting to note that this was not necessarily the function of women in the science fiction community ouside of the texts. Even though there were not many female members of the interwar science fiction community – either as consumers or producers – women did participate in discussions in the pulp magazines and publish their own science fiction stories in them. However, most of them only used their initials in order to disguise their gender.[51] Women of that period were just as much creatures of their own culture as men and thus did not push to turn over a new leaf in regard to the gender dynamic in science fiction, but instead followed the narrative path of their male counterparts. Therefore, it can be assessed that women did have access to the science fiction community in the interwar years but they did not push for an enhanced female presence in the community and science fiction.

Considering that the space societies of the 1930s were the precursors of the space age lobby, women seemingly had access to this particular element of the basis for the American space community. However, the science fiction based AIS eventually wanted to do more experimental work on rockets and spaceflight, and started to shift its standards regarding membership and course of action accordingly. In April 1934, the American Interplanetary Society rebranded itself and adopted a new name – *the American Rocket Society* (ARS). According to the society, the name change was a means to be taken more seriously by the academic, engineering and

49 Cheng, p. 112.
50 Attebery, p. 82.
51 Cheng, p. 117.

aeronautical community, but did not intrinsically imply that ARS had given up on the "interplanetary idea."[52] However, ARS started to be more selective regarding its membership because they began to be more concerned about how the actions of individual members would reflect upon the society. Being enthusiastic about space was no longer enough to become a member of ARS. Starting in the mid to late 1930s active members needed a scientific background in order to participate.[53] This prerequisite was a further obstacle to the participation of women due to women's difficult access to and exclusion from the natural sciences. However, it was not an insurmountable hurdle, especially in respect to the fact that ARS was in the end not very important for the physical development of astronautics and space exploration in the United States.[54] It produced almost nothing that was later used in the American space effort. Even Goddard, the father of American rocketry, was not able to leave his mark on present-day American rocketry. He was "on a branch that died."[55] The postwar American space program was mainly built with German technology. The vision for American space exploration, however, was provided by the German and American rocket societies alike.

2.1.2 The bifurcated birth of astronautics – rocketry

The Verein für Raumschiffahrt and the American Rocket Society were not the only space advocacy groups worldwide or even in their respective countries before the end of World War II. Quite a few space minded societies were founded in the 1930s all over the world.[56] Especially the Soviet Union had very active space societies. Along with Germany, the Soviet Union was the most energetic and innovative place to be regarding rocket research due to the many Soviet

52 Winter, *Prelude to the Space Age. The Rocket Societies 1924–1940*, p. 83. The society kept its new name up until its merge with the Institute of the Aerospace Sciences to form the American Institute of Aeronautics and Astronautics.

53 Cheng, pp. 289–90.

54 See chapters four, five, and six.

55 Theodore von Káraman and Lee Edson, *The Wind and Beyond* (Boston: Little Brown, 1967), p. 240. See also: Cheng, p. 298; Tom D. Crouch, p. 70.

56 There were the Cleveland Rocket Society, the Peoria Rocket Association, the International Cosmos Science Club, the Yale Rocket Club, the Junior Rocket Society, the Westchester Rocket Club, the Bernard Smith's California Rocket Society, Robertson Youngquist's Massachusetts Institute of Technology Rocket Research Society and the Springfield Society for Rocket Experimenters in the United States. In Germany there were other astronautical organizations like the

innovators and the government's and respected scientists' openness regarding space research.[57] The *zeitgeist* had paved the way for the imagining and engineering of the rocket in these two very different nations.[58] The Soviet interwar space societies laid the groundwork for the later space age successes and also produced the leaders of the Soviet space program like Sergei Korolev, despite obstructions in the form of the Stalinist purges in 1938 and 1939 and the resulting death and/or imprisonment of society members such as Korolev and Valentin Glushko.[59] Politics, culture, economy and technology laid the foundation for space travel in the Soviet Union and Germany despite the countries'

Verein für Fortschrittliche Verkehrstechnik e.V. (Society for Progress in the Technology for Traffic), the *Gesellschaft für Raketenforschung* (Society for Rocket Research), the *Studiengesellschaft für Raketen e.V.* (Study Society for Rockets), the *Raketenforschungsinstitut* Dessau (Dessau Rocket Research Institute), the *Deutsche Versuchsanstalt für Luftfahrt* (German Research Institute for Aeronautics), the *Gesellschaft für Raketenforschung Hannover* (Society for Rocket Research Hannover), and the *Gesellschaft für Weltraumforschung e.V.* (Society for the Exploration of Space). There were also other societies spread out in Austria, Argentina, Great Britain, Holland, France, and Japan. Winter, *Prelude to the Space Age. The Rocket Societies 1924–1940*, pp. 49–50, 99–111; Cedric Giles, 'The Rocket Societies', *Astronautics*, 60 (1944), 12–13 (pp. 12–13); Burrows, pp. 55–57.

57 The Central Bureau for the Study of the Problem of Rockets, the All-Union Association for the Study of Interplanetary Communications, The Gas Dynamics Laboratory, the Moscow based Group for the Study of Reaction Motion, the Leningrad based Group for the Study of Reaction Motion, and the Rocket Research Institute, which had consumed all the other rocket research societies due to Stalinist bureaucracy, can attest to that.

58 Tom D. Crouch, pp. 41–45.

59 Winter, *Prelude to the Space Age. The Rocket Societies 1924–1940*, p. 64. For more information about the Soviet rocket societies, the Soviet space program and its reception in the Soviet Union, see: Yevgeny S. Shchetinkov, 'Development of Winged Rockets in the USSR 1930–1939', in *First Steps toward Space*, ed. by Frederick C. Durant and George S. James (Washington, D.C.: Smithsonian Institution Press, 1974), pp. 247–57; Anatoli Blagonravov, 'Spaceship Designer Sergei Korolev', *Space World*, 1968, 9–11; Michael Stoiko, *Soviet Rocketry. Past, Present, and Future* (New York: Holt, Rineheart and Winston, 1970); G.A. Tokaty, 'Foundations of Soviet Cosmonautics', *Spaceflight*, 10 (1968), 335–46; James T. Andrews and Asif A. Siddiqi, *Into the Cosmos: Space Exploration and Soviet Culture* (Pittsburgh: University of Pittsburgh Press, 2011); Asif A. Siddiqi, *Challenge to Apollo: The Soviet Union and the Space Race, 1945–1974* (Washington, D.C.: National Aeronautics and Space Administration NASA History Division, 2000); Cathleen Lewis, 'The Red Stuff: A History of the Public and Material Culture of Early Human Spaceflight in the U.S.S.R' (George Washington University, 2008); Ilina Kohonen, 'The Space Race and Soviet Utopian Thinking', ed. by David Bell and Martin Parker (Malden: Blackwell Publishing/The Sociological Review, 2009), pp. 114–31; Slava Gerovitch, 'Creating Memories. Myth, Identity, and Culture in the Russian Space Age', in *Remembering the Space Age*, ed. by Steven J. Dick (Washington, D.C.: National Aeronautics and Space Administration

different outlooks in all these areas. Even though the values and beliefs of the two nations were quite divergent, there are striking similarities between the German and Soviet space pioneers. In both countries, the societies provided an experimental test environment for postwar space achievements. In both countries, the societies started out in the early 1930s pursuing the dream of space travel with the means of rocketry but were prevented from following their vision by 1935 because the respective armies utilized the rocketeers for militaristic purposes. In both countries, the early post war design and launch teams included several interwar society members, and in both countries a society member ascended to become an exceptional leader in the opening of the space age – while Korolev was this person for the Soviet Union, Wernher von Braun filled the same role for Germany and the United States.[60]

The engineering student Wernher von Braun started out as a member of the VfR. Although he was not one of the ten founding members, he became one of the most enthusiastic members in 1930, three years after the society was founded.[61] Like other space enthusiasts around the globe, von Braun had also been inspired by the fictional works of Jules Verne and Kurt Lasswitz as well as Hermann Oberth's technical book *Die Rakete zu den Planetenräumen* (The Rocket to the Planet Spaces). The interwoven concepts of hard science and science fiction would become one of von Braun's most successful means to make spaceflight a reality in the postwar United States. However, in the space societies of interwar Germany science fiction did not play an important part, unlike its American counterpart. But just like in the United States, women were not an integral part of the space society. Although women were not specifically excluded from joining the VfR and one of the founding members actually was a woman, neither contemporary records nor the secondary literature show the participation of women in the society's favorite past time – the construction and testing of various rockets. Von Braun, on the other hand, took part in numerous experiments at the Raketenflugplatz (rocket airfield) in Berlin. The VfR always operated financially on the brink of collapse, especially after the consequences of the Great Depression deepened. Therefore, the society always had to be on the lookout for donors. However, the society's members were not particularly interested in public funding because

External Relations, 2008), pp. 203–36; James J. Harford, *Korolev: How One Man Masterminded the Soviet Drive to Beat America to the Moon* (New York: John Wiley and Sons, 1997).

60 Winter, *Prelude to the Space Age. The Rocket Societies 1924–1940*, p. 64.

61 Founding members were: Max Valier, Willy Ley, Johannes Winkler, Georg Lau, Alfons Jakubowicz, Gerhard Guckel, Walter Neubert, H. Sauter as well as Herbert Fuchs and Theodor Fuhrmann, both clergymen, and the only woman, Hedwig Bernhard. Winter, *Prelude to the Space Age. The Rocket Societies 1924–1940*, p. 35.

most of them were convinced that bureaucracy and government red tape would only meddle with their work. So when the German military showed interest in liquid propellant rockets, they turned to the VfR but were not impressed with what they witnessed. Army Captain Walter Dornberger concluded that "rocketry was a sphere of activity beset with humbugs, charlatans, and scientific cranks, and sparsely populated with men of real ability."[62] He was impressed with von Braun, though, whom he considered energetic, shrewd and knowledgeable. Von Braun became the first VfR member to sign a contract with the Army to work on the new German rocket program in October 1932. Only the machinist Heinrich Grünow accompanied him to the Army's secret testing facilities in Kummersdorf and from 1937 on in Peenemünde. Between 1934 and 1937 fellow VfR members Klaus Riedel, Kurt Heinisch, Helmut Zoike, Hans Hüter and Hermann Oberth joined von Braun and Grünow at Peenemünde after the government's crackdown on civilian rocket experiments. Although Oberth was the superior rocket theoretician, von Braun had established himself as the chief rocket scientist because he had been the only one to recognize early on that only the government had enough money to realize the dream of space travel.[63]

Von Braun left the comfort and familiarity of the German and international spaceflight community in order to pursue his goal of building rockets for the German military, which did not want him to continue his involvement with the VfR but to concentrate fully on his military contract. His desire for individual distinction as well as his eagerness to achieve his goal led von Braun to subordinate everything else. According to Willy Ley, von Braun was not interested in politics. Apparently von Braun considered the Weimar Republic "no good" and the Nazis "ridiculous."[64] He was an opportunist and took his chance to build large scale rockets despite the knowledge that his creation would be weaponized. After all, his employer was the German Army. Von Braun and his team created the infamous *Vergeltungswaffe 2* (Vengeance Weapon 2), better known as the V2, and previously known as *Aggregat 4* (A4) before Propaganda Minister Joseph Goebbels changed its name. The full-scale liquid fuel rocket was the first manmade object to pierce the Kármán Line and enter space.[65] Equipped

---

62 Frank H. Winter, *Rockets into Space* (Cambridge: Harvard University Press, 1990), p. 46.

63 Winter, *Prelude to the Space Age. The Rocket Societies 1924–1940*, pp. 51–54; Tom D. Crouch, pp. 72–77.

64 Willy Ley, 'Count von Braun', *Journal of the British Interplanetary Society*, 6 (1947), 154–56 (p. 154).

65 Tom D. Crouch, p. 82.

with warheads, the V2 was used to target major European cities such as London and Antwerp. Starting in 1944, 3,200 rockets were launched, mainly to terrorize civilians. Even though working for one's government in times of war can be seen as a patriotic duty, von Braun and his rocket team not only started to work for the German Army before the war started but also made use of all equipment and possibilities with which the Nazis provided them – including forced labor. Concentration camp inmates from five different camps built 5,975 rockets on von Braun's orders during World War II. Most of the rockets were assembled by inmates of the concentration camp Mittelbau-Dora in the underground facilities of the Mittelwerk Company in Kohnstein.[66] Von Braun's rocket team certainly did not establish the concentration camps, but neither did they object to the utilization of concentration camp inmates in order to get closer to their dream of space travel. One member of the rocket team, Arthur Rudolph, even suggested the use of forced labor and the other rocketeers did not disagree. Their visits to the V2 production and assembly facilities can attest to that. Von Braun himself visited Mittelwerk several times.[67]

It is important to note the German rocket engineers' background because many of them would establish themselves as visionaries within the core of the American postwar space community. The German rocketeers pursued their careers during the Third Reich, an era shaped by war, genocide, forced labor, and supposedly 'traditional German' values such as racial superiority and the ideal of women being stay at home moms of Aryan children – concepts they would most likely bring with them to the United States.

The German rocket team did not end up as a pillar of the American space community by chance; as early as January 1945 von Braun was

66 Before 1944 the production of the V2 was accomplished at different construction sites. The inmates of Buchenwald concentration camp worked at Peenemünde while the inmates of Dachau concentration camp worked in Friedrichshafen in the Zeppelinwerke. Meanwhile the inmates of Mauthausen concentration camp labored in Vienna and the inmates of Sachsenhausen concentration camp worked for DEMAG-Panzer in Falkensee. Rainer Eisfeld, *Mondsüchtig. Wernher von Braun Und Die Geburt Der Raumfahrt Aus Dem Geist Der Barbarei* (Fernwald: zuKlampen, 2012), p. 106 et seqq.; *Konzentrationslager Mittelbau-Dora: 1943–1945. Begleitband Zur Ständigen Ausstellung in Der KZ-Gedenkstätte Mittelbau-Dora*, ed. by Jens-Christian Wagner (Göttingen: Wallstein Verlag, 2007), p. 49 et seqq.

67 Michael J. Neufeld, *The Rocket and the Reich: Peenemunde and the Coming of the Ballistic Missile Era* (Washington, D.C.: Smithsonian institution Press, 1995), p. 187; Tom D. Crouch, p. 90. Von Braun became a member of the Nazi Party (NSDAP) in 1937 and joined the ranks of the SS in 1940 after a personal invitation from Heinrich Himmler, the commander of the SS (*Schutzstaffel* – Protection Squadron). He received three promotions in the SS and was a major by the end of the war.

preparing for a life after World War II. The rocket team had realized that Germany would lose the war and they wanted to surrender only to the Americans because they "despise[d] the French", were "mortally afraid of the Soviets" and did "not believe that the British [could] afford" them.[68] By the end of the war in Europe, on May 8, 1945, von Braun and his team as well as their families had crossed Germany and were captives of the U. S. Army in Bavaria. On July 19, 1945, Operation Overcast was initiated by the U.S. Joint Chiefs of Staffs to recruit German scientists and engineers in order to remove their know-how and talent from Germany, but especially to remove them beyond the reach of the Soviet Union. The scientists and engineers were brought to the United States and assigned to the different branches of the U.S. military under the codename *Project Paperclip*. Von Braun and his team were sent to Fort Bliss, Texas, to work with the U.S. Army.[69]

2.1.3 Post-war astrofuturism

NASA was not founded until 1958, but the development of the American space community continued right after World War II was over. It had started with the science fiction based American Interplanetary Society, which had become the American Rocket Society during the interwar years. The society kept its new name up until it merged with the Institute of the Aerospace Sciences in 1963 to form the American Institute of Aeronautics and Astronautics (AIAA).[70] The American Rocket Society, and later AIAA, flowered into a great organization. In 1950, ARS had more than 17,000 members in its various regional chapters spread out all over the U.S. and grew to become AIAA, the world's largest technical

68 Dieter Huzel, *Peenemünde to Canaveral* (Englewood Cliffs: Prentice Hall, 1962), p. xx.

69 Wayne Biddle, *Dark Side of the Moon. Wernher von Braun, the Third Reich, and the Space Race* (New Yok: W. W. Norton & Company, 2012); Bob Ward, *Dr. Space: The Life of Wernher von Braun* (Annapolis: Naval Institute Press, 2005); Michael J. Neufeld, *Von Braun: Dreamer of Space, Engineer of War* (New York: Alfred A. Knopf, 2007); Neufeld, *The Rocket and the Reich: Peenemunde and the Coming of the Ballistic Missile Era*; Ulli Kulke, *Weltraumstürmer. Wernher von Braun Und Der Wettlauf Zum Mond* (Berlin: Quadriga, 2012); Annie Jacobsen, *Operation Paperclip: The Secret Intelligence Program That Brought Nazi Scientists to America* (New York: Back Bay Books, 2015); Eisfeld; Monique Laney, *German Rocketeers in the Heart of Dixie. Making Sense of the Nazi Past During the Civil Rights Era* (New Haven: Yale University Press, 2015); Gainor, pp. 70–88.

70 Winter, *Prelude to the Space Age. The Rocket Societies 1924–1940*, pp. 82–83.

society dedicated to aerospace. AIAA was established before NASA existed and continued to grow alongside NASA. Today it has 30,000 individual members from 88 countries, and 95 corporate members.[71] Besides AIAA, new space societies were founded from coast to coast, even after the establishment of NASA. Wernher von Braun founded the National Space Institute (NSI) in 1974, and Carolyn and Keith Henson founded the L5 Society in 1975 to promote space colonization. In 1987 the two organizations merged to become the National Space Society (NSS).[72] To this day they promote their vision of "people living and working in thriving communities beyond the Earth, and the use of the vast resources of space for the dramatic betterment of humanity."[73] International space societies and meetings also started up again after World War II. From 1950 on, the International Astronautical Federation Congresses attracted great numbers of participants, demonstrating the international interest in space travel and the need for and viability of private space societies.[74]

Groups fond of science fiction also continued and newly established themselves in the U.S. after World War II, such as the Hydra Club, an organization of science fiction writers, editors, artists and fans. Among its members were several women, for example, Judith Merril, who was the chairwoman of Hydra's Permanent Membership Committee and one of the earliest female science fiction authors; Inga Pratt, who was an artist and book illustrator; Bea Mahaffey, who was an editor for *Science Stories* and *Universe Science Fiction*; Katherine MacLean Dye, who was a science fiction writer; Margaret Bertrand, Evelyn Harrison, Lois Miles Gillespie, who were all fierce science fiction fans; and Olga Ley and her husband Willy Ley, the former VfR members who had fled national socialist Germany in 1935.[75] According to an article in *Life* magazine in 1951,

71 <http://www.aiaa.org/AboutAIAA/> [accessed 17 February 2016]; Winter, *Prelude to the Space Age. The Rocket Societies 1924–1940*, p. 117.

72 Linda Billings, 'Overview: Ideology, Advocacy, and Saceflight: Evolution of a Cultural Narrative', in *Societal Impact of Spaceflight*, ed. by Steven J. Dick and Roger D. Launius (Washington, D.C.: National Aeronautics and Space Administration Office of External Relations History Division, 2007), pp. 483–500 (pp. 488–490); Michaud; McCray.

73 'NSS Statement of Philosophy', *NSS* <http://www.nss.org/about/philosophy.html> [accessed 17 February 2016].

74 Winter, *Prelude to the Space Age. The Rocket Societies 1924–1940*, p. 117.

75 Gerald Jonas, 'Judith Merril, 74, Science Fiction Editor and Writer', *The New York Times*, 1997 <http://www.nytimes.com/1997/09/17/arts/judith-merril-74-science-fiction-editor-and-writer.html>; Judith Merril, 'The Hydra Club', *Marvel Science Fiction*, November (1951), 128–29; Dave Kyle, 'The Legendary Hydra Club', 2000 <http://www.jophan.org/mimosa/m25/kyle.htm>

there were as many as two million science fiction fans in the United States in the early 1950s and quite a few of them were female. *Life* magazine dubbed them 'fenne.' The article also pointed out that many of the consumers and producers of science fiction were scientists such as Isaac Asimov, Eric Temple Bell (aka John Taine) or Robert Richardson, and that all the authors like Robert Heinlein or Ray Bradbury who did not have a professional scientific background absorbed more scientific literature in the course of their research than some practicing scientists.[76] Thus, American science fiction continued postwar to harbor and nourish space enthusiasts of various backgrounds, be they women, German refugees or German scientist warmongers like Wernher von Braun, mixing the science and science fiction communities.

Von Braun had realized that only through the harnessing of public opinion through the science fiction community could the dream of space travel be accomplished, because without the public in their corner, America's space pioneers would not receive enough money from Congress to make space flight a reality. At the first Annual Symposium on Space Travel at the Hayden Planetarium in New York on October 12, 1951 organized by Willy Ley and planetarium director Robert Coles, von Braun and Fred Whipple of the Harvard Observatory convinced two writers for *Collier's* magazine of the importance of space travel for the future of the world and the United States in particular. *Collier*'s published eight articles which appeared between March 22, 1952 and April 30, 1954, complete with illustrations by Chesley Bonestell, Fred Freeman, and Rolf Klep. The first article announced to the world that it

> may be one of the most important scientific symposium ever published by a national magazine. It is the story of the inevitability of man's conquest of space. What you will read here is not science fiction. It is serious fact. Moreover it is an urgent warning that the U.S. must immediately embark on a long-range development program to secure for the West 'space superiority'. If we do not, someone else will. That somebody very probably would be the Soviet Union.[77]

The *Collier*'s articles inspired other magazines such as *Time, Look* or *This Week* to also pick up the topic of space travel. In the course of the *Collier*'s articles von Braun rose to fame. Working for the Army by day and moonlighting for the

[accessed 18 February 2016]; Judith Merril and Emily Pohl-Weary, *Better to Have Loved: The Life of Judith Merril* (Between the Lines(CA), 2002); Damon Knight, *A Century of Science Fiction* (New York: Simon & Schuster, 1962), p. 151; Jared Scott Buss, 'Willy Ley, the Science Writers, and the Popular Reenchantment of Science' (University of Oklahoma, 2014).

76 Winthrop Sargeant, 'Through the Interstellar Looking Glass', *Life*, 1951, 127–40.

77 Cornelius Ryan, 'Man Will Conquer Space Soon', *Collier's*, 1952, pp. 23–40 (p. 23).

promotion of spaceflight by night, von Braun additionally wrote several books and collaborated with Walt Disney in 1954. Disney wanted to publicize its magic kingdom via a television show on *ABC*, which would introduce the park's main themes of Fantasyland, Frontierland, and Tomorrowland. Von Braun was contracted to advise and star in the Tomorrowland section, turning him into one of the most recognizable figures of the up and coming space age.[78] It was a win-win situation for von Braun. He was able to push for the realization of space travel and reached millions of young Americans, while Disney provided von Braun with American patriotic cachet. He needed the latter because from the very beginning space flight was sold to the public as the last frontier, the new frontier, and the final frontier – a continuation of American manifest destiny from the American West right into space.[79] According to Fred Whipple, von Braun was "one of the best salesmen of the twentieth century" because he could convert anyone into a supporter of space flight.[80] Therefore, DeWitt Douglas Kilgore determined that von Braun's most important contribution to spaceflight was not as an engineer but as a politician who could sell the dream of travelling to space to different audiences using different angles; from manifest destiny, to an ever-expanding capitalism, to the gospel of space.[81]

Because von Braun was so flexible in the framing of his vision, using any available support to make spaceflight a reality, Kilgore assumes that there was no core to his vision. With or without a core, von Braun and his fellow members of the German rocket team functioned as the catalyst for postwar astrofuturism in the United States.[82] Although the roots of astrofuturism were international, the commitment to space and especially the space race turned it into an American phenomenon, ironically with Germans at its center. Because of the popularity of American astrofuturism, which was embedded in nineteenth-century Euro-American thinking about imperialism and utopian speculation, and which projected the American cultural narrative of frontier pioneering, continual progress, manifest destiny, free enterprise, rugged individualism, and a right to life without limits into space, the rhetoric of space advocacy had to comply with the set parameters of astrofuturism irrespective of the changing social, political,

78 Tom D. Crouch, pp. 116–122; Makemson, pp. 3–18.

79 Kilgore, p. 59; Benjamin, pp. 39–40; McCurdy, *Space and the American Imagination.*

80 Makemson, p. 9.

81 Kilgore, p. 57.

82 Kilgore, pp. 49–57.

economic, and cultural context in the United States during the Cold War.[83] The goal of astrofuturism, to leave Earth behind and conquer new space in the depths of space in order to realize the full potential of the American project, had a grave impact on the twentieth-century American space community with the National Aeronautics and Space Administration at its center. Astrofuturist ideas dictated the community's course of action and provided a conception of its membership.

2.1.4 The Sputnik shock

Even though astrofuturism was on the rise in postwar America, there was another school of thought regarding space exploration courting the public's as well as the government's attention. While von Braun and his team dreamed of humans flying to space, quite a few scientists argued that robotic space exploration would be much more sensible due to its reduced costs and absence of danger to human life. President Eisenhower was one of their main supporters. He preferred a cheaper space program of satellites and automated probes based on science and technical arguments. Eisenhower thus approved a plan to orbit an artificial satellite as part of the International Geophysical Year (IGY) in 1957. However, on October 4, 1957 the Soviets launched the world's first artificial satellite, Sputnik I, as their contribution to the IGY 1957, beating the Americans to the historical record. Sputnik took most Americans by surprise and had a massive, Pearl Harbor like effect on American public opinion. Most people felt that a technological gap had developed between the Soviet Union and the United States, and demanded a technological sprint in order to keep up with the Soviets. The fear that the Soviets would control space, winning the higher ground in a future war, provided the impetus for the creation of NASA, as well as increased aerospace spending and research, and placed the United States in a space race with the Soviet Union.[84]

---

83 Billings, 'Overview: Ideology, Advocacy, and Saceflight: Evolution of a Cultural Narrative', p. 438; Kilgore, pp. 1–11; Taylor E. Dark, 'Reclaiming the Future: Space Advocacy and the Idea of Progress', in *Societal Impact of Spaceflight*, ed. by Steven J. Dick and Roger D. Launius (Washington, D.C.: National Aeronautics and Space Administration Office of External Relations History Division, 2007), pp. 555–72.

84 Roger E. Bilstein, *Orders of Magnitude. A History of NACA and NASA, 1915–1990* (Washington, D.C.: Government Printng Office, 1989); Stephen J. Garber and Roger D. Launius, 'A Brief History of the National Aeronautics and Space Administration', in *National Aeronautics and Space Administration (NASA): Background, Issues, Bibliography*, ed. by C.V. Anderson (New York: Nova Science Publishers, 2002), pp. 3–12.

Despite Eisenhower's fear and warnings that a space race would force the United States into becoming a technocracy, neglecting constructive, level-headed scientific progress and ignoring democratic ideals in favor of a dramatic space opera, the American public led by politicians of both parties pushed toward a space race. The public and politicians were convinced that the winners of the space race would be the ones dominating the world and therefore could not understand and were very discontent with Eisenhower's space alternative. The call for a strong American presence that could beat the Soviets in space reached near-hysterical proportions which Eisenhower could not comprehend. After his tenure as president he remarked that the space race had "diverted a disproportionate share of our brain-power and research facilities from equally significant problems, including education and automation."[85] He also asked his fellow Americans:

> Why the great hurry to get to the Moon and the planets? We have already demonstrated that in everything except the power of our booster rockets we are leading the world in scientific space exploration. From here on, I think we should proceed in an orderly, scientific way, building one accomplishment on another, rather than engaging in a mad effort to win a stunt race.[86]

During his presidency, however, Eisenhower had to confine the Sputnik-induced panic and react to opportunists who tried to capitalize on the Sputnik Shock. Scientists, astrofuturists, and the military tried to use Sputnik to get more funding, members of Congress (mainly Democrats) rallied for more political support, status and power on behalf of Sputnik, and future president Lyndon B. Johnson used space policy evoked by Sputnik as a political stepping stone.[87] Thus, Eisenhower agreed to a limited response to the Soviet space advance. He pushed for a National Defense Education Act in 1958, providing federal loans to college students majoring in STEM (Science, Technology, Engineering, and Mathematics) subjects. He also reluctantly agreed to the formation of a civilian space agency by signing the National Aeronautics and Space Act of 1958, which provided NASA with a broad mandate to explore space for "peaceful purposes for the benefit of all mankind."[88] NASA began

85 Dwight D. Eisenhower, 'Why I Am a Republican', *Saturday Evening Post*, April 11, 1964, p. 19.

86 Dwight D. Eisenhower, 'Are We Headed in the Wrong Direction?', *Saturday Evening Post*, August 11, 1962, pp. 19–25 (p. 24).

87 Kay, p. 65; McCurdy, *Space and the American Imagination*, pp. 54–75.

88 *National Aeronautics and Space Act of 1958*, Public Law 85–568, 72 Stat., 426, RG 255, National Archives at College Park, M.D. (hereafter NARA II); Alison Griffith, *The National Aeronautics and Space Act: A Study of the Development of Public Policy* (Washington, D.C.: PublicAffairs Press, 1962).

operating on October 1, 1958, incorporating the National Advisory Committee for Aeronautics (NACA) with an annual budget of $100 million, 8,000 employees, three major research laboratories and two smaller test facilities. Eisenhower also forced the armed services to lay their competition in the field of space exploration and utilization to a rest and cooperate in order to produce a rocket capable of competing with the Soviet Union.[89]

2.1.5 Military space exploration

Although NASA was established, NASA did not become the only American space agency. There were even periods when NASA was not even the biggest space agency. When measured by funding, the Department of Defense (DoD) repeatedly surpassed NASA's capacities. For example, in 1988, the DoD's space budget was twice as much as NASA's budget.[90] The DoD had been interested in space long before the dawn of the space age spearheaded by NASA. The different branches of the U.S. military had planned to place objects and people into space starting in the mid-1940s. The U.S. Navy's Bureau of Aeronautics (BuAer) had plans to orbit an artificial Earth satellite in October 1945, but project costs exploded and BuAer invited the U.S. Army Air Forces (USAAF) to join the project and share the conquest of the new frontier as well as the $8 million. However, General Curtis E. LeMay, deputy chief of staff for USAAF Research and Development, declined and put the people at Project Research and Development (RAND) to work coming up with an exclusive USAAF satellite plan. BuAer wanted to force the USAAF to join forces and requested an official hearing on the satellite subject before the Aeronautical Board of the War Department. Although the panel usually encouraged cooperation between the different military branches, in this case the board suggested that the two military services fine-tune their satellite proposals instead of ordering the Air Force to take part in a joint venture. As it turned out, both

89 Richard M. Filipink Jr., *Dwight Eisenhower and American Foreign Policy during the 1960s: An American Lion in Winter* (Lanham: Lexington Books, 2015), p. 12; Roger D. Launius, 'History of Civil Space Activity and Spacepower', in *Toward a Theory of Spacepower: Selected Essays*, ed. by Peter L. Hays and Charles D. Lutes (Washington, D.C.: National Defense University Press, 2011); Roger D. Launius, 'Eisenhower, Sputnik, and the Creation of NASA: Technological Elites and the Public Policy Agenda', *Prologue: Quarterly of the National Archives and Records Administration*, 28 (1996), 127–43.

90 Kay, p. 4 footnote 5. For more information about the intertwined military and civilian space see: Wilson W.S. Wong and James Fergusson, *Military Space Power. A Guide to the Issues* (Santa Barbara: Praeger, 2010).

proposals became victims of budget constraints and, more importantly, intraservice rivalry – a common theme of the U.S. military.[91]

Intraservice rivalry was also evident during the International Geophysical Year in 1957. The Navy, Air Force and Army submitted proposals for a satellite and launch vehicle. A committee, headed by Homer Stewart of the Jet Propulsion Laboratory (JPL) and Assistant Secretary of Defense Donald Quarles, rejected the USAF's proposal and the Explorer program, which had been proposed by the Army Ballistic Missile Agency (ABMA) headed by von Braun, in favor of Project Vanguard, a program developed by the United States Naval Research Laboratory (NRL). The satellite bid went to the Navy because it promised a larger satellite payload. The Navy also did a better job at making the satellite look civilian, which was very important to President Eisenhower because he favored a science based space exploration that emphasized inquisitiveness and technology.[92]

Having lost the satellite bid, von Braun and his team were worried that they would lose the cutting edge of the long-range rocket business. While the development of the Vanguard rocket suffered several setbacks, the Army's Jupiter-C rocket was successfully tested several times. The DoD nevertheless stood by its decision that Vanguard was the only official U.S. satellite program, even as it worried that the Army Ballistic Missile Agency could try to upstage the Navy's Vanguard, the official U.S. satellite program at this time, with an 'accidental' satellite launch. However, after Sputnik II carried the world's first space traveler, a dog named Laika, into space on November 3, 1957, the Pentagon allowed the Army to also try to launch a satellite. Thus, all missiles from all military branches were tested at Cape Canaveral, the missile test center of the Air Force also known as Missileland or Earthstrip No. 1, but Vanguard remained the United States' official program. After several launches were pushed back, a Vanguard test launched on December 6, 1957 and malfunctioned. The explosion of the Vanguard TV-3 rocket, although just a test, was treated by the U.S. media as a disaster of epic proportions and belittled by the Soviet media, which always instantly reported on U.S. space program malfunctions. The next Vanguard launch had been scheduled for January 2, 1958 but was cancelled due to a fuel leak, giving von Braun and his team the opportunity to leave their mark on American history with the backup launch of their Jupiter C missile and its payload satellite Explorer I.[93] However, that did not put an end to the intraservice space rivalry.

---

91 Tom D. Crouch, pp. 104–115.

92 R. Cargill-Hall, 'Civil-Military Relations in America's Early Space Program', in *The U.S. Air Force in Space, 1945 to the Twenty-First Century* (Washington, D.C.: USAF History and Museum Program, 1998), pp. 17–32.

93 Makemson, pp. 19–39; Tom D. Crouch, pp. 133–150.

The Air Force and Navy quickly rejected an Army dominated space program and all of them volunteered to house and command a unified space program. However, the founding of an entirely new organization or the placement of a structured space program within existing organizations like the Atomic Energy Commission, the National Science Foundation, or the National Advisory Committee for Aeronautics were also considered by the authorities.

While the National Advisory Committee for Aeronautics turned into NASA and the DoD's interest in space did not vanish, the military space efforts had to undergo several changes because President Eisenhower transferred all non-military space projects to NASA, including the Navy's and the Army's satellite programs, as well as ARPA's plans for a lunar probe and a single-chamber rocket engine. Furthermore, ABMA's operation in Huntsville was folded into NASA against the protests of General John Bruce Medarison on July 1, 1960, turning von Braun and his team into NASA men. The Air Force space efforts did not become a part of NASA because Secretary of Defense Neil McElroy had assigned all military space projects to the Air Force. The National Aeronautics and Space Act of 1958 (P.L. 85-568) also clarified that all military space activities were still part of the Department of Defense's domain.[94]

Thus, civil space and military space efforts were separated from each other but co-existed within the same community. Moreover, there were several cross-over points between the two areas of space utilization; a constant flow of talent and ideas was exchanged. For example, NASA's first astronauts were drafted exclusively from the military but did not perform military jobs. However, the Air Force did not give up on its vision of military astronauts flying to space on defense missions. The USAF followed the recommendations of Edward Teller and several other scientific visionaries who did not specifically speak out in favor of a military mission, but who concluded that the first nation to accomplish human spaceflight would receive national prestige. According to Teller and his group, the Air Force could develop an orbital spacecraft with a human on board within the next two years, and the Air Force started the Man-in-Space-Soonest (MISS) program in 1957.[95] Because of the $133 million price tag and lack of a clear military mission, MISS was derailed in the Department of Defense until President Eisenhower yet again referred a military project to NASA.

94 Tom D. Crouch, pp. 148–153; Erin C. Hatch, 'The National Aeronautics and Space Administration (NASA): History and Organization', in *National Aeronautics and Space Administration (NASA): Background, Issues, Bibliography*, ed. by C.V. Anderson (New York: Nova Science Publishers, 2002), pp. 13–42 (p. 15).

95 David N. Spires, *Beyond Horizons: A Half Century of Air Force Space Leadership* (Peterson Air Force Base: Air Force Space Command, 1997), pp. 75–76.

MISS became a part of NASA's Project Mercury in August 1958 and the Air Force served as an advisor providing liaison personnel for NASA's Mercury program.[96]

The Air Force put its ambitions regarding a human military spaceflight mission on hold after the MISS setback. While cooperating with NASA on the Mercury program, the USAF developed the X-20 Dynasoar, a winged recoverable military spacecraft which was supposed to be launched atop a Titan IIIC rocket in 1961. The military plane could not carry a large payload but was intended to fly at the edge of space to reach targets anywhere in the world for long-range bombardment. Defense Secretary Robert S. McNamara canceled the program in 1963 and started a new program, the Manned Orbiting Laboratory (MOL) to "determine military usefulness of a man in space."[97] However, the military space station that would have been launched into space aboard a Titan IIIM rocket was scrubbed due to numerous technical and funding problems in June 1969.[98]

The next Department of Defense project intended to launch military astronauts and a payload into orbit was NASA's Space Shuttle project. The Space Transportation System (STS) in the form of a reusable delta-winged orbiter spacecraft with a compartment large enough to accommodate a flight crew of ten astronauts and a 15 by 60 foot cargo bay with a 45,000 ton carrying capacity was intended to fly all U.S. government payloads into low Earth orbit – including military ones. To prepare for the military parts of the Shuttle missions, 32 military astronauts, drafted from Navy and Air Force personnel, were trained between 1979 and 1986 as a part of the Manned Spaceflight Engineer program, founded by the Air Force. However, the Shuttle project could not deliver on its 'space truck' promise. The STS was neither cheap nor reliable. After the explosion of the Challenger

---

96 David N. Spires and Rick W. Sturdevant, ' ' ... to the Very Limit of Our Ability ... ': Reflections on Forty Years of Civil-Military Partnership in Space Launch', in *To Reach the High Frontier: A History of U.S. Launch Vehicles*, ed. by Roger D. Launius and Dennis R. Jenkins (Lexington: University Press of Kentucky, 2002), pp. 470–501 (p. 475); Roger D. Launius and Dennis R. Jenkins, *Coming Home: Reentry and Recovery from Space* (Washington, D.C.: National Aeronautics and Space Administration, 2012), pp. 48–49.

97 Department of Defense, *Air Force to Develop Manned Orbiting Laboratory*, news release, December 10, 1963, < http://www.nro.gov/foia/declass/mol/6.pdf> [accessed 11 August 2015].

98 Roy F. Houchin III, 'Interagency Rivalry: NASA, the Air Force, and MOL', *Quest: The History of Spaceflight Magazine*, 4 (1995), 40–45; Roy F. Houchin III, 'Air Force-Office of the Secretary of Defense Rivalry: The Pressure of Political Affairs in the Dyna-Soar (X–20) Program, 1957–1963', *Journal of the British Interplanetary Society*, 50 (1997), 162–268.

in 1986 all commercial and military payloads were canceled on future Shuttle missions, turning yet another military space mission into a failure. The desire for a human military space mission did not end there. The Department of Defense started work on a single-stage-to-orbit (SSTO) vehicle for military purposes, but like the previous plans this one fell through as well in 1994 due to technological and financial difficulties. To this day, the Department of Defense has not laid its quest for military astronauts to a rest. The potential positive effects of having military personnel in space are in fact confirmed by Department of Defense studies, according to which military astronauts could be useful in the future of human space exploration and control, using their warfighting, problem-solving and decision-making abilities to accomplish the martial elements of future space missions.[99]

NASA and the Department of Defense continuously cooperate but also compete in the area of space exploration and development, attesting to a difficult but fruitful interrelationship between the civil and military space programs. Despite the Department of Defense's longing for its own human spaceflight program, a continuous two-way flow of talent and technology between the civil and military spaceflight operations developed and illustrates the fluidity and flexibility, but also the top-dog-mentality, within the space community.[100] Thus, besides science fiction and space societies, the space-oriented part of the military must also be added to the space community, accommodating changing ideas, administrations and egos while presenting space-minded people with several openings into the space community, and possibly also creating other options for women to take part in the space community outside of NASA.

99 Launius, 'History of Civil Space Activity and Spacepower', pp. 215–222; Curtis Peebles, *High Frontier. The U.S. Air Force and the Military Space Program* (Washington, D.C.: Air Force History and Museum Program, 1997). See also: C.V. Glines, 'DoD: NASA's Silent Partner,' Armed Forces Management, May 1969, Box GH 51, General Reference, JSC. For more information on Cold War Science History see: *Cold War Science and the Transatlantic Circulation of Knowledge*, ed. by Jeroen van Dongen (Leiden: Brill Academic Publications, 2015); *Science and Technology in the Global Cold War*, ed. by Naomi Oreskes and John Krige (Cambridge: MIT Press, 2014); Audra J. Wolfe, *Competing with the Soviets: Science, Technology, and the State in Cold War America* (Baltimore: Johns Hopkins University Press, 2013); Corinna R. Unger, 'Cold War Science: Wissenschaft, Politik Und Ideologie Im Kalten Krieg', *Neue Politische Literatur*, 51 (2006), 49–68; Margaret Pugh O'Mara, *Cities of Knowledge: Cold War Science and the Search for the Next Silicon Valley* (Princeton: Princeton University Press, 2005).

100 Karsten Werth, *Ersatzkrieg Im Weltraum. Das US-Raumfahrtprogramm in Der Öffentlichkeit Der 1960er Jahre* (Frankfurt: Campus Verlag, 2006), pp. 86–94.

2.1.6 NASA's narrative

The National Aeronautics and Space Administration was a rushed, political creation, demanded and designed by a large number of people due to the Sputnik Shock. NASA was written into law through the National Aeronautics and Space Act of 1958 (PL 85-568), also known as the NASA Act. Despite fear of the Soviet Union being the trigger for the NASA Act, the act itself turned out to be a well-balanced and well-thought out legal instrument, not just for competing with the Soviet Union, but also for conducting civilian research and development in aeronautics and space.[101] In order to achieve the goal of improving and expanding human knowledge of space and space travel in peaceful cooperation with other nations and U.S. agencies as set out in the NASA Act, NASA needed to find and attract talent and partners for their various programs and primary business areas. NASA needed people to do research in space and Earth sciences, including such specializations as aeronomy, ionospheres, lunar and planetary studies, fields and particles, meteoroid studies, and Earth observations. Furthermore, NASA needed personnel to conduct research in life sciences and systems, the study of life and how to live in space, including psychological and physiological studies, neurobiology, exobiology, biochemistry, radiobiology, and man-machine systems. There were also other major groups such as fluid and flight mechanics, materials and structures, propulsion systems, flight systems, measurement and instrumentation systems, data systems, experimental facilities and equipment, and research piloting in which NASA needed qualified personnel.[102]

Even though NASA's research focus areas were all of a scientific nature, there were other jobs at NASA besides being a professional scientist or engineer. There were actually four major job categories at NASA:

(1) Professional Scientific and Engineering – including astronauts as well as scientists and engineers engaged in test piloting, aerospace research, development, operations and related work such as development and operation of specialized facilities and equipment to support NASA's programs for which a college degree or the equivalent and specialized training and experience were required

101 *National Aeronautics and Space Act of 1958*, Public Law #85-568, 72 Stat., 426, RG 255, NARA II.

102 Johnson Space Center, *NASA Facts. Lyndon B. Johnson Space Center. NASA Careers*, ca. 1970s, Box 18, women in space, Astronauts and Space Suits, Center Series Poindexter Reference Files, JSC.

(2) Technical Support – including scientific and engineering aids, technicians, drafting, photography, illustrating, salaried shop superintendents, quality assurance specialists, production planning and inspecting positions
(3) Clerical and non-professional administrative – including secretarial, specialized, general clerical and administrative specialist positions, which require clerical training and experience or specialized non-professional experience in supply, fiscal, procurement and related activities
(4) Professional administrative – including management positions in research and development administration in such activities as financial management, contracting, personnel, security, administration, law and public affairs for which a college degree or the equivalent and specialized training and experience were required[103]

Considering that astronauts served as the representatives of NASA and the space community, it has to be recognized that this particular job constituted a very small percentage of NASA's workforce. However, it is still the most recognizable job at NASA and thus also the job with the greatest publicity outlet, which can be either beneficial or disadvantageous for NASA's operation, depending on the astronauts' behavior and interactions with the public and the media. This has not always been the case. Before the selection of the first astronaut corps in 1959, space engineers and entrepreneurs like Wernher von Braun had been the faces of American space exploration.

The fame of the astronauts and their accompanying influence in the space community and with the public did not sit well with quite a few scientists and engineers.[104] The creators of the space vehicles and technologies did not necessarily appreciate being left out of the public picture and did not want their creations to be meddled with by the passengers – the astronauts. Captain Robert C. Truax (USN), member of the American Rocket Society, employee of the USAF's space program and Aerojet General, as well as the founder of Truax Engineering and the Volksrocket, a reusable space vehicle for the American people which he was never able to finish, wrote in his unpublished memoire that he wanted the public to know that the American space pioneers were not all of German descent, but also homegrown like him

103 Federal Women's Program Office of the NASA Office of Equal Opportunity Programs, *Women at Work in NASA*, 1975, Box 18, women in space, Astronauts and Space Suits, Center Series Poindexter Reference Files, JSC.
104 See, for example: Donald D. Arabian, interview by Kevin M. Rusnak, February 3, 2000, pp. 12–13, Edited Oral History Transcript, JSC Oral History Project, JSC.

and "as American as the log cabin" in which he was raised.[105] Furthermore, he emphasized that his goal was to put "the Spam back in the can (where it belongs)" because

> the real space pioneer is not the man who climbs into his trusty rocket and blasts off into the blue. The test pilots and the astronauts did not start the development of rockets or the conquest of space. They were the necessary evaluators of the final product, but the dreamers, the scientists, and the engineers were the ones responsible for getting us to the brink of space. Yes, even the skilled workmen who translated the paper of the engineers into working hardware made contributions at least equal to those who tested the final product: All of these also had 'the right stuff' and deserve recognition at least equal to the few who rode the pillars of fire to fame, and sometimes, fortune. Creation, it seems to me, deserves more applause than mere evaluation. If it were only a question of public acclaim, I would not bother, but fame has, on occasion, given the test pilots and astronauts influence on the space program disproportionate to their qualifications, and has led to some fatal (and I do mean fatal) mistakes in our space program.[106]

The tension between the scientists and astronauts was not the only point of friction in the space community. Besides the scuffle between civilian and military space development as well as private vs. public space visions, the tussle between human and robotic (formerly known as manned vs. unmanned) space exploration is the most contested topic in the space community up to this day. The two very different schools of thought both wanted to explore the space beyond Earth's atmosphere, but the origins of their desire to explore were quite different and their strategies for going about it were thus quite different as well. The scientists preferring space exploration with robotic probes preferred this route because they were solely interested in advancing science and accumulating knowledge about the universe. They wanted to know more about the planets, stars, galaxies, intergalactic space, matter, energy, the smallest subatomic particles and the technologies they needed to explore the universe. The scientists and engineers favoring a robotic space program were not generally against humans in space, but were against rushing that development to meet political goals instead of scientific objectives. The supporters of robotic space exploration would have been content to use the minimum equipment necessary to get the job done and keep expenses low. Instead, they frequently had to place their

105 Robert C. Truax, *Draft of an Autobiography*, undated, Truax Autobiography (Draft), 19959, HRC; Michaud, pp. 254–255; Chris Dubbs, *Realizing Tomorrow: The Path to Private Spaceflight* (Lincoln: University of Nebraska Press, 2011), pp. 32–60.

106 Robert C. Truax, *Draft of an Autobiography*, undated, Truax Autobiography (Draft), 19959, HRC.

scientific experiments on human missions, which was often a frustrating experience for them because the equipment costs were far greater than the costs for the same sort of equipment in a robotic program.[107] This group of scientists could also follow their dream of robotic space exploration, but had to accept that their part of the space budget was always tied to the human program with a Faustian implication. If the overall space budget increased, their cut increased as well; therefore, they had to publicly promote the human spaceflight program even though most of them regarded it as a "voracious circus act" with one huge pitfall.[108] Namely, the scientists were worried that if the human space program were shuttered due to people dying, the robotic program would die with them. Even though "it took years and years of effort to get that point across," eventually the advocates of robotic exploration understood that the human program "was a national decision that involved more than science and since it was going to happen anyway, it made sense to do the best science possible on it" and use the human program to their advantage.[109] Thus, scientists lobbying for their research and an increase in funding often used the narrative arguments of the human spaceflight program to back their scientific arguments in order for non-experts to understand the importance of their work.[110]

The narrative argumentation came straight from President Kennedy. With his 1961 moon speech he not only set the gold standard regarding presidential leadership in space, but he also set the tone and style of expressing the American space rationale.[111] Using a president's "power to define" Kennedy did not solely rely on political, military, scientific and economic arguments, but supported his space reasoning with comfortable metaphors from the past which people could understand and relate to.[112] Due to the American frontier experience, or rather the myth of it, the

107 W.M. David Compton, interview with Homer S. Newell, Jr., April 21, 1976, 4495, HRC.
108 Burrows, p. 437.
109 W.M. David Compton, interview with Homer S. Newell, Jr., April 21, 1976, 4495, HRC.
110 Kaufman, p. 107 and 114.
111 John F. Kennedy, Address before a Joint Session of Congress, 25 May 1961, <http://www.jfklibrary.org/Asset-Viewer/xzw1gaeeTES6khED14P1Iw.aspx> [accessed 8 January 2015]; Roger D. Launius and Howard E. McCurdy, 'Epilogue: Beyond NASA Exceptionalism', in *Spaceflight and the Myth of Presidential Leadership*, ed. by Roger D.; McCurdy Howard E. Launius (Champaign: University of Illinois Press, 1997), pp. 221–50 (p. 221).
112 David Zarefsky, *President Johnson's War on Poverty: Rhetoric and History* (Tuscaloosa: University of Alabama Press, 1986), p. 1; McCurdy, *Space and the American Imagination*, pp. 1–10; Kaufman, pp. 4–40; Launius and McCurdy, p. 223.

history of human expansion and the history of imagining space travel, it was relatively easy for Kennedy and NASA to justify human space travel because it was "self-justifying."[113] The notion of Frederick Jackson Turner's frontier thesis, that the frontier experience had formed the American character, was so deeply rooted in the mind of the American people that it was a powerful and familiar image. The use of the frontier thesis was a smart move on the part of Kennedy and NASA because it targeted the pride and hearts of Americans. Reasoning based on this powerful, emotional basis could withstand all kinds of logical attacks and could even survive decades of disappointment during the struggle to make the moon landing a reality.[114] By framing issues in terms non-experts could understand and reducing complex decisions to simple scenarios, such as the struggle between good and evil at a new frontier, the public was more engaged in reaching the set goal.

However, the public also needed people in this space endeavor onto which they could project their image of American values like they did with the frontiersmen of the American West. Thus, the astronaut was "invented."[115] The astronauts were the personification of the American space program and therewith the American space community. They combined hero and celebrity status and gave the American space program an "aura of competence" and a mythic quality such that all Americans wanted them to succeed.[116] The astronauts were presented as having the same qualities as the men of the frontier. They were trail-blazing, young, ruggedly handsome individuals who were also brave, honest, devote, patriotic family

113 'Interview with Michael D. Griffin, Administrator', in *NASA at 50. Interviews with NASA's Senior Leadership*, ed. by Rebecca Wright, Sandra Johnson, and Steven J. Dick (Washington, D. C.: Office of Communications Public Outreach Division NASA History Program Office, 2008), pp. 1–23 (p. 15).

114 Frederick Jackson Turner, 'The Significance of the Frontier in American History', in *Major Problems in the History of the American West*, ed. by Clyde A. Milner II (Lexington: D.C. Heath and Company), pp. 2–21; Margaret Walsh, 'The Frontier and the West: Realities, Myths and the Historians', in *The American West. Visions and Revisions* (Cambridge: Cambridge University Press, 2005), pp. 1–18; Matthias Waechter, *Die Erfindung Des Amerikanischen Westens. Die Geschichte Der Frontier-Debatte* (Freiburg: Rombach, 1996); McCurdy, *Space and the American Imagination*, p. 4; Werth, p. 69 and 143–149.

115 Michael L. Smith, 'Selling the Moon: The U.S. Manned Space Program and the Triumph of Commodity Scientism', in *The Culture of Consumption: Critical Essays in American History, 1880–1980*, ed. by T.J. Jackson Lears and Richard Wightman Fox (New York: Pantheon, 1983), pp. 175–209 (p. 199).

116 McCurdy, *Space and the American Imagination*, p. 88; Matthew H. Hersch, 'Return of the Lost Spaceman: America's Astronauts in Popular Culture, 1959–2006', *Journal of Popular Culture*, 44 (2011), 73–92.

men who were willing to give their lives to fight evil forces in a hostile environment.[117]

Despite the similarities of the 'old' and 'new' frontier, there are two main differences. Unlike the American West, space is limitless and therefore cannot be conquered because there is no final destination. Furthermore, the astronauts were basically just coming along for the ride in the capsules, operated by Mission Control, which severely distinguishes them from the men who mastered the perils of the American West with no, or at least less, assistance. Yet, in order for the space frontier narrative to work, the basic elements of the typical frontier story needed to be fulfilled; there always has to be a hero adventurer. To turn the 'spam in the can' into the hero adventurer, NASA started to change its messaging. In its change of language NASA started to label its astronauts as pilots, suggesting they were in control of their capsules. Furthermore, NASA started to call the capsules 'spacecraft' at the end of 1961, indicating that their engineering feats produced sophisticated modes of transport instead of just a 'can,' and that such a complex vehicle could not be operated by mere 'spam', but only by skilled men, the astronauts. In addition, the narrative requires an evil antagonist in every frontier story and the Soviet Union was already the perfect enemy. Last but not least, it is important for the story to conquer an identifiable hostile location that has not been conquered before. In appointing the moon to be this particular location, Kennedy gave Americans the missing piece of their new frontier story.[118]

Although the scientific importance of landing men on the moon has been second-guessed from the day Kennedy tasked Americans with the moon challenge, the rhetorical value has always been recognized despite later objections from New Western Historians.[119] However, the narrative framing of American space exploration in the context of the Cold War contest in combination with the frontier thesis never made the majority of the public or the four committees overseeing NASA in the Senate and the House question the sensibility of human spaceflight.[120] Even though the argumentation

---

117 Susan Faludi, *Stiffed. The Betrayal of the American Men* (New York: William Morrow, 1999), p. 462; McCurdy, *Space and the American Imagination*, pp. 91–92.

118 Kaufman, pp. 32–42.

119 Patricia Nelson Limerick, 'Imagined Frontiers: Westward Expansion and the Future of the Space Program', in *Space Policy Alternatives*, ed. by Radford Byerly (Boulder: Westview Press, 1992), pp. 249–73 (p. 252).

120 The House Committee on Science and Astronautics, the Senate Committee on Aeronautical and Space Sciences, and the Appropriations Committees of both House and Senate. Kaufman, p. 94. It has to be noted that the Soviet Union used a similar rhetoric and narrative for their own space program. See: Asif A. Siddiqi, 'Spaceflight in the National Imagination', in

for human spaceflight has changed over time because NASA needed to modify its image according to the changing of the political environment in order to secure funding and support from the public and Congress, the agency never entirely neglected the narrative form of the argument and consistently used its astronauts as public spokespeople.[121]

Because astronauts have been the acclaimed heroes of every space-related story, there is a mythic quality attached to them. They set the standards for everything regarding 'space.' According to public opinion there is no more prestigious job in the space community; thus, women's participation in the space community is mainly measured by their membership in the astronaut corps. This cannot come as a surprise considering the important role they were ascribed within the space community narrative. The explorer-frontier narrative was already implemented with NASA's first high profile human spaceflight program, Project Mercury.

Project Mercury ran from 1958 through 1963. The goal of Mercury was for an American to reach Earth orbit before the Soviet Union and return safely to the U.S. Even though Mercury conducted 20 unmanned developmental flights, some even carrying animals on board, and six successful manned flights, the first human to reach Earth orbit was a Soviet. Cosmonaut Yuri Gagarin, aboard Vostok 1, beat the American astronauts to the punch on April 12, 1961. Seven American astronauts had been selected to become the new space frontier heroes. These astronauts were collectively known as the Mercury Seven.[122] Alan Shepard was the first of them to be launched into space, making a suborbital flight on May 5, 1961. Gus Grissom followed with his own 15 minute suborbital flight in July.[123] Soviet cosmonaut Gherman Titov was launched on a day-long orbital flight in August, 1961, setting another spaceflight record for the Soviet Union. The American astronauts also eventually reached Earth orbit. On February 20,

---

Rembering the Space Age, ed. by Steven J. Dick (Washington, D.C.: National Aeronautics and Space Administration, 2008), pp. 17–36.

121 Byrnes.

122 LCDR (later RADM) Alan Bartlett Shepard, Jr., USN; Capt. (later Lt. Col.) Virgil Ivan 'Gus' Grissom, USAF; Maj. (later Col.) John Herschel Glenn, Jr., USMC; Lt. (later CDR) Malcolm Scott Carpenter (1925–2013), USN; LCDR (later Capt.) Walter Marty 'Wally' Schirra, Jr., USN; Capt. (later Col.) Leroy Gordon 'Gordo' Cooper, Jr., USAF and Maj. Donald Kent 'Deke' Slayton, USAF.

123 Every Mercury spacecraft was given a name ending with a '7' by its pilot: Carpenter – Aurora 7; Glenn – Friendship 7; Schirra – Sigma 7; Shepard – Freedom 7; Slayton was grounded from flying due to an erratic heart rhythm (idiopathic atrial fibrillation) diagnosis in 1962. He became Coordinator of Astronaut Activities in 1963 instead and was eventually allowed to fly again in the Apollo-Soyuz Test Project in July 1975.

1962, John Glenn orbited three times around the Earth. Ultimately, both nations each sent six people into space, but the Soviets were able to set more records. For example, the cosmonauts had spent more total time in space than the astronauts. The Soviet Union also sent a female cosmonaut, Valentina Vladimirovna Tereshkova, into space on June 16, 1963. She spent more time in space than all the U.S. astronauts together.[124]

Although Project Mercury started out slow and had to face several obstacles like the loss of Grissom's capsule, the astronauts gained popularity.[125] People around the world were interested in their missions and their private and family lives, and followed their every move in tabloids and newspapers as well as on radio and TV. Mercury laid the groundwork for Projects Gemini, and Apollo, the two follow-up projects that would bring the United States to the moon. Project Gemini was conducted between 1961 and 1966 in low Earth orbit and had the objective of perfecting space rendezvous and docking maneuvers needed for the manned lunar landings of Project Apollo. Furthermore, Project Gemini was launched to practice all important techniques essential for Apollo's moon mission, such as extra-vehicular activities (EVA) in order to be able to work outside of the capsule.

Apollo could not have embarked on its moon mission without the fundamental experiments of Mercury and Gemini. While Mercury was a one-man project, Gemini was already able to carry two astronauts in one spacecraft and Apollo was built for three astronauts. In order to fill up all these seats on the spacecraft, several new astronaut classes were recruited.[126] Even though new astronauts had been brought into play, the original seven were still the most

---

124 Siddiqi, *Challenge to Apollo: The Soviet Union and the Space Race, 1945–1974*; Donald K. Slayton and Michael Cassutt, *Deke! U.S. Manned Space: From Mercury to the Shuttle* (New York: Tom Doherty Associates Book, 1994); Colin Burgess, *Selecting the Mercury Seven: The Search for America's First Astronauts* (Chichester: Praxis Publishing, 2011); John Catchpole, *Project Mercury. NASA's First Manned Space Programme* (Chichester: Praxis Publishing, 2001); Loyd S. Swenson Jr., James M. Grimwood and Charles C. Alexander, *This New Ocean. A History of Project Mercury* (Washington, D.C.: National Aeronautics and Space Administration, 1966); Matthew H. Hersch, *Inventing the American Astronaut* (New York: Palgrave Macmillan, 2012), pp. 9–40; Carpenter and others; Colin Burgess, *Liberty Bell 7: The Suborbital Mercury Flight of Virgil I. Grissom* (New York: Springer-Praxis Books, 2014); Colin Burgess, *Sigma 7: The Six Mercury Orbits of Walter M. Schirra, Jr.* (New York: Springer-Praxis Books, 2016); Slava Gerovitch, *Soviet Space Mythologies: Public Images, Private Memories, and the Making of a Cultural Identity* (Pittsburgh: University of Pittsburgh Press, 2015).

125 His capsule Liberty Bell 7 sank after splashdown and could only be retrieved in 1999.

126 David J. Shayler, *Gemini. Steps to the Moon* (London: Springer, 2001); David M. Harlan, *How NASA Learned to Fly in Space. An Exciting Account of the Gemini Missions* (Burlington: Apogee Books, 2004).

recognizable and loved astronauts in the eyes of the public. Only the frontier heroes who actually made it to the moon for the very first time with Apollo 11 on live TV were able to surpass their fame. Project Apollo ran from 1961 to 1972 and peaked with said moon landing of Apollo 11 on July 20, 1969 when astronauts Neil Armstrong and Edwin 'Buzz' Aldrin landed their Lunar Module (LM) on the lunar surface, while Michael Collins remained in the Command/Service Module (CSM) orbiting the moon. Despite the major setback of the excruciating death of the entire Apollo 1 crew and the near death experience of the Apollo 13 crew, Project Apollo was able to set several major human spaceflight milestones for the United States and the entire world.[127]

Even today, it is the only human mission beyond low Earth orbit, making Americans the only people to have ever left Earth orbit. Furthermore, Apollo 8 was the first spacecraft with humans on board to orbit another celestial body. While Apollo 11 undertook the grand achievement of landing men on the moon and returning them safely back to Earth, it was not the only mission to the moon. Subsequent to Apollo 11, there were five successful manned missions to the moon and ten more astronauts had the opportunity to walk on another heavenly body and collect important scientific data. The astronauts, for example, gathered 842 pounds of lunar rocks and soil which were distributed among scientific facilities all over the U.S. Originally, there were supposed to be three more Apollo missions for extended lunar geological and astrophysical exploration, but these missions were cancelled due to budget cuts.[128]

"That's one small step for [a] man, one giant leap for mankind."[129] Neil Armstrong's famous words, uttered when he became the first human being to set foot on an extraterrestrial body are known worldwide. A member of the

127 The Apollo 1 crew perished on the ground in an electrical fire in a pure oxygen atmosphere on board the Apollo 1 cabin during a test for their planned February 21, 1967 launch. Apollo 13 was not able to land on the moon due to an oxygen tank explosion which knocked out the CSM's propulsion and life support. In this case, the crew was able to use the LM as a 'lifeboat' and safely returned to Earth on April 17, 1970. Project Apollo also lost two astronauts in air crashes during training, and during the Gemini program three astronauts died in training jet crashes, including the primary crew for Gemini 9.

128 Roger E. Bilstein, *Stages to Saturn. A Technological History of the Apollo/Saturn Launch Vehicles* (Washington, D.C.: National Aeronautics and Space Administration, 1996); Burrows; Siddiqi, *Challenge to Apollo: The Soviet Union and the Space Race, 1945–1974*; Gene Kranz, *Failure Is Not an Option. Mission Control from Mercury to Apollo 13 and Beyond* (New York: Simon & Schuster, 2000); Michael Collins, *Carrying the Fire: An Astronaut's Journeys* (New York: Cooper Square Press, 2001).

129 Neil Armstrong in: Garber and Launius, p. 6.

human species walking on the moon was a memorable and life-changing event for a variety of people in and outside of the U.S. and could not be replicated with NASA's robotic missions in terms of positive public feedback for the space community. Scientific accomplishments could not compete with the monumental victory of the moon landing, also known as the narrative climax of the epic heroic space frontier saga. Nevertheless, during its first twenty years of existence, NASA designed several scientifically important robotic missions to collect data about the universe. NASA, for example, sent robotic probes to the moon (Ranger, Surveyor, and Lunar Orbiter), to Venus (Pioneer Venus), to Mars (Mariner 4, Viking 1 and 2), and to the outer planets (Pioneer 10 and 11, Voyager 1 and 2). Furthermore, NASA also worked on remote-sensing Earth satellites for information gathering (Landsat satellites for environmental monitoring) as well as applications satellites for communication and weather monitoring (Echo 1, TIROS, and Telestar).[130]

2.1.7 The role of the aerospace industry

It becomes apparent that there are a lot of areas in the robotic and human spaceflight operations in which one could participate besides performing as a hero astronaut. NASA had various positions for people with different educational backgrounds, and so did NASA's business and science associates. NASA Administrator James E. Webb acknowledged the contributions of its partners and collaborators in a speech at St. Mary's College[131]:

> We have not made these advances, these steps forward, alone. We at NASA have had the benefit of the aerospace industry, which receives 90 cents out of every NASA dollar. We have had the benefit of the universities of the Nation, who supply the space program's most vital single ingredient – brainpower.[132]

The third-highest-ranking person at NASA, associate administrator Homer E. Newell, affirmed Webb's praise for the intellectual and scientific support of the universities in the United States. As the person responsible for NASA's university programs, he knew about the importance of a close relationship between NASA and the scientific community because "universities are one of the

130 Garber and Launius, p. 4.

131 Following Dr. T. Keith Glennan as administrator, James Edwin Webb served as the second NASA administrator from February 14, 1961 to October 7, 1968.

132 James E. Webb, *Speech at St. Mary's College*, January 29, 1964, GH 101, Speech Transcripts 1964, Speech Bureau Bulletins, General Reference Series, JSC.

most important sources of effort in the programs we are carrying out."[133] Therefore, NASA started its university work programs already in 1959 with an expenditure of $3.6 million for study contracts alone. Measured dollar-wise NASA's university involvement appears relatively minor, but the impact of the universities on NASA's program is comparatively extensive.[134] Although NASA had good relations with the universities in the country, it was not easy to cultivate these relationships because it was nearly impossible for NASA to keep such a high number of constituencies satisfied with their role in the space program. Dr. Thomas K. Smull, Director of the Office of Research Grants and Contracts, even admitted to "a few rough spots here and there" because the universities, the independent sources of brainpower in American society, had to at the same time become NASA's contractors and all of them wanted to participate in the agency's mission and receive the funds provided by it. Most importantly, however, the university researchers wanted to see their ideas and visions become a reality. Therefore, it was hard for NASA to not get caught in the middle of competitive university research groups.[135]

Especially in the 1960s, university research played an important role in NASA's affairs. University participation was usually contracted via the Office of Research Grants and Contracts, and some universities were more successful than others in securing NASA contracts. NASA generally did not give funds to universities which were hoping to build up a new space research department, opting instead to fund universities with space related research already underway, which gave certain universities an edge. While some universities like the Massachusetts Institute of Technology (MIT) or the California Institute of Technology (CalTech), which up to this date manages the Jet Propulsion Laboratory (JPL) for NASA, were able to reel in major space research funding, other universities went away empty-handed. At a time when universities started to handle their affairs in a very business-like manner and fought each other for financial support, NASA was frequently criticized for not paying enough attention to the recommendations of the scientific community. However, according to NASA employees, the criticism usually came from university research groups who

133 Edward C. Ezell, Interview with Homer E. Newell, Jr., 25.05.1977, 4495, HRC.

134 Thomas K. Smull, *NASA-University Relations*, 2141, HRC; Edwin P. Hoyt, *The Space Dealers. A Hard Look at the Role of American Business in Our Space Effort* (New York: The John Day Company, 1971), p. 84.

135 Thomas K. Smull, Interview by Arthur Levine, August 1960, 2141, HRC; Thomas K. Smull, *NASA-University Relations*, 2141, HRC. Also see: Edward C. Ezell, Interview with Homer E. Newell, Jr., May 25, 1977, 4495, HRC; Homer E. Newell, *Beyond the Atmosphere: Early Years of Space Science* (Washington, D.C.: National Aeronautics and Space Administration, 1980).

were unable to exert their influence to modify NASA's program. In order to avoid scientific quarreling between the universities, NASA had an advisory committee system (from the science community, for example the National Research Council) in place. The committee met every three or four months and also held special meetings to sort out which proposals made the most sense for NASA.[136]

Thus, the researchers at universities were serving double duty, as members both of the scientific community and of the space community. Even though only some researchers were able to build a more profound relationship with NASA, all of them held positions which could also have been used as a door opener for women into the space community. Most importantly, the universities were always the prelude for every woman who wanted to be a scientist in the space program. However, while universities educated the women in their scientific disciplines, they also set the tone for the social and cultural structures women were faced with when entering the space community.

"What has 8.7 million parts and was put together by 300,000 people?" The Saturn V rocket and the Apollo capsule.[137] Thus, another interesting opening for women into the space community, at least when measured by the sheer number of opportunities, were the other business associates of NASA in the space industry, since the major part of the NASA team actually consisted of the numerous contractor companies spread out all over the United States. At NASA's height in the mid-1960s, it relied on 411,000 employees, of which 376,700 were working for NASA's 20,000 contractors and sub-contractors.[138] However, most of NASA's prime contractors were

136 John Pitts, Interview with Dr. Clark T. Randt, August 14, 1979, 1761, HRC; James E. Webb, *Education in the Space Age, Address before the American Council on Education*, October 4, 1963, Box 101, General Reference, JSC; Thomas K. Smull, *NASA-University Relations*, 2141, HRC. For more information on universities in the Cold War see: Stuart W. Leslie, *The Cold War and American Science: The Military-Industrial-Academic Complex at MIT and Stanford* (New York: Columbia University Press, 1994); O'Mara; Andrew Jewett, *Science, Democracy, and the American University: From the Civil War to the Cold War* (Cambridge: Cambridge University Press, 2012); *American Higher Education Transformed, 1940–2005: Documenting the National Discourse*, ed. by Wilson Smith and Thomas Bender (Baltimore: The Johns Hopkins University Press, 2008).

137 C.V. Glines, 'DoD: NASA's Silent Partner', *Armed Forces Management*, May 1969, Box GH 51, General Reference, JSC.

138 David Meerman and Richard Jurek, *Marketing the Moon. The Selling of the Apollo Lunar Program* (Cambridge: MIT Press, 2004), p. 36; Janne van Nimmen, Leonardo C. Bruno and Robert L. Rosholt, *NASA Historical Databook, 1958–1968, Volume I, NASA Resources* (Washington,

situated in only three States: California (42.7 percent), New York (10.2 percent), and Louisiana (7.2 percent).[139]

Even though NASA spent 90 cents out of every NASA dollar for contractors in the space industry, the space business constituted only ten percent of the overall aerospace industry, out of which the space industry grew at the end of World War II. Because of the close relationship between the aviation and space industries, most of the space business practices come from the aerospace industry.[140] However, NASA officials had to meet internal requirements and needed to conform to the guidelines of four other agencies, which made dealing with the space industry considerably harder. When dealing with contractors, NASA needed to factor in the policies of the Bureau of the Budget, which monitored NASA personnel ceilings; the U.S. Civil Service Commission, which defined and supervised position classification as well as federal standards for human resource administration, including Equal Employment Opportunity; the U.S. General Accounting Office, which investigated NASA's conduct and use of funds pertaining to contractors and R&D programs; and the Department of Defense, which assisted NASA with its experience in managing large-scale projects.[141]

NASA generally preferred to buy its needed items from private companies even if it could have manufactured them itself, because most such items were only the means to build the final product and it would not have been very economical to try to match the already existing research for a piece of hard- or software. Furthermore, NASA's in-house staff was employed not as manufacturers but as planners. They established the goal, guidelines, timeline, parameters, and cost estimates for the space program and needed to keep track of all strands of the program. The in-house NASA employees thus performed the tasks that could not be contracted out. That is why NASA generally favored

D.C.: National Aeronautics and Space Administration, 1976), p. 106. The numbers for the size of NASA's workforce during its peak employment slightly vary in different outlets. NASA Deputy Administrator Robert Seamans, for example, presented slightly different figures for the third quarter (Jan.-Mar.) of FY 1966 to Congress. He counted 420,000 contractor and civil service employees at the beginning of that year. See: House Science and Astronautics Committee, 1968 NASA Authorization, 90th Cong., 1st sess. (1967), p. 111.

139 Thomas P. Murphy, *Science, Geopolitics, and Federal Spending* (Heath Lexington: Lexington Books, 1971), pp. 182–186, tables 6–14.

140 James E. Webb, *Speech at St. Mary's College*, January 29, 1964, GH 101, Speech Transcripts 1964, Speech Bureau Bulletins, General Reference Series, JSC. Edwin P. Hoyt, pp. 4, 20, 88. See also: Joan Lisa Bromberg, *NASA and the Space Industry* (Baltimore: Johns Hopkins University Press, 1999), pp. 16–44.

141 Arnold S. Levine, *Managing NASA in the Apollo Era* (Washington, D.C.: National Aeronautics and Space Administration, 1982), pp. 107–108.

buyer-seller relationships over establishing its own research centers, with the exceptions of Bellcomm, which was eventually dissolved, and JPL, which is still operated by CalTech.[142] Because most of NASA's employees do not work for NASA directly, but are part of their contractors' workforce or even employees of NASA's contractors' subcontractors, NASA needed to establish a sensible procurement process from the very get-go. This was especially important particularly with regard to the fact that NASA awarded high paying contracts "that literally may mean the life or death of particular companies depending on whether they happen to be the ones selected; and this makes it very difficult to have everybody feel that they're getting helped by NASA."[143]

In terms of NASA's acquisition process, the agency operated similarly to the Department of Defense with one tremendous difference.[144] While the Department of Defense was able to sweep misjudgments and errors under the rug of secrecy, NASA's procurement decisions were out in the open, having to withstand the wary eyes of the public. The first step in the acquisition process was always a procurement request by the NASA personnel in charge. If one could buy the requested item off the shelf, it was an easy matter. NASA put out ads in the *Commerce Business Daily*, the newspaper of the Department of Commerce, and the businesses which manufactured the desired items could put in their bid. However, when it came to experimental spaceflight almost nothing could be bought over the counter but had to be built specifically.[145] In order to allocate those multimillion dollar contracts to an external company, NASA first checked the possible contractors to see who had the capabilities and capacities to do the job. The companies which complied with the prerequisites were then asked to submit a proposal and NASA selected the proposal which fit their needs best.[146] Considering NASA's extensive use of contractors, there were many

---

142 Arnold S. Levine, pp. 65–67.

143 Thomas K. Smull, Interview by Arthur Levine, August 1960, 2141, HRC.

144 See: Armed Forces Procurement Act (AFPA).

145 Edwin P. Hoyt, pp. 81–82.

146 Edwin P. Hoyt, pp. 49–51. In the case of Project Mercury, for example, NASA asked the following companies to submit proposals for the rocket engine: Aerojet General Corp.; the aircraft gas turbine division of General Electric Co.; Bell Aircraft Corp.; Pratt & Whitney aircraft division, Thiokol Chemical Corp.; Rocketdyne of North American; and the Wright aeronautical division of Curtiss-Wright Corp. In the end, the bid went to North American Aviation. For the Project Mercury spacecraft, NASA also asked qualified manufacturers to submit proposals. Twelve bids were handed in. While Winzen Research Inc. submitted an incomplete proposal, the other eleven bids were reviewed. Avco Manufacturing, Douglas Aircraft, Lockheed Aircraft, Republic Aviation, the Martin Company, Northrop Aircraft, Convair, Chance-Vought Aircraft,

opportunities besides working directly for NASA which women could have used to gain access and execute agency within the space community.

NASA relied heavily on its contractors not only for the construction of hardware, but also in terms of public relations and promotion of its space mission. Because NASA had a relatively small public relations team consisting of 146 employees in 15 locations that was scarcely able to keep up with public interest in the 1960s, the contractors supported NASA's PR efforts. They also put out press kits and even maintained information centers at Cocoa Beach, America's space coast, to inform the public about their role in the space mission – in the space community.[147] This was another opportunity for women to establish themselves in the space community.

2.2 Mapping NASA

Not just NASA's contractors are spread out all over the United States; NASA as a decentralized agency is spread out as well. Thus, NASA's administrative structure is incredibly complex because NASA needed to keep track of its thousands of employees and the work they were performing while always keeping the goal for their missions in mind. Considering that NASA developed out of the small research organization NACA into one of the largest federal agencies within five years of its creation, NASA needed to develop explicit administrative and local, on-site structures for its employees and their families.

NASA's nine main field centers, four affiliate facilities, and one federally funded R&D center (FFRDC) were all administered by NASA headquarters (HQ) in Washington, D.C. HQ was responsible for establishing policies, defining requirements and objectives, coordinating the field centers, and assessing the performance of the field centers in regard to the agency's programs. Because HQ was the authority in civilian space and aeronautics activities, it was in charge of long-range strategic planning,

Grumman Aircraft, and North American Aviation were one-upped by McDonnell Aircraft, which won the bid. The Mercury space capsule was thereupon produced by McDonnell Aircraft, which then had to employ 600 subcontractors, like, for example, Garrett AiResearch which built the environmental control system for the capsule. Edwin P. Hoyt, pp. 90, 100–101; Catchpole, pp. 138, 153–159, 197; Swenson Jr., Grimwood and Alexander, p. 137. For NASA's contractors after Mercury see: Bromberg, pp. 45–190.

147 Meerman and Jurek, pp. 36–37. See also: Makemson.

external advocacy, internal and external communication, program formulation, and resource allocation. The head of HQ and all of NASA was the administrator. The administrator received advice from the several program and staff offices, which were responsible for planning, directing, and managing NASA's current and future R&D programs; he also oversaw workforce and procurement activities and compliance.[148]

The National Advisory Committee for Aeronautics (NACA) brought five centers into NASA's operation. They became the initial nucleus of NASA.[149] During the first seven years of NASA's existence six more centers were added.[150] Each center has its own specialty and purpose within

148 The Office of the Administrator included the Administrator, the Deputy Administrator, the Associate Deputy Administrator, the Assistant for Commercialization, the White House Liaison, the Chief Engineer, the Chief Information Officer, the Chief Scientist, the Chief Technologist, and the Chief of Staff. Other HQ offices: Office of Space Flight, Office of Aero-Space Technology, Office of Space Science, Office of Life and Microgravity Sciences and Applications, Office of Earth Science, Office for the Chief Financial Officer, Office of Headquarters Operations, Office of Equal Opportunity Programs (did not exist at the founding of NASA but was added later, see: Chapter seven), Office of Human Resources and Education, Office of the General Counsel, Office of Procurement, Office of External Relations, Office of Management Systems, Office of Small and Disadvantaged Business Utilization, Office of Legislative Affairs, Office of Public Affairs, Office of Safety and Mission Assurance, Office of the Inspector General, Office of Policy and Plans. See: Hatch, pp. 33–34.

149 Ames Research Center (ARC) in California, Dryden Flight Research Center (DFRC; previously known as Flight Research Center and since 2014 called the Armstrong Flight Research Center) in California, Langley Research Center (LaRC) in Virginia, Lewis Reseach Center (LeRC; previously known as the Aircraft Engine Research Laboratory, Flight Propulsion Research Laboratory, and Lewis Flight Propulsion Laboratory and since 1999 called the John H. Glenn Research Center) in Ohio, and Wallops Flight Facility (WFF; nowadays operated by GSFC) in Virginia.

150 George C. Marshall Space Flight Center (MSFC) in Alabama, Goddard Spaceflight Center (GSFC) in Maryland, Lyndon B. Johnson Space Center in Texas (JSC), previously known as the Manned Spacecraft Center (MSC), John F. Kennedy Space Center (KSC; previously known as Launch Operations Center (LOC)) in Florida, John C. Stennis Space Center (SSC; previously known as Mississippi Test Facility, NASA Mississippi Test Operations, and National Space Technology Laboratories) in Mississippi, and the Jet Propulsion Laboratory (JPL) in California, which to this day is the only field center that is not federally owned. Each NASA center has been assigned a geographic region of responsibility: ARC: Alaska, Northern California, Hawaii, Idaho, Montana, Nevada, Oregon, Utah, Washington, Wyoming; AFRC: Arizona, Southern California; JPL: California; GRC: Illinois, Indiana, Michigan, Minnesota, Ohio, Wisconsin; GSFC: Connecticut, Delaware, Maine, Maryland, Massachusetts, New Hampshire, New Jersey, New York, Pennsylvania, Rhode Island, Vermont; WFF: eastern shore of Virginia, Maryland & Delaware; JSC: Colorado, Kansas, Nebraska, New Mexico, North Dakota, Oklahoma, South Dakota,

NASA's various projects and programs. Ames Research Center (ARC), Dryden Flight Research Center (DFRC), Langley Research Center (LaRC), and Lewis Research Center (LeRC) were primarily concerned with aerospace technology, while Johnson Space Center (JSC), Kennedy Space Center (KSC), Marshall Space Flight Center (MSFC), and Stennis Space Center (SSC) were mainly in charge of human spaceflight operations and systems. Goddard Space Flight Center (GSFC) and the Jet Propulsion Laboratory (JPL), on the other hand, principally dealt with space and Earth science. Furthermore, NASA worked with various affiliate centers.[151]

Even though the field centers had to take directives from HQ, they were not always on the same page regarding specific personnel, scientific and political issues. Sometimes there was even "sort of a mini-war brewing" between HQ and the centers. An unidentified source at JSC, for example, reported that there were quarrels between HQ and JSC over the composition of Space Shuttle crews in the 1980s: while HQ wanted at least four women on the flights, JSC decided two were enough.[152] Disputes like these show that NASA cannot necessarily be regarded as one uniform entity, but rather that different NASA centers at different locations had different ideas about the space program and community.

Looking at a map of the United States, it becomes clear that NASA not only had a national but also a local and regional impact in specific areas of the United States, thus providing the space community with a locality of sorts, a place where one could go if wanting to gain access to the space community. There are three main focal regions where NASA's and its contractors' facilities are situated. California, the East Coast, and the South. Especially the new NASA facilities that were not bequeathed from the National Advisory Committee for Aeronautics, such as Johnson Space Center, Kennedy Space Center, Stennis Space Center and Marshall Space

Texas; KSC: Florida, Georgia; LaRC: Kentucky, North Carolina, South Carolina, Virginia, West Virginia; MSFC: Alabama, Arkansas, Iowa, Louisiana, Missouri, Tennessee; SSC: Mississippi; NASA HQ: Washington, D.C., U.S. Territories, International. See: <http://www.nasa.gov/content/nasa-center-assignments-by-state> [accessed 11 March 2016].

151 For Example: the Goddard Institute for Space Studies (GISS; part of GSFC and Columbia University Earth Institute) in New York City, Michoud Assembly Facility (MAF; organizationally part of MSFC) in Louisiana, Moffett Federal Airfield (partially managed by ARC; since 2014 leased to Google for 60 years) in California, Wallops Flight Facility (WFF) in Virginia, and White Sands Test Facility (WSTF; administered by JSC) in New Mexico. Today, there is also the Independent Validation and Verification Facility (IV&V) which was established in 1993 in West Virginia. See: Hatch, pp. 33–42; Sol Levine, *Your Future in NASA* (New York: Arco-Rosen Career Guidance Series, 1971), pp. 46–76.

152 Thomas O'Toole, 'Space Shuttle Friction is worsened', *Washington Post*, March 16, 1983.

Flight Center, all took up residence in Dixieland. Lyndon B. Johnson (LBJ) in particular, in his position as Senate majority leader, and later vice president as well as Chairman of the National Aeronautics and Space Council (NASC), lobbied for the new NASA facilities to be set up in the South in order to support his home state of Texas and establish a federal presence in the former Confederacy States in order to fight racial discrimination and establish civil rights. He wanted to use NASA as a fertilizer to spur the growth of technology and economy in that particular region and to lead an activist federal government transforming the South from a structurally lagging region into a "modest down-home version of the Massachusetts Miracle", with NASA as the provider of the much needed jobs.[153] He was convinced that that there was an inexorable link between Southern poverty and Southern racism and that by solving one the other one would be solved as well.[154]

The NASA centers and the aerospace industry actually did lead the Southern states into a period of economic growth. However, contrary to LBJ's objective of overcoming poverty and racism, the NASA centers in the South initially contributed to the rise of racial and social segregation because the type of personnel the states in the South could mainly provide was former farmers, which NASA was not looking for.[155] The "brightest jewel in the New South's economic and industrial 'second Reconstruction'" was Huntsville.[156] The changes in Huntsville can be considered as part of the rise of the military-industrial complex during the Cold War, the so-called Gun Belt.[157] Huntsville transformed from a small town into Rocket City, and it was not the only locality in the South where NASA changed the local cultural landscape solely through its presence in the region. Houston became synonymous with Mission Control; it transformed into Space City. The city's baseball team changed its name from the

153 Joe P. Dunn and Howard Lawrence Preston, *The Future South. A Historical Perspective for the Twenty-First Century* (Urbana: University of Illinois Press, 1991), p. 90; Richard Paul, 'How NASA Joined the Civil Rights Revolution. Integration Came to the Nation's Space Agency in the Mid-1960s', *Air & Space Magazine*, 2014 <http://www.airspacemag.com/history-of-flight/how-nasa-joined-civil-rights-revolution-180949497/> [accessed 20 October 2016].

154 Gregory Hooks, 'Guns and Butter, North and South: The Federal Contribution to Manufacturing Growth, 1940–1990', in *The Second Wave: Southern Industrialization from the 1940s to the 1970s*, ed. by Philip Scranton (Athens: University of Georgia Press, 2001), pp. 255–85 (p. 264); Richard Paul and Steven Moss, *We Could Not Fail. The First African Americans in the Space Program* (Austin: University of Texas Press, 2015), pp. 5–6, 68–87 and 222.

155 Laney, pp. 50–51.

156 Paul and Moss, p. 74.

157 Laney, p. 50.

Colts .45s to the Astros and renamed its stadium the Astrodome. The new basketball team which came to Houston in 1971 was named the Rockets. Cape Canaveral in Florida came to be known as the Space Coast. The presence of 'the future,' of space, inspired the people living in these areas, which is also reflected in the space inspired (re-)naming of businesses on the Space Coast, in Space City or in Rocket City.[158]

Another local aspect of the space community was the areas where NASA and space contractors' personnel set up camp with their families to settle down. The construction and growth of NASA centers and contractors' sites created a development boom close to NASA production sites. Along the Florida space coast, the population of Brevard County, which harbored Kennedy Space Center in Florida, climbed from 23,000 to 239,000 between 1950 and 1968. Space workers had relocated their families to build local communities around NASA facilities, including housing, infrastructure, healthcare needs, schools, consumer needs, and government infrastructure. "Cocoa was a town rising out of the swamp, a veritable city of the future being built out of concrete and hope."[159] The same happened at the other NASA facilities, especially the newly founded ones in the South. In Space City, for example, so called 'space burbs' started to spring up around JSC,[160] among them the newly developed suburbs of Clear Lake City, Timber Cove, El Lago Estates and Nassau Bay where space workers established a close-knit community. The wry name insiders and journalists gave the space community on the ground was 'Togethersville'.[161]

Togethersville included not only the people who were directly working for the American space mission, but also their spouses and children because they were almost as affected by the space program as the people who were involved with the American space mission. On the one hand, the spouses and children of the space workers were taken care of by fellow space spouses in the local space community. The space families were at the instant disposal of each other in cases of everyday assistance, slumps, and overwhelming tragedy. On the other hand, the space spouses

---

158 Roger E. Bilstein and Mary Amelia Taylor, 'Aerospace', in *The New Encyclopedia of Southern Culture: Volume 22: Science and Medicine*, ed. by James G. Thomas Jr. and Charles Reagan Wilson (Chapel Hill: University of North Carolina Press, 2012), pp. 29–33 (p. 32); Laney, pp. 43–70; Benjamin, p. 10; Paul and Moss, p. 70 and 77.

159 Benjamin, pp. 7–9, quote on p. 9.

160 Koppel, p. 171; Lynn Spiegel, 'White Flight', in *The Revolution Wasn't Televised: Sixties Television and Social Conflict*, ed. by Lynn Spigel and Michael Curtin (New York: Routledge, 1997).

161 Dora Jane Hamblin, 'The Fire and Fate have Left Eight Widows', *Life*, January 26, 1968, pp. 60–64.

and children were under considerable pressure because of their status as members of the space community. They needed to make the life of the space worker in their family as comfortable and easy as possible in order to spare that family member stress and problems not connected to the space program. Thus, the local space community could "be a warm and loving family or just a great gelatinous blob which close[d] around and smother[d] the individual."[162] This indicates that the families of the space workers were also a part of the space community, at least so as long as the individual was involved with the American space mission somehow. If the space worker of the family left NASA or passed away, though, their families had to leave the space community at some point because they were no longer considered a part of it.[163]

Thus, women often became a part of the space community through their spouses working in space-related jobs – if they wanted to or not. Women married to a NASA engineer, scientist, administrator or astronaut had no choice but to become a member of the local space community. This compulsory membership was especially significant for the astronauts' wives. Due to the public interest in everything relating to the astronauts, the women could not elude the space community. However, even the space community membership – nationally and locally – of the astronauts' spouses derived from their ties to their space working spouses. Interestingly, it appears that the compulsory membership did not necessarily extent to the husbands of women working in space-related jobs. The spouses of the space women were allowed to have their own identity and community alliances.

2.3 A mandala network of overlapping communities

Considering that this book wants to find out what the sociocultural coding of masculinity and femininity in the space community can tell us about the (changing) mindset and cultural as well as social structures of American society during the Cold War, it is about time to establish what exactly the space community is in order to find out to what extent women participated and men reacted to female participation in the community. Thus far, it has become obvious that quite a few people felt attached to the space community in one way or another, but they did not necessarily assess their standing similarly.

162 Dora Jane Hamblin.
163 Dora Jane Hamblin.

Some people felt closer to the center of the community than others, which would explain terms like 'the inner circle' or 'the extended family'. Furthermore, the space community was perceived by different groups as either a place, or as identity and belonging, or as ideology and policy.

It becomes apparent that the American space community was not centralized, but rather a network of overlapping, interacting regional and transregional imagined as well as imaginary communities.[164] Furthermore, it appears that since the foundation of NASA, the space agency has constituted the center of the space community network. The regional space burbs would not even exist without NASA since they only sprang up because of the space agency, while most of the transregional space societies' aspirations and dreams rested upon NASA during the Cold War era due to the agency's unique role of taking American civilians to space.[165]

Every community needs its own symbols to demonstrate community membership to insiders and outsiders. Without symbols – imagery, customs, habits, rituals and boundary marking processes – there are and could be no communities.[166] Symbols describe and help to construct a community, and the space community also developed its own imagery, customs, habits, rituals and boundary marking processes to mark membership in the community, for instance the signature NASA emblems.[167] Although the space community did not make totalizing claims on its members like the imagined national communities, space community members did proudly and willingly identify as community members even in the time they spent outside of the space community. NASA personnel, for example, flaunted their NASA badges, a symbol identifying them as members of the space community, while running errands, and NASA

164 Zygmunt Baumann, 'Utopia with No Topos', *History of the Human Science*, 16 (2003), 11–15; Blackhawk, pp. 137–144; Phillip Wegner, *Imaginary Communities. Utopia, the Nation, and the Spatial Histories of Modernity* (Berkeley: University of California Press, 2002); Benedict Anderson, p. 6.

165 James M. Potter and Mark D. Varien, 'The Social Production of Communities. Structure, Agency, and Identity', in *The Social Construction of Communities. Agency, Structure, and Identity in the Prehispanic Southwest* (Lanham: AltaMira Press, 2008), pp. 1–17 (pp. 3–4); Uriya Shavit, *The New Imagined Community. Global Media and the Construction of National and Muslim Identities of Migrants* (Eastbourne: Sussex Academic Press, 2009), pp. 13–23.

166 Blackhawk, p. 125.

167 Anthony P. Cohen, *The Symbolic Construction of Community* (London: Routledge, 1985); Joseph R. Chambers and Mark A. Chambers, *Emblems of Exploration. Logos of the NACA and NASA* (Washington, D.C.: National Aeronautics and Space Administration Office of Communication, 2015).

contractors sported NASA logos on their brochures to show off their community membership status.[168]

Like the national community, the space community also existed in the minds of its individual members in differing variations; therefore, it is impossible to determine membership with accuracy. However, this book considers all people who worked, lived, and associated within the realm of NASA's imagery and conducted themselves according to the space agency's customs, habits and rituals as members of the space community. This definition includes all people working for NASA and its contractors as well as their families. As long as they were able to receive a pass for the gates, they are considered community members, regardless of the position they held within the community.[169]

"Trying to study community is like trying to scoop up jello with your fingers. You can get hold of some, but there's always more slipping away from you."[170] Although this is also the case for the space community, this book nevertheless attempts to do just that. Not only is 'space' an umbrella term for many fields of activity, but the space community is also an umbrella for several types of communities. One needs to envision the space community as a network of overlapping communities in a type of mandala arrangement. As the centre of the space community network, NASA provided direction and course of action as well as government approval to the other communities in the network. The other space communities are arranged around the NASA center with several areas of contact and interaction. For example, the military space community and the contractor space community overlap with each other and NASA.[171] However, they are

---

168 Mary Wiegers, 'Space Community: Emphasis on Excellence', *Washington Post*, September 13, 1969; Meerman and Jurek, pp. 36–37.

169 Even though citizen space societies are also considered part of the space community to some extent within this definition, due to length constraints of this book it was not possible to analyze to what extent such societies were a part of the community and what kinds of roles women were able to assume in these groups. However, the Hypatia Cluster or the Committee for the Future and their interactions with NASA and the public would make interesting and insightful topics for research. Hubbard; Michaud; Dubbs; McCray; Peter Dickens and James S. Ormrod, *Cosmic Society: Towards a Sociology of the Universe* (London: Routledge, 2007), pp. 8–9. See also Barbara Marx Hubbard's and the Hypatia Cluster's files in the NASA Historical Reference Collection: Barbara Marx Hubbard, 15114, and Hypatia Cluster, 16386, HRC.

170 Marcia Pelly Effrat, 'Approaches to Community. Conflicts and Complementaries', in *The Community: Approaches and Applications*, ed. by Marcia Pelly Effrat (London: Collier Macmillan, 1974), pp. 1–32 (p. 1).

171 Like the space community culture around NASA, the military community's imagery, customs, habits, and rituals are also at a cultural level "between the organizational and national culture." The imagined military community had produced its own symbols and culture, and the

not wholly superimposed upon one another; rather, some areas are congruent with each other. The local space communities also partially superpose the space community center that is NASA, as well as the contractor space community. The citizen pro-space community, made up of space, rocket, and science fiction societies, also overlaps with NASA. They provide an anchor for all American citizens who want to get involved with the topic of space but do not necessarily work in this area. This particular community is quite important for the space community network because it provides an outlet for enthusiasm and welcomes every person who wants to join the space community, unlike the other communities in the network. They include the American public in the community network. The public's involvement and interest is something the space community needs and actually depends on, but which most of the communities in the network cannot effect. Michael A.G. Michaud elaborated:

> Most of the widespread support for space activity is inchoate and has no outlet for expression. Enthusiasts need rallying points: organizations, meetings, magazines. The citizen groups bring together the pro-space constituency and give it some coherence, a set of focal points for activism. Publications and conferences play the major role in creating a nervous system for the pro-space community. They encourage self-recognition by the space advocacy.[172]

Furthermore, it has to be kept in mind that the mandala type space community network is situated within American culture and politics and plays a vital role in that country's domestic and foreign policy agendas; this was especially true during the Cold War. Considering that the space community network also intersected with the Media Age, the Technotronic Era, as well as Mid-Century Modernism, and even established its very own era – the Space Age – the space community network represents the ideal window into twentieth century American society. The space community network, hereafter simply called the space

---

people working on space issues within the military still exhibited the military imagery and behaved according to military customs, habits, and rituals, marking these people as members of the military community. All people are members of several communities; however, there is a gradation of the importance of different membership statuses to individual identity. Thus, through expressing and acting out military symbols above space symbols, military space explorers were members of the military community first and members of the space community second. Joseph L. Soeters, Donna J. Winslow and Alise Weibull, 'Military Culture', in *Handbook of the Sociology of the Military*, ed. by Giuseppe Caforio (New York: Springer, 2003), pp. 237–54 (p. 238).

172 Michaud, p. 312.

community, a community where technology, military, science, culture, media, politics, and patriotism in the United States intersect, constitutes an optimal sample of the mindset and social as well as occupational and gendered structures of American society during the Cold War and is therefore a representative, albeit extraordinary, community to research.

3 Media space in the 1960s, 1970s, and 1980s

From the moment NASA was founded in 1958, the space community was in the public eye. Most Americans, but also an international audience, was interested in the American advances in space and the people in the community who helped to make them happen. The public attention was a good thing for the space agency because public backing was necessary to justify NASA's activities and exorbitant spending. This is why NASA courted the American people and the media to gain their favor. Therefore, one of the major risks for the space agency, besides dead or misbehaving astronauts, lacking or insufficient financial investments, and operation-delaying failures, was negative publicity or deficient public interest in the space community, its activities, and representatives.[1] Because the space agency was dependent on a complaisant American public, NASA and the space community had to persist in public. Staying in the public's favour over the decades could entail different things, like granting access to journalists or participating in public events. Sometimes, however, the space agency needed more drastic measures to appease emerging sullenness, fury, or overwhelming requests on the part of the public, like an adjustment of its public image, or its organizational structure – including its gender belief system and gender performances.

As it is the goal of this book to explore various social spaces in the space community, to study what the sociocultural coding of masculinity and femininity in the space community say about the state of mind, cultural patterns, and social spaces of American society during the Cold War, the media space must also be taken into consideration. Although not a physical space and certainly not a space exclusive to the space community, it is a socially produced space dedicated to discourse, a discourse which often picked out members and activities of the space community as a central theme of discussion.[2] Considering that the people participating in the exchange of information and opinions do not share the same physical space, the public needed to rely on the media for information, including information about the space community. This dependency can be problematic due to the editorial decisions that are made for every single

1 Byrnes, pp. 2, 151–152; Hersch, *Inventing the American Astronaut*, pp. 131–158.

2 Gerard A. Hauser, *Vernacular Voices: The Rhetoric of Publics and Public Spheres* (Columbia: University of South Carolina Press, 1999); Alastair Hannay, *On the Public* (New York: Routledge, 2005); Richard Butsch, *Media and Public Spheres* (Basingstoke: Palgrave Macmillan, 2007); Jürgen Habermas, *Strukturwandel Der Öffentlichkeit. Untersuchungen Zu Einer Kategorie Der Bürgerlichen Gesellschaft* (Berlin: Suhrkamp Taschenbuch Wissenschaft, 1990).

https://doi.org/10.1515/9783110629828-003

story released via newspapers, TV, and radio. The essential questions, which subjects are important to discuss, what constitutes news, and how to package the issues, are editorial decisions and are always answered before the information in the media reaches the public.[3] Because most news stories can be presented from multiple point of views and influence the way the public perceives certain issues, news stories are always published with an issue frame, "the central organizing idea or story line that provides meaning to an unfolding strip of events and weaves a connection among them."[4]

This is why the relationship of women to the media and the portrayal of women in the media have been complicated. Traditionally, the media space, has been presided over by white, upper-middle-class men who promoted and framed the issues in the news according to their view of the world and urged women to stay in the private sphere. This is why the media often tried to play "a stabilizing role in American society" and was "usually leery of social movements," when 'the other,' for example women and people of color, tried to push their framing of certain political, cultural, and social issues into the public, including the media space.[5] Although, the news media could appear less biased due to their matter-of-fact style of reporting, the news outlets did frame certain issues from gendered angles. Especially in the 1950s and 1960s, the traditional-gender-roles-frame was quite popular, while in the 1960s and 1970s feminism- and anti-feminism-frames were both used. A 'women at work' frame started to become popular in the 1960s and remained so in the 1980s.[6]

The framing and stereotyping in the media was also applied to issues concerning the space community. It is, therefore, the goal of this chapter to take a closer look at the space community in the media space in order to study the sociocultural coding of masculinity and femininity in the treatment and portrayal of the women who were or wanted to become part of the space community, and to assess to what extent cultural, political, and social changes in American society led to a possible shift in or resistance to the framing of the issues in the media, and ultimately in the policies and gender belief system of the space community.

---

3 Nayda Terkildsen and Frauke Schnell, 'How Media Frames Move Public Opinion: An Analysis of the Women's Movement', *Political Research Quarterly*, 50 (1997), 879–900 (p. 879).

4 William Gamson and Andre Modigliani, 'The Changing Culture of Affirmative Action', in *Reseach in Political Sociology*, ed. by Richard Braungart (Greenwich: CT: JAI Press, 1987), pp. 137–77 (p. 143).

5 Flora Davis, *Moving the Mountain: The Women's Movement in America since 1960* (New York: Simon&Schuster, 1999), p. 109; Susan G. Douglas, *Where the Girls Are: Growing up Female with the Mass Media* (New York: Times Books, 1994), p. 11.

6 Terkildsen and Schnell, p. 884.

3.1 Astronettes and the cosmonette in the 1960s

With the establishment of NASA, and certainly with the introduction of the first astronauts on the cosmic stage, many Americans almost immediately started to ponder the possibility of women in space. However, ideas about if and how women should be included in the American space effort differed greatly. The notion of female astronauts was startling and amusing while also intriguing and puzzling.[7] Especially the presentation of the performance of female pilots who underwent the physical and psychological tests the Mercury astronauts had to pass fanned public discussion in the early 1960s before Valentina Tereshkova orbited Earth for the Soviet Union.[8] Those women had been tested by the de-facto medical department of NASA, the Lovelace Clinic in Albuquerque. Dr. Randolph Lovelace and his associates wanted to find out how capable women were of spaceflight for possible use in future space stations, although they did their testing in 1960 and 1961 without NASA's permission. 13 out of 19 women passed the tests. The women later became known as the Mercury 13, even though they had never been part of NASA's Mercury program. The name was given to the women by the media in the 1990s, which is incredibly misleading because they had never identified as a collective in the 1960s. Considering that the women never even met during their space durability tests, their lack of a common name – a common identity – which may have offered them a better public platform is understandable.[9]

Of all those accomplished female pilots who had been tested, Geraldine "Jerrie" Cobb and Jane Hart were the ones who lobbied most prominently for female

7 Frank Rhoades, column, *San Diego Union*, September 29, 1960, unprocessed Jerrie Cobb Collection, Schlesinger Library, Radcliffe Institute, Harvard University (hereafter Cobb/Schlesinger).

8 Bettyann Holtzmann Kevles, *Almost Heaven. The Story of Women in Space* (Cambridge: MIT Press, 2006), p. 11.

9 Karen Gibson, *Women in Space: 23 Stories of First Flights, Scientific Missions, and Gravity-Breaking Adventures* (Chicago: Chicago Review Press, 2014), pp. 13–40; Martha Ackmann, *The Mercury 13. The Untold Story of Thirteen American Women and the Dream of Space Flight* (New York: Random House, 2003), pp. 1, 77–80; Margaret A. Weitekamp, *Right Stuff, Wrong Sex. America's First Women in Space Program* (Baltimore: The Johns Hopkins University Press, 2004), p. 41 and 63–135; Jerrie Cobb and Jane Rieker, *Woman into Space* (Englewood Cliffs: Prentice-Hall, 1963). Frances Bera, Virginia Holmes, Patricia Jetton, Georgiana McConnell, Joan Merriam (Smith), and Betty Miller did not pass the test. Jerrie Cobb, Rhea Hurrle Allison (Woltman), Myrtle Cage, Jan Dietrich, Marion Dietrich, Wally Funk, Sarah Gorelick (Ratley), Jane Hart, Jean Hixson, Irene Leverton, Bernice Steadman, Gene Nora Stumbough (Jessen), and Jerri Sloan (Truhill) passed the test.

astronauts at NASA and in the media. Cobb, in particular, was adamant in her pursuit of spaceflight. She was able to present herself and the cause of female astronauts on various public platforms, such as space symposia, parades, and media reports. Cobb's image cultivation was so good that she was sometimes already praised as the "first woman astronaut," for example at an event in Pittsburgh in 1960, where she rode a Cadillac with the banner "Pittsburgh welcomes first woman astronaut Miss Jerrie Cobb."[10] Her momentum in the media reached such a height that NASA Administrator James Webb declared Cobb to be a consultant to NASA at a conference on the Peaceful Uses of Space in Tulsa in May 1961. For NASA it was a public relations move to reel her publicity into NASA's corner, to demonstrate that the space agency did care about American women and their part in the country's space effort. The appointment was supposed to muzzle Cobb's very public suggestions and comments about women in space. Instead, she prepared a strategy paper and continued to publicly speak on the topic of women in space. Cobb actually wanted to work with NASA, but received neither feedback on her strategy paper nor other instructions from the space agency, and had no clear idea about the nature of her consultant position. In December 1961, the space agency fired Cobb and ordered her not to speak out publicly anymore and refer to herself as a NASA consultant. Cobb, however, insisted that she never received a written termination letter and continued to publicly lobby for female astronauts. Cobb and Hart eventually managed to meet Vice President Lyndon B. Johnson and get the Committee on Science and Astronautics of the House of Representatives to hold a hearing about the selection criteria for astronauts in July 1962. However, neither the Vice-President nor Congress supported the women in their quest to fly to space for their country, although they had proven in the tests that their mental and bodily performance was as good as that of their male counterparts.[11] Unlike the men at NASA,

---

10 See image on the cover or Fig. 3.1.

11 James E. Webb, *Remarks by James E. Webb, Administrator National Aeronautics and Space Administration First National Conference on the Peaceful Uses of Space Tulsa, Oklahoma,* May 26, 1961, 000336, HRC; Jerrie Cobb, *Strategy Paper*, June 15, 1961, 000336, HRC; Jerrie Cobb, *Letter to James E. Webb*, March 30, 1962, 000336, HRC; *Hearings before the Special Subcommittee on the Selection of Astronauts of the Committee on Science and Astronautics U.S. House of Representative,* 87th Congress, 2nd Session, July 17 and 18, 1962, No. 9, U.S. Government Printing Office, Washington, D.C., 1962; Weitekamp, *Right Stuff, Wrong Sex. America's First Women in Space Program*, pp. 136–155; Ackmann, pp. 123–136; Kevles, pp. 13–17; Jerrie Cobb, *Jerrie Cobb: Solo Pilot* (Sun City Center: Jerrie Cobb Foundation, 1997); David J. Shayler and Ian A. Moule, *Women in Space. Following Valentina* (Chichester: Praxis Publishing, 2005), pp. 87–92.

Congress, or the White House, the men and women who mused about female astronauts in the media had a bigger palette of reactions and fantasies.

Fig. 3.1: Jerrie Cobb in Pittsburgh, 1960, Ed Walsh Studio, Pittsburgh, in: Jerrie Cobb Collection, Schlesinger Library, Radcliffe Institute, Harvard University.

3.1.1 Cartoonesque depictions of female astronauts

To most Americans, the idea of women in space appeared to be plain funny. The thought of female astronauts seemed ridiculous in the face of all the stereotypes associated with women on Earth, let alone in space. For this reason, aspiring female astronauts were for the most part not taken seriously. They were often not given a chance to explain their virtues, qualities, and skills for the job. Instead, they were confronted with common stereotypes of female attributes and characteristics in written but also pictorial form in order to make fun of the idea of women in space or space-related work environments. It was an effective way of framing these ambitious women in their traditional gender roles and therewith affirming the masculinity of the job of the astronaut. Making light of Cobb and the other women who attempted to break the gender barrier often shut down a balanced discussion before it even started, like Robert C. Ruark did in his column:

> I believe that the capsule in which Miss Jerrie Cobb, the first incipient astronautress, is to be placed is somewhat less commodious than the first-class lounge of the Queen Elizabeth, so it figures this eaglette is going to have to cut down on distaff indispensables. What, I wonder, will be eliminated as dispensable from that cavernous catchall, the handbag? […] Certainly Miss Cobb will have no audience to which she may complain that she has nothing to wear, once she is actually in flight. But saying that she smacks into the moon or some other inhospitable planet, it's a cinch she is not going to feel her best in her Buck Rogers suit. Will they slip a little black Hattie or a near Dior or something bouffant with a heated shimmie into the do-it-yourself kit, so that our heroine may hold up her head among the space people? Who, I wonder, will pin the suddenly sundered shoulder strap? Who will go with the lady to the astral little girls' room, there to gossip and giggle and refurbish the face? Who will double park the space ship while she searches for a drug store? Who will zip the zippers, hook the eyes, inspect the stocking seams and locate the lost gloves?[12]

Ruark's column includes almost all female stereotypes which were used to portray women in space as nothing more than an amusing whimsy. Women were depicted as dizzy creatures, way less sophisticated and adept than men because they supposedly have no understanding of technology. Who would want a female astronaut when she can't even double park, uses hairpins to solve engineering problems, or might accidentally push the wrong button and ruin the American space effort?[13]

Another, and probably the most prominent, joke about women in space consisted of cracks targeting the stereotype of women's concern their appearance.[14] Women were depicted as vain and simple minded creatures who would turn space suits into fashion accessories or pass the long hours of spaceflight with the application of make-up or the joy of being weightless, meaning not having to worry about ones weight.[15] Therefore the press asked readers: "[C]an you imagine a woman locked up for weeks or months in space with no telephone to jabber over all day long? But perhaps they'll line the inside of the spaceship with mirrors and she can be happy admiring herself."[16] Others were concerned with hygiene in space. One (female) journalist raised the question

---

12 Robert C. Ruark, *Lady in Space? Ha-Ha and additionally Hee-Hee. Not to mention Ho-Ho*, unknown newspaper, ca. 1960–1963, Cobb/Schlesinger.

13 Robert Ruark, 'Let the Girls, bless 'Em, Cut Loose in Space!', *Washington Daily News*, August 8, 1962.

14 See comics.

15 *Sex and Space*, unknown newspaper, ca. June 1963, Cobb/Schlesinger; Robert Ruark, 'Let the Girls, bless 'Em, Cut Loose in Space!', *Washington Daily News*, August 8, 1962; 'The Craftier Sex is Cleared for Space', *Los Angeles Times*, August 28, 1960.

16 Reader's comment in Douglass Welch's column. Douglass Welch, 'Girls best the men, again', *Pittsburgh Post-Gazette*, September 16, 1960.

how female astronauts would like to wear "the same space suit without a bath or a change of clothes for six weeks?" Her argument was that "[b]oys don't mind this sort of thing as a rule, but it's different with girls. And those with space ambitions should start now figuring out ways and means of remaining dainty unbathed."[17] The conclusions reached by this line of thinking were mostly the same: women should stay on Earth and stick to blow drier helmets at the hair salon instead of space helmets.[18] Not only women's supposed preoccupation with their appearance was targeted to ridicule the idea of women in space, but all other available female stereotypes were used to make fun of the idea as well, such as women's need to gossip, female driving skills, excessive spending while shopping, female appreciation for beautiful but supposedly useless objects, or her alleged desire to become a celebrity.[19] This is why columnist John Crosby asked: "How's a girl to have a garden when there's no air for the flowers to breathe?" or "How are you going to yell at the children with one of those masks on your face all the time?"[20]

Another way to deny the determined female pilots credibility and respectability in the technical field of space flight was to portray them as sex objects. The press frequently quoted snide comments from male, mostly anonymous, NASA employees like: "Well, there IS an allowance of 125 pounds for recreation equipment aboard the Apollo spacecraft."[21] Yet another way to not take women in the field of space travel seriously was to imagine them in space only as something else than a NASA astronaut – the other. Women could only conquer space when they were depicted as, for example, aliens or witches. This portrayal is especially interesting because it uses the stereotype of powerful women being out of this world. While the witches were usually portrayed as evil and unfeminine, the female aliens were usually depicted as sexy with a surprising twist. These depictions were still practiced in the 1970s and 1980s.[22]

---

17 Dorothy Roe, 'Girls, Space Travel might be bit rough', *Orlando Sentinel*, March 18, 1965; Kevles, p. 37.

18 *First Blonde in Space? Should a woman take the risks?*, unknown newspaper, ca. 1960–1963, Cobb/Schlesinger.

19 See Fig. 3.2–3.5. More cartoons can be found in the Jerrie Cobb Collection, Schlesinger Library, Radcliffe Institute, Harvard University.

20 Albert Perry, *Woman's Place in Outer Space*, 1962, 008994, HRC.

21 United Press International, *Scoffs, Bravos Aim At Woman in Orbit*, unknown newspaper, Cobb/Schlesinger; 'A Mrs. In the Missile?', *Los Angeles Times*, September 7, 1958, p. 12. See also Fig. 3.12.

22 See political cartoons in: *Publishers Newspaper* Syndicate, October 27, 1963; *Chicago Tribune*, January 19, 1964; *Bold*, Vol. 8, October 1972; *Airman*, September 1980; *Daily News*, August 4, 1976. See Fig. 3.6–3.7.

Fig. 3.2: Washington Daily News, March 14, 1963.

Fig. 3.3: The Evening Bulletin, September 15, 1960.

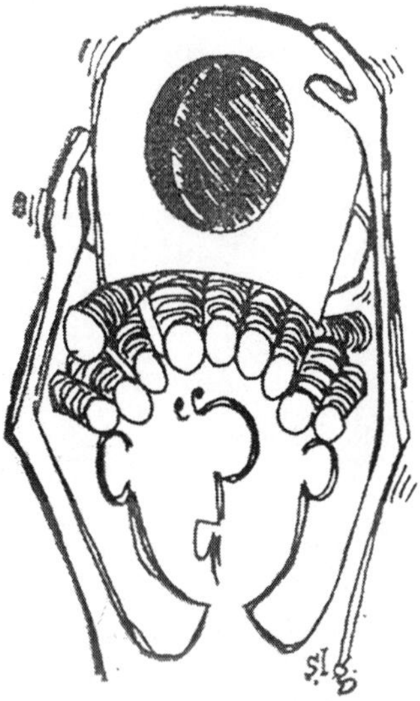

Fig. 3.4: Fort Lauderdale News, March 27, 1963.

Fig. 3.5: The Phoenix Gazette, August 22, 1960.

Fig. 3.6: Publishers Newspaper Syndicate, October 27, 1963.

Fig. 3.7: Bold, Vol. 8, October 1972.

3.1.2 The female astronaut body's depiction as a disruptive force

In addition to the ridicule spacefaring women were faced with in the media, the idea of women in space was also often bluntly rejected without any sense of ridicule or humor. A lot of American men and women were convinced that being an astronaut was a man's job and that women should stay away from it. One NASA employee was quoted stating that the thought of women in space "makes me sick to my stomach."[23] Others implied that female astronauts were not intended in God's creation and that women should stick to the gender performance of their sex: "With equal reverence I would suggest that maybe the Creator would be grateful to Jerrie if she got on with the job of being a woman – right here on earth."[24] Furthermore, women were regarded as a complication to the space effort for various reasons. According to many Americans, women in any workspace would lead to the decline of sexual morality. This held true for the workspace 'space,' as well. Cobb would sexualize the apparently hitherto sexually neutral work environment 'space,' and her willingness to fly to space with a man led one commentator to describe her as "astronaughty."[25] According to one newspaper caricature, even the 'man in the moon' would be distracted if a woman showed up in space, let alone the male astronauts. The disruptive power of the female body would tease the virile men's libidos and hinder successful, effective mixed-sex expeditions. A healthy, masculine man, like an astronaut, could not be expected to also deal with the distraction of female sexuality during work hours in their risky jobs in the hazardous environment of space. This argument was still occasionally applied in the 1970s, by, for example, astronaut Michael Collins who though that women on a long-term flight would be a "psychologist's nightmare" because of possible "sexual problems," amongst other issues like "jealousy" or "bickering." While this point of view was not contested by women during the 1960s in the media, women in the 1970s refrained from being at fault for men's supposed lack of self-control.[26]

---

23 United Press International, *Scoffs, Bravos Aim At Woman in Orbit*, unknown newspaper, ca. post June 1963, Cobb/Schlesinger.

24 *First Blonde in Space? Should a woman take the risks?*, unknown newspaper, ca. 1960–1963, Cobb/Schlesinger.

25 *No. 1 Space Gal Seems A Little Astronaughty*, unknown newspaper, ca. 1960–1963,Cobb/ Schlesinger. Some newspapers even joked about the the need for sex segeregated space programs. See Fig. 3.9.

26 Susan Watters and Patricia Anstett, 'Women well-suited for Voyages in Space but. . .they're lacking in 'spacemanship'', *Chicago Tribune*, August 12, 1976; 'Women's Place in Outer Space', *Washington Post*, March 17, 1968; Elaine Tyler May, p. 164.

Others felt that women impinged too deeply into the men's sphere in general, wanting to become involved in politics, sports, the military, industry and aviation. Spaceflight was just one of many fields those opponents of female astronauts argued against female involvement in without individual reflection on the specific case at hand.[27]

Another area of concern regarding women in space was the complication of space exploration on account of female biology. Women would or could need special facilities on Earth and in space, as well as special treatment concerning their supposed biological "weakness" of menstruation and their child bearing ability.[28] In this regard, Dr. Constantin D. Generales was quoted: "We must bear in mind, especially in view of future long-term flights of several month, the inborn weakness women have 12 times a year. During that time they suffer emotional instability. It wouldn't be a factor if the woman was just a passenger, but if she is given responsibility it would be significant."[29] Even the physicians who were part of the team that tested the Mercury 13 did not see a role for women in the near future because of the stigma attached to menstruating women. In a medical paper they published, they elaborated that the tests of the women had not been correlated with the menstrual cycle, during which women allegedly suffered a change in peripheral vision, a lack of attentiveness, nausea, and cramps. Because of it, women were also supposedly prone to mental illness and demonstrated an increased potential for violence. Thus, during menstruation, women would not be able to follow rigid schedules, or perform all of their work hours with the same vigor as men.[30] Yet another concern of a biological nature was that women lacked the physical strength and stamina to do some of the astronauts' jobs, like space-walking.[31]

27 See Fig. 3.8.

28 See Fig. 3.9.

29 Claire Wallace, 'Spacewoman up in the air over Red Tape', *New York World-Telegram and Sun*, June 21, 1963. See cartoons.

30 Johnnie Betson and Robert R Secrest, 'Prospective Women Astronaut Selection Program. Rationale and Comments', *American Journal of Obstetrics & Gynecology*, 88 (1964), 421–23.

31 Anne V. Thompson, 'Equality of Sexes Ends Where Outer Space Begins', *Tulsa Daily World*, October 1, 1967; 'Outer Space Out of Bounds to Women for Good Reasons', *Star*, April 14 1959; Joe Ceraso, 'The Question: Should there be Women Astronauts?', *Boston Herald*, August 27, 1962.

Fig. 3.8: Phoenix Gazette, ca. 1960–1963.

Fig. 3.9: The Daily Oklahoman, July 19, 1962.

3.1.3 Approaches of imagining female astronauts

Contrary to the minority who flat out rejected women's participation in space exploration, there were quite a few who generally supported the idea of women in space – just not at this point in time. A lot of NASA officials were often quoted as being open to the idea and suggested that women would fly to space at an indefinite time in the future. The *New York Post*, for example, wrote about a NASA official who had indicated that women would eventually be given a place in space. "We haven't decided yet," he said, "what we're going to do with that second seat in the Gemini."[32] This example gives a foreseeable timeframe; however, this was not the case most of the time. NASA officials usually said something like: "the future use of women astronauts is possible, but at what time in the future programs is another matter entirely" or "I have always believed that women can and will make a contribution to the space program."[33] Some such opponents of female astronauts even named tangible reasons for not including women in the astronaut corps right away. In a speech before the General Federation of Women's Clubs, NASA Administrator James Webb explained to the female audience that he hoped to increase women's participation "as this national effort accelerates", but he also pointed out that at the moment the qualifications to become an astronaut were "more readily met by men than by women." In particular, men in the military who already had extensive jet aircraft flight experience, scientific and engineering educations, and an already established security clearance were most apt to become astronauts.[34] This rationale was often repeated by NASA and Congressional representatives in public and private. Wernher von Braun, for instance, was convinced that women would fly to space as soon as the skill prerequisites, like the jet test pilot prerequisite, changed. He actually told Cobb women have a place in space and eventually assume it. [...] Besides I see no reason for not including a little beauty along with

32 William McGaffin, 'Female Space Barrier', New York post, March 19, 1962.

33 Hiden T. Cox, letter to Jerrie Cobb, January 31, 1962,Cobb/Schlesinger; Wernher von Braun, letter to Jerrie Cobb, July 20, 1962, Cobb/Schlesinger. See also: Robert R. Gilruth, letter to Jerrie Cobb, April 17, 1962, Cobb/Schlesinger; Rita Rapp, interview by KTRK, November 29, 1962, CD-G (cassettes), Center Series, JSC.

34 James E. Webb, *Address before the General Federation of Women's Clubs*, June 27, 1962, p. 12, unprocessed Ruth Bates Harris Papers, Moorland Spingarn Research Center, Howard University (hereafter Harris/MSRC); *'No Women Astronauts? U.S. can't find one'*, *Evening Bulletin*, October 20, 1966; 'Women are unable to make grade in Astronaut Test', *New Haven Register*, January 7, 1966; Joseph D. Atkinson and Jay M. Shafritz, *The Real Stuff. A History of NASA's Astronaut Recruitment Program* (New York: Praeger Publishers, 1985), p. 37.

technical competence."[35] California Congressman George Paul Miller was also convinced that the test pilot qualification would be dropped or the field of test piloting would eventually be opened up for women after the pioneering phase.[36] The argument that women could become astronauts 'someday' when they were able to meet the requirements was by far the most popular and was still used in the early 1970s.[37]

The famous pilot Betty Skelton, who was successfully tested for female abilities in spaceflight by *Look* Magazine, had a different explanation why women could not participate in the American space effort just yet: she felt that a "training program for women space pilots is not as important as the whole space plan."[38] In the face of the space race, a lot of Americans could understand where Skelton was coming from. They wanted to win the space race as Americans and could not understand why women would make such a "fuss" about themselves in view of the imminent Soviet threat, when there were already enough trained men in the armed forces who could get the job done right away. Powder Puff Derby Winner Mary Ann Noah and Jacqueline Cochran, another famed female pilot, also made public statements in this vein.[39] Yet another reason for holding back women from spaceflight was that some people were convinced that only 'manly' women should participate in space exploration. Another Powder Puff Derby winner, Mrs. Charles Roberts, for example declared that Jerrie Cobb "is a very soft spoken person, blonde, very feminine – and I would think the woman chosen should have more of a brash way about her."[40] The attitudes of these accomplished female pilots demonstrates that women, even those who were successful in the field of aviation, were not necessarily of the same opinion regarding women in space, much less supportive of the idea. The opinions of men and women varied extremely, which shows that the postwar discourse

---

35 Wernher von Braun, letter to Jerrie Cobb, May 9, 1962, Cobb/Schlesinger; Anne V. Thompson, 'Equality of Sexes Ends Where Outer Space Begins', *Tulsa Daily World*, October 1, 1967.

36 Marion Dietrich, 'No Skirts in Space?', *Oakland Tribune*, October 29, 1966.

37 Frank Macomber, 'Can Space Lady be far behind', *San Jose Mercury*, July 26, 1972; Donald K. Slayton, answers to various letters from girls/women who wanted to become astronauts, 1970–74, 008998, HRC.

38 United Press International, *Scoffs, Bravos Aim At Woman in Orbit*, unknown newspaper, ca. post June 1963, Cobb/Schlesinger.

39 Adele Durrell, *The Question: Should there be Women Astronauts?*, unknown newspaper, August 27, 1962, 008998, HRC; D. Brainerd Holmes, *Views on Women Astronauts*, June 7, 1962, 013080, HRC; Jim Maloney, '4 Being considered as Spacewomen', *Houston Post*, February 11, 1965; Ackmann, pp. 161–163; Weitekamp, *Right Stuff, Wrong Sex. America's First Women in Space Program*, pp. 140–144.

40 Audrey Ross, 'Space Ladies' in the Future?', *Phoenix Gazette*, August 22, 1960.

about women was much more complex than the stereotype of the 1960s women might suggest.[41] This was also the case for the discourse about women in space in the early 1960s.

There were actually a lot of people who supported the notion of women in space – the majority being men. Like the opponents of female astronauts, the proponents also had varying reasons for supporting women in space. Some believed in Jerrie Cobb and Jane Hart's warning that the Soviet Union would soon launch a woman into space and wanted an American woman to reach space first, in order to not give this 'first' away to the Russians. The news frequently reported Cobb's and Harts warnings that the Russians

> may soon launch a Mongolian woman into orbit ('They are a small, hardy race used to high altitudes'), while the first space-bound U.S. female may be a chimpanzee. 'There's a $1,000,000 budget for a place called Chimp College, New Mexico,' said the angry Jerrie, 'where at least one female, named Glenda, is taking astronaut training.'[42]

The most prominent argument in support of space bound women was science based. According to scientists from numerous medical fields, among them the 'father of the H-bomb,' Edward Teller, women were expected to be at least as, or even more capable, of space flight than men because women usually weigh less and need less oxygen and food than men. Hence, they would be able to go up in lighter capsules or survive longer than men on the same supplies on space missions. It was even pointed out that because "women's reproductive organs are internally located, they should be able to tolerate higher radiation levels."[43] Contrary to the opponents of women in space who used female anatomy against them, the proponents used the female body as a point in the women's favor. Even their reproductive organs, which threatened some men with the plight of menstruation, were suddenly an advantage for others due to their internal location. Moreover, they also assessed the working of the female psyche favorably. Dr. Donald N. Michael, for example, pointed out that "women appear to have

---

41 Meyerowitz, pp. 1–18.

42 'People', *Time*, November 23, 1962, p. 26; William McGaffin, 'Female Space Barrier', *New York Post*, March 19, 1962; 'People', *Time*, March 23, 1962; Leo O'Hara, 'The Question: Should there be Women Astronauts?', *Boston Herald*, August 27, 1962.

43 'From Aviatrix to Astronautrix', *Time*, August 29, 1960, p. 41; 'A Lady proves she's fit for Space Flight', *Life*, August 29, 1960, pp. 72–76, (p. 73); Erik Bergaust, 'Is Space a Place for the Ladies?', *This Week*, September 2, 1962, p. 12; Susan Okie, *Well-Wishers watch Sally Ride take new Step for Womankind*, June 19, 1983, 008987, HRC; Mohammed A. Rauf Jr., 'Woman Scientist Sees Definite Need for Women in Space', *Sunday Star*, June 24, 1962. Also see: Weitekamp, *Right Stuff, Wrong Sex. America's First Women in Space Program*, pp. 64–65; Kevles, p. 7.

greater capacity to put up with routine and boredom,"[44] while E.E. Clark emphasized that women can

> concentrate on simple maneuver with no side thoughts on the scientific problems involved. In simulated space flights, we find that women usually catch on to the space rendezvous controls faster than men. They usually come alongside the other vehicle in their first or second try. Somehow they get the manual feel of space driving immediately.[45]

Douglas Welch was also convinced that women were "a good deal quicker mentally than men" and they had "more wit and perception of wit than men." This brought him to the conclusion that women would make great astronauts given the same education, the same encouragement, and the same opportunity. Speaking directly to his fellow male Americans, he told them: "[F]ella, they'll do anything you can do and often better."[46]

Other scientists went a step further and suggested specific types of women to become astronauts. Col. John Stapp, director of the Air Force's aeromedical laboratory at Wright Field, could especially see single, passive women in space because in psychological tests men were only able to stay 42 hours in a test chamber, while a woman stayed for 5 days and only wanted to leave because she ran out of food. According to him, these women would make better astronauts because men "start worrying about their families and become aggressive and irrational."[47] Other scientists suggested that minority women, such as Inuits and Aborigines, in particular would make the best astronauts, along with schizophrenics and extreme introverts. One researcher felt that a "female midget (Inca or Acara Indian) from the Andes Mountains of Peru would probably make an ideal astronaut once trained" because "these women are small; already adapted to rarified air, and they normally consume very little food." Yet, another scientist, Professor Alan Kekwick, head of a team of medical space researchers at Middlesex College, believed that "the best space traveler would be a short, fat woman. It seems the small, chubby girls are best adapted to convert food into energy, although they may eat comparatively little."[48]

Not just women's minds and bodies were assessed as an advantageous element for spaceflight, but also their emotions, which were constantly

44 Erik Bergaust, 'Is Space a Place for the Ladies?', *This Week*, September 2, 1962, p. 12.
45 Erik Bergaust, 'Is Space a Place for the Ladies?', *This Week*, September 2, 1962, pp. 12 and 14.
46 Douglass Welch, 'Girls best the men, again', *Pittsburgh Post-Gazette*, September 16, 1960.
47 'Yup! Gals May Rule Space, Too', *Baltimore News Post*, January 13, 1959.
48 Donald Cox, 'First Travelers into Space may be Fearless Females', *Pittsburgh Press*, July 21, 1961.

mocked by those opposed to female astronauts. Some scientists were convinced that precisely because women were more emotional, they were less prone to psychosomatic illnesses, while men were more likely to suffer from psychosomatic illnesses because "little boys are taught that it is 'manly' to hold their emotions in tight check."[49] For other supporters of space bound women, the question if women could do the astronauts' job was not even sensible anymore because women were clearly the superior sex in every way. They told their readers:

> What difference does it make in actual achievement as to whether the pilot is an astronaut or an astronautess is not clear to us. If anything the achievement of a man is greater than that of a woman, for it is a known fact that the feminine element is stronger in our day, lives longer, is sick less often, and can endure more. So if a man can orbit, anyone knows a woman can.[50]

Although the science based arguments were pretty strong, the opponents of female astronauts did not accept this line of argumentation that women were better suited for spaceflight due to their bodyweight. The American capability of building rockets which could carry the weight of men made that argument obsolete in their eyes. They also reacted similarly to the arguments of Cobb and Hart at the hearing in 1962. If women could only do the job as well as men – not better – there was no advantage in adding women to the existing astronaut corps.[51] Because the field of psychology was not yet very prominent in the early 1960s and NASA was also wary about it, the pro-women-argument based on psychology also failed to take off.[52]

Yet another approach to endorsing women in space was the proclamation of women's rights there as well as on Earth. Cobbs quest for female astronauts was sometimes already considered in the light of the women's movement, calling her "space suffragette."[53] Some people already felt in the early 1960s that women were just as entitled as men to fly to space and that they should have a future in space exploration because they were just as important in the 'take-over generation'. *Life*, for example, listed Jerrie Cobb as one of 100 Americans who

49 'The Weaker Sex?', *Newsweek*, July 1, 1963.

50 *Sex and Space*, unknown newspaper, ca. June 1963, Cobb/Schlesinger.

51 Weitekamp, *Right Stuff, Wrong Sex. America's First Women in Space Program*, p. 146; Ackmann, p. 167.

52 Patricia A. Santy, *Choosing the Right Stuff. The Psychological Selection of Astronauts and Cosmonauts* (Westport: Praeger Publishers, 1994), pp. XIII–XIX, 1–24.

53 Phyllis Nibling, 'Best Foot Forward. Celestial Intrigue', *Denver Post*, June 18, 1963. See Fig. 3.10.

Fig. 3.10: Unknown Newspaper, ca. 1960–1963, in: Jerrie Cobb Collection, Schlesinger Library, Radcliffe Institute, Harvard University.

would be important in this "new breed of American" – the "take-over generation," which is "filled with purpose" and "thinks on a scale that often scares his elders. He demands responsibility, not because he craves authority but because he can get the job done. He is, at this moment in history, starting to take over our destiny."[54]

Although, some people, especially women, could not understand why a woman would even want to go to space, they thought that if there were women who actually wanted to do so, they should be given a chance in order to see how they would hold up to the challenge.[55] Senator Maurine Neuberger, for example, added to the discourse: "I don't see why a woman can't do it as well as

54 'The Take-over Generation. Introducing the Generation. A Foldout Gallery: Young Leaders of the Big Breakthrough. A Red-Hot Hundred', Life, September 14, 1962. See comics.

55 Audrey Ross, 'Space Ladies' in the Future?', *Phoenix Gazette*, August 22, 1960; United Press International, *Scoffs, Bravos Aim At Woman in Orbit*, unknown newspaper, ca. post June 1963; Mrs. Mortimer Stuart, 'The Question: Should there be Women Astronauts?', *Boston Herald*, 008998, HRC; Jim Maloney, '4 being considered as Spacewomen', *Houston Post*, February 11, 1965.

a man, but I also don't know why any woman would want to [...] I'm perfectly happy to be here on earth."[56] Yet another popular argument in favour of female astronauts, or at least not against them, was that American spaceflight technology was so dependable and well-constructed, even a women could use it. What did it matter then if the 'spam in the can' was a woman or a man? In her strategy paper, amongst other things, Cobb argued that including a woman in the Mercury program would "prove the safety, dependability and simplicity of our spacecraft systems. If a woman can do it – it must be safe and simple!"[57]

Fig. 3.11: Chicago Tribune, October 5, 1962.

56 *Lady 'Astro' Hopefuls Blast NASA Brushoff. Red Girl's Success Don't Shift U.S. Policy Despite 'Equal Rights'*, unknown newspaper, May 16, 1963, Cobb/Schlesinger.
57 Jerrie Cobb, *Strategy Paper*, June 15, 1961, 000336, HRC. See also Fig. 3.11.

3.1.4 Valentina Tereshkova's historic spaceflight in the American public discourse

After Valentina Tereshkova orbited the Earth in June 1963, the discourse about women in space changed in the United States. Tereshkova, as the first woman to ever fly in space, became the linchpin of all arguments in support as well as in opposition of female astronauts. Quite a lot of Americans were upset and disappointed that yet another 'first' had been lost to the Soviets and asked NASA and the government why American women, like Jerrie Cobb or Jane Hart, who could have flown first were still only warming up the bench. Although most Americans resented this Soviet first because they were anti-Soviet, not necessarily pro-women, they complained nonetheless that Cobb's "theme, and oft-repeated one, was the need to train women as astronauts and put one into orbit. 'If we don't, the Russians will', she told them prophetically. Jerrie, of course, was right. No one was more furious than she was when Russia's Valentina Tereshkova beat her to the launch and chalked up 48 orbits."[58]

Fig. 3.12: Unknown Newspaper, June 1963, in: Jerrie Cobb Collection, Schlesinger Library, Radcliffe Institute, Harvard University.

---

58 Clare Booth Luce, 'U.S. Team is still warming up the Bench', *Life*, June 28, 1963, pp. 28–33; Leonard Bertin, '20 Years a Pilot, wants to be an 'Astronette'', *Toronto Daily Star*, September 17, 1963; Clare Booth Luce, 'She orbits over the Sex Barrier', *Life*, June 28, 1963, pp. 28–33, (p. 32); *Lady 'Astro' Hopefuls Blast NASA Brushoff. Red Girl's Success Don't Shift U.S. Policy despite 'Equal Rights'*, unknown newspaper, May 16, 1963, Cobb/Schlesinger; Karl F. Rolvaag, *Report to the People*, July 11, 1963,Cobb/Schlesinger; *Women in Space Given Support. NASA Scolded*, unknown newspaper, ca. post June 1963, Cobb/Schlesinger; 'Jubel in Moskau – aber Amerikas Fliegerinnen sind eifersüchtig', *Bunte*, 1963, Cobb/Schlesinger; Atkinson and Shafritz, p. 95; Weitekamp, *Right Stuff, Wrong Sex. America's First Women in Space Program*, p. 163; Kevles, p. 36. See also Fig 3.12.

Many expressed not only their anger but also their composure, because they felt that Tereshkova's flight had attested to their already existing belief in female astronauts. Senator Clinton P. Anderson let the world know that he was "not surprised. There isn't any reason why women can't go up into space." He also promised that the U.S. "will train some women astronauts for sure."[59] His sentiment was further reinforced by Minnesota Governor Karl F. Rolvaag who pointed out that "[a]s surely as day follows night, the time will come when America will launch its first lady astronaut. The Russians have already done so and I don't believe that their orbiting lady has half the qualifications of Miss Cobb."[60] Senator Jon Stennis also wanted the United States to catch up to the Soviet Union when he said: "I am sure we have women capable of duplicating the flight."[61] Like their political representatives, many Americans wanted American women to have the opportunity to prove that the U.S. was just as capable and even more so. Therefore Blake Green called upon all American women to take action:

> The Russians, dern' em, have done it again. [. . .] Woe unto us. We're not only behind in the space race, but in Women's rights. And that, let me tell you, is just awful. [. . .] How can you expect to win influence among the newly emerging nations of the world when our country is steeped in unequal opportunity for women? Now if I know you, you've been apathetic about this whole thing. Why right now I bet you're cooking breakfast, getting the kids dressed for church and trying to get a roast in the oven when you should be out beating the bushes, ringing doorbells and applying all kinds of pressures against this male-dominated society. Why do you think the Russians got ahead of us? Because of female folly on our parts, because we haven't cared enough, because we've sat back for years with a 'let John do it' attitude when you know darn well we should have been saying 'let Mabel do it.'[62]

Blake Green and other journalists started to give thought to women's rights in the United States in comparison to the Soviet Union and came to the conclusion that women were much more valued in the Soviet Union "because Communism preaches and, since 1917, has tried to practice the inherent equality of men and women."[63]

---

59 *Lady 'Astro' Hopefuls Blast NASA Brushoff. Red Girl's Success Don't Shift U.S. Policy despite 'Equal Rights'*, unknown newspaper, May 16, 1963, Cobb/Schlesinger; United Press International, *Scoffs, Bravos Aim at Woman in Orbit*, unknown newspaper, ca. post June 1963, Cobb/Schlesinger.
60 Karl F. Rolvaag, *Report to the People*, July 11, 1963, Cobb/Schlesinger.
61 United Press International, *Scoffs, Bravos Aim at Woman in Orbit*, unknown newspaper, ca. post June 1963, Cobb/Schlesinger.
62 Blake Green, *Equality? Never!*, unknown newspaper, ca. post June 1963, Cobb/Schlesinger.
63 Clare Booth Luce, 'She orbits over the Sex Barrier', *Life*, June 28, 1963, pp. 28–33, (p. 31). Also see: Jerrie Cobb's response letter to Gary Wormser about an article in the Russian Newspaper *Izvestia* about her book *Solo Pilot* published on July 6, 1963, post June 1963, Cobb/Schlesinger; 'Russia: Women Are Different', *Time*, June 28, 1963.

Most Americans, however, did not think of the Soviet Union as a country of equal rights, but as a country where women had to labor hard because they were compelled to do so in order to support their families. These Americans assumed that Soviet women were craving American luxuries and would love to live an American housewife's life with no job outside of the house, but with a husband able to support her and buy her pretty clothes and technological household helpers. For these Americans the domestic fulfilment of women and the success of the American way of life were fundamentally intertwined. This is why women in particular argued that "[n]o number of space flights can free the average Soviet woman from household drudgery as could a good old clothes dryer, dishwasher or efficient diaper service. These things are not available in the Soviet Union."[64] However, they recognized at the same time that Tereshkova's success would give the Soviet Union a competitive edge in the fight for the hearts and minds of possible allies in the Cold War: "The peoples of the world are not going to indulge in any very refined analysis of the difference between the status of women in the free world and the Communist world when they have the picture of Valentina Tereshkova leading her sex into space."[65] While some opponents and proponents of female astronauts started to question women's right in general, other pursued a more direct approach and started to challenge the space agency's selection requirements for astronauts. They especially wanted to know why American astronauts had to be test pilots when an "unqualified" Russian woman could do the same job with less training.[66]

This line of argumentation, or rather questioning, opened a new door for female astronaut hopefuls. Although NASA did not take Tereshkova's flight seriously, the discourse opened up more favourably for women. Especially Jerrie Cobb, once again, used Tereshkova's international momentum to appeal to NASA and the American public. She usually pointed out that she and Tereshkova had a lot in common despite their differing beliefs because they both wanted to serve their countries by being the first woman in space. She then proceeded to congratulate her on "the splendid accomplishment of her mission" and used her compliments as an argument on her behalf.[67] She tried to convince the men at

64 Audrey Topping, 'First in Space but not Femininity', *New York Times*, June 30, 1963; Vera Bacal, 'the latest Fashion News – from Moscow', *New York Times Magazine*, November 20, 1960, p. 125; Laville, 'Gender and Women's Rights in the Cold War', p. 526; Llinares, p. 84.

65 *New York Herald Tribune*, May 17, 1963, in: Robert L. Griswold, '"Russian Blonde in Space": Soviet Women in the American Imagination, 1950–1965', *Journal of Social History*, 45 (2012), 881–907.

66 'Russia: Women Are Different', *Time*, June 28, 1963.

67 Jerrie Cobb, *letter to Gary Wormser*, post June 1963, Cobb/Schlesinger.

NASA and the public that Tereshkova's flight was not a publicity stunt because the Soviets were "coldly scientific about everything they do in space."[68] According to Cobb, Tereshkova had also proven that the test pilot requirement in the U.S. was completely unnecessary and should be removed.[69] Therewith, she intensified her call for American women in space. She lobbied: "Call it anything you like – call it Project Femini – but let's not wait for Project Gemini."[70]

While some parts of American society interpreted Tereshkova's flight as a huge success for women around the world and a game changer for the American female astronaut wannabes, others, like the men in power at NASA or the military, assessed Tereshkova's flight as not scientifically significant and merely a Soviet publicity stunt.[71] Lt. Gen. Leighton I. Davis, commander of the US Missile Test Center in Cape Canaveral, for example, knew that he risked drawing "the ire of the feminine people" when he classified Tereshkova's orbits as "merely a publicity stunt."[72] Other NASA employees were more polemical in their response because they assumed that Tereshkova must have flown in 'the chimp mode,' meaning that she had basically been a passenger in the capsule while an automatic system piloted for her. Walter C. Williams, NASA director of the Manned Space Center, for example, was convinced that Tereshkova's flight was a far cry from an accomplishment: "There is no reason why Russia shouldn't orbit a woman. We could do the same. We have orbited a chimpanzee with an automatic system, for example."[73] Because most NASA officials believed that Tereshkova's flight was solely for propaganda purposes, they felt even more validated in their belief in NASA's astronaut prerequisites and believed their point of view as confirmed when the Soviet Union stopped sending women to space after Tereshkova. Only four years after Tereshkova's flight, they gloated that "[e]ven the Russians no longer seem interested in having astronettes."[74] The Americans also ridiculed the supposed Soviet

68 *She Tried to Swap Places with Chimp*, unknown newspaper, 1963, Cobb/Schlesinger.

69 George Natanson, 'Leading Woman Flier Hits U.S. Space Policy. Talks in Mexico City with USSR's first Woman in Orbit, says she's ready to go', *Los Angeles Times*, October 17, 1963.

70 *She's ready for space, but NASA says No*, unknown newspaper, October 16, 1963, Cobb/Schlesinger.

71 *Lady 'Astro' Hopefuls Blast NASA Brushoff. Red Girl's Success Don't Shift U.S. Policy despite 'Equal Rights'*, unknown newspaper, May 16, 1963, Cobb/Schlesinger; Weitekamp, *Right Stuff, Wrong Sex. America's First Women in Space Program*, p. 162; Kevles, p. 36. See cartoons.

72 United Press International, *Scoffs, Bravos Aim at Woman in Orbit*, unknown newspaper, ca. post June 1963, Cobb/Schlesinger.

73 Llinares, p. 80.

74 Anne V. Thompson, 'Equality of Sexes Ends Where Outer Space Begins', *Tulsa Daily World*, October 1, 1967. For more information on Soviet Women Cosmonaut Flight Assignments see:

publicity stunt by facetiously claiming that if they wanted to, they could easily outperform the Soviets in space propaganda by marrying Cobb off to an astronaut on national television, just like Tereshkova and her cosmonaut husband Adrian Nikolayev.[75]

With comments like these circulating, the Americans wanted to point out their scientific integrity once again. The depiction of Tereshkova as a feminine woman, contrary to the more common depiction of Soviet women as graceless, shapeless, and sexless, helped serve this purpose. Even though Tereshkova was sometimes portrayed as asexual and described as "no astro-beauty," the anti-astronettes usually emphasized her femininity and beauty to negate any technical competence on her part, whereas the pro-astronette lobby highlighted her femininity in order to demonstrate that highly competent women could be the ultimate Cold War weapon – in the propaganda sphere, certainly, but more so as reinforcements for the workforce to build up national productivity and win the Cold War.[76] Thus Tereshkova was often described by the media as "plumpy pretty" with a "winning smile" and very feminine because she had "a white dove embroidered – not laminated or stencilled, mind you – on her space suit." This "unexpected touch of femininity" made her the "most potent propaganda projectile," a "heroine who will appeal to both men and women."[77]

The thought of any kind of sexual activity in space was awakened with a female presence in there, and the fact that that Tereshkova directed the thoughts of quite a few people toward sex was another reason to not take her seriously. While some people considered it intriguing and others in bad taste, neither opinion promoted the idea of a serious, efficient female spacefarer in the American and international public.[78]

Bert Vis, 'Soviet Women Cosmonaut Flight Assignments 1963–1989', *Spaceflight*, 41 (1999), 474–80; Rex D. Hall, Bert Vis and David Shayler, *Russia's Cosmonauts* ([New York]: PraxisPublishing Ltd., 2005), p. 125; Shayler and Moule, pp. 43–68.

75 Allan Cromley, 'A Weakness in U.S. Space Program: Where is Husband-Wife Team? Lone Bachelor, Jerrie Candidate', *The Daily Oklahoman*, November 8, 1963, p. 24.

76 'The Hawk and the Gull', *Newsweek*, June 24, 1963; Griswold, pp. 881–882 and 889–893; Julia Richers, *Die Erste Kosmonautin: Valentina Tereškova Und Der Transkontinentale Geschlechterkampf Im Kalten Krieg* (Zürich: Gender in Trans-it, 2009), p. 239.

77 Phyllis Nibling, 'Best Foot Forward. Celestial Intrigue', *Denver Post*, June 18, 1963; Griswold, p. 893. See also political cartoons in: Fig. 3.13; Unknown Newspaper, July 1963, Box 6, Current News 1962/1963, James E. Webb Papers, John F. Kennedy Presidential Library and Museum (hereafter JFKPLM).

78 See political cartoon: Waite in *London Daily* Sketch, June 1963, Box 5, Current News 1962/1963, July 1963, James E. Webb Papers, JFKPLM.

Fig. 3.13: Bill Mauldin, 1963, in: Library of Congress Prints and Photographs.

However, it wasn't just Americans who levelled accusations of using women for space propaganda; in the name of the Soviet Union, Tereshkova herself claimed that the United States had used Jerrie Cobb for the same purpose. In the Soviet media, Tereshkova declared that the United States had never planned on giving women equal rights on Earth or in space; she claimed that NASA had appointed Cobb to become a consultant without ever planning on giving her a flight assignment. Although she complimented Cobb on being a "daring and courageous woman", she also publicly pitied her because she was put "in such an awkward position."[79] According to Tereshkova, the "American woman pilot hoped to be first in space orbit. She became first in propaganda orbit."[80] At the 56th conference of the International Aeronautical Federation, both Tereshkova and Cobb were guests of honor. In the presence of the American media, Tereshkova was able to ask, or rather tease, Cobb in person about her role in the U.S. space program.[81]

79 'We have different Wings, Jerrie! American Publicity and Soviet Reality. The Story of a Book and its Author', *Izvestia*, July 6, 1963.

80 Roy P. Stewart, *Jerrie Gets in Soviet Paper. Book on Flying Sets Reds Off on Propaganda*, unknown newspaper, ca. post June 1963, Cobb/Schlesinger.

81 George Natanson, 'Leading Woman Flier Hits U.S. Space Policy. Talks in Mexico City with USSR's first Woman in Orbit, says she's ready to go', *Los Angeles Times*, October 17, 1963. See also Fig. 3.14.

Fig. 3.14: Jerrie Cobb, Alberto Branif and Valentina Tereshkova at the 56th conference of the International Aeronautical Federation in Mexico, October 1963, Jesus Martinez Photography, Jerrie Cobb Collection, Schlesinger Library, Radcliffe Institute, Harvard University.

Tereshkova's accusations are quite interesting, considering that she herself was used as a propaganda tool by the Soviet Union. The Politburo had decided that Tereshkova's femininity and status as a working Soviet woman had to be emphasized to highlight the superiority of Soviet society and therewith dwarf the achievements of the male U.S. astronauts. The woman who had spent more time in space than all the American astronauts together had to be called 'Miss Tereshkova.' She also had to wear feminine civilian clothes despite her military rank of junior lieutenant to mark her as a member of the female sex. Soviet propaganda transformed Tereshkova into a cultural ambassador for Soviet women and girls, into an international symbol of gender equality, as well as into an embodiment of the traditional Soviet qualities of a mother and wife.[82]

82 Sue Bridger, 'The Cold War and the Cosmos: Valentina Tereshkova and the First Woman's Space Flight', in *Women in the Khrushchev Era*, ed. by Melanie Ilic, Susan E. Reid, and Lynne Attwood (New York: Palgrave Macmillan, 2004), pp. 203–21 (pp. 229–231); Richers, pp. 240–241; Roshanna P. Sylvester, 'She Orbits over the Sex Barrier. Soviet Girls and the Tereshkova Moment', ed. by James T. Andrews and Asif A. Siddiqi (Pittsburgh: University of Pittsburgh Press,

3.1.5 Gendered framing of the discourse

Tereshkova's flight resulted in a great many accusations and much finger pointing by both supporters and opponents of women in space, both of which groups felt reassured in their initial point of view. It does not seem as though Tereshkova's flight converted enough people to opposing views to have a significant impact on the previously existing landscape of opinions. However, for a brief time, the discourse about female astronauts was able to get to the front page. The discourse about women in space, before and after Tereshkova's flight, always consisted of similar elements, as both supporters and opponents of spacefaring women used essentially the same ingredients to build up an argument for or against women in space. The same was also the case for the discussion surrounding Tereshkova's flight. All stories about female astronauts emphasized the woman, not the astronaut. The female body took center stage in all the press coverage, either to sexualize or problematize the topic of women in space or to present the female body as the solution to space flight, and the language used to describe female astronauts was intentionally feminized. The media coined a variety of new words for this new breed of women: moon maid, astronautess, astro-eve, lady astronaut, spacewoman, astronautrix, astronette, spacegirl, space suffragette, astrodoll, spacegal, astrotrix and astrodame[83] are just some of the examples. The same was done for Tereshkova. She was called cosmonette and the like.[84] If the press did not feel the need to invent a new term, they at least pointed out that they were writing about a woman astronaut or a female cosmonaut.

2011), pp. 195–213 (p. 197); Roshanna P. Sylvester, '"Let's Find Out Where the Cosmonaut School Is". Soviet Girls and Cosmic Visions in the Aftermath of Tereshkova', in *Soviet Space Culture: Cosmic Enthusiasm in Socialist Societies*, ed. by Eva Maurer and others (Basingstoke: Palgrave Macmillan, 2011), pp. 121–38.

83 'Shuttle open to astrodames. Space Shuttle won't bypass women, but will cut down program costs', *Miami News*, May 1, 1971; 'Moon Maid's Ready', *Oklahoma City Press*, August 24, 1960; *Sex and Space*, unknown newspaper, ca. June 1963, Cobb/Schlesinger; Wernher von Braun, *letter to Jerrie Cobb*, May 9, 1962,Cobb/Schlesinger; Lyndon B. Johnson, *letter to Jerrie Cobb*, April 23, 1962, Cobb/Schlesinger; Ann Marshall, 'Spacewoman's hopes crushed by tragedy', *Nairobi East African Standard*, June 8, 1967, Technical Files Wally Funk, United States Air and Space Museum (hereafter NASM); 'From Aviatrix to Astronautrix', *Time*, August 29, 1960, p. 41; Anne V. Thompson, 'Equality of Sexes Ends Where Outer Space Begins', *Tulsa Daily World*, October 1, 1967; Joan McCullough, 'The 13 Astronauts Who Were Left behind', *Ms.*, 3 (1973), 41–45 (p. 41).

84 Griswold, p. 896. See also Fig. 3.2–3.13.

The women in question themselves did not feel the need for a new job title. Jerrie Cobb, for example, felt that "the word is neuter and doesn't take a feminine ending."[85] Still, they had to tolerate the media's depiction if they wanted to make their goal of flying to space more public. Considering that the word 'sexism' was non-existent at that time and that the women, like the space program as a whole, needed the media's and the people's favor, they had to accept, sometimes even embrace, the deliberate accentuation of their femininity in the media. The discourse about femininity was presented as fact and the media presented not only the Mercury 13's skills to the world, but mainly fixated on their measurements, hairstyles, fashion and other supposed feminine qualities such as their housewives qualities.[86] *Fleetway* Magazine, for example, wrote about Cobb that one

> might expect her to be a hard-faced, strident masculine woman. But, in fact, the first American woman to qualify to be literally out of this world is a soft-voiced, dimpled blonde, who often walks and flies barefoot, and loves poetry, cooking and her dachshund pup. Some people might think these feminine qualities are more suited to life on earth than to whizzing around the universe in a space capsule.[87]

Similar descriptions were also published about Tereshkova. Besides the domestic themes and gendered language, quite a few articles about Tereshkova and the American wannabe astronauts were published as 'soft news' in the women's section of the papers. Sometimes the feminizations of the articles about those women are interpreted as hate speech.[88] However, while some opponents of those women certainly used the feminized language to put the aspiring astronauts and cosmonaut down, those who supported Cobb and the other women supposedly emphasized their femininity for a different reason. In a time when women were expected to always look beautiful and feminine, the space dreaming women had to look as feminine as possible in order to be taken seriously as members of their gender and in their ambitions. Masculine looking women would not have stood a chance in an environment which

85 Carolyn Bell Hughes, 'New Tune: Lady on Moon', *Washington Post*, September 6, 1961.

86 Donald Cox, 'First Travelers into Space may be Fearless Females', *Pittsburgh Press*, July 21, 1961; 'From Aviatrix to Astronautrix', *Time*, August 29, 1960, p. 41; Richard Taylor, 'Let me tell you', *San Diego Independent*, June 19, 1960; Ann Marshall, 'Spacewoman's hopes crushed by tragedy', *Nairobi East African Standard*, June 8, 1967.

87 *First Blonde in Space? Should a woman take the risks?*, unknown newspaper, ca. 1960–1963, Cobb/Schlesinger.

88 Llinares, p. 68.

forced its women to look this certain way. Had prospective spacefaring women been identified as masculine, they would not have been regarded as true representatives of their sex because they were not performing the gender assigned to it. Performing the 'wrong' gender might have even have led them to be considered lesbians, which would have forced them to the margins of society. At the very least, they could not have been regarded as prototypical American women – a propaganda prerequisite for becoming an astronaut. Therefore, their supporters needed them to be identified as feminine in order for them to be a part of and to represent mainstream American society. However, one has to agree with Dario Llinares when he points out that "[i]t is striking how the rejection of female astronauts did not rely overtly on the denial of their strength, character, biology or even their emotional instability, but rather on the ideological gender roles that the social order interpellates."[89] Careful attention to the arguments of those opposed to women in space shows that the men at NASA, along with the majority of the American public, knew that the limitations for women in space had not a biological but a social origin. Recall, for example, John Glenn's admission during the hearings before the Special Subcommittee on the Selection of Astronauts of the Committee on Science and Astronautics that the selection of the astronauts in 1962 is based on the American social order.[90]

One of the main problems for prospective spacefaring women was that American women did not stand together as a united front in their wish to alter the social fabric of the United States. The most rigorous opponents of the astronaut hopefuls were, in fact, often other women.[91] Despite dissonance within the sexes, the media often framed the coverage of Tereshkova or Cobb within the context of the "age-old battle" between the sexes and sometimes even assessed their call to put more women in space as the continuation of the women's movement.[92] Phyllis Nibling of the *Denver Post*, for example, though that

> [i]t looks like a new battle of the sexes is brewing, and some women are already in orbit. One can imagine latter-day Suffragettes – perhaps called Astronettes – picketing

89 Llinares, p. 74.

90 *Hearings before the Special Subcommittee on the Selection of Astronauts of the Committee on Science and Astronautics U.S. House of Representative*, 87th Congress, 2nd Session, July 17–18, 1962, No. 9, U.S. Government Printing Office, Washington, D.C., 1962., p. 67.

91 United Press International, *Scoffs, Bravos Aim at Woman in Orbit*, unknown newspaper, ca. post June 1963, Cobb/Schlesinger.

92 The Weaker Sex?', *Newsweek*, July 1, 1963; *Sex and Space*, unknown newspaper, ca. June 1963, Cobb/Schlesinger; William McGaffin, 'Female Space Barrier', *New York Post*, March 19, 1962.

> Cape Canaveral and staging sit-ins on launching pads. They will be led by Jerri Cobb and Mrs. Philip Hart in the roles of the Amelia Bloomer and Susan B. Anthony of the Space Age.[93]

The aspiring astronauts were not interested in a battle of the sexes, though. Cobb was often quoted saying: "This is not a battle of the sexes – we're talking about human beings, not just men and women" or "[t]his isn't a battle between the sexes, it's a battle of Red Tape. I've talked to everybody, and no one wants to take the responsibility to do what we must."[94] The women did not want to be seen as troublemakers. They did not want to create a new social order, nor were they feminists, despite pushing for a cause the women's movement would applaud just a couple of years later. They just wanted to fly and to serve their country. In the tradition of the female pilots of the 1950s, Cobb and most of the other women wanted to succeed within the existing social conventions.[95] For this reason, they colluded with coverage which emphasized their feminine nature by always accentuating their feminine side themselves through feminine clothes and make-up. Of all the women who wanted to fly to space in the 1960s, only Jane Hart complained about the prevalent sexism in the space community in the early 1960s. She explained that "a great many men" desired to "keep us in our place, whatever that is."[96] She would later join the National Organization for Women (NOW) as a member of the board.[97]

3.2 Lady engineers and astrowives in the 1960s

Tereshkova's historic flight changed the public discourse not just about female astronauts, but also about male astronauts. When the Mercury 7 had been introduced to the world, they had been presented as clean-cut, modest, hardworking, all-American family men. Similar to their Soviet counterparts, they were supposed to be ordinary heroes. Their masculinity of these military men was not especially emphasized because it appeared to be intrinsically

93 Phyllis Nibling, 'Best Foot Forward. Celestial Intrigue', *Denver Post*, June 18, 1963.

94 *She Tried to Swap Places with Chimp*, unknown newspaper, 1963, Cobb/Schlesinger; Claire Wallace, 'Spacewoman up in the air over Red Tape', *New York World-Telegram and Sun*, June 21, 1963.

95 Weitekamp, *Right Stuff, Wrong Sex. America's First Women in Space Program*, p. 55.

96 William McGaffin, 'Female Space Barrier', *New York Post*, March 19, 1962.

97 Ackmann, p. 176.

evident. Stories about their virility, like adultery and boozing, were in fact suppressed in the public sphere, including the media. However, with the appearance of Cobb and the question of women in space, NASA and the media started to highlight the masculinity of the pilot astronauts, something which was not done for the scientist astronauts, who were selected in 1965 and 1967.[98] The fortified emphasis on masculinity seemed necessary because Cobb and her fellow female pilots had "de-stabilized many of the iterative bodily performances enacted through NASA that prescribed binary gendered assumptions."[99] This masculinity framing was especially pursued after Tereshkova further proved that neither a penis nor a test pilot license was necessarily required for a successful space flight, though the portrayal of military pilot astronauts as the epitome of masculinity would not peak until almost 20 years later, in 1979, with the publication of Tom Wolfe's novel *The Right Stuff*.[100]

While the socialist world enjoyed the American gender debates after Tereshkova's historic feat upset the American conception of masculinity and femininity, another facet of the discourse about women in space became more

98 Margaret A. Weitekamp, 'The "Astronautrix" and the "Magnificent Male": Jerrie Cobb's Quest to Be the First Woman in America's Manned Space Program', in *Impossible to Hold. Women and Culture in the 1960s*, ed. by Avital H. Bloch and Lauri Umansky (New York University Press, 2005), pp. 9–28 (p. 22 and 25); McCurdy, *Space and the American Imagination*, pp. 88–91; Hersch, 'Return of the Lost Spaceman: America's Astronauts in Popular Culture, 1959–2006', pp. 75–84; James Spiller, 'Nostalgia for the Right Stuff: Astronauts and Public Anxiety about a Changing Nation', in *Spacefarers. Images of Astronauts and Cosmonauts in the Heroic Era of Spaceflight*, ed. by Michael J. Neufeld (Washington, D.C.: Smithonian Institution Scholarly Press, 2013), pp. 57–80 (pp. 57–67); Asif A. Siddiqi, 'Cosmic Contradictions. Popular Enthusiasm and Secrecy in the Soviet Space Program', ed. by James T. Andrews and Asif A. Siddiqi (Pittsburgh: University of Pittsburgh Press, 2011), pp. 47–76 (p. 72); Brian O'Leary, *The Making of an Ex-Astronaut* (London: Michael Joseph Ltd, 1970); Sharon Begley, 'Down to Earth', *Newsweek*, 1996, pp. 30–36 (p. 36); John Hockey, 'No More Heroes. Masculinity and the Infantry', in *Military Masculinities: Identity and the State*, ed. by Paul Higate (Westport: Praeger, 2003), pp. 15–26 (p. 17); David J. Shayler and Colin Burgess, *NASA's Scientist-Astronauts* (Chichester: Praxis Publishing, 2007); Hersch, *Inventing the American Astronaut*, pp. 41–102.

99 Daniel Sage, 'Giant Leaps and Forgotten Steps: NASA and the Performance of Gender', *Sociological Review*, 2009, 146–63 (p. 157).

100 Tom Wolfe, *The Right Stuff* (New York: Farrar-Straus-Giroux, 1983); Margaret Lazarus Dean, 'Warrior and Worriers: Risk, Masculinity, and the Anxiety of Individuality in the Literature of American Spaceflight', in *Spacefarers. Images of Astronauts and Cosmonauts in the Heroic Era of Spaceflight*, ed. by Michael J. Neufeld (Washington, D.C.: Smithonian Institution Scholarly Press, 2013), pp. 203–25; Spiller; Llinares, pp. 126–146.

pronounced in the public space.[101] NASA placed an exaggerated emphasis on the contributions of its female scientists and engineers as a response to the calls for female astronauts. Even though NASA had already done so before Tereshkova's historic flight, the space agency increased this type of framing after Tereshkova's accomplishment had fueled the discussion about women in space in June 1963. NASA personnel participated in several symposia panels and entire space symposia specifically catering to women, the agency authored various essays about its notable female scientists, and prominent NASA personnel and engineers delivered numerous speeches.[102] From the administrator to the astronauts to NASA's notable female scientists, they all gave public talks addressed to mixed and all-female audiences about the topic of women in the space community.[103] No matter the speaker or author, all of them generally applauded women's wits and capabilities in their speeches. James E. Webb, for instance, especially appreciated that women were always effective forces for the support of space exploration and generally "of any worthwhile human endeavor."[104] After complimenting the women's "genuine interest" in the space program, the speakers and authors usually moved on to explain why women could not become astronauts at the current time, but that they were still valued members of the space community. To prove their point, they always cited examples of the contributions of notable female scientists like NASA's chief of its Astronomy and Solar Physics Branch,

101 Richers, p. 240.

102 NASA Office of Public Information, *Women finding their Place in the Sun in America's challenging Space Program*, 1960s, 008998, HRC; Sol Levine, pp. 123–144. For example: the Women's Space Symposium in Los Angeles, February 22, 1962; the Mills College Symposium: Women's Place in Space Science in Oakland, October 228, 1966; Women in Aviation and Space Symposium at JSC, March 12 and 13, 1980. For more information see: *Mills College Founders Day Program*, Mills College Archive; Marion Dietrich, 'No Skirts in Space?', *Oakland Tribune*, October 29, 1966; Women's Space Symposium in Los Angeles, February 22, 1962. For more information see: Jerrie Cobb, *Space for Women?*, speech at the First Women's Space Symposium, February 22, 1962,Cobb/Schlesinger; Dee Dickson, 'Girl Students urged to be Space Pioneers', *Los Angeles Herald-Examiner*, February 23, 1962, Cobb/Schlesinger; *Women in Aviation and Space Symposium*, March 12–13, 1980,008987, HRC.

103 James E. Webb, *Address before the General Federation of Women's Clubs*, June 27, 1962, Harris/ MSRC; Alan B. Shepard, *Women in the Space Age*, speech presented at the Officers' Wives Club, May 11, 1965, 008998 HRC; Catherine Dryden Hock, *Space, the next Goal*, speech presented at the Student-Parent Night, Society of Women Engineers, February 24, 1965, 008998, HRC.

104 James E. Webb, *Address before the General Federation of Women's Clubs*, June 27, 1962, p. 1, Harris/MSRC.

Nancy Grace Roman. Dr. Roman was generally described as a "woman doer."[105] She herself was one of NASA's most frequented speakers on this particular topic and Roman used herself and other female scientists as examples in her addresses and essays, too.[106]

3.2.1 Framing female scientists as important for the American space effort

The agency was additionally very accommodating to media outlets which asked for information about NASA's most renowned female scientists. NASA's generosity with information led to various articles in newspapers and magazines about the women in the labs.[107] The main tenor of the articles was to reject the criticism of the U.S. space program, the charge that "we have no women 'up there,'" and to show that "we have a lot 'down here' who are helping push America to new heights in the peaceful competition for space leadership."[108] One integral part of this framing was to emphasize that most female scientists did not even desire to become astronauts because their work on Earth was just as exciting and important.[109] Another argument frequently pre-

105 A White House term for a woman of intelligence and achievement. Stems from President Johnson's emphasis on a wider role for women in government. NASA Office of Public Information, *Women finding their Place in the Sun in America's challenging Space Program*, 1960s, 008998, HRC.

106 Nancy Roman, *The Role of Women Scientists in the Space Program*, speech at Marymount College, Tarrytown-on-the-Hudson, January 31, 1963, MC 575, Box 160, Folder 2036, Betty Friedan Collection, Schlesinger; Richard l. Callaghan, *letters to Senator Gordon Allott*, August 19, 1965 and October 13, 1965, 1821, HRC; Zoe von Ende, 'NASA Astronomer: Women not barred from Space Program', *Denver Post*, March 18, 1966; Nancy G. Roman, *Careers for Women in Astrophysics*, May 16, 1966, 1821; Nancy G. Roman, *Visit to Sullins College, Bristol, Va.*, report to FGE/Speaker Scheduling Officer, Office of Public Affairs, February 19, 1969, 1821, HRC; Nancy G. Roman, *letter to Mrs. A. Dorothy Berger*, September 8, 1966, 1821, HRC; Nancy Roman, *Space Age Exploration and the Role of Women*, ca. early 1960s, MC 575, Box 160, Folder 2036, Betty Friedan Collection, Schlesinger.

107 'Women Aid Astronauts', *Telegram Bridgeport, Conn.*, June 26, 1963; 'The Supporting Cast: Women Scientists Back Up Space Shots', *Washington Daily News*, September 11, 1964; Joseph L. Myler, 'Girl Experts in Space Age!', *New York Herald Tribune*, June 16, 1963; Mary Lou Loper, 'Woman Explores Science's Frontier', *Los Angeles Times*, August 7, 1960; Judith Cass, '150 Hear Space Talk by Woman', *Los Angeles Times*, October 22, 1965.

108 'A Woman's Space in Science', *Goddard News*, July, 1963, pp. 3–5.

109 NASA Office of Public Information, *Women finding their Place in the Sun in America's challenging Space Program*, 1960s, 008998, HRC.

sented in these articles pointed out that women were very welcome to join the space community workforce because women's abilities were "this great untapped resource" which could give the U.S. the competitive edge over the Soviet Union in the space race.[110] It was usually noted by male and female NASA representatives that the agency only employed its workers on a merit basis and that every qualified American could therefore join the space program, irrespective of their sex. Thus, the "requirements for women in its space sciences, engineering, administrative and management jobs are exactly the same as they are for men."[111] Of course, journalists often questioned the female scientists about the truth of this statement. Reporters wanted to know why there were comparatively few women at work in the space agency and if the women had experienced any sex-based hostility or discrimination at the workplace. Female NASA scientists always assured their interlocutors that they were treated fairly at NASA and that it was American women themselves, not the agency, who were responsible for their underrepresentation. According to Edith Reed, a physicist in the astrophysics branch at Goddard, most women were more interested in preparing for marriage than a scientific career. Sometimes they also pointed out that one should account not just for the 160 professional aerospace technologists and 70–80 professional mathematicians at NASA, but also for the skilled female scientists and engineers in the space community, the space industry which worked for NASA.

However, women often admitted that there still were some sources of discrimination in "the sciences" in general. Barbara Lunde, a physicist in the stabilization and control branch at Goddard, for example, admitted that "women haven't yet gained complete equality in the sciences" and although there "may be some justified reasons," "most are unjustified, based on traditions and customs." While her African American mathematician colleague Dorothy Hoover merely disclosed that "women may be underestimated," Virginia Zanner, a systems analyst in the physics branch, was a little bit more specific. She felt that there was no barrier for women as long as "a woman works as hard as a man and doesn't expect any extra privileges," but she also revealed that there were some men "who sometimes find that 'they just can't work with women.'"[112] In

110 Marion Dietrich, 'No Skirts in Space?', *Oakland Tribune*, October 29, 1966.

111 NASA Office of Public Information, *Women finding their Place in the Sun in America's challenging Space Program,* 1960s, 008998, HRC.

112 'A Woman's Space in Science', *Goddard News*, July, 1963, pp. 3–5. See also: Blair Justice, 'Her Career in in Science. They have a common Trait: 'Stick-to-Itiveness'', *Houston Post*, February 7, 1965; Toni House, 'Woman Space Launcher Comes Home', *The Evening Star*, December 24, 1970.

the late 1960s and early 1970s, allusions like these usually led to the question if the female scientists were active in the women's movement. However, it has to be emphasized that successful working women in the late 1960s were constantly asked about their opinions about the women's movement no matter the actual reason for the interview. Just like the aspiring female astronauts in the early 1960s, most of the female scientists and engineers in the late 1960s also distanced themselves from the women's movement. They wanted to be recognized for their hard work and technical skills and did not want to give the impression that they had gotten their jobs because of anything other than that. Goddard Space Flight Center project manager Marjorie Townsend, for instance, did "not fancy herself as a leader of women's liberation" but attributed her professional success to her engineering skills and education.[113]

Most of these types of articles were written by women, most likely because they were often published in the 'woman's pages' of the newspapers or written for women's magazines. One article was basically like the next, the pattern they followed always the same, the tenor always upbeat and positive. The women were introduced to the readers and applauded for their part in the American space effort. There basically were no other portrayals. Clearly, the variety of opinions was by far not as broad as had been the case for the articles about the women who wanted to become astronauts. The portraits of notable female scientists, mathematicians, and engineers usually did not pay that much attention to the professional training, education, and work assignment of these women. Instead, they focused on the fact that these successful "female NASA employees were women first and foremost."[114]

Local newspapers mostly used the women who worked for NASA installations in the area they covered for their articles. However, there were some women who were featured more often than others. Dr. Nancy Roman, Laurel van der Wal, Barbara Lunde, Jaylee Burley, Melba Roy, Dr. Jocelyn Gill, Dr. Evelyn Anderson, Eleanor Pressly, Marjorie Townsend, Dr. Marcia Neugebauer, Harriet Malitson and Ann Bailey, in particular, were featured on a regular basis, most likely because they were among the most successful female scientists in the space community or because they made themselves more accessi-

113 'Spacecraft Launch Boss is a Mother', *Huntsville Times*, December 15, 1970.

114 Jennifer Ross-Nazal, *Women Scientists at NASA*, paper presented at Women, Gender, and Science, a Conference at Suny New Paltz, October 22, 2005, p. 5, 008988, HRC.

ble to journalists than women like JSC's Rita Rapp, who sometimes just ignored reporter's questions.[115] Their intelligence was most certainly one reason these women were often featured in the media, but another reason was their looks. Most articles not only emphasized the women's "calm mastery over a surprising range of space skills," but highlighted their beauty, clothes, and marital status.[116] The physiologist and food subsystems manager Rita Rapp was, for example, described as a "girl technical genius," and as "an unassuming, soft spoken woman" with "smoothly brushed hair [...] lustrous with strands of silver-blond," who was "slender and looked immaculately tailored in her green and white tweed dress."[117] Dr. Jocelyn Gill, NASA's Chief of in-flight sciences, was called an "attractive, cheery-voiced woman astronomer", and Barbara Lunde who had just submitted two patents for her inventions was described as an "attractive 26-year-old aerospace engineer-housewife."[118] Descriptions like these exist for every single female scientist featured in a speech or the media because the journalists wanted to show that 'beauty' and 'brains' do go together, that women do not necessarily have to lose their femininity as a lab worker in the space community. Eloise Engle directly approached the issue in her article for Dodge Magazine:

> The 'helpless, swooning dumb blondes' were given a third look as they moved into top career jobs. The reason for the first and second looks was obvious; the third look was caused by the 'startling' revelation that these women had completely shattered the old myth that beauty and brains don't mix. Nowhere in today's world is this fascinating combination more evident than in the latest feminine maneuvers in the field of Space and the Space Sciences. Quite literally there is no area here that is completely devoid of the rustling petticoat and the soft, albeit authoritative, soprano voice, but just the same they're dames.[119]

To further highlight the engineers' and scientists' status as members of the female sex, they were also always called 'lady engineer,' 'lady scientists,' 'lady

115 Juli Daves, 'She feeds Astronauts', *Today*, October 10, 1968, p. 1E-2E. See also Fig. 3.15–3.19.
116 Sara Azrael, 'Women Behind the Spacemen and the Satellites', *Sun Magazine*, May 17, 1964, pp. 19 + 21; NASA Office of Public Information, *Women finding their Place in the Sun in America's challenging Space Program*, 1960s, 008998, HRC. See Fig. 3.15–3.19.
117 Juli Daves, 'She feeds Astronauts', *Today*, October 10, 1968, p. 1E-2E.
118 'Woman Scientist to make Flight', *Evening Star*, June 29, 1963; James E. Webb, *American Progress and Goals in Space*, speech at the Inventors' Congress and Space Symposium, October 30, 1964, GH 101, General Reference Series, JSC.
119 Eloise Engle, 'Brainy Gals help push America's race-into-space', *Dodge Magazine*, April 1962, pp. 4–7.

astronomer' or received other gendered occupational titles.[120] The readers were basically never allowed to forget that these scientists were feminine women first.

3.2.2 Gendered framing of the reportages about the female scientists

The articles about these accomplished women often read like a mixture of first grade report cards and dating profiles. The women are praised for their scientific work with statements like 'good job' or 'fine job,' just like in a teacher's report card. The articles also usually list the women's hobbies, similar to dating profiles. In an article about the MSFC physicist Ann F. Whitacker, one can, for example, read that "Ann likes all aspects of her job" and that "she plans to make this a career." The article went on to report that

> ' [r]eading is one of her favorite pastimes but Ann also likes the outdoor life. She enjoys swimming, fishing and hunting. Her supervisor, Keith Demorhorst, says Ann has done a 'fine job' in every task assigned. Basically quiet but not shy, she is well liked by her fellow workers, most of whom are men, he said. 'She never gets upset,' Demorest said, 'even though she catches a lot of kidding, That's the way it is, though, when you're outnumbered.[121]

In another example, a different article referred to Roman as "young, 36, strawberry blonde" and a "peppery, bespectacled young lady" who does an "outstanding job" at work but who "also cooks, sews, and reads a great deal, too."[122] Homemaking skills are often especially highlighted in these types of reportages.[123] Although the women's cooking skills were frequently praised, it was also pointed out when women were lacking in this department, as was the case for Marjorie Townsend, whose son did not think that his mother was a good cook.[124] The coverage of the female scientists also liked to make

120 'The Supporting Cast: Women Scientists Back Up Space Shots', *Washington Daily News*, September 11, 1964; 'The Sky's the Limit for Lady Astronomer', *Newport News Times Herald*, July 19, 1962. The same was also done for female engineers and scientists who did not necessarily work in the space community: Amy Sue Bix, 'From "Engineeresses" to "Girl Engineers" to "Good Engineers": A History of Women's U.S. Engineering Education', in *Women in Engineering: Pioneers and Trailblazers*, ed. by Margaret E. Layne (Reston: ASCE Press, 2009), pp. 11–34.

121 'Ann Whitacker is a Physicist in P&VE's Materials Division', *Marshall Star*, March 24, 1965.

122 'The Sky's the Limit for Lady Astronomer', *Newport News Times Herald*, July 19, 1962.

123 Lillian Levy, 'Career Girl 1959. A Space Career. She doesn't want to be launched', *Sunday Star*, May 31, 1959.

124 Toni House, 'Woman Space Launcher Comes Home', *The Evening Star*, December 24, 1970.

comparisons between the women's professional work and their homemaking skills. Goddard engineer Marjorie Townsend, who supervised the work of 30 engineers, was, for example, described as a "pert housewife who looks as though she were about to demonstrate a cake on a TV commercial."[125] The work of the women who researched in scientific fields related to food or hygiene in space was particularly frequently compared to "housekeeping on a cosmic level," most likely because the articles were usually written by the magazines' food editors.[126] Phyllis Riely, biochemist and specialist in life support and waste management for the Pall Corporation, Mary V. Klicka, space food researcher for the U.S. Army Laboratories in Natick, Mass., Beatrice Finkelstein, senior research nutritionist for the Whirlpool Corp., and Rita Rapp were often confronted with the stereotype of being "chef de cuisine for America's astronauts" and working in a space "test kitchen."[127] However, Rapp specifically pointed out that the food she helped to develop, was "not just whipped up – it's engineered."[128] On rare occasions, the women were able to turn the kitchen metaphor around. Mary Hann, mathematician at GSFC, for instance, pointed out that "sex makes no difference" for scientific work, just like it did not matter in the kitchen because "some of our best cooks are men."[129]

Another important aspect of most of the articles were references to marriage and motherhood – the ultimate proof of femininity in American society, still in the twentieth century. The majority of the reportages pointed out which of the women were married and how many children they had. The point of these mentions was probably to show that a woman could work outside of the home without compromising her femininity, that she could be a scientist and a homemaker. The women themselves actively participated in this kind of portrayal. Barbara Lunde, for example, publicly disclosed that "she would like to have children herself but she will not quit her job."[130] ARC's Dr. Evelyn Anderson told women in the

---

125 Sara Azrael, 'Women Behind the Spacemen and the Satellites', *Sun Magazine*, May 17 1964. See also Fig. 3.18.

126 Dolores Alexander, 'She helps Astronauts live cleanly in Space', *Newsday*, June 7, 1967.

127 NASA Office of Public Information, *Women finding their Place in the Sun in America's challenging Space Program*, 1960s, 008998, HRC; Dolly Hlava, 'Apollo 15 Space Food – A Tase of Home', *Colorado Springs Sun*, September 22, 1971, p. 7; Nancy L. Ross, 'Rehydrated Shrimp Dish Replaces Shoe Leather', *Washington Post*, 1968, Box 8, Food Systems (R. Rapp Files), Center Series, JSC; Dolores Alexander, 'She helps Astronauts live cleanly in Space', *Newsday*, June 7, 1967.

128 Dolly Hlava, 'Apollo 15 Space Food – A Tase of Home', *Colorado Springs Sun*, September 22, 1971, p. 7.

129 'A Woman's Space in Science', *Goddard News*, July, 1963, pp. 3–5.

130 Sara Azrael, 'Women Behind the Spacemen and the Satellites', *Sun Magazine*, May 17, 1964.

audience of one of her speeches at a symposium on women in space to "choose both, marriage and career."[131] Nancy Roman also urged young women who wanted to concentrate solely on homemaking "to take at least an elementary course in a science, so that you can have some idea of what is going on" in the world, even when they did not plan on becoming scientists. In favor of her argument, she also pointed out that these young women "might marry a scientist, and will want to know how to talk about things that interest your husband and his friends."[132] JSC's Ivy Schneider, "a 5-feet 2-inches bundle of energy" also highlighted that "there are 'working mothers' in science and mathematics," and also emphasized that "certainly there is nothing in this field that keeps a woman from being feminine."[133] The concept of women in space in relation to femininity was also directly addressed by Lunde, who was convinced that women working in space science or flying to space would not lose their femininity. According to Lunde, "[f]emininity is a misty term. Women will remain women, and space colonies will need them for bearing babies." She also pointed out that the female deities of Rome and Greece who lived in space "were the most feminine ever conjured by mankind."[134] These examples show that most of the female scientists supported their very feminized portrayal in the public space by bring up the topic of femininity, motherhood, and marriage themselves. Just like the aspiring female astronauts in the early 1960s, it was important to the women in the labs of the space community in the 1960s to be perceived as women, too. Because of it, their achievements often took a back seat in the articles about them. Only on rare occasions did the reporter notice this fact. In an article for the *Christian Science Monitor*, Dorothy Leavitt, for instance, pointed out that Roman was an "unassuming, well organized lady, inclined to minimize her achievements and the role she is playing in the evolution of the Space Age."[135]

More often than not, however, the editors and reporters already had an idea about the article they wanted to publish before even meeting the female scientists. For example, when Lucy Rogers Baggett, reporter for the *National Business Woman*, requested an interview with one of the notable women

131 Marion Dietrich, 'No Skirts in Space?', *Oakland Tribune*, October 29, 1966.

132 Ruth Wagner, 'Astronomer Down to Earth', *Washington Post*, July 26, 1962.

133 Blair Justice, 'Her Career in in Science. They have a common Trait: 'Stick-to-Itiveness'', *Houston Post*, February 7, 1965.

134 Mohammed A. Rauf Jr., 'Woman Scientist sees definite Need for Women in Space', *Sunday Star*, June 24, 1962.

135 Dorothy Leavitt, 'Space Woman sets Sights on Moon', *Christian Science Monitor*, July 18, 1960.

working for the Apollo program, she seems to have had a certain storyline in mind. Although she emphasized that the first consideration in picking the woman should be "the importance of the contribution the woman is making to this vital activity of our times," she also revealed that "[w]e always hope for youth and photogenic qualities."[136] Her specifications show that the stories about the hard working, feminine, and beautiful woman scientists were often created before the interviews actually happened. An even more explicit example is the anecdote of Dorothy B. Lee, an aerospace engineer at Langley and JSC. She remembered that a reporter once asked her if it is true that "women working with men have to think like a man, work like a dog, and act like a lady?" Lee told the reporter that she agreed with that statement; however, when the article came out in the Sunday paper, the reporter had quoted his own question as a statement of Lee's.[137] The reporter basically put the words in her mouth because he (or she) apparently wanted the article to go in a certain direction.

Most of the articles in the mainstream media exclusively portrayed white female scientists. The few female scientists of color who worked in the space community in the 1960s were usually featured in minority magazines like *Ebony* or *Jet*. The tenor of these reportages was similar to that of those in the mainstream magazines, though. Just like her white colleagues, Yvonne Clark, an African American associate professor of mechanical engineering at Tennessee A&I State University and NASA contractor, was for example praised for "her feminine tidiness and attention to detail" in *Ebony*.[138] There were also already some articles and booklets which portrayed the work and life of women like Dorothy B. John, an African American mathematician, and Dr. Tsu-tzu Tsai, an Asian American chemist, both employed at Wright-Patterson Air Force Base.[139] The minority woman who was featured by far the most in the mainstream reportages about female scientists in the space community was Goddard mathematician Melba Roy. In the magazine articles, Roy was not portrayed as beautiful like her white

136 Lucy Rogers Baggett, *letter to Robert C. Seamans*, November 18, 1966, 008998, HRC.

137 Dorothy B. Lee, interview by Rebecca Wright, November 10, 1999, p. 17, Oral History Transcript, NASA Johnson Space Center Oral History Project, JSC. Apparently, JPL's Macie Roberts had often told her supervisees in the computing pool exactly that mantra. See: Nathalia Holt, *Rise of the Rocket Girls: The Women Who Propelled Us, from Missiles to the Moon to Mars* (New York: Little Brown, 2016), p. 256.

138 'Tennessee State's Lady Engineer. Professor Splits Time between Teaching, Space Project Work', *Ebony*, 1964, pp. 75–82 (p. 75).

139 Mary Finch Hoyt, *American Women of the Space Age* (New York: Athenum, 1966).

colleagues, whose bodies and skin color probably better fit the American ideal of female beauty and femininity in the 1960s. In an article in which Barbara Lunde is described as "trim and piquant" and Jaylee Burley as "pretty," Roy is, for example, only characterized as "pleasant" and "mild mannered."[140] Still, it has to be highlighted that women of color were already included in these reportages, not only in the text but also in the accompanying pictures. NASA as well as the mainstream media usually only started to stress the contributions of these women in the 1970s, in order to make the space agency look like a model equal opportunity employer.[141]

However, from today's perspective, it is quite interesting that Melba Roy was featured the most in the mainstream media, because she is not as famous nowadays as some of her other African American colleagues, like mathematician Katherine Johnson, who received the Presidential Medal of Freedom in 2015 and is the main character in the book and Hollywood movie *Hidden Figures*.[142] Then again, the same phenomenon happened with their white colleagues. In 2016, Margaret Hamilton received the Presidential Medal of Freedom. She had been a software engineer for Charles Stark Draper Laboratory at MIT, which developed the in-flight software for the Apollo missions. In 2015 a picture of her, standing next to the stacked-up code she and her MIT team had written for the Apollo project, went viral. This is quite interesting because Hamilton had not been one of the frequently featured female scientists in the news reports in the 1960s.[143]

140 Sara Azrael, 'Women Behind the Spacemen and the Satellites', *Sun Magazine*, May 17, 1964; Nancy Roman, *Space Age Exploration and the Role of Women*, ca. early 1960s, MC 575, Box 160, Folder 2036, Betty Friedan Collection, Schlesinger. See also Fig. 3.16, Fig. 3.17, Fig. 3.19., Fig. 5.3.

141 Federal Women's Program Office of the NASA Office of Equal Opportunity Programs, *Women at work in NASA*, 1979, Box 18, women in space, Center Series – Poindexter Reference Files, JSC; Rosen, p. 303; Margaret W. Rossiter, *Women Scientists in America. Vol. 3. Forging a New World since 1972* (Baltimore: Johns Hopkins University Press, 2012), p. 180.

142 Margot Lee Shetterly, *Hidden Figures: The Untold Story of the African American Women Who Helped Win the Space Race* (London: Harper Collins, 2016).

143 A.J.S. Rayl, 'NASA Engineers and Scientists-Transforming Dreams Into Reality', *NASA*, 2008 <https://www.nasa.gov/50th/50th_magazine/scientists.html> [accessed 9 January 2017]; Maia Weinstock, 'Scene at MIT: Margaret Hamilton's Apollo Code', *MIT News*, 2016 <http://news.mit.edu/2016/scene-at-mit-margaret-hamilton-apollo-code-0817> [accessed 9 January 2017]. See also pictures in: 'Women behind the Spacemen and the Satellites', *Sun Magazine*, May 17, 1964.

Fig. 3.15: Eleanor Pressly in: 'Women behind the Spacemen and the Satellites', Sun Magazine, May 17, 1964.

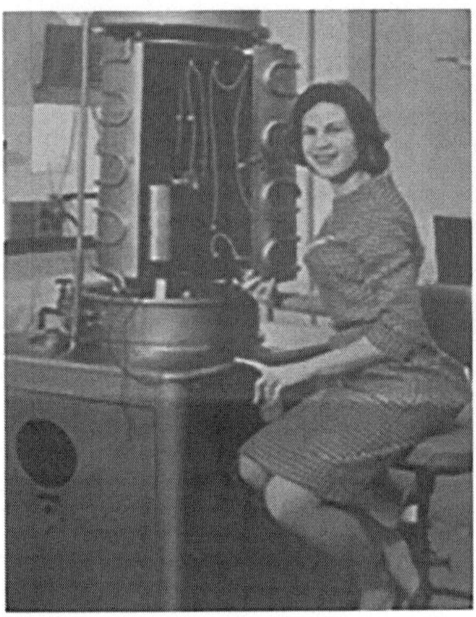

Fig. 3.16: Barbara Lunde in: 'Women behind the Spacemen and the Satellites', Sun Magazine, May 17, 1964.

Fig. 3.17: Melba Roy in: 'Women behind the Spacemen and the Satellites', Sun Magazine, May 17, 1964".

Fig. 3.18: Marjorie Townsend in: 'Women behind the Spacemen and the Satellites', Sun Magazine, May 17, 1964.

Fig. 3.19: Jaylee Burley in: 'Women behind the Spacemen and the Satellites', Sun Magazine, May 17, 1964.

3.2.3 Framing astronauts' wives as important for the American space effort

Women needed to fulfil several demands regarding their role in American society during the Cold War in the late 1950s and early 1960s. While women were invited to assume a more active role in the public sphere in order to beat the Soviet Union in the fierce Cold War competition, the pursuit of a private role as a housewife was promoted, too. Being a homemaker for a hard-working, patriotic husband and nice, healthy children demonstrated the national progress and superiority of the American system over the Soviet Union, where women were required to perform manual labor, a type of work which according to American stereotypes made Soviet women ugly and masculine.[144] Both kinds of roles were promoted in the public and private spheres of the space community. The women working as scientists and engineers performed the active part, outside of the home, in public, to support the American space program and to ultimately win the space race. The other women in the space community were the wives of all the men employed by NASA and the adjacent space industry. These women were as much involved in the space program as their spouses, because by engaging in such

144 Laville, 'Gender and Women's Rights in the Cold War', p. 533; Douglas, p. 45; Griswold, pp. 881–886; Elaine Tyler May, p. 159.

activities as housework or child rearing they kept their husband's plates clear of anything that had nothing to do with space exploration. They were subordinates of the space program, too, just as their husbands were. One wife of a NASA engineer even made clear that NASA had become "our god, our government and our goal."[145] Thus, they supported the United States in the Cold War at the home front, in the private space. Among these women were the wives of the astronauts.

When they had married their husbands, they had married ordinary military men and had expected the life of a military wife. A life with a clear structure wherein the men go off to war and the women care for the children at home. A life relatively isolated from the rest of society but very involved within the tight-knit military community due to the restless lifestyle a career in the military demanded. A life where a husband can be ordered to move to another base from one day to the next and the entire family has to follow the order.[146] Instead, they had become the wives of celebrities. Although they were still military wives and still lived on the same wage, being an astronaut's wife was more similar to the life of a rock star's, a sports star's, or a politician's wife, but with the constant fear of hazard and death.[147] The wife of the first American to reach Earth orbit, Annie Glenn, explained: "As the wife of a famous astronaut, I had to deal with being constantly in the public eye. I had to deal with the press. And if this wasn't hard enough, I had to deal with a severe handicap. A stuttering problem I have had since childhood."[148] While the other wives did not have to deal with a handicap like Annie Glenn, they all had to deal with the public obsession with their husbands – and with them. From the very first day the astronauts had been introduced to the public in 1959, the media and the American people wanted to know more about their private lives. While the work of the astronauts was exciting and dangerous, their family lives appeared to be as ordinary as the next and as American as apple pie. Thus, the American people were able to marvel at the bravery and extraordinary job of their space heroes on the one hand, but also to identify with the spacemen on a personal basis through their family life on the other

145 Mary Wiegers, 'Space Community: Emphasis on Excellence', *Washington Post*, September 13, 1969.

146 Harrison and Lalibeté, pp. 36–37; Soeters, Winslow and Weibull, p. 237; Deborah Harrison, 'Violence in the Military Community', in *Military Masculinities: Identity and the State*, ed. by Paul Higate (Westport: Praeger, 2003), pp. 71–90 (pp. 78–79).

147 James Burke, *Project Apollo*, interview with Jim Lovell, Dave Scott, Charlie Duke, Jack Schmitt, Gene Cernan, Jim McDivitt, aired July 20, 1979, conducted May 19, 1979, 1959, p. 91, HRC; Richard N. Billings, 'The Right Stuff? Tom Wolfe got the Facts right, but he got the Story wrong', *Washington Post*, October 1983, 10984, HRC; Koppel, p. 364.

148 Myra McPherson, 'The unsinkable Annie Glenn', *Washington Post*, February 23, 1984.

hand. The central person in an astronaut's private life was always the wife. This is why the wives, too, became a focus point in the public sphere. Unlike the wives of other NASA employees who could chose to what extent they wanted to get involved within the public space, for example the administrators' wives, the wives of the astronauts had no choice in the matter. The people were interested in everything regarding the wives: their looks, outfits, hobbies, children, parenting styles, composure, opinions, and their general 'Americanness.'[149]

Although the women had to deal with public scrutiny and intrusive representatives of the press from the get-go, NASA never officially briefed the wives about a media code of conduct before the very first astronauts were announced to the world. The public relations liaison Shorty Powers only informed the astronauts, though he did urge the men to prepare their wives for the media tornado. However, the astronauts did not take his warning too seriously and neglected to talk to their relatives. Thus, the women were left to figure out for themselves how to handle the press and how to deal with a life in the public eye.[150] If it had been up to President Eisenhower, the lives of the astronauts' wives would have been much different. He originally envisioned the NASA families as living in a secluded space city, just like the space families in the Soviet Union. Instead, the astronauts and their families were fair game for public consumption.[151] Reporters regularly waited in front of the women's houses to get a glimpse or a statement from them. Even when the wives tried to escape from the media frenzy, the reporters usually hunted them down. Rene Carpenter, for example, wanted to watch her husband's launch in peace from a beach in Florida, but the reporters and photographers found her to get a piece of the hot property of an 'astrowife.' Carpenter had made the mistake of telling Shorty Powers and he had tipped of the media about her whereabouts.[152] NASA actually welcomed the public interest in the wives because they made for positive headlines and were a good, personal link to the American public. Therefore, the NASA public

149 For example: Mary Beggs, wife of Administrator James Beggs, was very active at the space agency. She always publicly supported her husband and represented NASA with him for various major milestones. On the other hand, Fay Fletcher, wife of Administrator James Fletcher, only accompanied her husband when she was required to do so. Unlike Mary Beggs, she never spontaneously popped in at the office or organized public events at NASA. Josie A. Soper, interview by Rebecca Wright, April 19, 2006, pp. 37–38, Oral History Transcript, NASA Administrators Oral History Project, 19822, HRC.

150 Koppel, p. 7; Makemson, p. 44.

151 Koppel, p. 31.

152 Koppel, pp. 114–115.

relations team's orchestration of the multimedia content and portrayal of the space program included the astronauts' wives.

Especially *Life* Magazine set the tone for the early coverage of the wives because *Life* had purchased the exclusive rights to the astronauts' personal stories. Although the other media outlets were upset about the *Life* deal, *Life* had actually been the only magazine to bid on the personal story rights.[153] *Life*, however, still had to report on the astronauts and their families according to NASA's liking. Accordingly, the magazine depicted and even enhanced the all-American family angle and framed the astronauts as extraordinary ordinary American guys and their wives as faithful, beautiful, and feminine typical American women and mothers.[154] The astronauts and their wives were portrayed as the embodiments of American national identity. All the families were portrayed in the same way: wholesome, loving, patriotic, nuclear family units, a depiction that NASA wanted to transmit. The astronauts and their families knew this and played along because they were afraid that if they did not do so, the astronauts' careers would be at risk.[155] Marriage problems, addictions, or even curious quirks and imperfections were all swept under the rug of secrecy, leaving the media to eat up the display of the 'typical' American family, the basis of American values. Although the astronauts and their women were always portrayed in this homogenized way in the early 1960s, the public was already able to see behind the curtain of the perfect American family in the late 1960s. In 1971, McCandlish Phillips recognized in the *New York Times*, for instance, that the astronauts and their families ranged "over a fairly broad spectrum, from straight and square apple pie eaters to worldly men and women about town."[156] Earlier, in 1969, Maxine Cheshire had taunted the space agency about the separation of the Carpenters when she remarked that the "NASA image-makers are going to have to work overtime to maintain that All-American Boy wholesomeness for the history books."[157]

---

153 Makemson, p. 48; Kaufman, pp. 72–73; Robert Sherrod, 'The Selling of the Astronauts', *Columbia Journalism Review*, 12 (1973), 16–25 (pp. 16–18).

154 Loudon Wainwright, *The Great American Magazine: An Inside History of Life* (New York: Knopf, 1986), p. 271; Llinares, pp. 26–31.

155 Linda Gillan, 'No Place like Space. NASA testing Women Candidates', *Los Angeles Times*, October 27, 1977, HRC; Koppel, pp. 21, 148, 273.

156 McCandlish Phillips, 'Family Portraits of 3 Astronauts', *New York Times*, August 15, 1971.

157 Maxine Cheshire, 'Scott, Rene says: Separate Orbits', *Washington Post*, November 13, 1969, HRC; Shayler and Moule, p. 107.

3.2.4 Gendered framing of the reportages about the astronauts' wives

In the early 1960s, however, the perfect nuclear family took center stage in the media. The model of the bi-generational family unit, consisting of children, a mother who was the homemaker in the private sphere, and a father who was the breadwinner in the public sphere, was generally reproduced in the media, not just in case of the astrofamilies.[158] However, the presentation of the astronauts and their families as the paradigm for this way of life was ironic. The articles portrayed the women as courageous, faithful, and feminine. The wife of Scott Carpenter was, for example, described as

> BLONDE, bronzed, deep-dimpled, green-eyed and shapely. Rene Carpenter, 34, is by anyone's standard a dish. She is also self-possessed and wise. Wearing a Navy blue skirt and white middy blouse, and carrying a red scarf in her hand, she stepped before newsmen at Cape Canaveral after her husband's space voyage was over. Said she: 'I was dry-eyed the whole day.'[159]

Just like the female scientists and aspiring astronauts, the wives were described in gendered terms. Considering that the astronauts' wives were supposed to embody the typical American homemaker, the extreme emphasis on their femininity fits the picture. The press also took a lot of freedom with their coverage. If the wives dared to not fit the stereotype in the public space, the media often adapted the story to fit the women into the desired American family concept. The *Houston Chronicle*, for example, asked Jo Schirra if she would welcome back her husband Wally from space with 'steak and cake'. She replied that she never baked, which was a characteristic that most certainly did not fit the perfect housewife stereotype. The headline of the *Houston Chronicle* the next day nevertheless proclaimed that 'Astronaut's wife will bake a cake,' regardless of the fact that Schirra had no intention of baking her husband a cake.[160] This behavior shows that the urge to frame stories a certain way in the media was extremely strong.

Not only the media but also NASA fostered this kind of framing. For the *Life* photoshoot, NASA handed the women a guideline concerning their clothes and

---

158 Talcott Parsons, 'The American Family. It's Relations to Personality and the Social Structure', in *Family, Socialization and Interaction Process*, ed. by Talcott Parsons and Robert F. Bales (New York: Free Press, 1955), pp. 3–33; Isabel Heinemann, 'Introduction', in *Inventing the Modern American Family. Family Values and Social Change in 20th Century United States*, ed. by Isabel Heinemann (Chicago: University of Chicago Press, 2012), pp. 7–30 (p. 10).

159 'Nation: I've Been Thoroughly Checked Out', *Time*, June 1, 1962.

160 Koppel, p. 132.

appearance. They were instructed to wear proper pastel shirtwaist dresses, a clothing choice the women would not have made themselves because they wanted to look more fashionable, but which served NASA's and *Life*'s desire to portray the women as the nice mothers from next door. The women also wanted to wear pink lipstick in the pictures because the color was all the rage in those days. They were actually permitted to do so, but *Life* airbrushed the color to red without the women's knowledge. The patriotic color just better fit the framing of a patriotic wife of an American hero. Following the publication of the pictures in *Life*, Revlon actually launched a red lipstick called 'Moon Drops' – 'the lipstick to wear to the moon.' However, the product was not directly endorsed by the wives, although quite a few companies tried to win them, and the astronauts, too, of course, as their brand ambassadors. NASA did not allow the astronauts and their families to participate in profitable endeavors like this, and the media also condemned any commercial activities on the part of the astrofamilies because they would destroy the image of the ethical, unselfish American patriot.[161] *Life* particularly stressed the wholesome stereotypes of the American values, and its articles showed that the women were also in good pioneering spirit by supporting their husbands. Coverage stressed the wives' courage and their faith in their astronauts and NASA, because staying at home, staying behind, was just as hard as flying to space.[162] Other media outlets also highlighted the women's silent bravery, emphasizing that "[w]hatever they feel – the agony, the terror – they'll have to keep to themselves. Above all they realize that they cannot add to the awesome burden already heaped upon the shoulders of their men."[163] *Life*'s coverage was especially intriguing, though, because it also included a lot of pictures of the women in the privacy of their homes. Through the exclusive look into their homes, Americans felt especially connected to them. The home stories about the women also felt more personal because they were written by ghost writers who penned the articles on behalf of the wives. These home stories were specifically tailored for American women, both to sell the magazine to them and to get them excited about the space program by making them feel included.[164] *Life*'s

161 Koppel, pp. 31–52 and 126–128. See picture: Trudy Cooper, Betty Grissom, Marge Slayton, Louise Shepard, Jo Schirra, Annie Glenn and Rene Carpenter on the cover of *Life*, September 21, 1959.

162 James Burke, *Project Apollo*, interview with Jim Lovell, Dave Scott, Charlie Duke, Jack Schmitt, Gene Cernan, Jim McDivitt, aired July 20, 1979, conducted May 19, 1979, p. 91, 1959, HRC; Koppel, pp. 213, 348, 357; Kaufman, pp. 86–87; Shayler and Moule, p. 108.

163 Michael Pearse, 'The Astronauts' Wives: The special Way they help their Husbands', *Modern Screen*, September 1969.

164 Kaufman, p. 87; Koppel, p. 41.

portrayal of the wives as the supportive 'other' for the astronauts was a role most women could identify with, especially because the wives often also served as stand-ins for the American public. The inclusion of the wives in the story of the space program was very much embraced by American women, who were very interested in the wives and always wanted to know more about them.[165]

The space program was made to be witnessed by the public so that they could attest to the giant leaps made by the hero astronauts, and to be the exact opposite of the secretive Soviet space program. Therefore the astronauts needed to be watched. This made the public a part of the space program and imbued it with the very important role of the spectator. The wives were made into the model spectators; in pictorial coverage of rocket launches, one could often see them looking up into the sky to their husbands, the American heroes.[166] Through the selection of those specific images, a specific audience is constructed and associations are constructed on specific bodies, in this case white middle class women. For this reason, images of women started to symbolize the public reception in general. This also meant that other Americans who did not fit the archetype, namely minorities, were excluded from representations of the space program and placed behind the scenes.[167] The wives could also often be seen in their role as witnesses to their husbands' successes in space, when they served as the presenter of honors. They were the ones to pin medals to their husbands' chests or reveal commemorative plaques in public ceremonies. The women were always at their husbands' sides to bear witness to the tributes American society bestowed upon their spouses.[168]

Although NASA and the media loved to frame the astronauts in the midst of their nuclear family units alongside the wives in their starring roles as the perfect feminine housewives who marvelled at their spouses' brave, masculine risk-taking in space, the women needed to perform roles "outside

---

165 'Letters', *Time*, May 11, 1959.

166 See pictures: Rene Carpenter watches Scott Carpenters launch on the title of *Life*, June 1, 1962; Trudy Cooper and her daughters Jan and Cam watch Gordon Coopers Gemini 5 launch, August 21, 1965, Astronaut Biographies, Leroy Gordon Cooper Jr. Box 61, General Reference Series, JSC; Trudy Cooper and her daughters Jan and Cam, *Birmingham News*, May 12, 1963. See Fig. 3.20.

167 Jodi Dean, pp. 64–69; Llinares, p. 34; Heinemann, p. 13.

168 'Give Astronaut Pilot Wings to Cooper', *Chicago Tribune*, May 30, 1963. See also pictures: Annie Glenn and daughter Lyn Power unveil a commemorative plaque for the tenth anniversary of the flight of John Glenn in Friendship 7, February 22, 1972, 8318, HRC; Wife and mother of Alan B. Shepard unveil a plaque commemorating the first flight of an American into space at Cape Kennedy, May 13, 1971, Astronaut Biographies, Box 65, General History XI, General Reference Series, JSC. See Fig. 3.21-3.22.

of their heterosexual desire."[169] The astronauts were almost always at work and they had to travel a lot for their job. Beth Williams, for example, figured that she saw her husband "not more than 45 days in a whole year. [. . .] He'd be on his way to L.A. from the cape [Cape Kennedy]: I'd give him some clean clothes and he'd give me his laundry. Then I'd say 'goodbye.'"[170] Because their husbands were almost always absent in the women's daily lives, they had to perform all the roles and chores in the family and the house, including supposedly masculine roles and task such as cutting grass, fixing things around the house, or raising boys by themselves. All of the astronauts were "absentee fathers" who did not really participate in child rearing, not even for their male offspring.[171] Scott Carpenter later admitted that his kids mainly saw him on television as an astronaut: "I missed an awful lot of their childhood because of heavy flight and training schedules. I only see them occasionally now." Rene added that "[b]eing an astronaut may have meant fame for the men, but their wives and children all suffered. I don't know of one astronaut's family untouched by the space program. Fame always exacts its price."[172] That was especially true for the busy, famous astronauts, but the wives of less famous NASA men knew the problem all too well, too. One wife revealed that "[b]efore a shot, the tension is so great the kids are afraid to talk when their father is home. He is so pressurized and rushed for time he expects me to have whatever he wants – food, clean clothes – ready in a minute. Naturally, all of us feel the tension, and the minute he's out of the house the kids start fighting, whining and crying."[173]

The stress, the long hours and his travels left the wives to run the house by themselves and be mother and father to the children, therewith dissolving the gendered division of labor in the private sphere, even though they were the personification of the undiluted feminine housewife in the public. Dr. Katharine Kiehl, a paediatrician who worked with children from the space community, even remarked that "[a]nybody who can make it to the top in this area is almost

169 Sage, p. 152; Charlie and Dottie Duke, *Moonwalker* (Rose Petal Press, 2011), pp. 175–180.

170 Kingsley Wood, 'Space Lives on for 3 Widows', *Chicago Sun-Times*, June 16, 1969.

171 Frank Macomber, 'Can Space Lady be far behind?', *San Jose Mercury*, July 26, 1972; Mary Wiegers, 'Space Community: Emphasis on Excellence', *Washington Post*, September 13, 1969. Quite a few of the astronauts who have written autobiographies, have indicated the difficult times that their families had du endure because of their absence. Walter Cunningham, *The All-American Boys* (New York: Ibooks, 2003); Duke; Frank Borman and Robert J. Serling, *Countdown* (New York: William Morrow, 1988).

172 Lloyd Shearer, 'Scott Carpenter: for Astronauts – Fame equals Fortune', *Parade*, May 23, 1971.

173 Mary Wiegers, 'Space Community: Emphasis on Excellence', *Washington Post*, September 13, 1969.

necessarily a poor husband, poor lover and poor father."[174] Beth Williams, wife of Clifton Curtis Williams Jr. who died in a plane crash before he could fly to space, revealed that her life as a widow was not actually that different from before because her husband had been away most of the time anyway.[175] These revelations show that the gendered portrayal of the wives and their families' structures in the media in the early 1960s was far off from daily reality, a fact that they started to talk about in the public space after the "divorce bubble" burst in the community in the late 1960s. Jane Dreyfus (formerly Conrad), for example, thought the women looked like Stepford Wives, and Walt Cunningham complained that they had not been allowed to be human, that the men had to act like Boy Scouts.[176]

Because the husbands were almost never home, the women built a bigger support system to cope with their daily workloads. First of all the women, and their neighbours supported each other and helped one another out.[177] Moreover, the people working in the space agency also helped to move other family members to Houston, so the women would have some family in the area. This was, for example, done for Betty Grissom. Flight Director Gene Kranz' team assisted Betty's sister and her brother-in-law, Mary Lou and Albert Fosbrink, to relocate from Indiana and employed them both. The astronauts' secretaries also became honorary family members.[178] They developed close relationships with the wives and the families because they were the intermediaries between the astronauts and their families when the men were at work. Some of them also babysat for the astronauts' families, like Jamye F. Coplin who helped to look after the children of the Stuart A. Roosa family, the Ronald E. Evans family, the Joe H. Engles family, and the William A. Anders family.[179] Considering that the nuclear family concept was the family model being exalted in the public space, it is significant to realize that the women, who were public role models for this very concept, were dependent on people outside of their family

174 Mary Wiegers, 'Space Community: Emphasis on Excellence', *Washington Post*, September 13, 1969.

175 Kingsley Wood, 'Space Lives on for 3 Widows', *Chicago Sun-Times*, June 16, 1969.

176 Koppel, p. 357; Sherrod, p. 17; Shayler and Moule, p. 109; Holt, *Rise of the Rocket Girls: The Women Who Propelled Us, from Missiles to the Moon to Mars*, p. 216.

177 Janet M. Evans, interview by Jennifer Ross-Nazzal, August 7, 2003, pp. 3, 15, Edited Oral History Transcript, NASA Johnson Space Center Oral History Project, JSC.

178 Janet M. Evans, p. 11.

179 Maureen Bowen, interview by Rebecca Wright, June 21, 2007, p. 14, Oral History Transcript, NASA Johnson Space Center Oral History Project, JSC; Jamye Flowers Coplin, interview by Rebecca Wright, November 12, 2008, pp. 11–12, Oral History Transcript, NASA Johnson Space Center Oral History Project, JSC; Mary Wiegers, 'Space Community: Emphasis on Excellence', *Washington Post*, September 13, 1969.

units, like extended family, co-workers, and friends. However, the women did not just let anyone into the inner circle of their families and homes. In fact, they were very protective of their families due to the international fame of their husbands. Given the extreme hostility, devotion, and admiration the wives were faced with because of their famed status in the public, their closeness has to be assessed as self-protection.

The wives also had to deal with their husbands' groupies. Some of these women just admired their husbands from afar, commenting on their 'warm smile' or 'spellbinding eyes;' other women were more active. While one planned to build a shrine for John Glenn, others got involved in space citizen groups or started fundraising campaigns to help the space program. However, some women actually tried to entice their husbands away from them, especially when the latter had to travel for their jobs to Cape Canaveral. The wives dubbed these women 'cape cookies.' Not only did the women have to worry about their husbands' fidelity, but they also frequently received hate mail from those women who thought the astronauts deserved better wives. Furthermore, the women had to deal with their own admirers and members of the press, who made a sport of reporting about them, such as Norman Mailer.[180]

The female pilots who publicly lobbied to become astronauts were also on the women's minds, mainly because reporters frequently asked them their opinions regarding Cobb and the other aspiring female space travellers. As military wives, most of them generally disliked "women who work and desert their duties as wives" and they were also not in favor of women working as astronauts. Unlike Pat McDivitt, Annie Glenn, or Mailyn Lovell, Trudy Cooper could understand the female pilots' desire to become astronauts. She was a pilot herself and made her dream of flying in a Powder Puff Derby reality in 1971 after divorcing her famous husband. She also could have imagined herself as an astronaut.[181] Just as most of the astronauts' wives were critical of women

---

180 Richard N. Billings, 'The Right Stuff? Tom Wolfe got the Facts right, but he got the Story wrong', *Washington Post*, October 1983; David Braaten, 'Impromptu Poll shows Astronauts really send M'Lady into Orbit', *Sunday Star*, May 26, 1963; Charles Hillinger, '$1 Million Shrine for John Glenn', *Los Angeles Times*, March 7, 1963; George M. Low, *letter to Mabel Hodges*, October 9, 1973, Box 16, Chronological File 4, George M. Low Collection, Rensselaer Polytechnic Institute (hereafter Low/RPI); Gregory Donald H., *Designing the Space Age*, speech at Freeport Texas, Texas Society of Profession Engineers and Colorado Engineering Society in Denver, February 19 and 21, 1963, GH 51, General Reference Series, JSC; Mary Wiegers, 'Space Community: Emphasis on Excellence', *Washington Post*, September 13, 1969.

181 Mary Wiegers, 'Space Community: Emphasis on Excellence', *Washington Post*, September 13, 1969; Carl T. Rowan, *Blacks in Space – Rowan Report Number 214*, taped on July 9, 1976, on air July 22, 1976, transcript, Harris/MSRC; Koppel, pp. 22, 161, 353; Atkinson and Shafritz, p. 94;

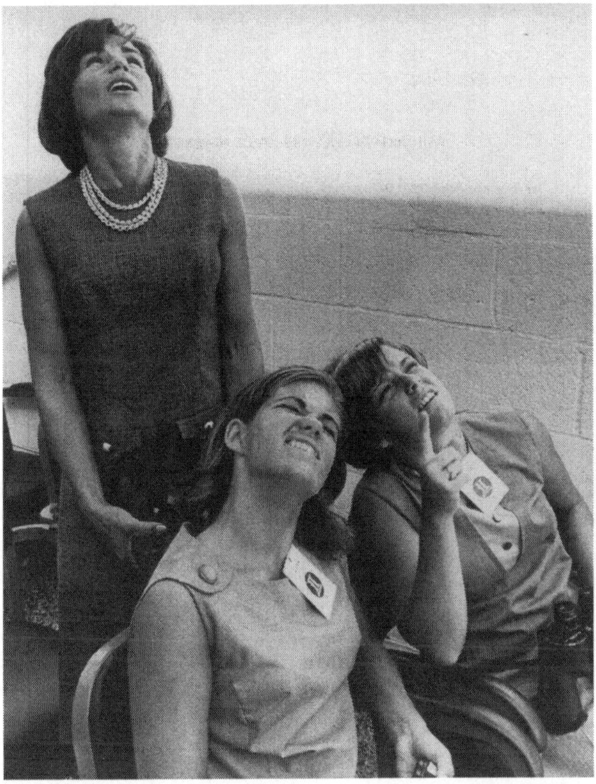

Fig. 3.20: Trudy Cooper and her daughters Jan and Cam watch Gordon Coopers Gemini 5 launch, August 21, 1965, NASA.

wanting to fly to space with their husbands, most of them were also sceptical about the women's movement in the late 1960s and early 1970s. Mary Irwin, for example, thought that the women of the movement have "missed the boat somewhere. Frankly, I like being a woman. And being treated like one."[182] Considering that the women's movement regarded the wives' way of life as undesirable and a relic of the American patriarchy, the wives'

Weitekamp, 'The "Astronautrix" and the "Magnificent Male": Jerrie Cobb's Quest to Be the First Woman in America's Manned Space Program', p. 23; Kevles, p. 72.

182 McCandlish Phillips, 'Family Portraits of 3 Astronauts', *New York Times*, August 15, 1971; Janet M. Evans, interview by Jennifer Ross-Nazzal, August 7, 2003, pp. 26–27, Edited Oral History Transcript, NASA Johnson Space Center Oral History Project, JSC.

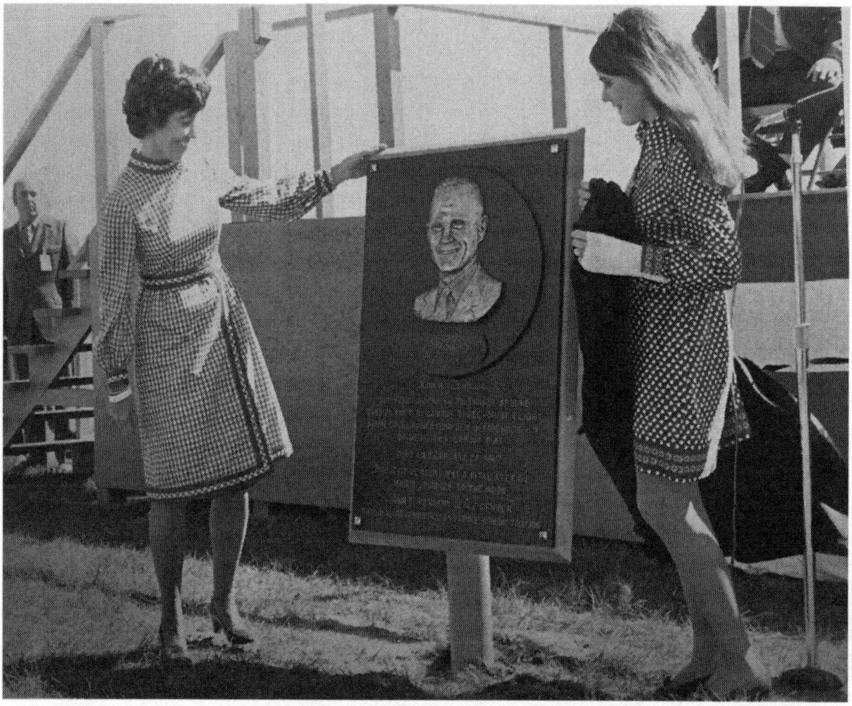

Fig. 3.21: Annie Glenn and daughter Lyn Power unveil a commemorative plaque for the tenth anniversary of the flight of John Glenn in Friendship 7, February 22, 1972, NASA.

defensiveness regarding the women's movement is understandable, especially because a lot of women in the late 1960s were afraid that the call for gender equality meant that women had to give up their femininity. Rene Carpenter was one of the few wives who did appreciate the women's movement, maybe because she had a new perspective after her divorce, just like Trudy Cooper. Indeed, Carpenter had never been the typical military wife because she found life on base too restricting and tried to escape from it. She started her own career around the time her husband became an astronaut. Her first journalistic endeavour was the column *A Woman Still* at the *Houston Chronicle*, which was eventually picked up by 35 other papers. She also guest starred in a TV show and worked as a TV commentator. Carpenter also went on the campaign trail with Robert Kennedy and in 1972 she started a feminist talk show named *Everywoman*.

Fig. 3.22: Wife and mother of Alan B. Shepard unveil a plaque commemorating the first flight of an American into space at Cape Kennedy, May 13, 1971, NASA.

However, pursuing a career outside the home and being involved with the women's movement was most unusual for the wife of an astronaut in the 1960s.[183]

183 Judith Martin, 'Astronaut's Wife grounded in July', *Washington Post*, July 21, 1968; Koppel, pp. 221–222, 266, 276, 355.

3.3 Libbers, glamournauts and a teachernaut in the 1970s and 1980s

Just like most of the female scientists and engineers, the majority of the astronauts' wives did not care for the women's movement. The women's movement, however, did care about them, especially about their portrayal in the media.[184] Dorothy L. Olaechea, who worked on NOW's media committee, complained to *CBS* for its remarks about two of the astronaut wives in a story which aired on July 27, 1973. The wives were shown looking into a space capsule while the commentator intoned that "... both ignored a wifely urge: they did not rearrange the furniture." Olaechea, incensed, asked *CBS*: "What, may I ask, was the purpose of this remark? Amusement, perhaps? Was there any constructive or instructive purpose behind it? [...] We are not amused."[185]

The women in the women's movement were not amused about a lot of things they had to witness in the media in the 1970s. The catcalling as well as the sexist and misogynistic comments about women in the media did not just magically disappear in that decade. A lot of women, and men for that matter, had to fight to change the content and language considered proper in the public space of the media. Feminists continuously challenged sex stereotyping and the portrayal of the traditional gender and family roles, as well as issues such as domestic violence, workspace equality, sexuality, rape, and reproductive rights, in the name of all American women. However, women of color for the most part did not feel represented by the women's movement, which mainly targeted issues of concern to middle-class white women. In the early 1970s, women's groups and coalitions started to petition newspapers, magazines, and radio and TV stations to draw public attention to the sexist and negligent portrayal of women in the media as well as the treatment of women in the media workspaces. Feminists felt so strongly about sex stereotyping because it not only acted as a catalytic agent for low self-esteem but also restricted opportunities for women by planting images about their limited capabilities in American heads – male and female.[186]

184 'Spacecraft Launch Boss is a Mother', *Huntsville Times*, December 15, 1970; McCandlish Phillips, 'Family Portraits of 3 Astronauts', *New York Times*, August 15, 1971.

185 Dorothy L. Olaechea, *letter to Roger Mudd*, July 28, 1973, MC 495, Carton 20, Folder 5, Wilma Scott Heide Papers, Schlesinger.

186 Davis, pp. 109–110; Rosen, p. 80; Allison Perlman, *Public Interests: Media Advocacy and Struggles Over U.S. Television* (New Brunswick: Rutgers University Press, 2016), pp. 65–93; Winifred Breines, *The Trouble Between Us: An Uneasy History of White and Black Women in the Feminist Movement* (Oxford: Oxford University Press, 2006).

The media outlets, however, often met the women's activism with ridicule, just as they had the ambitions of the female pilots who challenged the astronaut gender stereotypes in the early 1960s. They often portrayed the feminists as militant, crazy, hairy, man-hating 'dykes,' mostly because in the early 1970s the media liked to focus on the most radical women's liberationists, 'the libbers,' instead of on the moderate activists. The tussle between the women's movement and the media was not just a simplistic jab against women but a complex struggle between traditional values and new society-altering ideas, in this case between antifeminism and feminism. Other movements had to go through the same struggle in the beginning, though, and just like the movements before them, the media eventually started to lessen the pressure and began to deal with the feminists on a more serious note in the middle to late 1970s.[187] However, the media portrayal then often swung to another extreme, that of the superwoman who could do it all, a reporting style which the first female American astronauts had to deal with in the late 1970s and 1980s.[188] This is why feminists often resorted to writing their own news and history and started to create their own media outlets such as, for example, *Ms.* magazine founded by Gloria Steinem in 1971.[189]

3.3.1 Feminists' shaming of NASA

American society's ambivalence about women's role and the women's movement can also be seen in NASA's changing relationship to several groups belonging to the women's movement. NASA only tried to connect with the women's movement beginning in the late 1970s.[190] JSC, for example invited the National Organization for Women (NOW) to take part in the center's Federal Women's Week, named 'Awareness – Emerging Attitudes, Values, Beliefs.' The same center dispatched an employee to a

187 Davis, pp. 108–109; Douglas, pp. 12–13; Rosen, p. 80.

188 Rosen, pp. 295–296.

189 Mary Thom, *Inside Ms: 25 Years of the Magazine and the Feminist Movement* (New York: Holt, 1997); Amy Erdman Farrell, *Yours in Sisterhood: Ms. Magazine and the Promise of Popular Feminism* (Chapel Hill: University of North Carolina Press, 1998).

190 Harriett G. Jenkins, interview by Jennifer Ross-Nazzal, August 5, 2011, p. 30, edited Oral History Transcript, NASA Headquarters Oral History Project, JSC.

meeting with NOW women to show the film 'Stepping Stones to the Future,' which was supposed to encourage women and minorities to pursue engineering opportunities at NASA, and get the women's input on a prospective NASA film on 'Women in Space.'[191] However, some people at NASA, namely Deputy Assistant Administrator of the Equal Opportunity Office Ruth Bates Harris, had already recognized in the early 1970s that "[s]ex discrimination will be one of the leading equal rights issues of the seventies."[192] This insight is why Harris on several occasions emphasized that it was necessary to recognize the important role of the women's movement, which too many people "identify the women's right movement simply as a fad or with bra burning."[193] She also stressed that the women's movement was made up of several organizations and that people should refer to them accordingly. The misapplications of terms like "libbers" would evoke negative preconceived notions that kept men and women from making a stronger commitment to the women's movement, despite them profiting from it.[194] In a NASA guideline on contractor Equal Employment Opportunity (EEO) in 1972, Harris elaborated:

> It is extremely important that those of us in the compliance field evaluate feminist groups realistically. One step is to drop the terms "Women's Lib" and "Lib" from our vocabularies. The Women's Liberation Movement is a specific group of people; they do not refer to themselves as "Women's Lib" and neither should we. Another aspect is that these terms, "Lib" and "Women's Lib" conjure for many people a vision of frivolous [sic], bra-burning, radical lesbians. We have reached a level of sophistication that enables us to realize that all minority organizations do not speak for all minorities, that each one should be approached on its own terms. The same is true of feminist organizations.[195]

However, Harris' plea fell mainly on deaf ears in the space community in the early 1970s. It certainly did not help her cause that some women's

191 'Women at NASA –up, up, and away?', *NOW News Bay Area Chapter* 6/3 (1978), p. 1, PR-1 Carton 25, NOW Chapter Newsletter, Schlesinger; 'Johnson Space Center welcomes public to Federal Women's Week April 17–21', *NOW News Bay Area Chapter* 6/4 (1978), p. 4, PR-1 Carton 25, NOW Chapter Newsletter, Schlesinger.

192 Ruth Bates Harris, *Dissemination of Initial NASA Guidelines on Contractor Equal Opportunity for Women*, June 13, 1972, 18508, Box 1, Code u Chron Files 1972, HRC.

193 U.S. Department of Commerce (Bureau of the Census), 'Ruth Bates Harris urges strong commitment', *Census Bulletin* 15/8 (1975), Harris/MSRC.

194 U.S. Department of Commerce (Bureau of the Census), 'Ruth Bates Harris urges strong commitment', *Census Bulletin* 15/8 (1975), Harris/MSRC.

195 Ruth Bates Harris, *Dissemination of Initial NASA Guidelines on Contractor Equal Opportunity for Women*, June 13, 1972, 18508, Box 1, Code u Chron Files 1972, HRC.

organizations repeatedly shamed NASA in the public space for its sexist attitudes. NOW, especially, was very active in demonstrating against various activities of the space agency and its employees.

NOW was founded on October 29, 1966 with a founding conference in Washington, D.C. The feminist organization wanted to recast gender roles and demanded equality between the sexes. The women who were members considered themselves "part of the worldwide revolution of human rights now taking place within and beyond our national borders."[196] Similar to NASA, NOW had a headquarters and various local chapters all over the U.S, which enabled it to call upon women scattered all over the U.S. whenever a new protest was launched.[197] The organization used several protest strategies, including zap actions, sit-ins, and marches. One particularly popular form of consciousness raising was to encourage its members to write letters of protest to people in power, like CEOs, editors in chief, members of Congress, or, in the case of NASA, to the administrator. The women of the NOW chapters close to NASA installations were especially involved when it came to protesting the space agency. The public statements and behavior of the most visible members of the space community, the astronauts, were a popular target for such protests.

In the 1960s, astronauts embodied an innate masculinity that was presented, produced and legitimised by the media and generally accepted by American society.[198] Although the story of space exploration, a story by and about men, slowly started to change in the 1970s under the pressure of the women's movement, most of the American space heroes did not recognize the shift in the gender hierarchy that was taking place and continued their gender performance as the iconic space warriors to which women only existed in a peripheral, masculinity enhancing supporting role. So when feminists in the late 1960s started to question the existing gender roles and to point out sexism in American society, including the space community, they frequently clashed with NASA officials, especially astronauts.[199] For example, in 1973, the Apollo

196 Rosen, p. 78.

197 Richard L. Hughes, 'National Organization for Women', in *Culture Wars: An Encyclopedia of Issues, Viewpoints and Voices*, ed. by Roger Chapman and James Ciment (New York: Routledge, 2014), p. 457; Virginia Schomp, *American Voices from the Women's Movement* (New York: Marshall Cavendish Benchmark, 2007), pp. 101–102; Rosen, pp. 74–85.

198 Llinares, pp. 1–25; Hersch, *Inventing the American Astronaut*, pp. 9–74; McCurdy, *Space and the American Imagination*, p. 92.

199 Nita Ladewig, 'NITA and NASA: A Case History of Male Chauvinism in the Space Program', *NOW Acts*, 2 (1969), 16–19. In: Margaret W. Rossiter, *Women Scientists in America. Vol. 2. Before Affirmative Action 1940–1972* (Baltimore: Johns Hopkins University Press, 1995), p. 370.

17 astronauts came to the University of South Dakota for a Q&A and to exhibit a moon rock. There they encountered a women's group protesting the event with three banners, two of which read: 'Billions for space, pennies for daycare' and 'Where are the 13 women astronauts now?' Astronaut Pete Conrad had no sympathy for the women's complaints and demonstrated absolutely no grasp of the new gender awareness when he asked the protestors if they felt sexually deprived. The sexist comment was obviously not well received and prompted the South Dakota women to ask NOW to officially protest.[200] It would not remain the only incidence of feminists taking offense at a remark by an astronaut. James Lovell, for example, remarked in *Playboy* magazine that he definitely saw a future for women in space: "We fully envision that in the near future we will fly women into space and use them the same way we use them on earth – for the same purpose."[201] So in a time when women fought for equal participation in the workspace, Lovell, just like Conrad, met these hopes with ridicule, machismo, and sexism, indicating that the only reason for women to participate in space exploration was to provide relief from sexual tension and do the housekeeping. This statement was picked up by newspapers all over the U.S. and set off a public discussion in which feminists criticized the astronauts' behavior and mindset. This was a common tactic to get more publicity for one's agenda.[202] NOW also invited Lovell to come to the annual Barefoot and Pregnant Ball which was held in his 'honor.' They wanted to give Lovell an award for "the chauvinistic attitudes" he had "displayed towards women and their struggle for recognized equality and liberation," in order for him to recognize that the "women of the world" knew exactly where he stood "on issues that affect their lives."[203] The sarcastic award was one way of dealing with the astronauts' machismo in the public sphere. A lot of women within and outside of NASA, including Ruth Bates Harris, just criticized the of-

---

200 Catherine Dunlap, *letter to Wilma Scott Heide,* September 19, 1973, MC 495, Carton 20, Folder 5, Wilma Scott Heide Papers, Schlesinger.

201 *The Winners!!! . . .Lunar Landing Festival King (or Queen) Contest. . .*, 1973, MC 496, Carton 44, Folder 5, National Organization for Women Collection, Schlesinger.

202 Rossiter, *Women Scientists in America. Vol. 2. Before Affirmative Action 1940–1972*, p. 377.

203 Drury Kerik, C.C. Britton, The Honorable G. Cherry, K. Coolman, J. Frost, The Honorable E. Gordon, W.S. Heide, J. Lightfood, The Honorable L. Wilson, *letter to James Lovell*, October 22, 1973, MC 495, Carton 20, Folder 5, Wilma Scott Heide Papers, Schlesinger; Chase Untermeyer, 'Women's Unit asks Impeachment', *Washington Post*, May 29, 1974; Gayla Salinas, *Testimony of the National Organization for Women before the House Judiciary Subcommittee on Civil Rights & Constitutional Rights*, March 13, 1974, p. 4, Harris/MSRC.

fensive statement and demanded an apology from NASA.[204] However, the space agency did not issue any kind of comment regarding the astronauts' insensitive statements and remained stoically silent.[205] NASA Deputy Administrator George Low only explained to a female co-worker that Lovell was no longer employed by the agency, and that he therefore did not speak for NASA, which made a public disclaimer unnecessary. To the dismay of a lot of female NASA employees and feminists, the agency did find the time, within days, to deny the allegation that all astronauts had a vasectomy. So when the virility of the agency's heroes was threatened, the agency immediately came to the defense of the astronauts' honor as men; they did no such thing for the honor of the American women.[206]

Humor and ridicule were popular tools for dealing with such behavior from astronauts and other men. For example, when astronaut Michael Collins and science fiction writer Arthur C. Clarke publicly argued over who came up with the concept of breasts in weightlessness first, feminist activist Sharon Smith met the debate with sarcasm. Clarke had written a letter to *Time*'s Robert Sherrod in August 1974 to let the world know that he had "beaten Mike Collins to the NASA SUTRA" in the opening of chapter 11 of *Rendezvous with Rama*.[207] In his book, Clarke had elaborated about the distraction female bodies represented for men in space:

> Some women, Commander Norton had decided long ago, should not be allowed aboard ship; weightlessness did things to their breasts that were too damn distracting. It was bad enough when they were motionless; but when they started to move, and sympathetic vibrations set in, it was more than any warm-blooded male should be asked to take. He was

204 Ruth Bates Harris, *Minorities and Women into Space*, memo to Associate Administrator for Manned Space Flight, July 19, 1972, 18508, Box 1, Code u Chron Files 1972, HRC.

205 Jo DiMaggio, *testimony in front of the committee on the Judiciary U.S. House of Representatives Subcommittee on Civil Rights and Constitutional Rights*, March 14, 1974, Harris/Schlesinger.

206 *Statement of Diane Graham*, Department of Housing and Urban Development, Space, Science, Veterans, and certain other independent Agencies Appropriations for Fiscal Year 1975. Hearing before a Subcommittee of the Committee on Appropriations United States Senate 93rd Congress Second Session on H.R. 15572 an Act making Appropriations for the Department of Housing and Urban Development Space, Science, Veterans, and certain other independent executive Agencies, Boards, Commissions, Corporations, and Offices for the Fiscal Year Ending June 30, 1975, and for other Purposes, January 11, 1974, pp. 133–134, LoC; *Official Statement of the National Organization for Women regarding NASA*, January 24, 1974, Harris/MSRC.

207 Arthur C. Clarke, *letter to Robert Sherrod*, August, 21 1974, 006798, HRC.

> quite sure that at least one serious space accident had been caused by acute crew distraction, after the transit of a well-upholstered lady officer through the control cabin.[208]

Sherrod wrote back to Clarke the following month to tell him that Collins had sworn that he had "not read Rendezvous with Rama and that he arrived at the bosom concept on his own."[209] In his book *Carrying the Fire*, Collins had fantasized about women in space as well:

> Imagine a spacecraft of the future, with a crew of a thousand ladies, off for Alpha Centauri, with 2,000 breasts bobbing beautifully and quivering delightfully in response to every weightless movement . . . and I am the commander of the craft, and it is Saturday morning and time for inspection, naturally.[210]

Clarke and Collins proved that the idea of women in space, in science-fiction and in real life, was still not taken seriously as late as 1974. Female astronauts were still imagined as sexual objects. The astronauts apparently still needed to show off their virility, thus their masculinity, in this form. However, most women in the 1970s were no longer willing to be objects for male consumption and humorously invited the public to consider a paradigm shift in the gender hierarchy of America's social order:

> To my amazement, I found myself agreeing with Mike Collins and Arthur C. Clarke. Male astronauts certainly would be distracted by the sight of female astronauts' breasts bobbing weightlessly. The spacemen would not be able to perform their duties. It is not a case of cruel discrimination; it is simply a biological fact. As a woman, I have had to come to a very painful but inevitable decision. After all, the effectiveness of the space program must come first. We must exclude men from the astronaut program. We'll miss them.[211]

The women thus turned the biology-argument around and made it a reason to exclude men, not women, from space exploration.

Not only feminists but also adversaries of the women's movement made their opinions about women in space exploration publicly known in the 1970s and usually referred to one of the tests NASA conducted about and with women. In September 1972, for example, the space agency tested its newly developed waste containment system, essentially a space toilet, during parabolic flight sequences with the help of four female flight nurses from Wright-Patterson Air Force Base in order to see if the toilet could also properly be used by women

208 Arthur C. Clarke, *Rendezvous with Rama* (London: Gollancz, 1973).
209 Robert Sherrod, *letter to Arthur C. Clarke*, September 13, 1974, 006798, HRC.
210 Michael Collins, p. 392.
211 Sharon Smith, 'letter to the editor', *Time Magazine*, October 7, 1974.

during a future space mission.[212] Another experiment was conducted by Ames Research Center in the fall of 1973 for five weeks, with the goal of determining how a young female body, aged 23 to 35, would cope with a space environment onboard the Space Shuttle. The study involved a series of physiological tests with twelve flight nurse volunteers, including a period of two weeks of total bed rest to simulate the effect of weightlessness.[213] The tests, led by Ames' Dr. Harold Sandler, were expanded throughout the 1970s. Female and male volunteers up to the age of 65 were sought with the help of advertisements in local newspapers, on radio and TV as well as at the local universities and women's organizations in 1977. They were tested in order to find out how well "ordinary human beings," as compared to the "finely trained bodies" of astronauts, were able to live and work in a space environment.[214] Around the same time, in September 1973, *Ms.* published an article about the Mercury 13, called 'The 13 Astronauts who were left behind,' in which the author claimed that the space exploration capability tests undergone by the 13 female pilots in the early 1960s had officially been part of a NASA program. Despite not being true, the claim triggered an uninformed, emotional discussion between proponents and opponents of women taking part in space exploration, in which information about Ames's bed rest studies was mixed up with the *Ms.* article and the opinions of the debaters.[215] As had been the case in the 1960s, the range of opinions was broad. However, this time

---

212 Amy Foster, *Integrating Women into the Astronaut Corps. Politics and Logistics at NASA, 1972–2004* (Baltimore: The Johns Hopkins University Press, 2011), p. 120.

213 'Study of Women as Space Flight Candidates completed', *NASA News*, October 23, 1973, 008994, HRC; Ruth Bates Harris, *Harlem Princess. The Story of Harry Delaney's Daughter* (New York: Vantage Press, 1991), p. 263. Following women participated in the study: Felicia M. Benton, First Lieutenant USAFR; Frances F. Cappa, First Lieutenant USAFR; Sylvia K. DeJong, Captain USAFR; Carrol Lee French, Captain USAFR; Judith A. Keeton, First Lieutenant USAFR; Kathleen Queiser, First Lieutenant USAF; Bonnie L. Kultgen, First Lieutenant USAFR; Marcia R. Pack, Second Lieutenant USAFR; Evelyn Jean Parks, Captain USAFR; Lorraine Clare Schoen, First Lieutenant USAFR; Marina Van De Graaff, First Lieutenant USAFR; Lucinda Zimmer, Captain USAF.

214 'Women to begin Tests to assess Fitness for Space Travel', *NASA News*, April 7, 1977, 008994, HRC; Oceola S. Hall, *memorandum to Dr. Harriett G. Jenkins*, May 2, 1977, 18505, Box 2, Code u Chron Files 1977, HRC; Harold Sandler and David L. Winter, *Physiological Responses of Women to Simulated Weightlessness: A Review of the Significant Findings of the First Female Bed-Rest Study* (Washington, D.C.: National Aeronautics and Space Administration, 1978); Foster, *Integrating Women into the Astronaut Corps. Politics and Logistics at NASA, 1972–2004*, pp. 79–81; Shayler and Moule, pp. 147–151.

215 McCullough; Weitekamp, *Right Stuff, Wrong Sex. America's First Women in Space Program*, pp. 93–94.

around discussions about women in space were often connected to peoples' sympathy for or dislike of the women's movement.

Vernon A. Buford, for example, complained to NASA Administrator James C. Fletcher that NASA should not take the concerns of the women's movement to heart. He assumed that the bed rest studies at Ames were only conducted to appease the women's movement after the publication of the *Ms.* article:

> I would like to know just why the statement made by that ignorant magazine was taken so seriously. I myself did not read the article in that trashy publication, but read about the training program in TIME magazine. Now, I myself have had my fill of women's liberation. I am sick to my stomach of having to hear about how the women of the world have been treated unfairly since the dawn of time.[216]

He was not the only one who weighed in on the issue. Al Volker, for example, complained in a Miami newspaper that the news about women in space exploration was mainly ignored by the women's movement. The test of the twelve Air Force nurses, as well as Jerrie Cobb being made a laureate of the Harmon Air Trophy, a prize for outstanding services in aviation, "should have been hailed by women's libbers, but they weren't." He also complained that the tests of the flight nurses were unnecessary because the Mercury 13 had already set the criteria for women in space in the 1960s.[217] So while one man felt that the women's movement was too involved and had too much influence on NASA, another man was convinced that the women's movement was not nearly enthusiastic enough about these achievements for women in space. Volker was wrong: participants in the women's movement did notice the news about the testing of the nurses, but he was also right because the women of NOW, for example, were not overly excited about the tests. To them, the tests came late and were too small of a step to integrate women in human space exploration. They also did not like the way the media reported on the women of the Ames' study. In a letter to Senator John V. Tunney, Gloria Woody, for instance, remarked that she did not

> welcome the tone and quality of the image that is being projected. I do not recall mention of the male astronauts [sic] pastel underwear or that they wore lacy NASA jock straps and had their genitals and hormones monitored. I am really curious to know more about the testing of males in these newsworthy areas. I am more interested in seeing young women offered opportunity and training as jet pilots in the Navy and any other training that will lead to their being tested for NASA Command Astronauts. Not nurses, not stewardesses,

---

216 Vernon A. Buford, *letter to James C. Fletcher*, March 18, 1974, 008994, HRC.

217 Al Volker, 'Women in the News ignored', *Miami News*, September 21, 1973.

> not legitimatized courtesans, but serious scientific, exploring, challenging individuals on the very same ground as the men.[218]

Woody's letter clearly demonstrated what NOW and other women's organizations charged the space agency and the media with. They did not want to read sexist comments about nurses' underwear or marital status. They wanted to read news about a lift on institutional and structural discrimination within the space agency and the military, which thus far had provided the pipeline for NASA careers.

One of the most contested topics within the 1970s controversy about women in space was the theme of 'mothers.' While some of those who complained pronounced that women should not be tested because mothers have no business in space, others were upset that none of the twelve flight nurses tested at Ames was a mother.[219] NOW, on the other hand, lamented that when a prospective spacewoman was a mother, the fact was emphasized in the media, as in the case of the physicist Ann F. Whitacker of MSFC in Huntsville. Whitacker, Doris Chandler, Carolyn Griner, and Mary Helen Johnston, all scientists and engineers at Marshall Space Flight Center, were an all-woman crew who simulated a five-day Spacelab/Space Shuttle mission in December 1974. The scientists spent eight hours every day in the General Purpose Laboratory (GPL), a cylindrical structure similar to the ESA spacelab that was being developed for the Space Shuttle flights in the 1980s, at Marshall to conduct experiments as part of Marshalls' concept verification test (CVT). CVT was supposed to involve potential experimenters early in the development cycle to test the hardware and identify operational problems that might occur on an actual mission, and to see what could and could not be done by astronauts preforming science experiments in Earth orbit.[220] Although most articles hailed the women for their skills and as prospective astronauts, proclaiming that female astronauts are "no longer a science fiction fantasy," they also all discussed if the scientists' husbands and children were ok with them potentially flying to space.[221] All of

218 Gloria Woody, *letter to Senator John V. Tunney*, October 26, 1973, MC 495, Carton 20, Folder 5, Wilma Scott Heide Papers, Schlesinger.

219 'NASA Doctor feels trapped in the Middle of Controversy over Women in Space', *Sun*, October 8, 1973.

220 'Women Experimenters in simulated Space Test at Marshall Center', *NASA News*, December 13, 1974, 008995 HRC; 'Four Marshall Women Begin Five Day Laboratory Test', *Marshall Star*, December 18, 1974, pp. 1 and 4, HRC; Shayler and Moule, pp. 157–160. See Fig. 3.24.

221 Paula Bernstein, 'It's no longer a Science Fiction Fantasy', *Daily News*, December 8, 1975, HRC; Barry J. Casebolt, 'Women train here for possible Mission on Spacelab', *Huntsville Times*, December 17, 1974; Barry J. Casebolt, 'Females want Space Flight', *Huntsville Times*, December 21, 1974.

the women dreamt of becoming astronauts and eventually applied for astronauts' positions on Spacelab and the Shuttle, making it to the final rounds of the selection process.[222] However, the concept of a mother flying to space was extraordinary for a lot of Americans. While fathers flying to space was not newsworthy, mothers in space and in the space community were.

NOW called attention to this gender bias in the 1970s. They complained that the headline 'Mother among space mission crew finalist' in a *Houston Chronicle* article from 1977 was sexist because Whitacker was not solely portrayed as a scientists, like a man would have been described; instead, the article positioned her as a mother first, starting with: "The mother of a 7-year old girl …"[223] Even today, journalists like to frame women in traditional gender roles and never forget to remark who is a mother or a grandmother. The same gender-based framing happened to quite a few women in the space community, for example, electrical engineer Marjorie Townsend. In 1970, the headline of an article in the Huntsville Times proclaimed that the 'Spacecraft Launch Boss is a Mother' when she became the project manager of the Small Astronomy Satellite (SAS) project.[224] Other women in the space community, like the winner of the Teacher in Space program, Christa McAuliffe, and astronaut Anna Fisher, famous for being the first flying 'astromom,' were subjected to similar reporting, revolving around their status as mothers in the 1980s.[225] There were no exceptions, the media gender framed all women in all occupations, and space workers were far from the only ones. Israel's Prime Minister Golda Meir, for example, was continuously described as a mother and grandmother.[226] Although NOW was an ardent supporter of women having a career and children, feminists took the media and the space agency to task for their double standard. The simulation undergone by the Marshall scientists bore no relation to their gender; the fact that they were all women

---

222 Beth Russler, 'Huntsville Women may be first female Astronauts', *Birmingham Post-Herald*, November 29, 1977.

223 'CR Excercise: What's Wrong With This Headline', *NOW News Bay Area Chapter* 6/1 (1978), p. 4, PR-1 Carton 25, NOW Chapter Newsletter, Schlesinger; 'Mother among Space Mission Crew Finalist', *Houston Chronicle*, December 22, 1977.

224 'Spacecraft Launch Boss is a Mother', *Huntsville Times*, December 15, 1970.

225 Anna L. Fisher, interview by Jennifer Ross-Nazzal, May 3, 2011, p. 2, Edited Oral History Transcript, NASA Johnson Space Center Oral History Project, JSC; 'In NASA Training she carried a Stowaway', *Garden City Newsday*, October 25, 1984; Aimee Lee Ball, 'When Mom Is an Astronaut', *Congressional Record Daily Edition*, 131 (1985); Jennifer Ross-Nazzal, 'You've Come a Long Way, Maybe. The First Six Women Astronauts and the Media', in *Spacefarers. Images of Astronauts and Cosmonauts in the Heroic Era of Spaceflight*, ed. by Michael J. Neufeld (Washington, D.C.: Smithonian Institution Scholarly Press, 2013), pp. 175–202 (pp. 184–186).

226 Davis, p. 108.

was a coincidence.[227] They had not been trying to prove that women could live and work in space, like the Ames bed rest study, or to determine what women needed to work in space; they had tried to establish what kind of experiments any scientists, irrespective of the astronauts' gender, could be expected to perform in Spacelab or the Space Shuttle. Still, most reports did not emphasize that the women were scientists, but that these scientists were women, wives, and mothers.

The NOW women did not see the space community making any kind of effort to include women in space exploration. In their January 1971 newsletter, the Houston chapter of NOW elaborated on their feelings: "The launching of Apollo 14, history's largest phallic symbol, carrying male astronauts, with its flight preparation and control emanating from the Manned Spacecraft Center, is an almost-too-flawless symbol of the position of women in Houston."[228] They felt that women in the space community were only "space trash," whose job in space city was to support the men as sexual objects and maids. A popular opinion, which they even made into an image on the cover of their newsletter.[229] The feminists wanted more for the women in the space community: an equal place in space. Hence, NOW frequently raised awareness of the women in the space community. The feminist organization, for example, demonstrated in three different locations to honor the 54th anniversary of women's suffrage in 1974. One of the demonstrations was at NASA HQ to demand the hiring of female astronauts. Demonstrators brandished foot-shaped signs with red painted toe nails, upon which was the proclamation "One Giant Step for Humankind." They also announced that that they had placed NASA Administrator James Fletcher on its list of "FBI (Feminist Bureau of Investigation) Ten Most Wanted Men for Offense against Women."[230] The negative treatment of Fletcher was most likely connected to one of NOW's biggest space community related consciousness raising efforts ever in the public space – the case of Ruth Bates Harris.

Up to this day, NASA needs political support to function. The space agency is dependent on the media's positive portrayal of the agency itself and space exploration in general. This was even more so the case in the 1970s than today, because NASA needed to convince the American public that space exploration was still important even after the success of Apollo 11. However, by the 1970s the Cold War legitimation of the agency no longer resonated with the majority of the

227 Barry J. Casebolt, 'Females want Space Flight', *Huntsville Times*, December 21, 1974.

228 'Our Place in Space', *The Broadside*, January 1971, Pr-1 Carton 25, NOW Chapter Newsletter, Schlesinger.

229 See Fig. 3.23.

230 Judy Flander, 'NOW Banks too hard on one Zap', *Washington Star*, August 27, 1974.

American people and a majority of the public felt that the money could better be invested in other, Earth-based programs. Their concerns had shifted to other issues such as, for instance, gender equality. Thus far, the media and the American people had been mainly interested in the astronauts and the defeat of the Soviet Union, and the NASA administrators, just like the astronauts, did not grasp the change in public opinion and media focus right away.[231] This is probably why NASA Administrator James C. Fletcher did not even consider the possibility of a public reaction when he decided to fire the agency's African American Deputy Assistant Administrator for Equal Opportunity for being a "disruptive force" and a "divisive element" in 1973.[232] However, times had changed and when the news of Harris' termination broke, the majority of the American people, including the press, members of Congress, and civil rights as well as women's organizations were outraged and publicly demanded the reinstatement of Harris as well as significant change within the space agency regarding equal rights. Fletcher even received a bomb scare and harassment calls.[233]

Fletcher and his deputy, George M. Low, had underestimated the effect of their action in these gender and race sensitive times and, most of all, they apparently had not expected Harris to strike a chord with the American public and the media. In his personal notes, Low repeatedly complained that Harris was "trying to get all of the publicity that she can" and that she achieved her goal of good publicity while NASA mainly got bad publicity regarding the treatment of Harris and equal employment opportunity.[234] Considering that Harris was a media pro and her employment at NASA was terminated because she had

231 Byrnes, pp. 2, 114–115, 147–148; Howard E. McCurdy, *Inside NASA. High Technology and Organizational Change in the U.S. Space Program* (Baltimore: Johns Hopkins University Press, 1993), p. 124; Kim McQuaid, 'Race, Gender, and Space Exploration: A Chapter in the Social History of the Space Age', *Journal of American Studies*, 41 (2007), pp. 405–34 (pp. 408–409).

232 James C. Fletcher, *Memorandum to all NASA employees*, November 2, 1973, Harris/MSRC; Constance Holden, 'NASA: Sacking of Top Black Woman Stirs Concern for Equal Employment', *Science* 182 (1973), pp. 804–807. For an extensive chronicle about the termination of Ruth Bates Harris see: Kim McQuaid, '"Racism, Sexism, and Space Ventures": Civil Rights at NASA in the Nixon Era and Beyond', in *Societal Impact of Spaceflight*, ed. by Steven J. Dick and Roger D. Launius (Washington, D.C.: National Aeronautics and Space Administration Office of External Relations History Division, 2007), pp. 421–50; McQuaid, 'Race, Gender, and Space Exploration: A Chapter in the Social History of the Space Age'; Harris, pp. 253–282.

233 George M. Low, *personal notes*, October 29, 1973, Box 61, Folder 3, Low/RPI; Rossiter, *Women Scientists in America. Vol. 3. Forging a New World since 1972*, pp. 169–170.

234 George M. Low, *personal notes No. 108*, November 25, 1973, Box 67, Folder 5, Low/RPI; George M. Low, *personal notes*, October 29, 1973, Box 61, Folder 3, Low/RPI; George M. Low, *personal notes No. 107*, November 13, 1973, Box 67, Folder 5, Low/RPI.

Fig. 3.23: Title of the Broadside (Houston Chapter) 2/1 (1971), in: NOW Chapter Newsletter, Schlesinger Library, Radcliffe Institute, Harvard University.

a confrontational "take the bull by the horns and wrestle with him" approach to management and consciousness raising, it is quite surprising that NASA's top management was dumbfounded by Harris reaching out to the public and the resulting media interest in her story.[235] According to the speculations of Harris' co-workers, Fletcher had assumed that the termination of Harris would be "safe" because Harris, as an African American female, was "expendable" in terms of the impact on NASA management and the black community. Apparently Fletcher had stated that Harris' boss and target of her criticism, Assistant Administrator for Equal Opportunity Dr. Dudley McConnell, was able to get along with the NASA center directors better

235 Nathaniel Pilot, interview by Dr. Aaron Gillette, July 28, 1992, p. 41, Tape Transcript, HRC.

Fig. 3.24: From left to right: Mary H. Johnston, Ann F. Whitaker, Doris Chandler, and Carolyn S. Griner, November 13, 1974, NASA.

than Harris because he was "a man and an engineer." While co-workers did not understand "why, if a Center Director refuses to get along with a woman, it is a reflection on her," they did understand that Fletcher and Low had assumed that the termination of the highest ranking woman at NASA would go unnoticed within the black and feminist community because the agency had no immediate connection to it.[236]

However, Fletcher had underestimated NASA's symbolic status as a place to reach for the stars – metaphorically and literally. The termination, a symbol demonstrating that reaching for the stars was off limits to minorities and women, especially women of color, invited the entire country to look critically at the space agency. Fletcher and Low had also underestimated Harris' ability

236 *Testimony of Andrea Diane Graham*, 1974, Harris/MSRC; *Testimony of Samuel Lynn*, 1974, p. 15, Harris/MSRC.

to weave her personal story of termination from the space agency into a bigger story about NASA as a space of social injustice.[237]

3.3.2 Gendered framing of the discourse about the first female astronauts

After Ruth Bates Harris war reinstated at NASA due to public pressure in 1974, the space agency's personnel decisions came into the focus of the media once again in the late 1970s. In 1969, NASA had ceased hiring new astronauts due to a lack of funding and of human space missions, and by the early 1970s, the agency had even been considering dismissing some of its astronauts. However, in 1977 NASA announced new astronaut positions for its new space transportation system, the Space Shuttle. All qualified Americans were invited to send in their application. Contrary to previous years, especially female and minority candidates were encouraged to apply. The space agency was looking for pilot astronauts and mission specialist astronauts, who were graduate school educated scientists. Even though the test pilot pipeline career in the different military branches had recently opened up for women, the mission specialist position and the increased emphasis on science and therewith scientist astronauts in the coming Shuttle era gave more women the opportunity to become an astronaut.

The Navy had abolished its separate women's services in 1972, the Air Force in 1976, and the Army in 1978, while the military academies were opened to women in 1976 through Congress. Still, it was only in 1988 that the Air Force allowed women into the test flight training program and NASA accepted its first female pilot astronaut in 1990 – the Air Force pilot Eileen Collins.[238] Before Collins, though, NASA selected 35 new astronauts in January 1978, including six women,

237 More detailed information about the case of Ruth Bates Harris can be found in chapter seven of this book.

238 Carolyn L. Huntoon, interview by Rebecca Wright, June 5, 2002, p. 10, Oral History Transcript, NASA "Herstory" Oral History Project, JSC; Francine D'Amico, 'Policing the U.S. Military's Race and Gender Lines', in *Wives and Warriors. Women and the Military in the United States and Canada*, ed. by Laurie Weinstein and Christie C. White (New York: Praeger, 1997), pp. 199–234 (p. 219); Kevles, pp. 171–172; Hersch, *Inventing the American Astronaut*, p. 176; Amy Foster, 'The Gendered Anniversary: The Story of America's Women Astronauts', *The Florida Historical Quarterly*, 87 (2008), 150–73 (p. 171); Dennis R. Jenkins, *Space Shuttle: The History of the National Space Transportation System – 3rd Edition: The History of the National Space Transportation System: The First 100 Missions*, 3rd edn (Forest Lake: Specialty Press, 2001); Shayler and Moule, pp. 178–180.

three African Americans and one Asian American, and the media went wild for the "ten interesting people" in the pool of "standard white guys." The women, in particular biochemist Dr. Shannon W. Lucid, surgeon Dr. Margaret Rhea Seddon, geologist Dr. Kathryn D. Sullivan, electrical engineer Dr. Judith A. Resnik, emergency physician Dr. Anna L. Fisher, and physicist Dr. Sally K. Ride, were the main focus in the mainstream media.[239] Keeping in mind that by 1978 the American public had witnessed the Civil Rights Movement and experienced the first couple of years of second wave feminism, and that the majority of Americans and media representatives had already been waiting for the announcement of women and minority astronauts, the extreme media focus on these new recruits is understandable.

Furthermore, the media, generally always interested in everything new and unprecedented, in the 1970s was particularly into 'first women' stories about women who were the first of their sex to perform a job previously closed to them. The experience of the first female astronauts with the media was thus not that different from that of other 'first women.' However, because the astronaut profession had not only been a male dominated domain but the epitome of masculinity, the media was especially curious about the female astronauts' take on the job. Some of the reporters were almost obsessed with the women's performance of femininity in a job originally marketed to be for virile, heterosexual men only. The reporters usually just grazed the topic of the women's extraordinary skills and concentrated on their appearance, emotions, marital status, motherhood, domesticity, their male co-workers' behavior toward them and the questions if she considered herself a 'libber' and if the women's movement helped her to get the new position. The focus of the

239 Kathryn D. Sullivan, interview by Jennifer Ross-Nazzal, May 10, 2007, pp. 36–37, Edited Oral History Transcript, NASA Johnson Space Center Oral History Project, JSC; Jeffrey A. Hoffman, interview by Jennifer Ross-Nazzal, April 2, 2009, pp. 12–13, Oral History Transcript, NASA Johnson Space Center Oral History Project, JSC; Sally K. Ride, interview by Rebecca Wright, October 22, 2002, p. 4, Oral History Transcript, NASA Johnson Space Center Oral History Project, JSC; J. Alfred Phelps, *They Had a Dream: The Story of African-American Astronauts* (Novato, CA: Presidio, 1994), pp. 65–66; Foster, *Integrating Women into the Astronaut Corps. Politics and Logistics at NASA, 1972–2004*, pp. 91–95; Atkinson and Shafritz, pp. 147–164; Foster, 'The Gendered Anniversary: The Story of America's Women Astronauts', pp. 158–159; Kevles, pp. 60–67; Jennifer Ross-Nazzal, 'Pioneering Women of Today: Space Travellers', *Sound Historian*, 2011, 23–32; Shayler and Moule, pp. 169–174. See also political cartoons: Have you considered sending up a lady astronaut and bobby pin? *Houston Chronicle*, May 16, 1973; *Washington Star*, February 1, 1978; *Washington Post*, February 23, 1975. See Fig. 3.25

Fig. 3.25: From left to right: Shannon W. Lucid, Margaret Rhea Seddon, Kathryn D. Sullivan, Judith A. Resnik, Anna L. Fisher, and Sally K. Ride, February 1979, NASA.

articles usually revealed unspoken gender assumptions and was intended to reassure the American public that these female astronauts were women still.[240] Just like Tereshkova, Cobb, and the female scientists of the 1960s, the first female astronauts were also bestowed with new occupational titles like 'astrowomen,' 'lady astronaut,' or 'glamournauts,' and were frequently referred to as 'girls,' 'ladies,' or 'lovelies,' not only by the media but also by their co-workers. It was frequently highlighted that the "glamournauts aren't just great lookers. They also know what they're doing."[241] Statements like these

240 Rosen, p. 303; Ross-Nazzal, 'You've Come a Long Way, Maybe. The First Six Women Astronauts and the Media', pp. 175–177; Foster, *Integrating Women into the Astronaut Corps. Politics and Logistics at NASA, 1972–2004*, pp. 94–95, 144–145.

241 Barry Hart, 'The Glamornauts', *Globe*, March 18, 1980; 'Space Shuttles will have mixed Crews. Here come the Astrowomen', *Equal Opportunity Current News*, November 18, 1976; Frederic Golden, 'Toward A New Frontier', *Time*, June 27, 1983.

repeated the gender assumptions found in articles from the 1960s, when reporters were astonished that the scientists had “a surprising range of space skills” besides being physically attractive.[242] Fifteen years and a women’s movement later, the reporters were nearly just as much amazed by the fact that “looks and books” do go together.[243]

Before Sally Ride was selected to become the first of the group to make a spaceflight and the focus concentrate on her as the first American woman to fly to space, the media was interested in all of the women. Especially “pretty” and “pert” Anna Fisher and Margaret Rhea Seddon attracted the public’s attention and the women also knew why.[244] Kathryn Sullivan, for example, explained that from the get-go, she guessed who the media would be interested in because of the media’s predictability. The press was interested in stereotypes and some of the women fit the bill better than others:

> They’ve got a cute blonde. They’ve got a cute kind of curly-haired, flirtatious, single gal. They’ve got two married gals, and of the two of them, Anna was slighter of build, younger. Rhea, Anna, and Judy are going to be their lightning rods and their icons. They’re very telegenic, photogenic. Sally might get in there; she fits the physiological stereotype, you know, slender. Shannon and I are taller, stockier. Shannon’s married. If you were going to line the six of us up, put our six photos up, and pick cover girl shots, I’m not a cover girl type. [. . .] It’s about fitting the mental models that the writers or the reporters have, and I’ve never really fit very well into people’s standard bins. [. . .] Too tomboy, too smart, too strong, too all those things; not the archetypal little girl.[245]

Sullivan was right. In the beginning the reporters were not that interested in the sporty tomboys Sullivan and Ride, and Resnik was not as approachable as Fisher, or as flashy as Seddon and thus also did not receive as much coverage. Resnik repeatedly declared that “my private life is private and I’ll keep it that way,” whereas Seddon and Fisher were more comfortable to give interviews, even when the reporters asked about their private lives.[246] Seddon just

242 Sara Azrael, ‘Women Behind the Spacemen and the Satellites’, *Sun* Magazine, May 17, 1964, pp. 19, 21; NASA Office of Public Information, *Women finding their Place in the Sun in America’s challenging Space Program*, 1960s, 008998, HRC.

243 Lillian Levy, ‘Career Girl 1959. A Space Career. She doesn‘t want to be launched’, *Sunday Star*, May 31, 1959.

244 ‘Space: Squabbling over Astro Turf’, *Time*, August 30, 1982.

245 Kathryn D. Sullivan, interview by Jennifer Ross-Nazzal, May 10, 2007, pp. 39–40, Edited Oral History Transcript, NASA Johnson Space Center Oral History Project, JSC.

246 James Schefter, ‘Space Chores awaiting the Touch of Women’, *Washington Star*, March 26, 1981; Margaret Rhea Seddon, interview by Jennifer Ross-Nazzal, May 20, 2010, p. 57, Edited Oral History Transcript, NASA Johnson Space Center Oral History Project, JSC; Anna L. Fisher,

accepted that the media would comment on her appearance and feminine qualities whether she liked it or not, so she did not worry about it and did not let it ruin her day.[247] Fisher understood the public's curiosity about the first female astronauts. She also felt that it was her duty to create positive publicity for NASA, therefore she was always open to sit for an interview, even when she was asked about her newlywed life.[248] Although Lucid's appearance did not appeal to the media as much as, for example, Fisher's looks, she was very interesting for the public because she was already a mother. Thus, every article about her identified her as a mother of three, just as articles about female scientists in the 1960 had focused on their domestic roles.

Considering that the twenty first century press still likes to document how women are able to combine work and motherhood, the media in the 1980s was, of course, especially interested in how women could balance the two roles while thousands of miles away from their children in Earth orbit.[249] Similar to today, women in the late 1970s and 1980s were considered the main caretakers of children. This is why the public was especially interested in how far the women could find their supposed biological fulfillment as mothers while having a full-time job without neglecting their children. Not only did Lucid have to render an account of her motherhood, but NASA, as her employer, was also asked how much consideration it had given to "her responsibilities to her children versus her responsibilities to the program." This question, posed by Bill Hines of the *Chicago Sun Times*, revealed American society's way of thinking about working mothers, because a question like that had thus far never been asked about the male astronauts and their responsibilities as fathers. It is interesting to note, though, that Hines was aware that his question might be criticized and evaluated as an unfair double standard for women, because he started his question off with a backhanded apology. He said: "Forgive me for what may seem like a 'male chauvinist pig' question." JSC Director Christopher

interview by Jennifer Ross-Nazzal, February 17, 2009, p. 18, Edited Oral History Transcript, NASA Johnson Space Center Oral History Project, JSC.

247 Margaret Rhea Seddon, interview by Jennifer Ross-Nazzal, May 21, 2010, p. 19, Edited Oral History Transcript, NASA Johnson Space Center Oral History Project, JSC.

248 Anna L. Fisher, interview by Jennifer Ross-Nazzal, February 17, 2009, p. 18, Edited Oral History Transcript, NASA Johnson Space Center Oral History Project, JSC.

249 For example: Lori Garver was asked during her entire time as NASA Deputy Administrator how to combine career and children and if it was hard to go back to work after maternity leave. Questions male members of the space community never have to answer. Lori Garver, 'Women, Innovation and Aerospace. Speech at George Washington University', *NASA*, 2012 <https://www.nasa.gov/pdf/632664main_Deputy_Administrator_Lori_Garver_womens.pdf> [accessed 17 November 2015].

C. Kraft let him know that her status as a mother had not been taken into consideration at all and NASA Administrator Robert A. Frosch emphasized that all of the astronauts, irrespective of their gender, were picked because of their motivation, academic background and training.[250]

Although Lucid's status as a mother did not matter to NASA, it persisted in the media. However, the focus was taken off of her after Seddon and Fisher gave birth to their 'astrotots' with their astronaut husbands, Robert 'Hoot' Gibson and Bill Fisher, in 1982 and 1983 respectively. Stories and pictures of the women and their babies were frequently featured in various magazines. Even though the babies also had astronauts for fathers, the focus was always on the mothers. The public wanted to know if they were good mothers or if their astronaut abilities might even make them better mothers. At least one organization in New York City must have come to the conclusion that Fisher was indeed a good mother, because they awarded her the Mother of the Year Award.[251]

3.3.3 Female astronauts and the women's movement

The public and the press just saw, honored, and assumed what they wanted to see according to the prevailing gender stereotypes in the United States in the 1970 and 1980s. For example, when Seddon and Ride got married, they did not take their husbands' names. Some coverage, however, just assumed that they did according to American custom and called them Margaret Reha Seddon Gibson and Sally K. Ride Hawley respectively.[252] Like the 1960s scientists and the astronauts' wives, the first female astronauts also had to realize that the media was often intent on writing preconceived stories, scrambling up statements to make them fit the desired narrative. Some newspapers even made up entire interviews.[253] Seddon, for instance, remembered a reporter who asked her: "Don't you think it's about time NASA took women?" She only replied with "Yes." However, the next day's paper bore

250 Space Shuttle Astronauts Press Conference, transcript, January 16, 1978, 008987, NASA HRC.

251 Anna L. Fisher, interview by Jennifer Ross-Nazzal, May 3, 2011, p. 2, Edited Oral History Transcript, NASA Johnson Space Center Oral History Project, JSC; Ball; Ross-Nazzal, 'You've Come a Long Way, Maybe. The First Six Women Astronauts and the Media', pp. 184–186; Ross-Nazzal, 'Pioneering Women of Today: Space Travellers', p. 28.

252 Ross-Nazzal, 'Pioneering Women of Today: Space Travellers', p. 27; Ross-Nazzal, 'You've Come a Long Way, Maybe. The First Six Women Astronauts and the Media', p. 185.

253 Jeffrey A. Hoffman, interview by Jennifer Ross-Nazzal, April 2, 2009, p. 13, Oral History Transcript, NASA Johnson Space Center Oral History Project, JSC.

the headline: "Astronaut says it's about time NASA took women."[254] Astronaut Jeffrey A. Hoffman remembered a similar lesson learned by Resnik about how the media worked. Resnik was asked whether she felt that the women's liberation movement was responsible for her getting selected as an astronaut. She replied that her own professional qualifications had gotten her the job; however, the headline the next day read: "New Woman Astronaut Says Women's Lib Not Necessary."[255] Resnik and the other first female astronauts felt that the "focus on women isn't fair. We're all going through the same thing. The women are not doing anything different than the men and don't deserve all this public attention and interest."[256]

The women did not want to be seen as 'women scientists' or 'women astronauts' but just as scientists and astronauts who were selected because of their credentials and not because of affirmative action or as a publicity stunt.[257] Their emphasis on their skills and hard work and not on their sex was consequential considering the tenacious rumor that NASA had "bowed to women's lib" and had to include women and minorities. Just like other first women, the first female astronauts thus tried to distance themselves from the women's movement and to portray themselves as 'one of the boys' who did not receive any preferential treatment.[258] This attitude seemed necessary, considering how the media highlighted their femininity and communicated to the public that men and women were fundamentally different. Keeping in mind that the women were the first of their sex

254 Margaret Rhea Seddon, interview by Jennifer Ross-Nazzal, May 20, 2010, p. 11, Edited Oral History Transcript, NASA Johnson Space Center Oral History Project, JSC.

255 Jeffrey A. Hoffman, interview by Jennifer Ross-Nazzal, April 2, 2009, pp. 12–13, Oral History Transcript, NASA Johnson Space Center Oral History Project, JSC. See also: Sally K. Ride, interview by Rebecca Wright, October 22, 2002, p. 6, Oral History Transcript, NASA Johnson Space Center Oral History Project, JSC; Kathryn D. Sullivan, interview by Jennifer Ross-Nazzal, May 28, 2009, pp. 95–96, Edited Oral History Transcript, NASA Johnson Space Center Oral History Project, JSC; Foster, *Integrating Women into the Astronaut Corps. Politics and Logistics at NASA, 1972–2004*, p. 147.

256 Donna Miskin, 'In the new Astronaut Class', *Washington Star*, November 19, 1978.

257 Anna L. Fisher, interview by Jennifer Ross-Nazzal, May 3, 2011, p. 8, Edited Oral History Transcript, NASA Johnson Space Center Oral History Project, JSC; Sara Sanborn, 'Sally Ride, Astronaut. The World is watching', *Ms.*, January 1983 11/7, pp. 45–52 and 87–88; Ross-Nazzal, 'You've Come a Long Way, Maybe. The First Six Women Astronauts and the Media', p. 176.Disler, p. xviii; Ross-Nazzal, 'You've Come a Long Way, Maybe. The First Six Women Astronauts and the Media', p. 176.

258 Don Kirkman, 'Women Astronauts', *San Francisco Chronicle*, July 7, 1974; 'People', *Time*, August 14, 1978; Rosen, pp. 304–306; Foster, *Integrating Women into the Astronaut Corps. Politics and Logistics at NASA, 1972–2004*, p. 151.

not just in a male-dominated sector but in an agency molded by military gender performance, where 'male' was accepted as the norm, the women preferred to expunge all things feminine, just like the first women in the military.[259] However, just like the military women, the female astronauts' gender performance could not be too masculine, otherwise there would be accusations of lesbianism, which was not widely accepted in the United States in the 1970s. Therefore the women's gender performance still had to portray the 'right stuff' – being heterosexual first and foremost. This might explain the media's obsession with the women's dating and love lives.[260] Maybe that is also why Sally Ride, the first American woman in space, waited until her death in 2012 to tell the world she was gay and had loved a woman for the past 27 years.[261]

Because the women did not ask for, and were not given, any special treatment, they did not "want to be portrayed as the cute little girls NASA [had] hired to be stewardesses on the Space Shuttle."[262] This is also why Resnik had refused to discuss her personal grooming rituals with the media and why Ride had refused to carry make-up with her on the Shuttle, smile for photographers on demand or accept a bouquet of carnations and roses when she returned from her historic space flight. None of her crewmates were gifted with flowers, but their wives were presented with bouquets. Thus, Ride refused the gendered American custom because it singled her out for her sex and not her accomplishment.[263] Even though the women preferred to be 'one of the guys,' they were indeed aware that they had benefitted from the "feminist movement and from women who went before us pushing hard for equal rights" and that their opportunity at the space agency was "in part because of the point in time at which we all came along, and that point in time in American history was certainly

---

259 Jeffrey A. Hoffman, interview by Jennifer Ross-Nazzal, April 2, 2009, p. 13, Oral History Transcript, NASA Johnson Space Center Oral History Project, JSC; Elgin Medea Brunner, *Foreign Security Policy, Gender, and US Military Identity* (New York: Palgrave Macmillan, 2013), pp. 24, 33.

260 Disler, p. 16; Ross-Nazzal, 'You've Come a Long Way, Maybe. The First Six Women Astronauts and the Media', pp. 183–184.

261 Alan Boyle, 'Why Sally Ride Waited until Her Death to Tell the World She Was Gay', *NBC News*, 2012 <http://www.nbcnews.com/science/science-news/why-sally-ride-waited-until-her-death-tell-world-she-f908942> [accessed 10 January 2017].

262 Donna Miskin, 'In the new Astronaut Class', *Washington Star*, November 19, 1978.

263 Frederic Golden, 'Mission Accomplished', *Time*, July 4, 1983; 'People', *Time*, August 14, 1978; Ross-Nazzal, 'Pioneering Women of Today: Space Travellers', p. 31; Ross-Nazzal, 'You've Come a Long Way, Maybe. The First Six Women Astronauts and the Media', pp. 193–194; Foster, *Integrating Women into the Astronaut Corps. Politics and Logistics at NASA, 1972–2004*, p. 146, ft. 15.

shaped by the Civil Rights Movement. It was shaped by the '60s in general, and shaped by the feminist movement, absolutely."[264] Still, they needed to highlight their individual talents and accomplishments to show that though the women's movement might have unlocked the door for them, it was their own hard work and professional skills which got them over the threshold. To them, "[c]laiming and demanding some respect for your individual talents and accomplishments is not synonymous with totally rejecting that social change teed up opportunities that you were able to take ahold of."[265] To most media representatives, on the other hand, their point of view was not acceptable because it did not fit into the preferred binary system of good and evil, women and men, or 'libber' and antifeminist. The media wanted the women to either be feminist heroes or 'anti-libber.' The women felt as if they were "damned either way" in answering the press' questions about the women's movement, and therefore tried to stay away from the subject, if possible.[266]

However, it was rarely possible, because women's groups and girls of all ages had declared them to be their heroes and role models and projected their expectations onto the astronauts, who in turn were very much aware of their responsibility and felt the pressure of it. Seddon, for instance, knew that "[p]eople watch [them]. There are those who think there is no way for women to succeed. Others think whatever you do reflects on women in general. And there are other who want to be future astronauts and are looking for you to open the door. I feel a great responsibility to do a good job."[267] Fisher was grateful to the women's movement but did not consider herself a part of it.[268] Ride, on the other hand, turned into the feminists' main idol because she was not only the first of the six women to break through the space ceiling, but also because she had been a member of NOW in college and revealed that she had seen the NASA ad at her

264 Anna L. Fisher, interview by Jennifer Ross-Nazzal, May 3, 2011, p. 8, Edited Oral History Transcript, NASA Johnson Space Center Oral History Project; Kathryn D. Sullivan, interview by Jennifer Ross-Nazzal, May 28, 2009, p. 95, Edited Oral History Transcript, NASA Johnson Space Center Oral History Project, JSC.

265 Kathryn D. Sullivan, interview by Jennifer Ross-Nazzal, May 28, 2009, pp. 95–96, Edited Oral History Transcript, NASA Johnson Space Center Oral History Project, JSC-.

266 Kathryn D. Sullivan, interview by Jennifer Ross-Nazzal, May 28, 2009, p. 96, Edited Oral History Transcript, NASA Johnson Space Center Oral History Project, JSC.

267 Donna Miskin, 'In the new Astronaut Class', *Washington Star*, November 19, 1978; Foster, *Integrating Women into the Astronaut Corps. Politics and Logistics at NASA, 1972–2004*, pp. 147–152.

268 Kevles, p. 67.

women's center in Stanford. Although she had not been very active in her NOW group, she apparently "warmly credit[ed] the Women's Movement with changes in NASA since the late 1960s" and had "no doubt about its effect on her own career."[269] Despite wanting to be a genderless member of the team and refraining from phrasing her accomplishment in gendered terms, like Tereshkova had done in the 1960s, Ride nevertheless fed into the feminists' portrayal of her.[270] She, for example, attended Gloria Steinem's fiftieth birthday party and accepted awards from women's groups. Ride also tickled their fancy when she revealed that, when she rocketed through the glass ceiling into space, her mother had exclaimed "God bless Gloria Steinem," an exclamation which gave legitimate credit to one feminist in particular and therewith honored the work of an entire movement.[271]

Even though the other women were never called "a lasting friend of women's rights" by a women's group, as Ride had been, they also were involved with feminist causes.[272] Resnik, for example, was one of the speakers at JSC's Federal Women's Week. She also honored the first female dean of her alma mater, the University of Maryland, in a speech in which she reminded the audience that women should stick together and urged them to continue breaking into male-dominated sectors with skill and talent to achieve equality. However, it was important to her for women to remember that "firsts are only the end of full equality, not the end itself."[273] Resnik underestimated her status as a role model for people previously excluded from the American space adventures, though. When she refused to be a "woman astronaut" or a "Jewish astronaut," in favour of just being an "astronaut," she denied Jewish women a role model and contributed to the concealment of the numerous and valuable contributions made to space

269 Sara Sanborn, 'Sally Ride, Astronaut. The World is watching', *Ms.*, January 1983 11/7, pp. 45–52 and 87–88; Kevles, p. 63.

270 Sylvester, 'She Orbits over the Sex Barrier. Soviet Girls and the Tereshkova Moment', p. 202; Valentina Tereshkova, 'Women in Space', *Impact of Science on Society*, 20 (1970), 5–12.

271 'People', *Time*, June 4, 1984; Several letters from WEAL to Sally Ride, 1983–1987, MC 500, Carton 8, Folder 53, Women's Equity Action League Records 1967–1990, Schlesinger; Michael Robertson, 'Feminist Push in Outer Space. Finding better Role Models than Miss Piggy aboard the Swinetrek', *San Francisco Chronicle*, July 4, 1983. See also Fig. 3.26

272 Several letters from WEAL to Sally Ride, 1983–1987, MC 500, Carton 8, Folder 53, Women's Equity Action League Records 1967–1990, Schlesinger.

273 Lon Rains, 'Astronaut lauds another Woman Pioneer', *Washington Post,* September 25, 1983; 'Johnson Space Center welcomes Public to Federal Women's Week April 17–21', *NOW News Bay Area Chapter* 6/4, April 1978, p. 4, PR-1 Carton 25, Texas – Bay Area Chapter, NOW Chapter Newsletters, Schlesinger; 'Outstanding Speakers spark Federal Women's Week at John Space Center', *NOW News Bay Area Chapter* 6/5, May 1978, p. 2, PR-1 Carton 25, Texas – Bay Area Chapter, NOW Chapter Newsletters, Schlesinger.

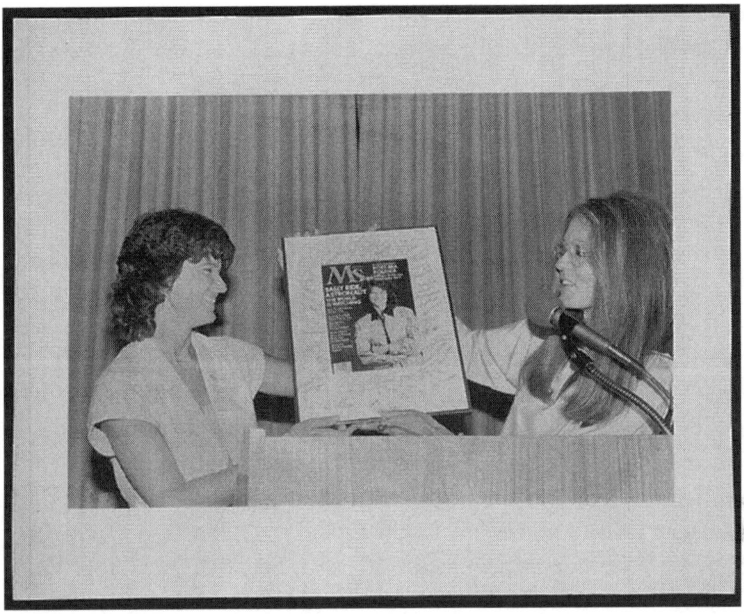

Fig. 3.26: Sally Ride accepts Ms. Magazine award at NY City Hall from Ms. editor Gloria Steinem, March 8, 1988, Bettye Lane Studio, Library of Congress Prints and Photographs.

exploration by female and Jewish members of the space community, since the American image of the prototypical astronaut was a white, protestant male.[274]

The first African American woman to be selected as an astronaut, in 1987, almost ten years after the first six women, portrayed herself very differently from Resnik and the other female astronauts or female soldiers. Dr. Mae Jemison wanted her accomplishments to be described in gendered and racial terms because she wanted people, and especially young African American girls, to know that dreams of going to space "can be true, and they're real."[275] To her a 'social first' was as worthy of being highlighted as the space firsts from the 1960s because they demonstrated how far American society had come and where it could go. To NASA, which had ignored the social aspect of their mission in the 1960s and early 1970s, diversity and equality had become valued assets in the public space, too, because they attracted renewed attention to the space program. The fact that women and people of color were not always well

274 Disler, p. xviiii; Kevles, p. 97.
275 Marcia Dunn, 'Black Woman on Shuttle. She'll be a First in Space. Ship also has Husband, Wife', *Sun*, September 8, 1992.

represented at NASA is nowadays acknowledged on the anniversary timeline on NASA's homepage, where space and social firsts are listed equally.[276]

3.3.4 The case of Sally Ride

As the first woman to fly to space for the United States, on June 18, 1983 as a crew member on Space Shuttle Challenger for STS-7, everyone wanted a piece of Sally Ride – the women's movement, the space community, the science community, the public, the media, the President of the United States, even the Soviet Union. With the news that the United States would send up its first female astronauts and would let her do equal work in space, even EVAs (Extra Vehicular Activity – a space 'walk'), the Soviet Union selected its own class of female cosmonauts, to try to prove yet again that the Soviet Union was first in space and in equality. Thus, Svetlana Savitskaya became the second Soviet woman to rocket into orbit, on August 19, 1982, before Sally Ride could fly for the United States, and she also became the first woman to perform an EVA on July 25, 1984, before Kathryn Sullivan.[277] This time around, however, the flight of a female cosmonaut did not ignite a broad public discussion in the U.S. about gender equality versus publicity stunt, as it had after Tereshkova's flight in 1963. The Americans were unanimously convinced that their way of integrating women into the astronaut corps was more sophisticated, professional, and most of all truly equal, and Sally Ride was the living proof. Most of the media outlets therefore mainly described Savitskaya's time in orbit as well as the celebration of the twentieth anniversary of Tereshkova's historic flight as Cold War propaganda and not a sign of gender equality in the Soviet Union.[278] Savitskaya confirmed this assumption years later, when she stated that the first female astronauts had changed Americans' mentalities regarding women in space. In her view, female astronauts were truly respected for their professional work in American society, while this was not the case in Russia. According to

---

276 Frederic Golden, 'Toward a New Frontier', *Time*, June 27, 1983; Miriam F. Williams, 'Reimagining NASA: A Cultural and Visual Analysis of the U.S. Space Program', *Journal of Business & Technical Communication*, 26 (2012), 368–89 (p. 379); Ross-Nazzal, 'Pioneering Women of Today: Space Travellers', p. 30.

277 Sally K. Ride, interview by Rebecca Wright, October 22, 2002, p. 23, Oral History Transcript, NASA Johnson Space Center Oral History Project, JSC; Shayler and Moule, pp. 199–208, 219–220.

278 Dusko Doder, 'Moscow Publicizes 2 Women who made Flights in Space', *Washington Post*, June 21, 1983; Robert L. Crippen, interview by Rebecca Wright, May 26, 2006, p. 57, Oral History Transcript, NASA Johnson Space Center Oral History Project, JSC; Thomas O'Toole, 'Female Astronauts walks in Space', *Washington Post*, October 12, 1984.

Savitskaya, "[i]n the domain of gender equality, we've got our work cut out for us in society [in Russia] as a whole and in space travel in particular."[279]

While Savitskaya was honored as a new female hero in the Soviet Union, Sally Ride became the symbol for gender equality and American frontier spirit in the United States. Although she tried to be a genderless member of her crew, she was always targeted in the public space and NASA and her teammates could not shield her from everything, even though they tried.[280] Even President Ronald Reagan singled her out in his public telephone chat with the crew, telling her that she had earned her seat on the Shuttle because she was "the best person for the job."[281] A well intentioned compliment, but one that nevertheless pointed out that she was different from her crewmates. It was not just the president who focused most of his attention on Sally; the reporters also did, sometimes even knocking over her fellow astronauts to get to her. Like the president, journalists also often complimented her or asked her questions which would never be put to a male astronaut. At one press conference, for instance, a reporter asked her if she thought she would have to cry when she was in orbit, in response to which Ride just complained that her male colleagues were never asked this type of patronizing question.[282] The crux of the matter was that most Americans, male and female, just assumed that men had naturally earned their spot on the Shuttle or that they would be manly enough to not cry in space, while women had to prove it. They were also forced to rely on the affirmation of their skills in the public space by 'experts,' who were usually the white men in power. After her successful flight, Ride became such an 'expert' herself and the other women, who flew after her, were compelled to accept her confirmation in the public eye. Sullivan's first space flight in 1984 was already Ride's second and reporters repeatedly asked Ride and her male crewmates: "Do you think Kathy can do this?"[283]

---

279 Svetlana Saviskaya, 'Gagarin Is Our Everything', *International Affairs: A Russian Journal of World Politics, Diplomacy & International Relations*, 57 (2011), 237–43 (p. 242).

280 Sally K. Ride, interview by Rebecca Wright, October 22, 2002, p. 14, 16, Oral History Transcript, NASA Johnson Space Center Oral History Project, JSC.

281 Frederic Golden, 'Mission Accomplished', *Time*, July 4, 1983; Frederic Golden, Sam Harris, and Jerry Hannifin, 'Space: Sally's Joy Ride into the Sky', *Time*, June 13, 1983.

282 Frederick H. Hauck, interview by Jennifer Ross-Nazzal, November 20, 2003, pp. 20–21, Oral History Transcript, NASA Johnson Space Center Oral History Project, JSC; John M. Fabian, interview by Jennifer Ross-Nazzal, February 10, 2006, pp. 47–48, Oral History Transcript, NASA Johnson Space Center Oral History Project, JSC; 'On the Record', *Time*, June 6, 1983. See also political carton: It's important people don't think I was picked because I am a woman, *Newsweek*, June 13, 1983.

283 Kathryn D. Sullivan, interview by Jennifer Ross-Nazzal, May 10, 2007, pp. 74–75, Edited Oral History Transcript, NASA Johnson Space Center Oral History Project, JSC. See also

While the majority of the reporters inconsiderately singled out Ride for her gender, others were already discussing this very behavior by their fellow Americans in the news in the early 1980s. Some just called it "old-fashioned chivalry," but others realized that it

> should come as no surprise that women are capable, that they are competent and involved. Yet this is still newsworthy, still the subject of some commentary. Maybe one day it won't be. For now, the commentary is to be expected, and the point does need to be made that after many years of breaking down barriers, women are not only involved in what used to be called 'a man's world,' they are also good at it to the same extent that men are good at it. The initial tokenism and what sometimes passed for enlightenment has given way to a situation where women are involved in integral ways, and that is an important development. [...] Having said all of this, it is difficult for your editor to draw a conclusion without sounding patronizing, and it is patronizing to comment on a woman's performance in view of the fact that the men around JSC have not been honored for jobs well done simply because they are men. At this time and place, however, it is appropriate to contemplate and even applaud the substantive contribution of women in our highly demanding technical atmosphere. The next great step will have come when editorials such as this are no longer written.[284]

Other reporters concluded that Ride and her fellow astronauts had to struggle with the same problem all first women in every occupation had to deal with: they were not allowed to just succeed for themselves, they had to take every step for womankind, too. This is why they were asked about their "opinions on everything 'female' – from fashion to feminism," and why most people figured it was necessary to offer their opinions about the women from fashion to feminism.[285] Unlike other first women, however, the astronauts were thrust in the limelight from the day they were selected and had to quickly figure out how to deal with the media and the publicity, with only little guidance from NASA regarding the sheer volume and gendered portrayal of their selection and personas. This is why the women tried to establish among themselves a consistency in their public performance as female astronauts, so no one could play them off against each other.[286]

---

political carton: It's important people don't think I was picked because I am a woman, *Newsweek*, June 13, 1983.

284 JSC Public Affairs Office, 'Feminine Perspectives', *Roundup*, November 2, 1984.

285 Frederic Golden, 'Toward A New Frontier', *Time*, June 27, 1983; Ellen Goodman, 'The Future of a 'First Woman'', *Boston Globe*, June 25, 1983.

286 Sally K. Ride, interview by Rebecca Wright, October 22, 2002, pp. 4, 23, 24, Oral History Transcript, NASA Johnson Space Center Oral History Project, JSC; Kathryn D. Sullivan, interview by Jennifer Ross-Nazzal, May 10, 2007, pp. 36–39, Edited Oral History Transcript, NASA Johnson Space Center Oral History Project, JSC; Kathryn D. Sullivan, interview by Jennifer Ross-Nazzal, May 28, 2009, p. 89, Edited Oral History Transcript, NASA Johnson Space Center

While the focus on gender issues in space exploration waned in the media in the mid to late 1980s, most Americans still had a lot of feelings about the first female astronauts in the early part of the decade.[287] Jeanne L. Crews, an engineer and astronomer at JSC, for example was proud and pleased when they selected the women and thought that all of the women were wonderful.[288] It was not only the women who worked in the space community who were excited for the first female astronauts; little girls and women all over the United States were happy, too. Some suggested that Sally Ride should be elected "Man of the Year," others felt that "Sally Ride's story was most uplifting," and others were grateful for the impact Ride had on the outlook of their daughters regarding space exploration and science.[289] Jerrie Cobb was apparently also "delighted for Sally Ride," according to her neighbor.[290] Tereshkova was pleased with Ride's historic flight, too. In a telegram to the American astronaut she congratulated her and told her that she was happy that a "third representative of the women of the planet, this time from the United States of America, finds herself in space."[291] Most people, especially women, were excited for the first female astronauts.

Keeping in mind that the voting majority of the U.S. was neoconservative and embraced the antifeminist rhetoric of Phyllis Schafly in the beginning of the 1980s, it cannot be surprising that not all Americans were overjoyed with Ride's historic time in orbit or her behavior as a female astronaut.[292] The woman who fought for ten years against the Equal Rights Amendment, Schafly herself, thought that Ride's selection as the first American woman in space was "fine," but she did not forget to mention that she herself and her daughter had "no desire to be an astronaut" and were glad to give Ride their places.[293] Other

Oral History Project, JSC. See also political cartoons about the female astronauts' fame in: *Chicago Tribune*, February 21, 1978; *Washington Post*, June 25, 1983.

287 Margaret Rhea Seddon, interview by Jennifer Ross-Nazzal, May 21, 2010, pp. 32–33, Edited Oral History Transcript, NASA Johnson Space Center Oral History Project, JSC.

288 Jeanne L. Crews, interview by Rebecca Wright, August 6, 2007, p. 39, Oral History Transcript, NASA Johnson Space Center Oral History Project, JSC.

289 M. Roy Brenner, 'letter to the editor', *Time*, December 12, 1983; Julie Horpedahl, 'letter to the editor', *Ms.*, February 1986 14/8, p. 4; Dr. Shirley Kyle, *letter to Gloria Steinem*, December 14, 1982, MC 568, Box 16, Folder 35, Ms. Letters, Schlesinger.

290 'Foiled Astronaut', *New York Times*, June 26, 1983.

291 *Hypatia Cluster Newsletter*, Fall 1983, 16386, HRC.

292 Donald Critchlow, *Phyllis Schafly and Grassroots Conservatism: A Woman's Crusade* (Princeton: Princeton University Press, 2005), p. 1.

293 *Susan Okie, 'Well-Wishers watch Sally Ride take new Step for Womankind', Washington* Post, June 19, 1983.

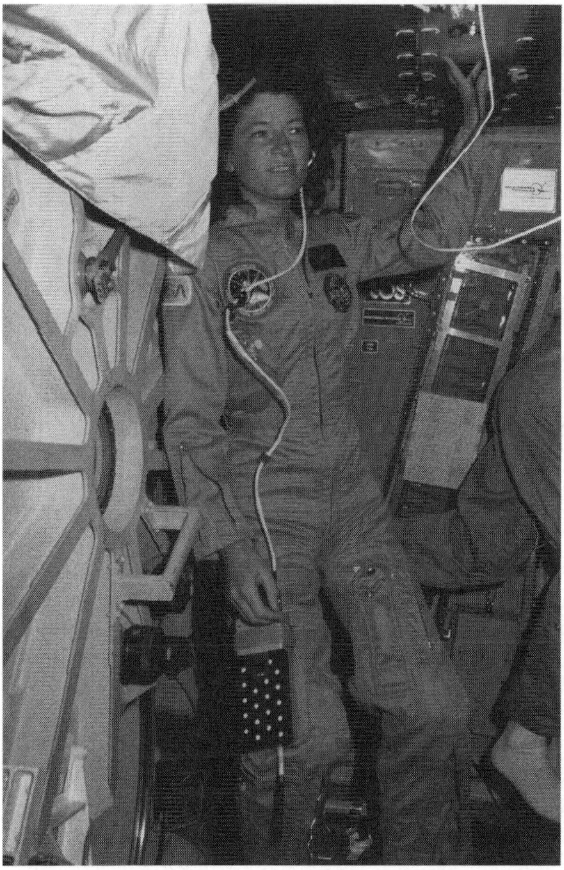

Fig. 3.27: Sally Ride on shuttle Challenger's middeck, STS-7, June 21, 1983, NASA.

women could not understand why Ride refused to wear make-up in space or accept the flowers she was offered upon her arrival back on Earth. This uncertainty is what prompted Lynn Hall to ask: "Now that women can do anything men can do, must they sacrifice being women?"[294] Her question shows that women in the early 1980s put an emphasis on femininity just as much as, maybe even more so, than men. Stella Peterson, in another example, did not like that Ride had refused to answer if she wanted to have children, because "you have a country that's in big trouble" when women don't want to become

294 Lynn Hall, 'letter to the editor', *Time,* July 25, 1983.

mothers. She also pointed out that she knew how "most of these astronauts are like" and therefore concluded that the astronaut lifestyle was not a "kind of a life for a lady to want to duplicate."[295] The topic of motherhood and being able to stay in the 'woman's world' was as important to Americans as to the media. Thus, people frequently applauded Fisher for her gender performance as a working mother because Fisher demonstrated that "the board room is not entirely incompatible with the nursery. The launch pad veteran can scramble on a playground trampoline with the best of them. Free of the stereotypical extremes that characterized the women's movement in its earlier stages, Astronaut Fisher demonstrates that you can excel in a man's world without abandoning the unique joys of a woman's world."[296]

3.3.5 The case of Christa McAuliffe

While the first female astronauts were generally admired for their extraordinariness, it was hard for 'average' Americans without a Ph.D. or test pilot experience to relate to their achievements. It was much easier for the public to connect with New Hampshire teacher Christa McAuliffe. In 1985, she was selected from among more than 11,000 applicants in NASA's Teacher in Space Program to represent American teachers in the space program and give lessons to American children from orbit to spark the kids' interest in science and space exploration. It was supposed to be the first special spaceflight participation project, followed by the Journalist in Space Program and an Artist in Space Program.[297] Christa McAuliffe was picked for her averageness, a fact of which she was very aware and frequently played into with remarks that showed the public that she was one of them, one of the run-of-the-mill. She, for example, stated that it was "mind-boggling" to her that electrical engineer Judy Resnik

295 Tom Zito, 'On the Cape riding high for Sally', *Washington Post*, June 18, 1983.

296 Lois A. Engel, 'Women's Accomplishments', letter to the editor, *Washington Post*, November 26, 1984.

297 Byrnes, p. 148; Dubbs, pp. 81–85; Margaret A. Weitekamp, 'Selling Education in the Shape of a Shuttle', *The Florida Historical Quarterly*, 87 (2008), 210–34 (pp. 217–220); Colin Burgess and Grace George Corrigan, *Teacher in Space: Christa McAuliffe and the Challenger Legacy* (Lincoln: University of Nebraska Press, 2000); Grace George Corrigan, *A Journal for Christa: Christa McAuliffe, Teacher in Space* (Lincoln: University of Nebraska Press, 2000); Robert T. Hohler, *I Touch the Future. The Story of Christa McAuliffe* (New York: Random House, 1986).

knew about the function of all the circuits in the Shuttle "and can figure all of this out."[298] Most Americans without the scientific background of the astronauts could relate to this statement. For this reason, the public embraced her as 'one of them' and the media idealised her ordinariness already before her space flight and even more so after. She was always described as a wife and mother, "as American as apple pie."[299] The *New York Times*, for instance, wrote that "[w]hat was worthy about Christa McAuliffe was her cornflower-blue ordinariness and an accessible style that set her apart from the professional astronauts with their arcane knowledge and career long acceptance of danger" because she thought of herself as 'just' a school teacher who wanted to inspire children.[300]

McAuliffe, often affectionately called "teachernaut," also publicly revealed insecurities about her skills and competence in comparison to the astronauts, but mainly to the other teacher applicants who were more accomplished than her. This is why Mopsy Strange Kennedy of the *New York Times* concluded that McAuliffe was ordinary in an extraordinary way because "[i]n a sense, what she offered were plain good qualities, so clear and straightforward that other ordinary people could imagine themselves in her place."[301] Many other media outlets repeated the same sentiment. For instance, Denise Kalette of *USA Today* figured that "teachers represent anyone who has looked skyward on a clear night and wished for the impossible."[302]

McAuliffe herself liked her portrayal and frequently demonstrated her ordinariness by positioning herself as a wife and a mother. After a fatal shooting in the high school at which she was teaching, she commented that "[i]t was like

298 Christa McAuliffe, 'I've got enough Courage. . .', *USA Today*, January 29, 1986.

299 Neil Chesanow, 'The Right Stuff. How an ordinary Teacher, Wife, and Mother became the first private Citizen in Space', *New Woman*, February 1986, pp. 3–7, 4.

300 Mopsy Strange Kennedy, 'Ordinary in an Extraordinary Way', *New York Times*, March 15, 1987. See also political cartoon: "I'm hoping to be in space the day the competency tests are given." in *Washington Times*, July 2, 1985.

301 Mopsy Strange Kennedy, 'Ordinary in an Extraordinary Way', *New York Times*, March 15, 1987; Lucy Keyser, 'Friends recall a Teachernaut', *Washington Times*, January 28, 1987.

302 Denise Kalette, "An ordinary Person' who lived the Dream', *USA Today*, December 23, 1986. She also referred to McAullife as "the first ordinary citizen in space" – a description which was correct. There were also a lot of articles calling her "the first private citizen in space" – which was not correct. While the 'ordinary' idealization is most certainly debatable, she was definitely not the first private citizen in space. Before her had flown private citizens Senators Jake Garn and Bill Nelson, an Arab prince, and German and Mexican engineers. Kathy Sawyer, 'Shuttle Liftoff delayed, but Teacher ready', *Washington Post*, January 26, 1986.

one of my children was hurt."[303] Acquaintances of McAuliffe often reinforced this type of depiction.[304] She thus stood for patriotism, family values and traditional feminine roles such as mother and educator of the young.[305] However, the "startlingly normal American," who probably served as the role model of the "averagenaut" in the TV show the Simpsons, was not just a modern version of the 1950s housewife, but also a self-proclaimed feminist.[306] The media judged McAuliffe's approach to feminism positively because she was not at all radical in her vision and still embodied all appreciated feminine virtues. According to a fellow teacher, "[s]he never sounded angry," she just "wanted women to do more, to learn more."[307] Her supposedly gentle approach to feminism was also reflected in her application to NASA. She claimed that

> [a]s a woman, I have been envious of those men who could participate in the space program and who were encouraged to excel in the areas of math and science. I felt that women had indeed been left outside of one of the most exciting careers available. When Sally Ride and other women began to train as astronauts, I could look among my students and see ahead of them an ever-increasing list of opportunities.[308]

Although McAuliffe considered herself a feminist and a role model, she usually preferred to perform traditional female roles, such as being a mother and a witness. She made herself into a spectator/witness with her proposed space project of keeping a diary in space to "humanize the technology of the Space Age" and help future historians with her eye-witness accounts "in their studies of the impact of the Space Age on the general population."[309] In short, she wanted to go to space to bear witness for the ordinary people. However, through McAuliffe the spectator in the 1980s was once again embodied by a woman. She basically in-

303 'Teacher mourns for Student', *Sun*, December 16, 1985.

304 Matthew L. Wald, 'Christa McAuliffe: Teacher, Wife and Mother who radiated Enthusiasm', *New York Times*, February 10, 1986.

305 Llinares, p. 94; Constance Penley, 'Spaced Out: Remembering Christa McAuliffe', *Camera Obscura: Feminism, Culture, and Media Studies*, 29 (1992), 179–213.

306 'Christa McAuliffe 1948–1986', *Time*, February 10, 1986; Michael Kranish, 'N.H. Teacher-Astronaut McAuliffe breaks the old Stereotype', *Boston Globe*, August 6, 1985; Hersch, 'Return of the Lost Spaceman: America's Astronauts in Popular Culture, 1959–2006', pp. 84–88.

307 'Christa McAuliffe 1948–1986', *Time*, February 10, 1986.

308 Christa McAuliffe, *Excerpt from Application of Christa McAuliffe to be the first private Citizen in Space*, congressional record, February 18, 1986, 1378, HRC.

309 Christa McAuliffe, *Excerpt from Application of Christa McAuliffe to be the first private Citizen in Space*, congressional record, February 18, 1986, 1378, HRC.

tended to perform the same act as the astronauts' wives in the 1960s, the only difference being that she was to observe from up-close in space on the Shuttle.[310]

After several scrubbed launches, Christa McAuliffe and her crew of professional astronauts on mission STS-51-L launched on January 28, 1986. The Space Shuttle orbiter Challenger exploded and broke apart 73 seconds into its flight, killing all of its seven crew members upon impact with the ocean below.[311] The *Challenger* accident changed the face of the space program and also severely shifted the discourse about women in space because "Christa McAuliffe's death opened cultural imaginings of space to the sacrifice and victimization of women."[312] Before the accident, NASA had hoped that the public would identify with McAuliffe and thus also identify with NASA. The space agency therefore also liked to emphasize the 'ordinariness' of McAuliffe, to connect with the American public. Through the inclusion of, and emphasis on, McAuliffe, NASA also tried to reverse the discourse with which it had historically rejected women. However, the depiction of McAuliffe as an everyday, feminine American woman before the accident had a profound impact on the discussions about civilians, especially women, in space after the Challenger explosion. Because McAuliffe had been portrayed as a symbolic mother figure to America's children due to her status as a mother and a teacher, the media primarily observed the death of a mother, not the death of an astronaut. The media assumed that McAuliffe, as "a nonprofessional, an innocent", had not been aware of the risks of spaceflight, unlike the other, career astronauts, including fellow woman Judy Resnik. Her death was not ascribed the same symbolic value as the death of the private citizen and mother McAuliffe, and thus Resnik's and her male colleagues' death did not seem as tragic, as "doubly poignant and unfair."[313]

---

310 Jodi Dean, p. 100.

311 For more information about the Challenger accident see: Mark Maier and James W. Messerschmidt, 'Commonalities, Conflicts and Contradictions in Organizational Masculinities: Exploring the Gendered Genesis of the Challenger Disaster', ed. by Mark Maier and James W. Messerschmidt, *Canadian Review of Sociology & Anthropology*, 35 (1998), 325–44; Julianne G. Mahler and Maureen Hogan Casamayou, *Organizational Learning at NASA. The Challenger & Columbia Accidents* (Washington, D.C.: Georgetown University Press, 2009); Diane Vaughan, 'Autonomy, Interdependence, and Social Control: NASA and the Space Shuttle Challenger', *Administrative Science Quarterly*, 35 (1990), 225–57; Ben Evans, *Space Shuttle Challenger: Ten Journeys into the Unknown* (New York: Springer-Praxis Books, 2007).

312 Jodi Dean, p. 101; Andrew J. Butrica, 'The Right Stuff: The Reagan Revolution and the US Space Program', in *Remembering the Space Age*, ed. by Steven J Dick (Washington, D.C.: U.S. Government Printing Office, 2008), pp. 121–34 (p. 133).

313 Lance Morrow, 'A Nation mourns: CHALLENGER Heroes. After three Days of communal Grieving, the Reagans join tearful Survivors in a Memorial to the seven CHALLENGER Heroes',

The media's accusation that McAuliffe had been unaware of the hazards of spaceflight was frequently repeated and seemed plausible due to NASA's marketing of the Space Shuttle as a dependable space truck and McAuliffe's refusal to call herself an 'astronaut' rather than a 'space participant.' The media repeatedly remarked that McAullife had never intended to become a space hero, therewith forgetting that Shuttle astronauts like Resnik also never wanted to be heroes *per se* but considered themselves scientists. However, the public was very conscious that McAuliffe, unlike Resnik, had not been a professional astronaut and thus "probably lacked what has come to be known, almost reverently, as 'the right stuff,'" although the media did not forget to emphasize that "she had plenty of good stuff on her own."[314] She was described as having a childlike sense of wonder, as well as pure and uncomplicated beliefs. The media often repeated the story that McAullife felt safer in the Shuttle than on the D.C. Beltway. According to the media, McAuliffe truly believed that spaceflight was super safe because she had once stated that a Shuttle flight "isn't the type of thing, I think, that anybody looks at with fear that there's going to be an accident."[315] Especially McAuliffe's brother fed the media's wrath. He repeatedly accused NASA of using McAuliffe for publicity and wanted the agency to abandon programs that would put 'everyday' people into space because he also felt that his sister had "thought it was more safe than it was."[316]

NASA was criticized not only in the American media but also internationally, with the Soviet Union's *Pravda* especially scolding NASA for putting an amateur into space. According to the Soviets, it was impermissible to use space as some sort of propaganda show.[317] This accusation on the part of the Soviet Union is especially interesting because some U.S. media outlets had alleged similar things about Tereshkova's flight in the early 1960s. Barbara Morgan, McAuliffe's backup, had a completely different assessment of the situation, though. According to her, McAuliffe had very much been aware of the risk of spaceflight. Furthermore, she did not appreciate that the professional astronauts' lives were valued differently because they were "no more expendable than the rest of us are."[318] Still, that was the sense one got in the media space. Some media outlets also decried the fact that most of the attention and grief

Time, February 10, 1986; Llinares, pp. 94–100; Constance Penley, *NASA/Trek. Popular Science and Sex in America* (New York: R.R. Donnelley & Sons, 1997), pp. 22–29.

314 Ron Arias, 'A Lesson in Uncommon Valor', *People*, February 10, 1986.

315 Ron Arias, 'A Lesson in Uncommon Valor', *People*, February 10, 1986.

316 'McAuliffe's brother blasts NASA', *Washington Times*, January 1987.

317 Serge Schmemann, 'Soviet Honoring Shuttle's Women', *New York Times*, February 2, 1986.

318 Barry E. DiGregorio, 'The Time is right', *Countdown*, September/October 1994, pp. 36–39.

was given to McAuliffe, although they did acknowledge why people tended to do so – namely because she

> was articulate and photogenic, a reminder to everyone of what classroom experiences were or should have been. Her presence made this Shuttle flight different from the 24 that had gone before. She would be our representative in space, carrying the proxy for those of us who cannot go because we are too old or too young or too timid, or too busy holding down jobs, raising families and worrying about the mortgage.

However, at the same time, it was also emphasized that

> the other six crew members were our representatives too. Like the teacher, their new colleague, they were citizens pursuing careers. Their jobs involved bigger risks and greater sacrifices than many others. But when they went home they found spouses and children, friends, phone calls to return, the ups and downs of daily living. When they wanted to relax, they played the saxophone or the piano, they painted oil pictures or jogged.[319]

Still, McAuliffe was given way more space in the public than any other of the deceased crew members. While all the grief stricken families of the crewmembers received (proposals of) generous donations, grants, and honors, none received more than McAuliffe. The National Education Association, for example, began to collect for a program that sought to honor McAuliffe by financing 'pioneering' projects by teachers and scholarships for education majors. Pennsylvania's Republican Senator Arlen Specter wanted President Ronald Reagan to name one of the Education Department's buildings after McAuliffe and New York's Democratic Congressman Gary Ackerman wanted to name January 28 as National Teacher Recognition Day.[320] McAuliffe was also nominated for a place in the National Women's Hall of Fame and posthumously received the Nile Gold Medal during the opening ceremonies of the 79th General Conference of the Federation Aeronautique International (FAI), the world's oldest aerospace organization, for her accomplishments and sacrifice in educating and motivating youth about space.[321] While Resnik would also have qualified for most of

319 Paul Gray, 'Seven Who Flew for All of Us. Diverse in Race, Sex and Aspiration, they embodied the Ideals of America', *Time*, February 10, 1986; Margot Hornblower, 'Honoring Judith Resnik. $8 Million Memorial Gym to be built in Jerusalem', *Washington Post*, March 24, 1986.
320 Ed Magnuson, 'Space: They slipped the surly Bonds of Earth to touch the Face of God. In 73 seconds, a new Era in Space Travel explodes into a searing Nightmare', *Time*, February 10, 1986.
321 'Personalities', *Washington Post*, February 14, 1986; 'Christa McAullife Teacher Astronaut receives (posthumous) World's highest Aerospace Education Award', *Aviation Space Magazine*, November 1986.

these honors, McAuliffe was usually the center of attention. People shared poems and even jokes about her to cope with their grief.[322]

Sometimes Resnik was remembered as well; the Soviets named two craters on Venus after her and McAuliffe.[323] The male crew members were rarely recognized with these kinds of honors, although Florida's Democratic Congressman Bill Nelson proposed that seven of the newly discovered moons of the planet Uranus each be named for one of Challenger's victims. Colorado Republican William Armstrong suggested something similar when he asked the Senate to name ten moons of Uranus after the Challenger crew and the three Apollo astronauts who died in the 1967 Apollo 1 fire. Democratic Representative Mickey Leland of Texas proposed that the entire Challenger crew should be posthumously awarded the Presidential Medal of Freedom.[324] Ultimately, however, the crew did not receive the Presidential Medal of Freedom and no moons of Uranus were named after them. Thus, although the men of the crew did receive some honors, the attention and homages remained concentrated mainly on the women, especially McAullife because she was not an astronaut but a wife, a teacher and a mother – one of the people – qualities her male colleagues could not offer, despite some being fathers.[325] In the eyes of the public, it was apparently not as tragic to lose a father as it was to lose a mother, therewith revealing that the social status of women was still very much associated with traditional gender roles. Because Resnik had no husband or children, her death did not appear as dreadful. The Justice Department even initially offered the Resnik family less compensation money than the other Challenger families, apparently because Resnik, unlike all the other crewmembers, left behind neither a spouse nor children. This provision was a particularly explicit demonstration that a female life has less value in the eyes of the American government and society when she is unmarried and childless.[326]

Not only McAuliffe, but other women as well, took center stage in the portrayal of and discourse surrounding the Challenger accident. Women, like the astronauts' wives, had always been used in pictures to witness the great deeds

322 Louis Stokes, *A Poem for the McAuliffe Family*, congressional record, March 17, 1986, 1378, HRC; Patrick Morrow, 'Those Sick Challenger Jokes', *Journal of Popular Culture*, 20 (1987), 175–84.

323 Serge Schmemann, 'Soviet Honoring Shuttle's Women', *New York Times*, February 2, 1986.

324 Ed Magnuson, 'Space: They slipped the surly Bonds of Earth to touch the Face of God. In 73 seconds, a new Era in Space Travel explodes into a searing Nightmare', *Time*, February 10, 1986.

325 Mary McGory, R'ight Stuff, Wrong Place?', *Washington Post*, January 30, 1986.

326 Michael Isikoff, 'U.S. refuses to settle with 3 Shuttle Families', *Washington Post*, February 19, 1988.

Fig. 3.28: From left to right: Barbara Morgan and Christa McAuliffe, January 1986, NASA.

of space men explorers; now women were also used as the faces of the Challenger accident. While McAullife's portrait was often featured as the face of innocence and the American dream, an ordinary person reaching for the stars, other women were often presented as the faces of grief.[327] After the Challenger tragedy most pictures depicted the exploding Shuttle and weeping, devastated women.[328] Most of the time it was the crewmembers' widows or female family members and children who were depicted. Although a lot of the stories focused

327 Lucy Keyser, 'Friends recall a Teachernaut', *Washington Times*, January 28, 1987; Mike Pride, 'There had been a Death in the Family', *Newsweek*, February 10, 1986; Jean Becker and Peter Johnson, 'It hurts a little more', *USA Today*, January 30, 1986.

328 For example, NASA employees Estelle Coleman and Bettye McNaughton, who were joined by 14,000 of their colleagues (not pictured) at Kennedy Space Center observing 73 seconds of silence at 11:38 a.m. one year after the tragedy: Mary Belcher, 'Reagan marks Challenger Anniversary', *Washington Times*, January 29, 1987.

on McAuliffe, the accompanying pictures usually did not feature her husband Steven McAuliffe, but mainly the widows. Men were mainly shown as supporting the women or gazing at the missing man formation, demonstrating that the gendered division in the cultural norms 'witnessing' and 'being strong and brave' was still very much in place in the late 1980s.[329]

While McAuliffe was hailed as "a special member of the NASA family," the public discourse surrounding her person before and after the Challenger accident demonstrates that the American public had not changed that much in their regard and valuation of female contributions to society.[330] Being a feminine wife and mother was just as important in the 1980s as it had been in the 1960s, but unlike the 1960s, professional women in space were widely regarded as the norm in the 1980s; the public acceptance of Judy Resnik's death can attest to that. Her death was not assessed to be as tragic as McAuliffe's in the media because she was regarded as a true member of the space community, a trained astronaut in a line with other (male) astronauts who had perished before and with her. This perception shows that the role of women in the space community, especially in the role of astronauts, had severely transformed in the eyes of the public in the previous three decades.

3.4 Media space and the space community

The historian Mark Byrnes hypothesized that NASA has had to repeatedly adjust its public image according to the always changing political and social environment of the United States, because all government agencies over the years have needed political support to function.[331] NASA is dependent on the public for rallying political support and the agency needs the media to portray them and space exploration in general in a good light in order to ensure that political support. Thus, NASA frequently courted the press with briefings, materials, video materials and access to good stories. Byrnes identified three distinct image periods NASA used in the twentieth century. According to him, NASA's

---

329 Sue Anne Pressley, 'Reagan bids Shuttle Crew Farewell', *Washington Post*, February 1, 1986; Kathy Sawyer, '73 Seconds of Silence for Challenger's Crew', *Washington Post*, January 29, 1987; D'Vera Cohn, 'Challenger Crew is honored at Arlington', *Washington Post*, March 22, 1987; Laura A. Kiernan, 'The Lights of America's Mourning', *Washington Post*, January 30, 1986; Laura A. Kiernan, 'Suddenly, the Celebration stopped', *Washington Post,* January 29, 1986.

330 Mel Levine, *Tribute to Christa McAuliffe*, congressional record, February 6, 1986, 1378, HRC.

331 Byrnes, p. 2.

image was molded along the lines of the 'nationalism' period in the 1950s and early 1960s, the 'romantic' phase in the late 1960s, and the 'pragmatic' era during the 1970s and 1980s. The media discourse surrounding women's participation in the space community can be better understood within this framework because it reflects the change of the cultural and gender norms in American society and media and its translation to the space community.

The nationalism period was marked by the Sputnik Shock and public anxiety about Soviet space hegemony. During this period, NASA had to market itself within a Cold War-defined framework emphasizing national pride, prestige, and strength through the use of symbols or patriotic names for their space capsules and programs. Accordingly, NASA picked its first astronauts suited to the 'nationalism' theme: 'all-American' type of men.[332] The media and the public never questioned the selection of these men. Still, there was a public discussion if women should be able to become astronauts, too. However, women had to face ridicule in written and cartoon form because the stereotypical American woman in the role of an astronaut in the early 1960s did not fit into the period's emphasis on nationalism.

The romantic era was more hopeful because the majority of the American public no longer feared Soviet space dominance to the same extent due to Kennedy's pledge of getting to the moon and the following American successes in space, like the successful completion of the Mercury and Gemini missions and the upcoming moon landing mission Apollo. In the romantic period, NASA put an emphasis on the exciting aspects of space exploration. Although NASA still positioned itself in opposition to the Soviet space program, the romantic era shifted the agency's image and accentuated romantic themes such as exploration, heroism, curiosity and the emotional reward of space exploration.[333] With the lessening of the imminent Soviet threat in space, the space agency had one reason less to restrict women in the space community. This is why NASA needed to present its female scientific personnel to the public in order to demonstrate in the media that women officially had a place in space, too.

The pragmatic period reflected American society's change of values during and after the civil rights and women's movements. Most Americans were not willing any longer to spend so much money in space for national prestige or romantic visions. They wanted to use the space money to better the social and living standards for all Americans. NASA therefore had to start to emphasize its practical, pragmatic assets and its central theme was that the space activities improve

332 Byrnes, pp. 8–46.
333 Byrnes, pp. 8–98.

life on earth in various direct and indirect ways through the technological stimulus, spinoffs from NASA inventions and a routine access to space.[334] The versatile use of the space shuttle was a key argument in the pragmatic phase; this line of argumentation also included the versatility of the space shuttle crew, opening the door for female astronauts at NASA. This decision was mainly embraced in the media space and gave NASA the political support it needed.

Looking from the outside in, it appears that the assessment of spacefaring women changed in the media over the decades in accordance with the changing cultural norms in U.S. society and NASA had to adjust its public image accordingly. The changes brought about by the civil rights and women's movement as well as the accomplishment of the moon landing changed the public discourse about women in the space community. However, this observation is not necessarily true for the treatment of women within the space community itself, because the media space was not the main place where women actually had to persist and prove themselves. The next chapters thus try to evaluate to what extent women had agency and were generally accepted within the workspaces of the space community away from the public eye, which might have presented other opportunities and prospects, especially with regard to the Cold War.

334 Byrnes, pp. 99–135.

4 Office space in the 1960s and 1970s

America's women were interested in the space program from the very beginning. Some ambitious female pilots even wanted to become astronauts and breach the space frontier by outperforming the Soviet space program. They dreamed of a spot at the very center of the space community with responsibility and esteem equal to that enjoyed by men. However, most American men were not willing to let these women into their prestigious community just yet. When the first female astronauts did join the men in the late 1970s, the male members of the space community had to get used to their presence. Astronaut John Creighton, for example, admitted that he had to change his view of women. Thus far he had considered them to be "social creatures" only and he had to start viewing them as "fellow workers."[1] Astronaut Richard M. Mullane echoed Creighton's assessment. According to him, there were a lot of men at NASA, including him, who did not know how to work together with women on a professional basis. In hindsight, he considered himself to be "a jerk" because he "didn't know how to act around them, telling jokes that probably were not appropriate to tell and just doing dumb things that were inappropriate and probably would have gotten me a prison sentence in this day and age now with sexual harassment and all that."[2] Fellow astronaut John M. Fabian said something similar about his first work experience with the female astronauts in an interview he gave 28 years after his initial work experience with the first female astronauts. He is convinced that he

> didn't do it all right, either. That was a slightly different era. It was an era in which you would take the centerfold of Playboy magazine and post it up on the back of your office door, and that was thought to be totally acceptable as long as it was the back of the door instead of the front of the door. And people hadn't yet thought of the word harassment. We were all learning. We were all learning in those days.[3]

Considering that the men were all still learning how to interact with women at the workspace in 1978 when the first female astronauts joined NASA, the logical assumption is that it must have been hard for the women to join this kind of male dominated work environment. However, according to all parties involved, that was not the case. Fabian, for example, felt that nobody "looked down

---

1 Donna Miskin, 'In the New Astronaut Class', *Washington Star*, November 19, 1978, C-5.

2 Richard M. Mullane, interview by Rebbecca Wright, January 24, 2003, transcript, p. 13, JSC Oral History Project, JSC.

3 John M. Fabian, interview by Jennifer Ross-Nazal, February 10, 2006, transcript, p. 72, JSC Oral History Project, JSC.

https://doi.org/10.1515/9783110629828-004

upon them, ever, in the program."[4] And the first American woman to break through the space frontier, Sally Ride, remarked that the men at NASA "didn't give us a hard time at all."[5] Similar remarks have often been repeated by the other females who joined the space community in the astronaut class of 1978.

It is interesting to know that the first female astronauts did not face a lot of opposition from their male colleagues in the workspace they shared, due to the latter's recognition that "women are human beings," too.[6] However, the six female space pioneers were by far not the first women to join the space community's workforce. By the time the first women joined the astronaut corps, NASA had already been in existence for almost two decades. Thousands of women had quietly joined the space community's workforce long before the very public entry of Mss. Ride, Seddon, Sullivan, Resnik, Fisher and Lucid. It stands to reason that the men of the space community also had to share a workspace and interact with these women long before the pioneering six burst onto the space scene.

The statistics of NASA and the space industry confirm the early participation of women in the workspaces of the space community. However, women, and minorities for that matter, do not specifically show up in NASA statistics from the 1960s. Coherent data pertaining to gender and race was only collected starting in the 1970s with Equal Employment Opportunity.[7] Even so, there are numbers mentioned in various NASA documents that can give us an idea of the involvement of women in the workspace of NASA and its contractors in the early 1960s. For example, in 1964 the Manned Spacecraft Center (now Johnson Space Center) in Houston employed 983 women as compared to 3,733 men.[8] MSC was not alone in employing a considerable number of women; a quarter of NASA's nationwide workforce was female.[9] Considering that NASA had 10,200 in-house employees in 1960 and 36,200 in-house employees during the agency's peak employment period in 1967, a fourth of NASA's 'manpower' being female seems quite a favourable percentage rate.[10] Astronaut Alan

4 John M. Fabian, interview by Jennifer Ross-Nazal, February 10, 2006, transcript, p. 74, JSC Oral History Project, JSC.

5 Sally K. Ride, interview by Rebecca Wright, October 22, 2002, transcript, p. 7, JSC Oral History Project, JSC.

6 Don L. Lind, interview by Rebecca Wright, May 27, 2005, transcript, p. 53, JSC Oral History Project, JSC.

7 Nimmen, Bruno and Rosholt.en.

8 Alan B. Shepard, *Women in the Space Age*, speech presented at the Officers' Wives Club, May 11, 1965, pp. 6–7, 008998, HRC.

9 Sylvia D. Fries, *The History of Women in NASA*, paper presented at Women's History Week, March 28, 1988, p. 7, 008995, HRC.

10 Nimmen, Bruno and Rosholt, p. 106.

Shepard was certainly convinced that this male-female ratio in the workspace was sound. In a speech presented at the Officers' Wives Club at Patrick Air Force Base in 1965 he told the attending women:

> Almost 1 to 4 – which is really not a bad average when you also compare the percentage of women who are seeking employment outside the home with the percentage of women who are not interested in employment outside the home. This is an average which holds true nationally – women constitute about one-fourth of the federal workforce – nearly 600,000 employees.[11]

Shepard continued to describe the triumph of the women in the workspace by

> Boost[ing] your ego as a woman, you may be interested in learning that the Federal Commission reported that the Post Office and the Treasury Departments submitted Information that women receiving the lower salaries outperformed the men! But the fundamental challenge for all of us is not to view women as a group apart – but to let the individual merit prevail.[12]

The statistics and facts Shepard presented show that there most definitely were women in NASA's workspaces long before the first female astronauts. However, his supposed laudation of women also revealed the crux of the matter with NASA's 25 percent female 'manpower:' most of the women at NASA worked in low-paying administrative and clerical jobs GS-7 and below. According to Shepard, they should have been content and proud in their low income jobs because they apparently were the better workers in comparison to the men. Additionally, when one looked at the female representation at NASA as a percentage of women who actually wanted to work, factoring out those women who did not even desire to join the workforce and preferred to be homemakers, their numbers at NASA were even more impressive. Shepard's number crunching is quite interesting considering that he claimed to be all about 'individual merit.' His statements provide an insight into the male mind regarding women in NASA's workforce in the 1960s: the diligent work women provided at NASA was treated as a matter of course and their payment was commensurate with their 'individual merit,' meaning that if they wanted to enhance their salaries, they should have performed more outstandingly at their jobs. To prove his point, he gave an example of three distinguished scientists at NASA whose stellar performance won them the Federal Woman's Award of the Civil Service

11 Alan B. Shepard, *Women in the Space Age*, speech presented at the Officers' Wives Club, May 11, 1965, p. 6, 008998, HRC.
12 Alan B. Shepard, *Women in the Space Age*, speech presented at the Officers' Wives Club, May 11, 1965, pp. 6–7, 008998, HRC.

Commission (CSC).[13] Although individual merit is most certainly important, it is questionable if a lack of individual merit was what really prevented women from earning more money in the space community in the 1960s. The question of what factors were, in fact, responsible will be more thoroughly debated now.

In order to get to the bottom of the gendered office dynamics in the space community in the 1960s, it is crucial to look at the statistics, which provide insight into the gender composition within NASA's workspaces. Only about three percent of the female 25 percent of NASA's employees worked as engineers and scientists in the mid-1960s.[14] Jennifer Ross-Nazzal estimates that there were approximately 990 female scientists and engineers in 1967.[15] Keeping in mind that there were only 146 professional engineers and scientists and 77 professional mathematicians at NASA in 1962, one can make out an upward tendency in the number of women in NASA's workspaces employed in science and engineering labs.[16] The upward trend of female scientists joining NASA continued uninterrupted despite declining personnel numbers in the 1970s and 1980s.[17] In 1987 eight percent of the women at NASA worked as engineers and scientists.[18] This development mirrored a national trend in the United States, or rather pushed the upward development from the space community to other areas in the United States. While in the mid-1960s women constituted three percent of all employed engineers, their proportion had been only 0.9 percent in 1960, even though 1.2 percent of all employed engineers had been women already in the 1950s. Furthermore, in a 1964 *Science* magazine survey of government jobs GS-12 and higher held by women from 1959 to 1963, the magazine measured a 790

---

13 Alan B. Shepard, *Women in the Space Age*, speech presented at the Officers' Wives Club, May 11, 1965, p. 7–9, 008998, HRC.

14 Catherine Dryden Hock, *Space, the next Goal*, speech presented at the Student-Parent Night, Society of Women Engineers, February 24, 1965, p.1, 008998, HRC; Sylvia D. Fries, *The History of Women in NASA*, paper presented at Women's History Week, March 28, 1988, p. 7, 008995, HRC; Sylvia D. Fries, *NASA Engineers and the Age of Apollo* (Washington, D.C.: NASA Scientific and Technical Information Program, 1992), p. 205; Donna Shirley, *Managing Martians* (New York: Broadway Books, 1998), p. 80.

15 Jennifer Ross-Nazal, *Women Scientists at NASA*, paper presented at Women, Gender, and Science, a Conference at Suny New Paltz, October 22, 2005, p. 4, 008988, HRC.

16 James E. Webb, *Address before the General Federation of Women's Clubs*, June 27, 1962, p. 14, Harris/MSRC.

17 NASA had 23,169 in-house employees in 1978. NASA's workforce decreased almost 30 percent from 1967 to 1978. Ihor Gawdiak and Helen Fedor, *NASA Historical Databook, 1969–1978, Volume 4, NASA Resources* (Washington, D.C.: National Aeronautics and Space Administration, 1994), p. 68.

18 Sylvia D. Fries, *The History of Women in NASA*, paper presented at Women's History Week, March 28, 1988, p. 7, 008995, HRC.

percent increase in computer sciences, a 137 percent increase in mathematics, and a 122 percent increase in the physical sciences. The enthusiasm and interest in space seemed to have sparked a female interest in engineering and science in workspaces all over the U.S., according to Catherine Dryden Hock.[19]

The emphasis on science and engineering when reviewing women working in the space community is rather significant because the main work of NASA and its contractors was space research and development. This is also how NASA perceived itself as an organization. Being a research and development organization, NASA's engineers and managers, who mostly had a science/engineering background as well, did not value clerical and administrative work the same way they did research and development. Although administrative work was the glue that held NASA as well as other space companies together through planning and developing programs, a majority of NASA employees just regarded the clerical and administrative workers as "support for the truly important work of the agency, handling the hardware."[20] Most women in the space community, however, entered the space workspace in exactly those areas and had to demonstrate and educate the engineers and scientists about the value, significance and difficulties of their work.[21] A letter from Willis B. Foster, the Director of Manned Space Science, appealing to the Associate Administrator for Space Science & Applications for more "Womanpower" in 1964, shows that women were an important and integral part of NASA's office spaces years before Mss. Ride, Seddon, Sullivan, Resnik, Fisher and Lucid joined the agency and the men had to learn how to work alongside those women.[22] In the letter Foster wrote: "We're proud of the headway we've made – little as it is – but I'm very concerned that unless we are given additional secretarial help we're going to be forced to a screeching halt, or more appropriately, we're going to be smothered in paper."[23] The letter not only shows that Foster needed more secretarial staff – i.e. women – to run his department smoothly, which was the main

19 Catherine Dryden Hock, *Space, the next Goal*, speech presented at the Student-Parent Night, Society of Women Engineers, February 24, 1965, p. 8, 008998, HRC.

20 Sylvia D. Fries, *The History of Women in NASA*, paper presented at Women's History Week, March 28, 1988, p. 7, 008995, HRC. See also: Emogene Greville, interview by Dr. Aaron Gillette, July 27, 1992, pp. 33–35, Tape Transcript, HRC; Linda Rogers, interview by Dr. Aaron Gillette, July 27, 1992, p. 38, Tape Transcript, HRC.

21 Linda Rogers, interview by Dr. Aaron Gillette, July 27, 1992, p. 38, Tape Transcript, HRC.

22 *Manpower (or rather, Womanpower)*, August 6, 1964, Rc. No. 22053, Folder 075-13, Apollo Series, JSC.

23 *Manpower (or rather, Womanpower)*, August 6, 1964, Rc. No. 22053, Folder 075-13, Apollo Series, JSC.

intent of his plea, but on another level it also demonstrated that women were an essential component of NASA offices from the very get-go.

Considering that not only Foster but all offices in but also outside of the space community relied heavily on their, mostly female, clerical and administrative staff to keep the office going, it has to be understood that these women contributed to the successes of the American space program just like their scientist/engineer bosses; maybe not to the same extent but they were just as involved in the workspace.[24] Thus, it is the goal of this section to examine NASA's offices in the 1960s to discuss the role of those women who did not work in the sciences in the R&D organization that is NASA, in order to get an insight into the social and cultural structures of this particular workspace in the space community and its relation to American society. Like other social spaces, NASA's office spaces in its various centers were perpetuated and reproduced on a daily basis through the policies, social conventions, and performances of the men and women who worked there. The next section will examine these spaces.

4.1 Female secretaries and scientists – a double standard

Considering that office dynamics were all in a state of upheaval because of the first female astronauts joining the workforce, despite the prior presence of women in NASA offices, it stands to reason that the men at NASA differentiated between the women working in administration and those working in science. A statement by Skylab astronaut Dr. Robert Parker seems to confirm this assessment. He was quoted in an article in a 1974 issue of the *Philadelphia Inquirer*, stating: "There was a time when the public was not ready for the thought of men and women going to the moon together. But times have changed. I don't foresee any problem about men and women getting along together on space missions. We're not talking about dizzy blonde secretaries, but reputable women scientists."[25] Apparently there were two types of women in the eyes of at least some of the men in the space community: the career women, esteemed and respectable, seeking success within the space community, and the female support staff, taken for granted and without ambitions of their own. However,

---

24 Gail Collins, *When Everything Changed: The Amazing Journey of American Women from 1960 to the Present* (New York: Little Brown, 2009); Lynn Peril, *Swimming in the Steno Pool: A Retro Guide to Making It in the Office* (New Yok: W. W. Norton & Company, 2011); Alice Kessler-Harris, *Out of Work: A History of Wage-Earning Women in the United States* (New York: Oxford University Press, 1982), pp. 303–304.

25 Howard Benedict, 'Women Astronauts in 1980s?', *Philadelphia Inquirer*, May 1974.

the administrative and clerical staff did not see themselves the same way and were very aware of their important contributions to American successes in space. NASA secretary Delores A. Parys complained to Parker in a letter: "I, along with many other secretaries in NASA and elsewhere, am offended by the reference to secretaries as 'dizzy' and/or 'blonde.' On the contrary, where would the space program have gone without the intelligent, responsible help of NASA's secretarial staff?"[26]

Aware that other members of the space community begged for more secretarial help, knowing the value of a well-run office, the secretarial and administrative staff of NASA knew the value of their work as well. They felt as much part of the space community as the astronauts or scientists. Josie A. Soper, a secretary who started working for NASA in Washington, D.C. right after high school in 1962, was very pleased with her contribution to the big picture – the United States' success in space. She even went so far as to say that she had "a right to be proud of the tiny, little piece that I've had in this over these years, just not to the degree that these rocket scientists do, but at least to some degree."[27] Although Soper did not attach as much importance to her work as she did to that of those who invented the space vehicles, she felt as much part of the space program as her fellow co-workers.

A lot of the women who worked for NASA in a clerical or administrative position did not end up there by chance because it was the best available job; they had chosen NASA as their employer because they wanted to be part of the space community. Jerri Bell, who started to work in procurement at Kennedy Space Center in 1962, for example, had heard Alan Shepard's flight on the radio and in that moment knew she just had to get involved in the space program.[28] As she recounted it, she was so excited to be part of the space program that she never really thought about trying to climb the career ladder to move into a more prestigious position.[29] Other women like Nancy Fee, who started to work as a secretary-stenographer for NASA at Ellington Air Force Base in Houston in the 1960s, initially had other career plans. She wanted to become an engineer but settled for a secretarial position because she was told by the people in her environment that as a woman she would not be able to find work as an engineer. Coping with the disillusionment she decided to work as a secretary in the

26 Delores A. Parys, *letter to Robert A. Parker*, November 13, 1974, 008994, HRC.

27 Josie A. Soper, interview by Rebecca Wright, April 19, 2006, p. 68, Oral History Transcript, NASA Administrators Oral History Project, 19822, HRC.

28 Jerri Bell, interview by Dr. Aaron Gillette, July 24, 1992, p. 4, Tape Transcript, HRC.

29 Jerri Bell, interview by Dr. Aaron Gillette, July 24, 1992, p. 7, Tape Transcript, HRC.

Engineering Directorate because she could earn "pretty good money" as a secretary-stenographer and still stay in the field she was initially interested in.[30]

Even though some men, like Skylab astronaut Robert Parker, publicly disrespected the women working in administrative and clerical positions at NASA as less capable and smart than the women who worked in the labs, the interpersonal relationships between the secretarial staff and the men in charge seems to have been good in NASA's office spaces. Emogene Greville, a secretary at Kennedy Space Center, remembered that she was always treated nicely and with courtesy. Furthermore, she felt that her job as a clerical person was respected and recognized as important. She loved working for NASA and being part of the community.[31] Jamye Coplin, who started working as a secretary in the astronaut office in 1966, also felt that her bosses, the astronauts, were "very sweet and very protective" and appreciated their considerateness toward her in her first weeks at work as an 18 year old nervous novice in the job.[32] Jerri Bell also only had good things to say about the astronauts. She never observed that the astronauts treated the office staff, i.e. women, differently.[33] Furthermore, she denied that NASA ever specified the gender for its job positions despite the obvious division in the workforce, with the majority of the men working in higher paying engineering and science careers and women mainly representing the secretarial staff, and its gendered practices.[34]

Emogene Greville, for example, had to change offices because she got engaged to a man who worked in the same department at KSC. He was not her boss, just another member of the team. Although she did not want to change offices, much less branches, she was nevertheless transferred according to NASA policy. The policy in question was most likely an unwritten rule instead of a formal policy but it affected the women at NASA nonetheless.[35] However, it appears, once again, that there was a double standard for women who worked in the office and the women who worked in the labs and in space. Couples such as Donald M. Packer and his wife Irene were allowed to work together at the Naval Research Laboratory as a research physicist team for Skylab's Ultraviolet

30 Nancy L. Fee, interview by Rebecca Wright, November 29, 2011, p. 1, Oral History Transcript, NASA Johnson Space Center Oral History Project, JSC.

31 Emogene Greville, interview by Dr. Aaron Gillette, July 27, 1992, pp. 31–32, Tape Transcript, HRC.

32 Jamye Flowers Coplin, interview by Rebecca Wright, November 12, 2008, p. 6, Oral History Transcript, NASA Johnson Space Center Oral History Project, JSC.

33 Jerri Bell, interview by Dr. Aaron Gillette, July 24, 1992, p. 16, Tape Transcript, HRC.

34 Jerri Bell, interview by Dr. Aaron Gillette, July 24, 1992, p. 6, Tape Transcript, HRC.

35 Emogene Greville, interview by Dr. Aaron Gillette, July 27, 1992, pp. 15–16, Tape Transcript, HRC. She did not remember if it was an informal or written-down policy.

Airglow Horizon Photography Experiment beginning in the late 1960s. They also operated together at Johnson Space Center during the mission as part of the science support team.[36] Astronaut couples such as Anna and Bill Fisher were also allowed to work together. Granted, that was in the late 1970s, but a differing handling of women who worked in the sciences and women who worked in administration nevertheless becomes apparent.

The main difference was most likely between 'having a job' and 'having a career.' The few women who worked in the space community as engineers and scientists were expected to strive for a career, while most secretaries were assumed to lack that particular ambition. The female clerical and administrative staff supposedly just worked to keep busy until they got married and left their position in the space community in order to take care of house and kids. Not only the men in the space community made this assumption, but most women did, too.[37] Thus, upper management did not expect women to reach the top of the career ladder in administration. Emogene Greville, for example, was convinced in hindsight that NASA leadership did not think women were able to fulfil job requirements at the upper management level and that this mentality only changed in the 1980s.[38] However, she also assumed that there was not really a planned resistance against advancing women in administrative positions, because the leadership could not even conceive of the idea of women climbing the career ladder since they lacked the education women nowadays have.[39] Linda Rogers, on the other hand, hypothesized in retrospect that the administrative side of NASA was just mirroring the technical side in regard to the employment, or rather advancement, of women.

There were only a handful of women in the sciences at NASA. By inference, there were also only a few women who were even able to start a career in that area. Deductively, according to Roger's mirror theory, only a few women out of the much larger group of secretarial personnel was able to pursue a career as well.[40] When Rogers herself had asked her manager at Kennedy Space Center when she could expect a promotion, her boss was surprised to hear that she wanted to advance in her position. He told her: "I didn't realize you were ambitious, Linda, that you wanted to, you know, further your career."[41] Her manager

36 *A family affair…*, NRL News Release, May 20, 1973, GH 85, Subject Files, General Reference Series, JSC.
37 Linda Rogers, interview by Dr. Aaron Gillette, July 27, 1992, p. 10, Tape Transcript, HRC.
38 Emogene Greville, interview by Dr. Aaron Gillette, July 27, 1992, pp. 22–24, Tape Transcript, HRC.
39 Emogene Greville, p. 45.
40 Linda Rogers, interview by Dr. Aaron Gillette, July 27, 1992, p. 43, Tape Transcript, HRC.
41 Linda Rogers, p. 10.

was not even all wrong in his assumption. When Linda Rogers started as a contract specialist at Kennedy Space Center she had no concrete career plans and just wanted to have a job; only after her divorce in 1968 did she start to want a career.[42] Josie Soper had a similar experience. She recounts: "What ended up happening to me personally was I ended up having a career, and I didn't really start out to have a career. I started out with a job, and that's where the women's movement, I think, really started making a big difference."[43] Unlike other women in NASA offices in the 1960s, Soper never felt that being a woman working in administration held her back in the space community. But she also admits that she was very happy in her job, did not aspire to "lofty heights" and thus did not "represent any threat to anybody, and other women have, and that's why they've been stunted."[44]

4.2 The implicitness of the secretarial staff

In order to pursue a career in the space community one had to be willing to move around the country, or the world, for one's career, but that career step was considerably harder for those secretaries who wanted to get ahead for several reasons. For one, they often had young kids at home and a husband who had a career of his own and was not able and/or willing to move or look after the children when his wife was away on business trips. Furthermore, managers often assumed women were not willing to travel or move without checking with the particular woman in question. Daisy B. Fields, who started working in the formulation, development, and administration of the employee-management relations branch at NASA in 1960 and became one of the founding members of the Federally Employed Women (FEW) in 1968, recounted a time when she almost was not hired for a job which required a lot of travel because she had kids and a husband. Some of the managers had just assumed that she would most likely not be able to travel and that if she were willing to travel she would expect the male members of her team to carry the luggage for her.[45]

42 Linda Rogers, pp. 5–8 and 45.

43 Josie A. Soper, interview by Rebecca Wright, April 19, 2006, pp. 44–45, Oral History Transcript, NASA Administrators Oral History Project, 19822, HRC.

44 Josie A. Soper, p. 75.

45 Daisy B. Fields, interview by Bernice Sandler, April 8 and 9, 1983, pp. 88–89, Women in the Federal Government Project Transcript, Schlesinger; Josie A. Soper, interview by Rebecca Wright, April 19, 2006, p. 65, Oral History Transcript, NASA Administrators Oral History Project, 19822, HRC.

It has to be recognized at this point that the secretarial and administrative staff also had to travel alongside their managers and by themselves without it being mentioned in their job descriptions. They often performed extra tasks and worked overtime or took work home with them just like their fellow space careerists. However, their contributions were not necessarily perceived as career moves but as a matter of course; they had to do it for the team. Jamye Coplin remembers extremely full days. She was only rarely able to get out of the office before 6:00 p.m. and regularly came in on weekends, even though her job description required her to work only until 4 p.m. Monday to Friday and she did not receive paid overtime. The secretarial, clerical and administrative staff was expected to be part of the team effort like the scientists, engineers and astronauts. In fact, she only remembers getting paid extra one single time for putting in those long hours.[46] Everyone had to work those hours – including the women who were not thought of as having a career but as 'just doing their job.' They also took work home with them, just like their colleagues and bosses. Coplin, for example, recalls that maintaining the astronaut's manuals and checklists was time-consuming, because one had to be meticulous in making sure the right pages were pulled. Because she had so many assignments, she did not have time and quietude enough to do it during the day, so she piled the manuals up and took them home with her to work on at night while she was watching television.[47] Still, in the early 1970s, the work of the secretaries continued to be considered a matter of course. The women were not allowed to use NASA's stress lab, a fitness center with the latest exercise equipment as well as a sauna. The area was for men only because "[w]omen don't have stress," as Ruth Bates Harris, Deputy Director of Equal Employment Opportunity, was told "since they sit at their desks all day."[48]

The secretaries were also required to wear many hats and accomplish highly diverse tasks, unheard of for women in the 1960s and early 1970s, especially women who were supposedly not pursuing a career but just doing their job. These secretaries experienced considerable independence in their jobs. The women of the astronaut office, for example, drove cars by themselves to the Cape and back to Houston, an undertaking which was quite unusual for women at that time, let

46 Jamye Flowers Coplin, interview by Rebecca Wright, November 12, 2008, p. 13, Oral History Transcript, NASA Johnson Space Center Oral History Project, JSC.
47 Jamye Flowers Coplin, interview by Rebecca Wright, November 12, 2008, p. 10, Oral History Transcript, NASA Johnson Space Center Oral History Project, JSC.
48 *Ruth Bates Harris at NASA*, undated, Harris/MSRC; Harris, p. 257. The EEO office saw to it that women also received access to the stress lab in the early 1970s.

alone 'single ladies.' A Chevrolet dealership in Florida leased cars to the astronauts and the women would pick them up and bring them to the astronauts.[49]

The secretaries also experienced independence in their private space. Quite a lot of the female secretarial staff went to work for NASA right out of high school, which often required moving out of their parents' homes to be close to the center they worked at. It was often an unusual but transformative experience for these women to live on their own right after high school as single women.[50] On top of that, the women also performed the roles of travel agent, public relations specialist, private detective, mother hen, screener, liaison manager, babysitter and many more.[51] It was most certainly a difficult job. In particular because their managers expected them to perform all tasks immediately and they often had to say 'no' to their bosses because they were the ones in the offices who had an overview of the workload. They had to prioritize. Coplin, who worked in the astronaut office, even thinks that this was the most challenging part of her work because each astronaut thought that their work was the most important and their mission the most significant. Apparently they were not used to waiting in line and it was really difficult to explain to them the necessity of doing so.[52] The women also performed jobs that were most certainly not in their official job description but which they felt compelled to do because they were part of the NASA family, the space community. They often developed very close relationships with the wives and the families of their bosses because they were the liaison persons in the office between their bosses and their families. Some of the women even babysat for their bosses.[53]

4.3 Barriers to the pursuit of a career in administration

The main hat the women in the NASA offices had to wear was, of course, the secretary's hat, that of the work they were actually employed to do. Secretarial

---

49 Jamye Flowers Coplin, interview by Rebecca Wright, November 12, 2008, pp. 40–41, Oral History Transcript, NASA Johnson Space Center Oral History Project, JSC.

50 Nancy L. Fee, interview by Rebecca Wright, November 29, 2011, pp. 10–11, Oral History Transcript, NASA Johnson Space Center Oral History Project, JSC; Jamye Flowers Coplin, interview by Rebecca Wright, November 12, 2008, p. 47, Oral History Transcript, NASA JSC.

51 Josie A. Soper, interview by Rebecca Wright, April 19, 2006, p. 32, Oral History Transcript, NASA Administrators Oral History Project, 19822, HRC; Jamye Flowers Coplin, interview by Rebecca Wright, November 12, 2008, p. 50, Oral History Transcript, NASA Johnson Space Center Oral History Project, JSC.

52 Jamye Flowers Coplin, p. 50.

53 Jamye Flowers Coplin, pp. 11–12.

work in the 1960s was gruelling because there were no computers and they had to work with nonelectric typewriters. Everything had to be letter perfect; there were no mistakes allowed. If they made a single typo, they spent the rest of the day erasing it on all the different-colored carbon copies. The offices were run the military way, with strict rules and procedures that were supposed to be the same, no matter in which particular office one worked, in order to make sure they could all work well together.[54] Nancy Fee remembers that

> [y]ou didn't break chain of command. You could be a secretary in the branch and take care of sections, because they had sections, then they had branches, then they had the division. So a branch secretary could not go to the division, the branch secretary could not go to the directorate. The branch secretary went to the next level up. If she was over a section, she went to the branch. The branch then went to the division. The division then went to the directorate. You never went above what they called the command.[55]

Thus, secretarial work at NASA was manifold and challenging and only rarely acknowledged as a career. Historically, however, the position of the secretary was an emancipating career opportunity for women in the beginning of the twentieth century. It was a way for women to support themselves and gain independence in a job that was not an extension of women's domestic sphere, unlike the traditional female occupations of nursing and teaching, and also not hard manual labor in a factory.[56] Until the 1960s, it was mainly the men who got promoted and were able to pursue a career, but starting in that decade there were also a few women who rose from secretary to manager and were able to start promising careers in administration. One might expect that those secretary trailblazers would have helped each other out. However, the gender based fraternization men often exhibit failed to appear for the women – there was no work sorority. In many cases, women were divided and suspicious of each other, and this was true for NASA as well.

According to Bell, there was an intangible separation between the professional and non-professional people in the NASA workspaces, with women in particular being jealous of each other because some were able to get better jobs or move up in professional jobs while others who also aspired to create a career for themselves did not make it.[57] Coplin also remembered territorial behaviour by

---

54 Jamye Flowers Coplin, p. 11; Nancy L. Fee, interview by Rebecca Wright, November 29, 2011, p. 2, Oral History Transcript, NASA Johnson Space Center Oral History Project, JSC.

55 Nancy L. Fee, p. 16.

56 Sylvia D. Fries, *The History of Women in NASA*, paper presented at Women's History Week, March 28, 1988, pp. 1–2, 008995, HRC; Peril.

57 Jerri Bell, interview by Dr. Aaron Gillette, July 24, 1992, pp. 29 and 35, Tape Transcript, HRC.

the women in the NASA offices. According to her, the secretarial positions, in the astronaut office for example, were highly coveted jobs, and there was competition for them amongst the women. Coplin herself admitted that she used to want to be known as the best of the group of women.[58] Greville also remembered that a lot of the women worked against each other because there was such jealousy if a woman got a promotion or anything else the other women did not receive. They would talk behind each other's backs and say something unkind about her or why she got the promotion. According to Greville, this behaviour changed only as all women started to network together. They got to know each other and help each other out starting in the late 1960s and 1970s.[59]

Before that the NASA secretaries had their own network; however, the professional women did not participate in the days before the Federally Employed Women. The professional women were even advised not to associate and bond with the secretaries but to associate more with professional people, who at that time were still mainly men.[60] Thus, the women who moved up the administration career ladder in the space community had to become 'one of the boys' instead of forming a united front with their fellow female employees. The women who remained in their secretarial positions also had their difficulties with the women who were now their managers. They often were not so keen about working for a woman. Greville, for example, had a female supervisor at some point in the 1960s, but she did not like to work for her. According to her

> [w]orking for a woman was difficult. I mean, you just, for some reason, it was different from working for a man. [. . .] Because men seemed to not pay attention to as many small details, like she would maybe give one person a job, and then she would come along and take it back, or give it to somebody else, and we just felt like maybe she did not know exactly how to supervise.[61]

Fee, on the other hand, was not bothered by her female manager; however, she also admitted that female supervisors have a different leadership style. According to Fee "[t]hey're more focused on their agenda than getting off track like men do, because men will do that. They'll have a chain of thought, then it's like they get onto something else. Women that I've worked for, they are focused. 'Go down this path, I want to get this done, then we go to the

58 Jamye Flowers Coplin, interview by Rebecca Wright, November 12, 2008, p. 49, Oral History Transcript, NASA Johnson Space Center Oral History Project, JSC.

59 Emogene Greville, interview by Dr. Aaron Gillette, July 27, 1992, p. 53, Tape Transcript, HRC.

60 Emogene Greville, pp. 58–60.

61 Emogene Greville, pp. 18–19.

next.'"[62] Although a lot of the women did not work smoothly together and did not help each other out in their careers during the early 1960s, there actually were women who did. Judith L. Robinson, who worked in Space and Life Sciences at Johnson Space Center, was out to help and better all women irrespective of the job category they were in.[63] Another example was Mae Morris Walterhouse, who started working as a secretary in Center Director Kurt Debus' office in 1964 and was very active on behalf of her fellow female coworkers at Kennedy Space Center. Walterhouse recalled that she was "the 'Bella Abzug' of KSC. I ruffled feathers."[64] At some point her supervisor even told her that she needed to decide between advancing women and advancing her own position because she could not do both. She eventually decided to concentrate on advancing women and started working in the Federally Employed Women national leadership in the mid-1970s.[65]

Not surprisingly, the women of the NASA office spaces were not the only ones who had their difficulties with women pursuing a career in administration; a lot of the men were also uncomfortable with women pursuing a career at NASA and felt threatened in their status. Astronaut Alan Shepard, for example, stated: "So, while we at NASA believe in 'full partnership of women,' I think at this point I ought to tell you that I am a firm believer in equal rights for men, too!"[66] He not only voiced his concern about the retrenchment of male rights, but also expressed concern for the future of women's femininity should they continue to compete with men. He alerted women that

---

62 Nancy L. Fee, interview by Rebecca Wright, November 29, 2011, p. 18, Oral History Transcript, NASA Johnson Space Center Oral History Project, JSC.
63 Nancy L. Fee, p. 14.
64 Kay Grinter, 'Federal Women's Program has come a long way', *Spaceport News*, September 4, 2009, p. 7. Bella Savitsky Abzug was a democratic U.S. Representative (in office 1971–1973), lawyer, and a leader of the Women's Movement. She was nicknamed 'Battling Bella' because of her activist dedication to better the social and legal situation for women and minorities around the world but specifically in the U.S. Alan H. Levy, *The Political Life of Bella Abzug, 1920–1976: Political Passions, Women's Rights, and Congressional Battles* (Lanham: Lexington Books, 2013); Suzanne Levine Braun and Mary Thom, *Bella Abzug: How One Tough Broad from the Bronx Fought Jim Crow and Joe McCarthy, Pissed Off Jimmy Carter, Battled for the Rights of Women and Workers, Rallied Against War and for the Planet, and Shook Up Politics Along the Way. An Oral History* (New York: Farrar, Straus and Giroux, 2007); *Women in American History: A Social, Political, and Cultural Encyclopedia and Document Collection. Volume 4: Cold War America to Today*, ed. by Peg A. Lamphier and Rosanne Welch (Santa Barbara: ABC-CLIO, 2017), p. 87.
65 Emogene Greville, interview by Dr. Aaron Gillette, July 27, 1992, p. 41, Tape Transcript, HRC.
66 Alan B. Shepard, *Women in the Space Age*, speech presented at the Officers' Wives Club, May 11, 1965, p. 6, 008998, HRC.

> [t]he goal of women in this space-oriented society should not be 'to be like men', but 'to be more womanly', and, therefore, of greater value to themselves, to men, and to society – thus realizing their full potential as human beings. Men and women will learn to work, not for a man's world, not for a woman's world, but for a world dependent on both.[67]

Thus, the loss of femininity was a concern men had not just for the first women who wanted to become astronauts in the 1960s, but also for all the other women in the space community who wanted to further their careers and become scientists, engineers or administrative managers. Some of the men at NASA were intimidated by women who moved up the career ladder and captured positions in NASA office spaces which had heretofore been held exclusively by men. They, in turn, gave those women who were now their bosses a hard time. While most men accepted the fact that women started to move into professional positions, there were some who thought that women should live out their full potential as female human beings by staying at home with the kids rather than by being their bosses. In her career, Bell had to deal with a couple of these men, who thought it was "beneath their dignity to work for a woman."[68] Greville also remembered some resentment toward her becoming a supervisor in procurement: "It was hard for some of the men who I had been a peer with, I suddenly became their supervisor and I met with quite a bit of personal resistance. But whether that was widespread, I don't know. I mean there weren't that many women supervisors, so it couldn't be very widespread."[69] Even if the men who worked for a woman were not reluctant about working for her, the men in other NASA offices made sure to make fun of these men, joking about their masculinity. The men who worked for Greville, for example, were mockingly called the 'squaw men' by their peers at Kennedy Space Center.[70] So not only were women scolded on account of their breakout gender performance, but the men were scoffed at as well because of their insufficient gender performance; the women in charge lost their femininity while the subordinate men's masculinity was impaired. As a female facility manager, Fee had a similar experience. Her working relationship with the maintenance and the construction staff was a little rocky at first because they did not react to her directly but double-checked all her business decisions with other men in charge of other departments. The men in her crew let her know that she was not "one of the guys" but

67 Alan B. Shepard, *Women in the Space Age*, speech presented at the Officers' Wives Club, May 11, 1965, p. 17, 008998, HRC.

68 Jerri Bell, interview by Dr. Aaron Gillette, July 24, 1992, p. 36, Tape Transcript, HRC.

69 Emogene Greville, interview by Dr. Aaron Gillette, July 27, 1992, p. 40, Tape Transcript, HRC.

70 Emogene Greville, p. 41.

> a woman. They will work for you and do what you tell them, but they're going to go to the man to convey all the stuff. I had to learn to make them direct their statements and react to me. I had to show them, 'Hey, I know what I'm doing, do what I ask.' Over time they learned, and I learned, and they were great.[71]

Even if an office (or lab for that matter) full of men accepted a career woman in their midst, this did not necessarily mean that they would accept another one. Fields, for example, suggested a capable and educated woman to fill a vacant position in her office, but her superiors did not want to consider a woman because they already had one in the office, her. After telling her manager that times had changed and that they should integrate more women in the office staff, her boss told her: "We've got you, honey. We don't need anybody else."[72] Fields did not want to be the token woman of the office and did not appreciate her manager's answer because she regarded it as a putdown. Even though her boss claimed that he did not mean it that way, she knew that she was there for representation: "you know, the typical...you put the black receptionist out front so that you show you've got a minority" on your staff.[73] In the end Fields' assertive way, which was mostly perceived as 'pushy' by her male co-workers, paid off: the woman was hired and did an excellent job, and the men of her office had the greatest respect for her.

'Pushiness' being the keyword here says a lot about the office gender dynamics at NASA in the 1960s and early 1970s. A lot of the women who wanted to advance up the career ladder were hesitant to openly show their determination because they did not want to appear pushy or aggressive – qualities that would make others perceive them as unfeminine.[74] Their mantra was: "You're a female; don't be aggressive because you'll be accused of being pushy."[75] Thus, most

71 Nancy L. Fee, interview by Rebecca Wright, November 29, 2011, pp. 19–20, Oral History Transcript, NASA Johnson Space Center Oral History Project, JSC.

72 Daisy B. Fields, interview by Bernice Sandler, April 8 and 9, 1983, p. 89, Women in the Federal Government Project Transcript, Schlesinger. See also the case of chemist Marie Crowley at JPL: Holt, *Rise of the Rocket Girls: The Women Who Propelled Us, from Missiles to the Moon to Mars*, p. 88.

73 Daisy B. Fields, p. 89.

74 Women being described as 'pushy' is still an issue today. See for example: Nic Subtirelu, 'Beyond Bossy: More on Our Gendered Characterizations of Leadership and Authority', *Linguistic Pulse*, 2014 <https://linguisticpulse.com/2014/05/19/beyond-bossy-more-on-our-gendered-characterizations-of-leadership-and-authority/> [accessed 13 March 2016]; Olga Khazan, '"Pushy" Is Used to Describe Women Twice as Often as Men', *The Atlantic*, 2014 <https://www.theatlantic.com/business/archive/2014/05/pushy-is-used-to-describe-women-twice-as-often-as-men/371291/> [accessed 13 March 2016].

75 Daisy B. Fields, p. 90.

working women accepted the status quo. However, some women, like Fields, were not content with this situation. She was not willing to accept that status for herself, and did not want to see it for other women, so she kept on pushing.[76]

One of the main issues that bothered a lot of women in the space community's office spaces, in clerical and professional positions alike, was that they were often treated like a monolith, all built from the same mold, in office exchanges with their male co-workers. Although the women had different backgrounds, education, positions and characteristics, they were often not addressed as individuals, but only as a member of their gender.[77] However, this gender-based treatment gradually changed and by the mid-1970s

> the men, little by little, started to treat women more and more as individuals based on their own behaviour and their own characteristics, rather than women as a member of a class solely defined by their sex, which is how it was initially I think. Women were just women as a group, and then we sort of broke out, some of us broke out and started being treated as individuals more.[78]

4.4 Recreational activities for women

The women of NASA did not only interact with each other and their male co-workers in the office spaces, but also spent time together outside of work. Early on, NASA encouraged its workforce to participate in training and extra outside and indoor activities. Coplin remembered that she went to numerous training classes.[79] NASA offered various activities for all of its employees, either at its own locations or through cooperation with local providers. Activities were usually announced by management or in the centers' own publications put out by the centers, for example in the Johnson Space Center's *Roundup* or the Ames Research Center's *Astrogram* – each center had its own.[80] NASA's recreational and educational offers ranged from themed summer picnics and Christmas

---

76 Daisy B. Fields, p. 90.

77 Linda Rogers, interview by Dr. Aaron Gillette, July 27, 1992, p. 17, Tape Transcript, HRC.

78 Linda Rogers, p. 16.

79 Jamye Flowers Coplin, interview by Rebecca Wright, November 12, 2008, p. 46, Oral History Transcript, NASA Johnson Space Center Oral History Project, JSC.

80 Johnson Space Center had the *Roundup*, Ames Research Center had the *Astrogram*, Glenn Research Center had *Aerospace Frontiers*, Goddard Spaceflight Center had *The Goddard View*, Jet Propulsion Laboratory had the *JPL News*, Kennedy Space Center had *Spaceport Magazine*, Langley Research Center had the *Langley Researcher News*, Marshall Spaceflight Center had the *Marshall Star*, Neil Armstrong Research Center had the *X-Press* and Stennis Space Center had the *Lagniappe*.

dances to safe boating classes to a karate club to various softball teams.[81] There were also special offers aimed solely at NASA's female employees, for example a women's softball team named the 'Lunar Rovers.'[82]

Fig. 4.1: the Lunar Rovers, *Roundup*, July 20, 1973.

NASA was also concerned with the safety and health of its female employees. In 1967, Johnson Space Center (then MSC) held three showings of a film about breast self-examination to detect cancer. All female NASA and contractors' employees were encouraged to come and see one of the showings of the educational movie 'Self-Examination of the Breast' and to talk to a physician from the Occupational Medicine Dispensary afterwards. Charles A. Berry, MD, the Director of Medical Research and Operations, informed female employees that "[c]ancer can be self-detected in many instances and cured. The earlier the

81 *Roundup*, June 8, 1973; *Roundup*, May 25, 1973; *Roundup*, October 22, 1976; *Roundup*, September 14, 1973; *Roundup*, December 9, 1977. See picture of female NASA softball team in: *Roundup*, July 20, 1973.

82 'Lunar Rovers Seize Softball Championship', *Roundup*, July 20, 1973. See Fig. 4.1.

detection, the easier the cure."[83] NASA, at Johnson Space Center (then MSC) for instance, also provided its female employees as well as its contractors' female employees and wives of employees with the opportunity to go to a self-defense seminar. The goal of the lecture was to prepare women to cope with threats and danger. The instructor intended to give practical, common-sense suggestions with the goal of giving NASA's women a healthy respect for the unknown without fright or panic. Wesley L. Hjornevik, the Assistant Director for Administration, urged "every woman to attend" in his announcement.[84]

The NASA centers' newspapers also addressed less serious supposedly 'female issues,' such as, for example, the struggle against excess weight. Helen Ragsdale, the chief telephone operator in the Johnson Space Center Telecommunication Section who had been with JSC since 1962 and was also a Weight Watchers lecturer, gave advice on how not to indulge during the Christmas season and how to keep off the extra pounds.[85] Another popular activity offered to the women of NASA were so called 'Charm Clubs.' Jerri Bell, for instance, was a member of the Kennedy Space Center charm club in the 1960s. She fondly remembered that "it was lovely, everybody enjoyed it, enjoyed the activities and everything."[86] The Charm Clubs brought the women of NASA's offices together and organized luncheons and fashion shows. The members of the Manned Spacecraft Center (later JSC) Charm Club, for example, held a 'Blast into Spring' style show at the King's Inn in Clear Lake City in 1966.[87] However, the Charm Clubs, along with NASA's very own beauty pageants, just sort of disappeared in the 1970s and some former members, like Belll, were still wondering why two decades later.[88]

The changing gender dynamics in the NASA office spaces also changed the agency's manner of conduct toward its event offers for female employees. In some cases these events were discontinued, while in others they were able to continue under the guise of publicity events. The astronauts, for example, still made appearances at local festivals where beauty pageants were held. Apollo 10

83 Charles A. Berry, *For Women Only*, April 25, 1967, GR 1013, MSC Announcements NOS. 67-1 through 67–98, January 9, 1967 – June 30, 1967, Center Series, JSC.

84 Wesley L. Hjornevik, *Lecture for Women on the art of self-defense*, November 19, 1965, GR 1013, MSC Announcements NOS. 65–89 through 65–170, July 6, 1965–December 28, 1965, Center Series, JSC.

85 Helen Ragsdale, 'Helen Ragsdale tells how to tactfully avoid undesirable holiday baggage', *Roundup*, December 23, 1977.

86 Jerri Bell, interview by Dr. Aaron Gillette, July 24, 1992, pp. 25–26, Tape Transcript, HRC.

87 'Fashion Show', *M.S.C. News*, March 1966, GH 85, Subject Files, General Reference Series, JSC. See Fig. 4.2.

88 Jerri Bell, interview by Dr. Aaron Gillette, July 24, 1992, pp. 25–26, Tape Transcript, HRC.

Fig. 4.2: MSC Charm Club's 'Blast into Spring' at the King's Inn in Clear Lake City, News of MSC, March 1966.

Fig. 4.3: George M. Low and Miss America Jane Anne Jayroe, 1967, NASA.

astronaut John W. Young, for instance, was the star of the Spirit Festival in Decatur, along with the contestants of the Miss Spirit of America Beauty Pageant.[89] Meanwhile, George M. Low, Deputy Center Director of Manned Spacecraft Center received Miss America Jane Anne Jayroe in his office and showed her around the center in 1967.[90] Even today, NASA still teams up with Miss America contestants. Miss America Mallory Hagan and astronaut George Zamka, for example, shared a stage for the Department of Education's 'Let's Read! Let's Move!' campaign in 2013 and Nina Davuluri, Miss America 2014, visited NASA's Goddard Space Flight Center to celebrate Asian-Pacific American Heritage Month.[91]

Unlike NASA's ongoing friendly collaboration with the United States' beauty queens, their very own beauty pageants did not continue. Although NASA never held an agency-wide beauty pageant including women from all centers, some centers and contractors crowned their own beauty queens, from their office and lab spaces, between 1952 and 1973. North American Aviation, the Jet Propulsion Laboratory, the Lewis Space Center, and the Johnson Space Center each respectively crowned a 'Goddess,' 'Miss Guided Missile,' 'Queen of Outer Space,' 'Miss NASA,' and, 'Lunar Landing Festival Queen' out of a selection of their female employees. North American Aviation, for example, selected the "fairest of the fair" out of 13 young women in the Apollo Engineering Unit at a Christmas Party for the employees and their spouses after a contest in semi-formal attire in 1962.[92]

The Jet Propulsion Laboratory was the first NASA center to start and to end its beauty contest. In 1952, before NASA was even founded, JPL held an annual spring dance which climaxed in the election of 'Miss Guided Missile.' After JPL was added to NASA and transitioned from missiles to spacecraft in 1958, the beauty pageant transitioned as well and the title holder was renamed 'Queen of Outer Space' in 1959. Because the movie *Queen of Outer Space* starring Zsa Zsa Gabor was released the same year JPL joined NASA, one has to imagine that the rebranding of the contest was intended humorously. In the film, American astronauts landed on the planet Venus, which they found to be inhabited solely by beautiful women and

89 'Moonwalker highlights Spirit Festival', *Huntsville Times*, July 3, 1974.
90 See Fig. 4.3.
91 Katy Mersmann, 'Former Miss America Walks the Goddard Runway', *Goddard View*, 11 (2015), 8 (p. 8); 'Space and Exploration Are Focus of Let's Read! Let's Move! Event in Washington, D.C.', 2013 <http://www.nasa.gov/content/space-and-exploration-are-focus-of-lets-read-lets-move-event-in-washington-dc/#.V3EErKK2qE0> [accessed 27 June 2016]; 'Miss America 2013 Mallory Hagan as a NASA Fitness Instructor', 2013 <https://www.youtube.com/watch?v=JGocgsZR8Dg> [accessed 27 June 2016].
92 "Goddess' in Apollo Unit to be named', *Skywriter*, XXII(52), December 21, 1962, HRC.

Fig. 4.4: JPL's director Walter Pickering crowning Miss Guided Missile, Alane Hine, 1955, NASA.

their despotic queen. The beauty contest continued throughout the 1960s. The lavishness of the campaigns indicate that the competition was taken very seriously. Each candidate was sponsored by a manager and campaigned with baked goods, flyers and other campaign material such as handicraft papier mâché model rockets with their names on it, for one week before the queen was chosen by a vote of employees, upon which the winner of the contest was always crowned by JPL's director, Walter Pickering, and had several responsibilities just like her Miss America colleague. She had to host a luncheon, go to inaugurations and generally represent JPL. She was also the star on top of a decorated convertible in a parade held in her honor. The contest was discontinued around 1970. It is not quite clear why, though it is known that there were complaints from managers that it took too much time away from productive work. Moreover, the beauty contest was perceived as no longer fitting with the times.[93]

---

93 'Vinson named to JPL-ERC Honor Roll', *Lab-Oratory*, January 1968, p. 7; Franklin O'Donnell, *Explorer I* (Pasadena: California Institute of Technology, 2007) <https://pub-lib.jpl.nasa.gov/docushare/dsweb/Get/Document-2417>; 'Solid Propellant Chemistry Gains Fitzgerald Transducer', *Lab-Oratory*, January/Fe (1956), p. 7 <https://beacon.jpl.nasa.gov/mystery-photo/

Fig. 4.5: Alane Hine, Miss Guided Missile of JPL 1955, christens the new transducer unit of the Fitzgerald Apparatus. Hine was very involved in JPL's Employee Recreation Club (ERC) and was honored for her "devotion, graciousness and contribution to the successful operation and progress of the ERC," as was her husband who also worked for JPL, Lab-Oratory, January 1968.

Lewis Research Center also held its very own beauty pageants from 1968 to 1973. The employees of Lewis elected 'Miss NASA' from amongst the ladies of its very own workforce. Similar to JPL's queens, Lewis Research Center's Miss NASA presided over parades and literally presented LeRC's prettiest side.[94]

---

page/6?mopt=1>; 'P-489C, Miss Guided Missile, 1955', *Jet Propulsion Laboratory*, 1955 <https://pub-lib.jpl.nasa.gov/docushare/dsweb/Services/Document-808> [accessed 27 June 2016]; Holt, *Rise of the Rocket Girls: The Women Who Propelled Us, from Missiles to the Moon to Mars*, pp. 78–79, 100–101, 212, 223. See pictures in *Explorer I*. See Fig. 4.4–4.5.

94 'Miss NASA Lewis Merri C. Fahnenbruck with Apollo Lunar Rock Display', *Glenn Research Center*, 1973 <https://archive.org/details/C-1973-1016> [accessed 27 June 2016]; 'Miss NASA Lewis Research Center and Lewis Research Center Band on Agena Trailer', *Glenn Research Center*, 1970 <https://archive.org/details/C-1970-2578> [accessed 27 June 2016]; 'Miss NASA in

Fig. 4.6: Lewis Research Center's Miss NASA, 1968, NASA.

There is also no information about why this beauty pageant was discontinued, but the reasoning must have been similar to JPL's. The public perception of beauty pageants was in the midst of changing and the centers most likely did not want their contest to become an issue of public debate. It seems like the shift started around the time of the 1968 protest against the Miss America pageant, during which feminists verbally attacked the women participating in the contest with songs and signs, unintentionally alienating the contestants and a lot of ordinary women. They also threw things like aprons, fake eyelashes and bras into a 'freedom trash can' which they had originally wanted to burn. However, they did not set fire to their freedom trash can because doing so would have constituted a fire hazard. Thus, the Miss America protest started the myth of bra burning although not a single bra was actually burned.[95]

Apollo 8', *Glenn Research Center*, 1971 <https://archive.org/details/C-1971-2942> [accessed 27 June 2016]. See Fig. 4.6.

95 Rosen, pp. 159–161; Bonnie J. Dow, 'Feminism, Miss America, and Media Mythology', *Rhetoric & Public Affairs*, 6 (2003), 127–49; Nell Greenfieldboyce, 'Pageant Protest Sparked Bra-Burning Myth', *Npr*, 2008 <http://www.npr.org/templates/story/story.php?storyId=

While the Jet Propulsion Laboratory and Lewis Research Center were putting a stop to their in-house beauty pageants, Johnson Space Center was just starting out. In 1973, the Johnson Space Center was holding an election for a 'queen' and her 'court' to reign over the Lunar Landing Festival activities in Houston.[96] The female JSC staff member who would become the 'Lunar Landing Queen' was supposed to be crowned at the Lunar Ball at the Albert Thomas Convention Center. Similar to the queens at the Jet Propulsion Laboratory and Lewis Research Center, her royal duties included cutting the ribbon of the NASA Space Show and Artifact Exhibit, participating in social events at Johnson Space Center and Houston such as a social tea at River Oaks Country Club and going to the Astroworld Grand Ball, and generally being an emissary for JSC.[97] There is no information available that would reveal why the managers in Houston thought that a beauty pageant would be a nice addition to the festivities. Most likely, they did not put that much thought into it to begin with. They clearly paid no attention to the concerns of the women's movement but must likely assumed that it would be a fun activity and a good way to get the space center involved with the activities in the city.

However, the election of Lunar Landing Queen Sharon Boniface did not go as planned. A group of anonymous feminists, most likely NASA employees given the closed environment, wanted to protest what they perceived as NASA's misogynistic practices. Overnight, they swapped the official ballot with a mock ballot that listed 45 male NASA employees who could become 'King or Queen' or one of the 'boys' of the court. The notification for the selection mockingly referred to the space agency's repeated, public affirmation of only employing accomplished personnel irrespective of their gender, when it announced that the "[s]election should not be based on merit. Vote for a pretty face & a good bod."[98] Some NASA men took offense and called in the center's

94240375&from=mobile> [accessed 2 April 2016]; Rosalyn Baxandall and Linda Gordon, *Dear Sisters: Dispatches From The Women's Liberation Movement* (New York: Basic Books, 2000), pp. 184–186; Lamphier and Welch, pp. 180–182.

96 See pictures of the Lunar Landing Queen and her court in: 'Lunar Landing Queen Chosen', *Roundup*, July 6, 1973; 'Space Research contributes to Fire Prevention Program', *Roundup*, October 12, 1973.

97 'Lunar Landing Queen Chosen', *Roundup*, July 6, 1973, pp. 1–2; 'Announcement', *Roundup*, July 20, 1973, p. 1; 'Space Research contributes to Fire Prevention Program', *Roundup*, October 12, 1973.

98 Wendy Haskell Meyer, 'The Decline and Fall of Lunar Royalty', *Texas Monthly* (1974), pp. 38–39 (p. 38).

security team. Inspector Glenn L. McAvoy investigated and all the ballots were collected. However, most NASA employees thought the mock election was hilarious. They reproduced the ballots and kept on voting for their favorite Lunar Landing King. The pageant saboteurs thus received enough votes. Despite the watchful eyes of the security police, the pageant saboteurs still managed to distribute the results of the mock ballot in the dead of night along with caricatures to match. It was announced that the employees of JSC had elected the famous feminist Gloria Steinem to be Lunar Landing King. Her court, composed of prominent male NASA employees, was announced as well.[99] The announcement further mockingly declared:

> The wide cross section of size, weight, color, religion, political views, sex, and individual deviances represented by the winners of this contest dispense, once and for all, the blatant charges that have been laid at the doorstep of Johnson Space Center – that its record of discriminatory practices is worse than any other NASA installation; worse than the aerospace community surrounding it; and possibly among the worst in the Federal Government. This election has, without question, countered these charges. Once again it has been demonstrated that the great aerospace team of courageous and creative individualists, who accepted – and are conquering – the challenges of the frontiers of space, are also willing to challenge and overcome the barriers to the attainment of tolerance and humanity. Just as we have demonstrated the peaceful uses of space, we will demonstrate peacefully for equality. We will, in fact, recognize that talented persons come in various shapes, colors, beliefs and life styles and we will allow them to cohabitate.[100]

The NASA men who won the racially and misogynistic charged beauty king election were not amused. Only one of them took it humorously, asking "when he was going to get a fitting for his dress."[101] The implicit accusations of the humorous mock election that Johnson Space Center was a discriminatory workplace was just ignored, despite clearly demonstrating the issues

---

99 The 'boys' of her court all received titles: Joe Atkinson was dubbed 'Black Revolutionary,' George Abbey was called 'Conservative Reactionary,' Jeff Warner received the title of 'Hippy Liberal,' Iou Wong became 'Inscrutable Oriental,' Larry Dietlein was named 'Southern Aristocrat,' Jose Perez was referred to as 'Latin Lover,' Gene Kranz was designated 'Yankee Stormtrooper,' and Pete Conrad was dubbed 'Stuntman & Human Cannonball.' *The Winners!!! . . .Lunar Landing Festival King (or Queen) Contest. . .*, 1973, MC 496, Carton 44, Folder 5, National Organization for Women Collection, Schlesinger.

100 *The Winners!!! . . .Lunar Landing Festival King (or Queen) Contest. . .*, 1973, MC 496, Carton 44, Folder 5, National Organization for Women Collection, Schlesinger.

101 Wendy Haskell Meyer, 'The Decline and Fall of Lunar Royalty', *Texas Monthly* (1974), pp. 38–39, (p. 39).

feminists were concerned with in the space community in the early 1970s. The winner of the genuine beauty contest, Sharon Boniface, and the ladies of her court also most likely did not think it was funny because the mock election made them look bad, too – in front of the men and the women of their own community.[102]

The objectifying practice of beauty pageants, and with it the race and gender imbalance in society and at Johnson Space Center in particular, was publicly criticized through the mock election. However, it was not the men of the mock election but the women of the actual contest who had to deal with the consequences. While the men were able to just sweep their 'pageant success' under the rug, the women had to keep going to fulfill the Lunar Landing Queen's obligations. Knowing that a number of their female, and possibly also their male, colleagues were critical of the beauty contest business, "[t]hey [still] had to buy their own dresses and no men came out to hustle them. It turned out to be a bummer for them."[103] The triumphant women of the Lunar Landing Festival contest must have felt degraded. Although Sharon Boniface and her court represented a women's purpose that the organizers of the mock election wanted to overcome, they still had to grin and bear it and ignore how some of their female colleagues perceived them. The caricatures that were enclosed in the proclamation of the mock election's winners revealed how some women in the space community felt about the role of women at NASA, of which the Lunar Landing Queen had become the exemplar. The caricatures portrayed the NASA women as, for example, a 'moon maid' or as a crowned queen on the moon about to be crushed by the foot of an astronaut who is taking the "giant step for mankind", while the male winners were derisively depicted in typical beauty pageant poses.[104]

Looking into NASA documents and publications, such as Johnson Space Center's *Roundup*, one cannot find any indication of what had happened surrounding the Lunar Landing Queen contest in the offices at JSC. The matter of the mock election and the attendant accusation of misogynistic and racist office dynamics was concealed and certainly not publically debated. The Lunar Landing Queen contest was the last time a NASA center initiated a beauty pageant

102 The ladies of the court: Mary Yarbrough, Dorothy Holloway, Marilyn Ross, Carol Brinkman, and Sharon Kemp.

103 Wendy Haskell Meyer, 'The Decline and Fall of Lunar Royalty', *Texas Monthly* (1974), pp. 38–39 (p. 39).

104 Anastasia Sams, *Several Lunar Landing Festival Queen Caricatures*, 1973, MC 496, Carton 44, Folder 5, National Organization for Women Collection, Schlesinger. See Fig. 4.7–4.14.

Fig. 4.7: JSC's Lunar Landing Queen Sharon Boniface and her court sitting on a lunar rover, Roundup, July 6, 1973.

for its female staff as a leisure activity offer. Starting in the early mid-1970s, NASA tried to be more sensitive toward its female and minority staff members during work and free time. However, the objectification of NASA personnel because of their looks did not end with the Lunar Landing Queen contest in 1973. Switching roles, twelve male employees of Kennedy Space Center posed for a sexy but clothed calendar in 1992. The "strictly, young, single, and sexy" models who worked for Kennedy Space Center or its contractors were all chosen by Shuttle software analyst Sheryl Farmer and three co-workers.[105] Jim Gentile, a

105 Beth Dickey, 'The Faces that launched a thousand Ships', *Air &Space* (Dec. 1991/Jan. 1992), p. 10. See. Fig. 4.15.

Fig. 4.8: Sharon Boniface, Lunar Landing Queen of JSC 1973, presents fire safety items, Roundup, October 12, 1973.

Rockwell employee, who posed in skin-tight zebra-striped trunks was happy that the calendar would "verify that not all engineers are pencil-necked geeks."[106] Unlike the female beauty contests of the 1950s, 1960s and early 1970s, the project started out as a charity project for adult literacy in Brevard County, turning objectification into a good cause and thus a good deed by the men who participated. Interest in the calendar was so great that it became a global marketing project with 5,000 calendars pre-ordered from all over the world, including Canada and the Philippines, turning it into "an inside source

106 Beth Dickey, 'The Faces that launched a thousand Ships', *Air &Space* (Dec. 1991/Jan. 1992), p. 10.

Fig. 4.9: Caricature referring to the Lunar Landing Queen elections, Anastasia Sams, 1973, National Organization for Women Collection, Schlesinger Library, Radcliffe Institute, Harvard University.

Fig. 4.10: Caricature referring to the Lunar Landing Queen elections, Anastasia Sams, 1973, National Organization for Women Collection, Schlesinger Library, Radcliffe Institute, Harvard University.

Fig. 4.11: Caricature referring to the Lunar Landing Queen elections, Anastasia Sams, 1973, National Organization for Women Collection, Schlesinger Library, Radcliffe Institute, Harvard University.

for eligible men."[107] Even though their co-workers took to singing 'Calendar Boys' every time one of the models walked past them, their act of objectification was perceived differently because they wanted to show that smart people could also be good looking. Unlike the women in the workspaces of the 1960s, they

107 Beth Dickey, 'The Faces that launched a thousand Ships', *Air & Space* (Dec. 1991/Jan. 1992), p. 10.

Fig. 4.12: Caricature referring to the Lunar Landing Queen elections, Anastasia Sams, 1973, National Organization for Women Collection, Schlesinger Library, Radcliffe Institute, Harvard University.

did not have to prove that they were smart in the first place. Furthermore, they willingly accepted to become an object of desire whereas the women in the 1960s did not yet have the vocabulary, and the women in the early 1970s were only in the process of establishing vocabulary, to describe that they felt objectified in the workplace.[108] Furthermore, they did it for a good cause in a way which ran counter to what was commonly expected of a serious NASA employee, while the women in the NASA beauty pageants had mostly just been

108 Beth Dickey, 'The Faces that launched a thousand Ships', *Air & Space* (Dec. 1991/Jan. 1992), p. 10; Fred R. Shapiro, 'Historical Notes on the Vocabulary of the Women's Movement', *American Speech*, 60 (1985), 3–16.

Fig. 4.13: Caricature referring to the Lunar Landing Queen elections, Anastasia Sams, 1973, National Organization for Women Collection, Schlesinger Library, Radcliffe Institute, Harvard University.

acting according to the general expectations of a female employee. Natalia Holt, however, assessed the beauty contests tradition at NASA differently, at least at the Jet Propulsion Laboratory. According to her, "the beauty contest was a result of JPL's progressive hiring practices" because "the competition was unintentionally highlighting the presence of educated young women working at JPL."[109]

109 Holt, *Rise of the Rocket Girls: The Women Who Propelled Us, from Missiles to the Moon to Mars*, p. 79.

Fig. 4.14: Caricature referring to the Lunar Landing Queen elections, Anastasia Sams, 1973, National Organization for Women Collection, Schlesinger Library, Radcliffe Institute, Harvard University.

The Lunar Landing Queen incident shows that at least some of the Johnson Space Center employees were fed up with the discriminatory treatment women and minorities received in the space community. It also reveals that there was an area of conflict between NASA personnel who identified with the values of the women's movement or women's liberation movement in the 1970s and the one's who did not. Unlike the first female astronauts, most of the women working in NASA offices on the ground had clear opinions about the women's movement and positioned themselves accordingly – at home and in the office. This produced a trouble spot for the relationships and internal exchanges between all the women and men working in NASA offices in the 1970s and 1980s.

Fig. 4.15: KSC's 'Calendar Boys' of 1992, Air & Space, December 1991, January 1992.

4.5 The establishment of women's movement chapters in the space community

It has to be emphasized at this point that there were not a lot of conscious ness-raising demonstrations such as the King of Outer Space mock election, typical for the women's liberation movement, at NASA. Female NASA employees usually turned instead to the methods of the women's movement to have their issues recognized. While the women's movement, represented by various organizations, amongst others the National Organization for Women (NOW) and the Federally Employed Women, wanted to create change through economic and legislative methods, the 'libbers' chose active demonstration over courtroom battles in order to raise awareness and transform the private and public lives of women. Although the organizations that made up the women's movement also wanted to recast gender roles at home as well as on the job and not just add women's rights to the canon, the different sides of the movement had varying ideas how to go about it. The women who were often derogatorily called 'libbers' were mainly young women under the age of thirty. They felt that organizations like NOW were elitist, but they still joined forces with them from time to time for specific protests. However, there was no permanent coalition. It is also important to note at this point that neither the women's movement nor the women's liberation movement fully addressed the grievances of women of color, working class women, or lesbians; both movements predominantly addressed the nuisances of white middle class women.[110] The same was true for NASA when women's movement issues first came to its centers in the early 1970s, due to the agency's predominantly Caucasian female workforce at that time.

Matters pertaining to women's rights were handled by NASA's Equal Employment Opportunity (EEO) Program, which was founded in 1970 and expanded over the years. The agency-wide Federal Women's Program (FWP), which started in 1974, was responsible in particular for the advancement and training of the female workforce.[111] However, before the EEO officers and the Federal Women's Program manager started to intensively work on women's grievances at NASA and its contractors, a handful of women in different NASA centers took charge and initiated their own groups to advance themselves and their fellow women at NASA. These women connected with the private, non-profit organization Federally Employed Women (FEW) and either joined FEW's regional chapters or, in the case of the women of Kennedy Space Center, founded their own. Federally Employed Women was founded in 1968 by thirteen federally employed women who recognized the need to create equal opportunities for women in federal employment, shortly after Executive Order 11375, which added sex-discrimination to the list of prohibited discriminations within the federal government, was issued. Although the Federal Women's Program had been established as a result of EO 11375 in 1967, FEW wanted to make sure that all federal agencies complied with the executive order and did not want to solely rely on a government function which could easily be dismantled by the government to enforce such compliance. Furthermore, the early FEW organizers thought it would be best to also have a private organization for the rights of federally employed women because it was not limited in its political scope like the Federal Women's

110 Rosen, pp. 75–87 and 276–290; Marc Stein, *Rethinking the Gay and Lesbian Movement* (New York: Routledge, 2012); Maylei Blackwell, *¡Chicana Power!: Contested Histories of Feminism in the Chicano Movement* (Austin: University of Texas Press, 2011); Breines; Lamphier and Welch, pp. 111, 118, 166–167, 255; Dorothy Sue Cobble, *The Other Women's Movement: Workplace Justice and Social Rights in Modern America* (Princeton: Princeton University Press, 2004).

111 *NASA Hearings before the Subcommittee on Civil Rights and Constitutional Rights of the House Committee on the Judiciary. Background Information*, Judiciary Committee, 93rd Congress, Subcommittee on Civil and Constitutional Rights, NASA Hearings Correspondence, Box 37 of HR. BX. 33, NASA 1974 Background Information for Members for Hearing, RG 233, NARA I.

Program, which was legally prohibited from lobbying Congress. The organization that intended to improve the status of women was thus not in competition with the Federal Women's Program but considered itself a partner. In addition, it was not a labor organization that represented employees directly, but an advocacy group that wanted to initiate and engage in discourse as well as lobby for legislative change and offer training to its members.[112]

The FEW Space Coast Chapter at Kennedy Space Center was founded in 1973. There were only 15 to 20 women who took part in the Federally Employed Women's group at Kennedy Space Center in the early days, among them Linda Rogers and Mae Walterhouse. According to Rogers the Kennedy Space Center FEW chapter "was a pretty struggling little outfit and we didn't have any time at work or anything to meet. We had to meet on weekends and at night and that sort of thing, you know, there were no activities that were sanctioned during the day."[113] Apparently, only the women working in the office spaces, professional, secretarial, administrative and clerical employees, took part in FEW early on. The women who worked in the labs and engineering departments shied away from associating with women's right groups such as FEW.[114] Like the first female astronauts, a lot of these women preferred to be 'genderless' members of their crews.[115] Moreover, the FEW women were seen as troublemakers by most of the male NASA engineers and scientists in the 1970s, even in the later part of the decade.[116] Therefore, the female engineers most likely also chose not to consort with the FEW women in order to avoid getting caught in the crosshairs, which would presumably have made their position among the men in the engineering department more complicated. Federally Employed Women was considered a "blackballed group" in the early 1970s, most likely because the change they wanted to implement in the space community was rejected at the time. Thus most people – men and women alike – tried to keep their distance from the women who were really active in FEW. Especially the NASA men thought that they were troublemakers, who were "trying to make women get out of their place."[117] FEW was too radical for a lot of NASA women who felt threatened

---

112 'What Is FEW' <https://www.few.org/about-us/what-is-few/> [accessed 28 June 2016]; 'FEW History' <https://www.few.org/about-us/few-history/> [accessed 28 June 2016].

113 Linda Rogers, interview by Dr. Aaron Gillette, July 27, 1992, p. 12, Tape Transcript, HRC.

114 Linda Rogers, p. 13.

115 For more information about the attitude of women in science and engineering at NASA see chapter: *Lab space in the late 1950s and 1960s.*

116 Emogene Greville, interview by Dr. Aaron Gillette, July 27, 1992, pp. 26–27, Tape Transcript, HRC.

117 Emogene Greville, p. 26.

in their femininity and most likely also feared for the opportunities they had by emphasizing their female attributes. Quite a few of these women shared their male colleagues' perception of FEW's activities as "rabble-rousing;" they did not want to be associated with 'women's libbers.'[118] Bell, for example, explained that she was one of the women who had "never really been into this woman's lib stuff" because she "would rather be feminine than be, you know, going out for that women's right."[119] In her opinion, organizations like FEW were completely unnecessary because the work environment in the NASA offices was already as good as it gets regarding teamwork and gender and race equality. According to Bell,

> [i]n the early days, everyone was motivated, dedicated and really good team work, you know, there was no such thing as women, I mean if it was, I didn't notice it, as far as race, sex, anything, it was just one good happy team together. It was no, you know, women challenging another women or vice versa with this woman's lib and all that.[120]

Linda Rogers, who was one the FEW chapter founders at KSC, had a different explanation for the hesitant or even hostile attitudes of some NASA women who avoided FEW even though the group was rallying for their rights as well. She mused that all "women are individuals and some were very protective of their femininity and thought that feminism or women's libber type activity was anti-feminine, anti-woman."[121] Keeping in mind the reserved and deprecatory attitude of a good portion of people in American society and their negative and derogatory communication regarding the women's movement in the public space in the early 1970s, the hostile attitude against ideas of the women's movement within the space community cannot come as a surprise.[122] Considering the dismissive treatment the feminist often received in and outside of the space community, the women who organized the Federally Employed Women chapter at Kennedy Space Center were as cautious as their critical colleagues about their own activities. The FEW organizers were very sensitive to the fact that the men felt threatened in their masculinity and privilege because of their organization

118 Linda Rogers, interview by Dr. Aaron Gillette, July 27, 1992, pp. 10 and 21, Tape Transcript, HRC.

119 Jerri Bell, interview by Dr. Aaron Gillette, July 24, 1992, pp. 6 and 26, Tape Transcript, HRC.

120 Jerri Bell, p. 6.

121 Linda Rogers, pp. 21–22.

122 Terkildsen and Schnell; Deborah L. Rhode, 'Media Images/Feminist Issues', in *Feminism, Media, and the Law*, ed. by Martha A. Fineman and Martha T. McCluskey (Oxford: Oxford University Press, 1997), pp. 8–21; Lamphier and Welch, pp. 83–86.

asking for rights previously exclusively held by men. They were also very aware of their female colleagues' fears that their femininity might be impaired by the possible loss of attention in the NASA offices.

The FEW ladies sought to avoid being perceived as people who were only causing trouble. They therefore tried to be very strategic about their declarations and activities, such as the selection of their first president; they did not want to alienate their colleagues further.[123] In the eyes of the FEW women, their organization was the backing that women had been lacking before. The women in the group helped each other out and shared information and strategies on the job site.[124] Some of the FEW women, like Mae Walterhouse, even made a career out of advancing other women.[125] Thus, the fight for equality also opened up new career opportunities for women in and outside of NASA – career opportunities that some of the women who did not want to associate with FEW in the beginning begrudgingly witnessed. Bell, for example, claimed that FEW members got promoted faster.[126] If that was actually the case is almost impossible to retrace because there are no statistics for FEW members' promotions at NASA, but it does seem very likely considering the training sessions and seminars they attended. Further education and skill enhancement for the job was very much appreciated at NASA and often resulted in a promotion.

It has to be recognized that the apparent acceptance of FEW's own self-organized training sessions and seminars at NASA was a far cry from any advantage enjoyed by the Federally Employed Women members in the founding days of their group at Kennedy Space Center. The FEW seminars were not considered legitimate training by some people in NASA's leadership, and most of the time the FEW women were too shy to aggressively push for official acknowledgement of their events because they did not want to make "a fuss."[127] In some instances, however, they received backing. Kennedy Space Center's EEO officer, Willie D. Gaither, for example, complained about the management of Kennedy Space Center to EEO director Dudley McConnell at NASA headquarters in Washington, D.C. in 1973.[128] He claimed that Kennedy Space Center's management

123 Linda Rogers, interview by Dr. Aaron Gillette, July 27, 1992, pp. 10–11, Tape Transcript, HRC.

124 Linda Rogers, p. 11.

125 Emogene Greville, interview by Dr. Aaron Gillette, July 27, 1992, p. 41, Tape Transcript, HRC.

126 Jerri Bell, interview by Dr. Aaron Gillette, July 24, 1992, pp. 24 and 28, Tape Transcript, HRC.

127 Linda Rogers, interview by Dr. Aaron Gillette, July 27, 1992, p. 12, Tape Transcript, HRC.

128 Willie D. Gaither, *letter to Dudley McConnell*, April 12, 1973, 18508, Box 1, HRC.

exercised "its management prerogative in an extremely arbitrary, discriminatory and conservative fashion, [and] refused to allow the Space Coast Chapter FEW training seminar official training status."[129] Most of the women who wanted to attend the Federally Employed Women seminar to which Gaither was referring had to pay out of their own pockets. Two women actually did receive reimbursement, but they had only been notified about it the evening before the start of the seminar, having fully expected to pay for the seminar themselves like most of their colleagues. In contrast, a lot of NASA contractors as well as other agencies, such as Patrick Air Force Base, the Department of the Treasury, the Defense Department and even NASA headquarters, all paid their attending employees' expenses. Kennedy Space Center management furthermore made its position regarding -Federally Employed Women well known around the center, according to Gaither. Managers let the female employees subtly know that they did not approve of this particular seminar or of FEW in general, thereby implying that women who valued their careers at NASA should not attend. Kennedy Space Center's management did not even bother to send one of their own as a representative to the seminar, dispatching instead Ann Taylor, who was not part of the management team. In this way, Kennedy Space Center's management signalled its disinterest in seminars concerning the advancement of and equal rights for women.[130]

On the other hand, Kennedy Space Center's management encouraged its female employees to go to a one-day secretarial seminar in 1972 and 1973. On both occasions, the management purchased and distributed the tickets among the female staff, not just the secretaries, without asking who wanted to attend the seminar beforehand. The contents of this seminar were mainly about 'how to be a lady' and included lessons about grooming, pleasing the boss, reading the mail, and choosing the correct color of eye shadow, amongst other topics. Thus, many of KSC's female employees felt that George VanStaden, director of Kennedy Space Center, was in no way responsive to, much less supportive of, the Equal Opportunity Program at KSC.[131] Considering that Kennedy Space Center is situated in Florida, it is, perhaps, not surprising to learn that KSC management was not at the forefront of the women's rights movement. Not only was the center located in a Dixie state, but a lot of its employees were born and raised Southerners with conservative values. The Mexican-American mechanical engineer Ernie Reyes, for example, observed that "some of the southern folks, were still fighting the Civil War" when he

129 Willie D. Gaither, *letter to Dudley McConnell*, April 12, 1973, 18508, Box 1, HRC.
130 Willie D. Gaither, *letter to Dudley McConnell*, April 12, 1973, 18508, Box 1, HRC.
131 Willie D. Gaither, *letter to Dudley McConnell*, April 12, 1973, 18508, Box 1, HRC.

tried to describe the political and ideological atmosphere at Kennedy Space Center in the 1960s.[132] According to him, KSC eventually became a more cosmopolitan environment due to the influx of new people from other areas. Even then, however, the level of acceptance of a diverse workforce differed between departments, some of which were not exposed to the world as much as others.[133]

4.6 NASA's hiring and promotion system – female perspectives

Kennedy Space Center was not the only NASA center in a former Confederate state. Marshall Space Flight Center in Alabama was also very submerged in Southern culture.[134] Unlike Kennedy Space Center, however, Marshall Space Flight Center management was willing to send its female employees to professional training years before Kennedy Space Center. While KSC's managers started to accept the FEW training in the 1970s as useful and legitimate, Marshall's leadership encouraged its female workforce to train for high-level managerial and executive responsibilities as early as 1965 at management seminars.[135] Marshall, for example, offered its female employees a three day seminar at the University of Alabama's Huntsville Campus. The seminar's purpose was to train the female employees "to look beyond their specialty, if they are career bound, and to begin grooming themselves for high-level managerial and executive responsibilities."[136] The women were instructed in logical and practical approaches to supervisory problems, to analyze and resolve management problems, and taught coping mechanisms for situations which every executive manager will be confronted with eventually. The women were also encouraged to take part in a dialogue about women's role as managers and co-workers in the technological world of the 1960s. Furthermore, the seminar attempted to broaden their perspectives as civil servants and members of the community.[137]

132 Ernie Reyes, interview by Dr. Aaron Gillette, July 28, 1992, p. 31, Tape Transcript, HRC.
133 Ernie Reyes, pp. 31–34; See also: Paul and Moss.
134 Ernie Reyes, pp. 31–32.
135 Linda Rogers, interview by Dr. Aaron Gillette, July 27, 1992, p. 13, Tape Transcript, HRC; *Seminar for MSFC Executive Women*, leaflet from the University of Alabama, January 1965, 008998, HRC.
136 *Seminar for MSFC Executive Women*, leaflet from the University of Alabama, January 1965, 008998, HRC.
137 *Seminar for MSFC Executive Women*, leaflet from the University of Alabama, January 1965, 008998, HRC.

This shows that the women at different NASA centres as well as contractors' places of employment had differing career opportunities depending on the leadership and the manner of conduct of their male co-workers.

Although NASA headquarters issued behavioral guidelines regarding women and EEO for all its centers and contractors, their implementation was a long time coming in some cases. Inconsistent application of these guidelines might explain the different perceptions of career opportunities held by female office employees of different NASA centers and departments. Rogers, for instance, thought that NASA had a culture of not giving big promotions to women despite being one of the better places for them to work in civil service. However, she was also convinced that this situation was partially the women's fault because they furthered this type of culture by quitting after having a family.[138] Indeed, according to a Johnson Space Center survey in 1974, the main reason women left NASA was to become a full-time housewife or because their husband transferred away for his job.[139] Soper, on the other hand, felt that NASA had always been "on the cutting edge" as far as offering all types of opportunities, especially management opportunities, to women.[140] She even went a step further and claimed that she had never seen any discrimination against women at NASA and that NASA had been, from what she witnessed, "very, very, very good for women to have the opportunities."[141] Greville was less rosy than Soper in her perception of the opportunities given to women at NASA. She felt that lots of women attended evening college courses and sacrificed their own personal time in hopes of advancing in their job at NASA, yet were not promoted despite their college degrees. Extremely overqualified, they had to stay in their clerical positions.[142]

There are several cases that support Greville's claim. The U.S. District Court for Southern Texas, for example, ordered Johnson Space Center in 1977 to grant its employee Charlotte R. Smith a retroactive promotion with back pay because NASA discriminated against the woman with a master's degree by assigning her to perform routine clerical tasks despite her college education. Due to a physical

138 Linda Rogers, interview by Dr. Aaron Gillette, July 27, 1992, pp. 41 and 46, Tape Transcript, HRC.

139 Johnson Space Center, *EEO Progress Report for Second Quarter*, July 25, 1974, Box 1, Equal Opportunity 1974, Center Files, JSC.

140 Josie A. Soper, interview by Rebecca Wright, April 19, 2006, p. 72, Oral History Transcript, NASA Administrators Oral History Project, 19822, HRC.

141 Josie A. Soper, p. 74.

142 Emogene Greville, interview by Dr. Aaron Gillette, July 27, 1992, pp. 62–63, Tape Transcript, HRC.

handicap, Smith's supervisor at Johnson Space Center had made the assumption that she was not able to travel and that she could not perform lab work requiring manual dexterity. He supported his assumption on the basis that it was "obvious to the eye." Had he talked to Smith about her abilities he would have known that she had a driver's and pilot's license and that she was very much able to do lab work. She started to work in a lab after the court's decision and also began to pursue her doctorate. However, NASA appealed the case despite Smith's demonstration of skills in the lab.[143]

Another case of a woman NASA wanted to work in a position beneath her educational level transpired at Goddard Space Flight Center in 1974. Situated in Greenbelt, Maryland, close to the nation's capital, Goddard's misconduct, just like other misconducts of centers in the South, cannot be explained away by a 'Southern mentality,' which shows that gender discrimination was not necessarily connected to the location of a NASA center, but a general theme in the U.S. society of the 1960s and 1970s. In 1974, Goddard Space Flight Center experienced a reduction in force (RIF) and Barbara E. Lowrey was laid off from her GS-12 position as an AST flight mechanic. In return, she was offered a GS-4 supply clerk-typist position, which she declined. The Civil Service Commission recommended her reinstatement and the decision was still pending when she wrote a letter to NASA administrator James C. Fletcher in which she shared her story and her opinion of women in the NASA workforce. She wrote:

> I am a white female and it is my conclusion that being a woman cost me a grade or two and ultimately my job. It is, of course, difficult to separate other factors, known or unknown, from the discrimination issue and I can certainly think of many people who were non-discriminatory, but my net feelings is that there is no such thing as an equal opportunity for women at NASA. I feel that the work that I did was never given the same value that the work would have received if a man had done it.[144]

Lowrey felt discriminated against at NASA because she was a woman and she was also convinced that a male with an equivalent record would not have suffered such an adverse RIF action because men were not the victims of gender discrimination at Goddard. According to Lowrey, the discriminatory practices were a "condition of life" at Goddard.[145] Although gender discrimination in the workspace is difficult to pinpoint, she gave Fletcher various examples. For example, men got better assignments or 'helped' women with their projects, hence

143 'Unsung local heroine', *NOW News Bay Area Chapter* (1977) 5/8, p. 2, PR-1 Carton 25, Texas – Bay Area Chapter, NOW Chapter Newsletter, Schlesinger.
144 Barbara E. Lowrey, *letter to James C. Fletcher*, February 14, 1974, Box 33, 13694, HRC.
145 Barbara E. Lowrey, *letter to James C. Fletcher*, February 14, 1974, Box 33, 13694, HRC.

ensuring that the men ended up receiving the promotions while women often only received excuses. Like Rogers, Lowrey also found partial fault for the discrimination in the habits of her own gender: women were often "unconscious co-conspirators" because they did not present themselves as boldly confident and held themselves back, often choosing instead to conform to expectations in order to still appear feminine and to avoid making men feel threatened by female ability.[146] Accordingly, compared to her male colleagues, she appeared hesitant and not as competent to the supervisor who could have promoted her.

Although Lowrey did not unconsciously conspire with the other gender against her own interests, she felt that her work was not appreciated by her male co-workers and managers. She was frustrated by the feeling that her work did not count, but she had never considered filing a discrimination complaint, which in her mind took too long and could not produce results because it was "the refuge of the incompetent."[147] Like the women at Kennedy Space Center who wanted to take part in the Federally Employed Women seminar, she did not feel like there was a real commitment to EEO policies on the part of Goddard Space Flight Center management. Lowrey also claimed that secretaries were almost never allowed to advance to higher positions and that a RIF affected women the most because of veteran's preference and the fact that women had a lower grade structure than men, peaking at GS-11 versus GS-13 or GS-14 for men. Therefore, women were more vulnerable to "bumping" than men and thus their jobs were more likely to be abolished. According to Lowrey, there were also cases of direct discrimination by supervisors who believed that no woman should block a job that a man actually needed. Then there were cases of more subtle discrimination by supervisors. Perhaps some supervisors had preferred saving men's jobs because they had more sympathy for a fellow man facing the RIF. In such cases, the supervisors themselves might not even actually have realized they were discriminating. Lowrey also suggested that another reason women were more vulnerable to RIFs was the previous behaviour of a supervisor. If he had already discriminated against a female employee by giving the best job assignments to men, then the female worker was left with tasks of lesser importance and thus became more expendable than her male colleagues.

Lowrey also made several recommendations to remedy the gender discrimination at NASA. She suggested 10 steps that NASA should take, like a stronger EEO office at Goddard or a more thorough review of women's records for upward mobility. However, Lowrey also made two other interesting recommendations:

---

146 Barbara E. Lowrey, *letter to James C. Fletcher*, February 14, 1974, Box 33, 13694, HRC.
147 Barbara E. Lowrey, *letter to James C. Fletcher*, February 14, 1974, Box 33, 13694, HRC.

although this was not directly up to NASA, she felt that women should not be forced to change their names after getting married, and she proposed to create a seminar to educate managers on the subtle form of discrimination. This was not her first time suggesting this type of training for NASA managers; she had previously suggested just such a seminar, but was told that Goddard's management was not "ready" yet.[148] Lowrey was far from alone with her complaints about gender discrimination at NASA. Rogers, too, described the atmosphere at NASA as "paternalistic."[149] She complained that women had bleak chances of receiving a promotion and that all the positions of any power were held by men. Accordingly, all supervisors were men as well. Rogers connected this gender inequality to the composition of NASA's workforce, which included a lot of retired military people. According to Rogers, they brought their military mentality to NASA and merely tolerated the work women contributed, without really considering female employees as the equals of men in regard to their ability to grow in their jobs. This, in her view, was why women at NASA were not given assignments that would have helped them prove their eligibility for higher level work.[150] Rogers was also upset about the fact that women often found themselves training new male employees who were later promoted over them.[151]

Although not all female NASA employees groused about it because they did not mind, most of the women griped about NASA's promotion policies amongst themselves.[152] Greville also despised NASA's hiring and promotion practices in the 1960s and early 1970s. She recalled various cases of men being employed as clerks and typists; however, they only ever stayed for a couple of months before advancing to a better position, because male management looked out for them to get a better, professional position. Needless to mention, they did not put in the same effort for female employees.[153] Although Fee felt that NASA gave her every opportunity to personally grow, she was also not happy with the treatment of the agency's secretaries, namely the way in which the antiquated rules and regulations of the grading promotion system held secretaries back in their careers.[154] If a secretary got to the division level, she reached the peak of her career at a GS-6 and was not able to somehow grow further in her job. According to Fee, GS-6

---

148 Barbara E. Lowrey, *letter to James C. Fletcher*, February 14, 1974, Box 33, 13694, HRC.
149 Linda Rogers, interview by Dr. Aaron Gillette, July 27, 1992, p. 8, Tape Transcript, HRC.
150 Linda Rogers, p. 8.
151 Linda Rogers, p. 11.
152 Linda Rogers, p. 11.
153 Emogene Greville, interview by Dr. Aaron Gillette, July 27, 1992, p. 56, Tape Transcript, HRC.
154 Nancy L. Fee, interview by Rebecca Wright, November 29, 2011, p. 3, Oral History Transcript, NASA Johnson Space Center Oral History Project, JSC.

"was it. No ifs, ands or buts. Some of us worked on this, this was unfair. You had good quality people, but you were losing them because you weren't giving them the salary that they were entitled to."[155]

The system was eventually changed in the late 1970s, but women still struggled with the promotion system at NASA because there were also informal practices that kept them from advancing in their jobs.[156] For example, some of the promotions were not really competitive to begin with. In some cases, the position descriptions were specifically written with favored people, who management wanted to see in the job, in mind.[157] Like Lowrey had mentioned in her letter to Fletcher, Greville also observed that men were often allowed to take credit for the work of their female co-workers, which then often resulted in a promotion for the male colleague even though his female co-worker had done all the work.[158] Even though promotions were hard to get for women at NASA in the 1960s and early 1970s despite going to seminars and trainings, there apparently were some departments in which women were more likely to advance in their careers. According to Greville, the financial management sector was one area in which it was easier for women to rise to higher grades, in part because it did not require an AST education. Furthermore, there was not that much resistance compared to in other areas. However, a woman still had several men above her and also had to fiercely compete with a lot of other women, who also wanted to further their careers.[159] Another problem for women who wanted to advance their careers at NASA was their demanding workload. In order to get promoted and leave secretarial work, one had to take summer or night school classes while working full-time. Coplin, for instance, took several seminars and training courses offered by NASA but eventually had to quit the classes because her job did not allow her the time to enhance her skills.[160]

4.7 Minorities' standing in the office

As illustrated, the women in the NASA offices faced various types of discrimination due to their gender from their male supervisors and co-workers but also

155 Nancy L. Fee, pp. 13–15.
156 For more information see chapter: Equal Space in the 1970s.
157 Emogene Greville, p. 63.
158 Emogene Greville, p. 22.
159 Emogene Greville, pp. 22 and 32.
160 Jamye Flowers Coplin, interview by Rebecca Wright, November 12, 2008, p. 39, Oral History Transcript, NASA Johnson Space Center Oral History Project, JSC.

from members of their own gender in the 1960s and early 1970s. So far, however, only the office gender dynamics between white females and almost exclusively white males have been discussed. Although minority women faced similar problems to their white female colleagues in the NASA offices, they also had to battle other types of discrimination – especially from their white female colleagues. This latter group was often not even aware of the type of discrimination their fellow female African American, Asian American, and Hispanic co-workers had to deal with, much less that they were a source of discrimination themselves. In the letter Lowrey wrote to Fletcher, she described all the grievances of white women at NASA, and also explicitly mentioned that she did not know how severe the problems for African American female employees at Goddard Space Flight Center were. However, she was sure that black females faced at least the same amount of discrimination. This observation is quite interesting because she did not act on the assumption that minority women in fact faced more discrimination based on gender and race. Furthermore, Lowrey claimed, correctly, that Goddard's EEO goals were mainly achieved by bringing in black women for training at the GS-1 level.[161]

Ironically, it was Goddard which received NASA's first EEO award in 1981. The award went to "the NASA center which has implemented the most effective management strategies and has achieved the most positive affirmative action results."[162] However, this particular affirmative action strategy to pad the EEO statistics was not popular with just Goddard Space Flight Center, other centers had already used similar strategies in the late 1960s and 1970s. This practice ostensibly put a bull's eye on the minority women because they disturbed the insular, high-flying, head-in-space presidential mission of NASA's workforce. Coplin explained NASA's state in the late 1960s as follows. Despite the brewing social and racial unrest in the United States, "at the time I felt like I was very insulated and was not really a part of all that. It was going on and I was aware of it, but we were so focused on NASA's mission, and I was so involved that quite honestly, I didn't pay a whole lot of attention to things that were going on in other parts of the country."[163] However, starting in the late 1960s, NASA personnel, too, had to deal with the topic of racial equality due to the Civil Rights Act of 1964 and Executive Orders 10980 and 11375, issued in 1966 and 1967 respectively, which outlawed discrimination based on race, color, religion, sex, and national

161 Barbara E. Lowrey, *letter to James C. Fletcher*, February 14, 1974, Box 33, 13694, HRC.

162 *Goddard Center earns Equal Opportunity Award*, NASA News release, March 16, 1981, Box 4, 18507, HRC.

163 Jamye Flowers Coplin, interview by Rebecca Wright, November 12, 2008, p. 20, Oral History Transcript, NASA Johnson Space Center Oral History Project, JSC.

origin. NASA had to start employing minorities and most NASA centers, like Goddard, employed the majority of their minority employees in low grade secretarial and clerical positions, supposedly because not as much education was needed for the lower grades and they could not find enough educated minority personnel. However, it was hard to get minority clerical personnel because all government agencies needed to hire minorities, so NASA started to recruit them directly out of high school. The problem was that they often went to ill-equipped, segregated high schools and lacked certain skills in the beginning. Furthermore, they were also often not used to an integrated environment.[164]

A lot of the white NASA employees did not or did not want to recognize the adverse circumstances influencing the employment of the minorities. Greville, for instance, believed that the African American employees who were employed on a large scale in the late 1960s at Kennedy Space Center were less qualified. She was convinced that parts of the typing and shorthand tests were waived for the minorities. According to Greville, waiving those requirements was wrong and NASA should have employed better qualified people.[165] Bell found even more candid words about NASA's minority hiring practices in the late 1960s and early 1970s. According to her, the black employees could not read, they could not write, "they couldn't do a bloody thing."[166] She also claimed that the black employees were lazy, incapable and not even interested enough in the space program to watch a launch out the office window.[167] Emphasizing that her critique was not about race but solely about skill and ability, she went on to explain that they had it easier than their white female counterparts: "[I]f you are white, I would think you would have to qualify, you would at least have to have, you know, some sort of education and sort of be able to read and spell and things. And they couldn't do that, and yet they were hired on. And I never could understand that."[168] However, she never officially complained about her black coworkers since she felt that the work atmosphere did not allow it, because in the end the white employee would have had to warrant it.[169]

164 Nathaniel Pilot, interview by Dr. Aaron Gillette, July 28, 1992, pp. 12–15, Tape Transcript, HRC.

165 Emogene Greville, interview by Dr. Aaron Gillette, July 27, 1992, p. 50, Tape Transcript, HRC.

166 Jerri Bell, interview by Dr. Aaron Gillette, July 24, 1992, p. 17, Tape Transcript, HRC.

167 Jerri Bell, p. 18.

168 Jerri Bell, p. 19.

169 Jerri Bell, p. 22.

A lot of NASA employees, whites and minorities, were not as afraid as Bell and went ahead with their complaints. Tom Martin of the NASA headquarters Accounts Branch, for example, reprimanded his black secretary for her poor work and she immediately complained to NASA's EEO office.[170] Quite a few of the conflicts even made it to court. Four Marshall Space Flight Center employees, for instance, sued William R. Lucas, the director of MSFC, James Fletcher, NASA Administrator, Robert Hampton, the chairman of the Civil Service Commission (CSC), Ludwig Andolsek and Joyne B. Spain, members of the CSC, Bernard Rosen, Executive Director of the CSC, Irving Kator, Assistant Executive Director of the CSC, and Anthony S. Hudson, the EEO director at the CSC, for discrimination based on race and sex in 1975. The plaintiffs claimed that African American men and women were placed in low grades, that despite educational qualifications equal to those of their white co-workers, they were required to work in menial and clerical jobs. Furthermore, they felt that various jobs were "manipulated" in order to bar blacks and women altogether, or at least prevent their promotion.[171] The lawsuit filed against the NASA administrator and the Civil Service Commission by Minority Employees at NASA (MEAN) in the name of three black female employees in 1974 was quite similar. The plaintiffs demanded back pay, attorney fees and retroactive promotions because NASA's EEO program was a sham, according to the three black women. They alleged that minorities and females were only hired into dead-end positions and thus had less opportunities for promotions than white males.[172] It is obvious that minority females and males faced similar types of discrimination as white females. They were mostly forced into low grades without any chances to receive a promotion or switch jobs. In addition, minority women in the NASA offices also had to put up to the prejudiced belief of their white male and female co-workers that they lacked skills and were undeserving of their jobs in the first place. Thus, they faced double discrimination.

4.8 NASA's social reward system

Even though most of the white employees eventually changed their minds about the abilities of their minority co-workers (and about women for that

170 *Note* on March 7, 1973, Box 34, 013703, HRC.

171 *CA 75G0352 NE, in the United States District Court for the Northern District of Alabama North Eastern Division*, March 26, 1975, Harris/MSRC; Barry J. Casebolt, 'Job, Suits Filed against MSFC', *Huntsville Times*, March 27, 1975, p. 4.

172 *CA 74-1832, United States District Court for the District of Columbia*, December 10, 1975, Harris/MSRC; 'Blacks, Women at NASA sue', *Federal Ti.mes*, January 8, 1975.

matter), NASA was in need of a content workforce in these changing times.[173] One of the methods to unite and please the community was to reward them with incentives and awards. NASA and each center had its own medals and awards to help retain their employees and publicly praise them for their good work and attitude. As mentioned before, NASA, for example, established an EEO trophy in 1981 rewarding the NASA installation that had accomplished the most significant EEO progress.[174] Even before that, there had already been an Equal Opportunity Honor Medal which was rewarded to a deserving individual who had contributed to important EEO progress within NASA. However, this award was perceived as a 'second class citizenship' award in comparison to other Honor Awards. Furthermore, there had been complaints about the EO Honor Award because the recipients were mainly white males and it was not quite clear how the winner was decided in the first place.[175] Thus, racial and gender discrimination was a topic of concern even in an area which was supposed to award people who did not do that very thing – who did not discriminate because of gender and race but fought for an integrated workforce.

The recognition of their contribution to the space community through awards was very important to most employees, especially the ones in secretarial and clerical positions. Greville, for instance, was upset that she never received an award for her work, even though her boss at Kennedy Space Center had actually tried to get her one. She was convinced that clerical personnel at NASA never received awards despite their important contribution because if they gave an award to one secretary then they would have to give an award to all the secretaries.[176] Sensing the importance of secretarial work, some NASA centers did, however, start to give awards to secretaries. Johnson Space Center, for example, established the monthly Outstanding Secretary Award in 1973 to honor the "exceptional contributions to the effective operation of JSC through

173 Jerri Bell, interview by Dr. Aaron Gillette, July 24, 1992, p. 20, Tape Transcript, Oral History Transcripts, NASA HRC; Nathaniel Pilot, interview by Dr. Aaron Gillette, July 28, 1992, pp. 12–15, Tape Transcript, HRC.

174 *Goddard Center earns Equal Opportunity Award*, NASA News release, March 16, 1981, Box 4, 18507, HRC; Harriett G. Jenkins, interview by Jennifer Ross-Nazzal, August 5, 2011, p. 67, edited Oral History Transcript, NASA Headquarters Oral History Project, JSC.

175 Harriettt G. Jenkins, *Some Concerns about our Honor Awards*, Memo to Members of the Agency Honor Award Committee from Assistant Administrator for Equal Opportunity Program, February 2, 1977, Box 2, 18505, HRC.

176 Emogene Greville, interview by Dr. Aaron Gillette, July 27, 1992, p. 38, Tape Transcript, HRC.

Fig. 4.16: Christopher C. Kraft presents the Outstanding Secretary Award to Donna Sanders, Roundup, October 12, 1973.

professional competence and personal dedication."[177] The recipients received a handshake from the center director, a desk plaque and $100.[178]

Not only female employees were awarded. NASA and the centers, for example Johnson Space Center, had various awards for various contributions to the community, such as, for example, the NASA Exceptional Service Medal or the Suggestion Award.[179] Considering that there were so many types of awards for so many

177 'Outstanding Secretary Award to be Given', *Roundup*, August 3, 1973, GR 1104, Nos. 12.1 thru 12.26, General Reference Files, JSC.

178 'Outstanding Secretary Award to be Given', *Roundup*, August 3, 1973, GR 1104, Nos. 12.1 thru 12.26, General Reference Files, JSC. See Fig. 4.16.

179 All the JSC awards in 1978: NASA Exceptional Scientific Achievement Medal, NASA Exceptional Service Medal, NASA Equal Opportunity Medal, NASA Group Achievement Award, JSC Certificate of Communication, JSC Superior Achievement Award, JSC Equal Opportunity Award, JSC Group Achievement Award, JSC Certificate of Appreciation, Special Achievement Award, Cooperative Education Student Award, Invention Award, Tech Brief Award, Suggestion Award, Outstanding Performance Rating, Quality Increase Award, Superior Performance Award, Length of Service Award, Cost Reduction Award, and the Special Award. *JSC Honor Awards*, February 10, 1978.

types of employees, it is difficult to draw a conclusion whether or not the Outstanding Secretary Awards were a demonstration of paternalism or sincere appreciation on the part of NASA's male white leadership. It was definitely appropriate to acknowledge and even applaud the contribution of women working in administration in the technical atmosphere of NASA, but the way it was gone about appears to have been quite patronizing. Why weren't secretaries included in the consideration for awards like the Superior Performance Award? Why did they need their own category? Furthermore, a woman like Donna L. Sanders, who had received the Outstanding Secretary Award, was not thanked for her skill and hard work but for her "exceptional dedication, tact and cheerfulness."[180]

Unlike men, women were often requited for their charisma and sweet behaviour. In addition, an award was a small gesture pleasing the (possible) recipients without having to tackle the actual problems women faced in the NASA offices, such as not receiving promotions. Some of NASA's contractors used similar reward strategies, which might also have been patronizing instead of remunerating. North American Rockwell, for example, selected two women out of its 2,700 employees to watch the Apollo 16 launch live and they reacted in the cheery way women were expected to, which does not necessarily mean that they were not actually happy about it. Jennie Muncaster thanked her employer with the following words: "I never in my wildest dreams ever thought I'd get to see a moon shot [...] The fact that I was selected really puffed me up with pride. I'm very, very pleased. Words just can't express my feelings."[181] Helen Miller reacted similarly: "When my boss told me all I could say was, 'Oh my, oh my, oh my.' Then I gave him a big hug. He's so tall and I'm so short I had to reach up to his waist. I never thought I'd ever get to see an Apollo launch. I'm so excited I can hardly sleep."[182]

4.9 A sexual work atmosphere

A lot of women who worked in the NASA offices did not feel like they received special treatment one way or the other. Jerri Bell, for instance, was convinced that she had not been treated differently because she was a

---

180 'Donna L. Sanders gets Outstanding Award for October', *Roundup*, October 12, 1972.

181 *Two Women Employees of North American Rockwell to see Apollo 16 Launch*, 1972, Press Release, 080-16, Center, JSC.

182 *Two Women Employees of North American Rockwell to see Apollo 16 Launch*, 1972, Press Release, 080-16, Center, JSC.

woman.[183] However, she was very aware of the fact that some of her female co-workers deliberately used their feminine wile to get ahead in the office hierarchy. According to Bell, it helped to have sex with an important man to push one's career at NASA along: "Oh, yeah, it was a known fact in the old days. It was just something people don't talk about. But, you know. [...] Sort of consistently. I don't know how frequent. But it happened. A lot of cases that I won't say anything about. But there was a lot going on."[184] Bell was not the only one who noticed the "hanky-panky" conduct at NASA.[185] Several former female office employees remarked on the very sexual atmosphere in the NASA offices in the 1960s. Greville, for example, also claimed that there was a lot of sexual tension in the NASA offices when she had worked there. She even assumed that "[t]here were a lot of probably marriages broken up, and all that kind of thing, about it, during that period of time."[186] However, Bell noted that it was the woman's option if she wanted to participate or not participate in the "chasing" that was frequently happening in the offices.[187] Greville, on the other hand, experienced a case where the "fun games" turned into sexual harassment.[188] Her female supervisor and the other women in the office were verbally sexually harassed by their overall chief, but she did not know if that was just him or general practice at NASA.[189] How this sexual atmosphere came to be palpable is not known, but Greville thought the playful tension which could sometimes turn into unwanted attention was "because of people being attracted to each other, and more so than there is today, I think. There was a lot of things. Maybe just people were not used to working together, male and female, out in an environment like this, or something."[190]

One of the reasons the sexual atmosphere most likely developed was due to the party culture at NASA. Despite their all-American Boy Scout image, it was no secret that most of the astronauts in the 1960s liked to party to let off steam. There are several rumors of sexual impropriety, especially surrounding Gus Grissom and Alan Shepard and their love of space groupies, derisively called 'cape cookies' by their

183 Jerri Bell, interview by Dr. Aaron Gillette, July 24, 1992, p. 30, Tape Transcript, HRC.
184 Jerri Bell, p. 33.
185 Jerri Bell, p. 31, HRC.
186 Emogene Greville, interview by Dr. Aaron Gillette, July 27, 1992, p. 20, Tape Transcript, HRC.
187 Jerri Bell, p. 31.
188 Jerri Bell, p. 32.
189 Emogene Greville, p. 20.
190 Emogene Greville, pp. 20–21.

wives.[191] Bell acknowledged NASA's party culture and the importance of the astronauts in it. The employees apparently celebrated grand parties together after the launches. They called the events, which the press was not welcome to attend, the 'Mercury Club.'[192] Coplin also affirmed that they "partied hearty" at NASA.[193] The Johnson Space Center staff, for example, regularly mingled during happy hour at the Nassau Bay Hotel and the Kings Inn. The NASA employees also partied at the "darn good" launch and splashdown parties at Kennedy and Johnson Space Center, "where everybody wanted to be."[194] Maureen Bowen, a secretary at JSC, confirmed the sexy party atmosphere at NASA as well, when she stated that the NASA employees "worked hard and they played hard." She even admitted that it was "'probably a good thing that I was young and married to a policeman and had young children, because I wasn't able to participate in all those parties,' the splashdown parties, you know. I heard all about it the next day and I might have ended up in some kind of trouble. I always enjoyed the stories. Some of them were almost unbelievable. I see a difference nowadays."[195]

Like the NASA beauty pageants, the frisky parties appear to have ended with the Apollo era. Women at NASA apparently stopped participating in the sexual party atmosphere and started complaining about it. The splashdown party for Skylab 2 in the Fort McNair Officer's Club in Washington, D.C. in 1973, for example, did not go as planned because several employees had complained about the poster announcing the party. The poster depicted a caricature of a nude woman being held over a large champagne-filled bath tub by two men while several other men watched the spectacle.[196] People complained to Dudley McConnell, the Assistant Administrator for Equal Opportunity Programs at the time, but he could see nothing wrong with the poster. He actually thought the poster was rather "silly, childish, and unimportant" as well as to the point

191 Roger D. Launius, 'Heroes in a Vacuum: The Apollo Astronauts as Cultural Icon', *The Florida Historical Quarterly*, 87 (2008), 174–209 (pp. 191–192); James Schefter, *The Race: The Complete True Story of How America Beat Russia to the Moon* (New York: Anchor Books, 2000), p. 72; Koppel, pp. 58–67.

192 Jerri Bell, interview by Dr. Aaron Gillette, July 24, 1992, pp. 13–14, Tape Transcript, HRC.

193 Jamye Flowers Coplin, interview by Rebecca Wright, November 12, 2008, p. 48, Oral History Transcript, NASA Johnson Space Center Oral History Project, JSC.

194 Jamye Flowers Coplin, p. 48.

195 Maureen Bowen, interview by Rebecca Wright, June 21, 2007, p. 15, Oral History Transcript, NASA Johnson Space Center Oral History Project, JSC.

196 See Fig. 4.16.

Fig. 4.17: Poster for the Skylab II Splashdown Party, 1973, NASA.

because "she sure had a good pair."[197] He turned the complaint over to Jo Marie Di Maggio, the Federal Women's Coordinator, who had to contact the

197 F. George Drobka, *Memorandum for the Record*, June 29, 1973, Judiciary Committee, 93rd Congress, Subcommittee on Civil and Constitutional Rights, NASA Hearings Correspondence, Box 37 of HR. BX. 33, RG 233, NARA I.

Skylab people in order to request a poster switch. After her official complaint a replacement poster was used to announce the party.[198]

This incident shows that the semi-open display of sexual coquetry and sexism in the context of office parties was also abolished in favour of the gender equality values of the 1970s, as had been the fate of the beauty pageants and the charm clubs.

4.10 The switch to non-gender specific language

Another important matter which was changed in the course of establishing gender equality at NASA in the 1970s was the conscious switch to a less derogatory, gender neutral use of language at NASA. However, it was impossible for NASA to control the language of all the people in the space community and the media.[199] Thus, nicknames like 'glamornauts' or 'astrowomen' remained popular in the late 1970s and early 1980s in the media, showing the general state of mind in American society regarding gender equality.[200] Even within NASA offices, there were nicknames which were hard to get rid of because they had been a part of certain departments for years. People at Kennedy Space Center, for example, called the men who worked for Linda Rogers in the procurement department, the 'squaw men.'[201] Women were also often referred to as 'girls' or 'beauties,' shunting them into the less competent spheres of childhood and superficialities.[202]

While there was not much NASA could do about the informal use of gendered language, the agency tried to at least improve its official terminology, which had been a sore spot for a lot of women in the space community. The women of the Hypatia Cluster, a female pro-space citizen group, for instance, felt that terms like

198 F. George Drobka, *Memorandum for the Record*, June 29, 1973, Judiciary Committee, 93rd Congress, Subcommittee on Civil and Constitutional Rights, NASA Hearings Correspondence, Box 37 of HR. BX. 33, RG 233, NARA I.

199 For more information about gendered language in the media, see chapter three and Susan U. Philips, 'Language Ideologies', in *The Handbook of Discourse Analysis*, ed. by Deborah Tannen, Heidi E. Hamilton, and Deborah Schiffrin (Chichester: Wiley-Blackwell, 2015), pp. 557–75 (pp. 561–563); Talbot.

200 Barry Hart, 'The Glamornauts', *Globe*, March 18, 1980; 'Space Shuttles will have mixed Crews. Here come the Astrowomen', *Equal Opportunity Current News*, November 18, 1976, 008994, HRC.

201 Linda Rogers, interview by Dr. Aaron Gillette, July 27, 1992, p. 41, Tape Transcript, HRC.

202 Wernher von Braun, *letter to Jerrie* Cobb, May 9, 1962, *Cobb/Schlesinger; Apollo space suit A-7L at the International Latex Corp. (ILC) Plants*, picture caption, May 14, 1968, 008985, HRC.

'astronette' were "abomination[s]", and that "[a]n alternative [term] to 'manned' and 'unmanned' that's not impossibly awkward has yet to be found."[203] Aware of the criticism from within the space community and American society, NASA established a Task Force on Sex Discrimination in 1978 to review and revise the agency's regulations, directives and publications for gender-specific language. The agency published a gender-neutral language policy in NASA's Management Directives Handbook NHB 1410.12 on November 7, 1979. The language policy required the use of gender neutral formulations in all written NASA regulations, policies, programs, publications, and all other aspects of the agency's public information program. NASA publications were furthermore checked for possible derogatory visual representations of women. By 1982, under Executive Order 12336, NASA had changed the terminology of approximately 200 directives which had previously included gender-specific language. NASA's Management Instruction (NMI) on 'Power and Authority to Authorize or Approve Travel on Official Business and related Matters' (NMI 9710.1J), for example, used terms like 'spouse' or 'parents' instead of 'wife,' 'mother' or 'father' after the gender revision. NASA also replaced terms like 'men,' 'manpower,' 'man-hours,' or 'his' with terms like 'person,' 'human resources,' 'staff-hours' and 'his/her' respectively.[204] However, the language change with the greatest impact was the switch to 'robotic' and 'human' spaceflight instead of 'unmanned' and 'manned' spaceflight. Thus, NASA's official directive style guide now reads: "All references referring to the space program should be non-gender specific (e.g. human, piloted, un-piloted, robotic). The exception to the rule is when referring to the Manned Space Craft Center, the predecessor to the Johnson Space Center in Houston, or any other official program name or title that included 'manned' (e.g. Associate Administrator for Manned Space Flight)."[205]

4.11 Appropriate work attire for women

Another aspect that changed for the women in the NASA offices in the 1970s due to the shifting gender norms that the women's movement propagated in

203 *Hypatia Cluster Newsletter*, Fall 1983, 16386, HRC.
204 *Final Submission for the Fourth Quarterly Report to the Attorney General on Gender Discrimination under Executive Order 12336*, NASA, May 15, 1983, Box 3, 18506, HRC; *Quarterly Report to the Attorney General on Gender Discrimination under Executive Order 12336*, NASA, January 1983, Box 3, 18506, HRC.
205 Paul Dickson, *A Dictionary of the Space Age* (Baltimore: The Johns Hopkins University Press, 2009), p. 97.

American society was their choice of clothes. Before the end of the 1960s, women had to wear dresses. That was just a fact of life for them, although American women, including those at NASA, started to push the respectable dress boundaries in the 1960s. Coplin, however, remarked that they always still "dressed professionally. Now that said, the skirts were short, because it was the '60s. For some of us anyway, our skirts were a little bit shorter. . . ."[206] Even though no one at NASA "ever said there's a dresscode [sic]," a discussion surrounding appropriate work attire started to come up there in the early 1970s.[207] The women in the NASA offices had a smoldering dispute about the appropriate length of the dresses which were worn in the office and there was also "a big flap over pantsuits," which led Alberta C. Moran, the administrative assistant of Goddard Space Flight Center director Dr. John F. Clark, to issue a statement on the topic.[208] There is no indication that she wrote the circular, which was addressed to all the female employees at Goddard, in the name of her boss. Moran was obviously not a fan of the new work attire herself, because she reminded all her female co-workers, whom she called the "Goddard Gals," about the consequences of wearing a 'mini' instead of a 'midi' skirt to the office.[209] She was concerned that the display of a lot of leg could distract male co-workers and offend visitors from Far Eastern countries. She also remarked that a short skirt might not look flattering on an older co-worker: "Very often I wonder, when following a 'mature' woman in a mini, if she ever has looked at her rear view in a mirror."[210]

This not only shows that age- and body shaming were as popular then as they are today, it also reveals that the shaming was done by women first and foremost. Thus, women in the NASA offices, did not stand as a united front but antagonized each other, as they had over other topics surrounding the women's (liberation) movement. As for pantsuits, Moran remarked that they were acceptable as long as they were not "offensive to your boss and would not embarrass him when he has outside visitors. [. . .] However, bear in mind, if someone forgets to treat you like a lady – it was you who elected to wear the 'pants'."[211]

Like many other women in the NASA offices but also in other places of business outside of the space community, Moran was afraid of the loss of respect

---

206 Jamye Flowers Coplin, interview by Rebecca Wright, November 12, 2008, p. 45, Oral History Transcript, NASA Johnson Space Center Oral History Project,J SC..

207 Annie J. Easley, interview by Sandra Johnson, August 21, 2001, 13, Oral History Transcript, NASA Johnson Space Center Oral History Project, JSC.

208 Jamye Flowers Coplin, p. 45.

209 Alberta C. Moran, *Pants Suits*, September 28, 1970, 008988, HRC.

210 Alberta C. Moran, *Pants Suits*, September 28, 1970, 008988, HRC.

211 Alberta C. Moran, *Pants Suits*, September 28, 1970, 008988, HRC.

and femininity, qualities, which she and a lot of other women implicitly associated with male attention and reactions.[212] Because the pantsuit rattled so many cages, women who decided to wear them to the office did so very consciously and most of them could still remember the first time they wore their pantsuit to work decades later. Coplin, for example, remembered her first time very well even though she actually still preferred to wear dresses. For her, the pantsuits were just a nice change, while she assumed that women who were slightly older than her really took to them.[213] The dress code issue affected all female employees at NASA, no matter their occupation or racial and ethnic background; engineers and secretaries as well as white and black women faced the same challenges and prejudices. Annie J. Easley, who was an African-American computer at Lewis Research Center, remembered her first pantsuit office adventure very precisely, too:

> It was in 1970. In fact, I talked to my room supervisor about – because we'd started to wear pants in the outside world, and I said something to her about, "I have a pants suit. I'd really love to wear it to work." So we made a pact that we'd come to work the next day in pants, and it did cause quite a stir, but there was one woman who said, "I was just waiting for the first one to wear pants." I don't think she wore a dress ever after that. So that's a change. You know, we took the emphasis off what you're wearing. It's more like what you're actually producing.[214]

The discussions surrounding 'traditional' skirts, pantsuits and miniskirts in the NASA offices were rather serious because they symbolically showed the bifurcation of gender philosophies in American society in the social space of the NASA office. Thus, NASA, like American society at large, did not smoothly transition from passive beauty queens as the epitome of femininity to active pants wearing equality fighters. This shows that the office spaces in the various NASA centers were social spaces that mainly reproduced the social conventions and negotiations that male and female employees were familiar with from other spheres of their lives. It appears that the social changes which affected the gender dynamic of NASA's administrative and clerical workforce in the offices were therefore mainly dictated by the changing social conventions which were also

---

212 Brandon Marie Miller, *Dressed for the Occasion: What Americans Wore 1620–1970* (Minneapolis: Lerner Publications Co., 1999), p. 89.

213 Jamye Flowers Coplin, interview by Rebecca Wright, November 12, 2008, p. 45, Oral History Transcript, NASA Johnson Space Center Oral History Project, JSC.

214 Annie J. Easley, interview by Sandra Johnson, August 21, 2001, 13, Oral History Transcript, NASA Johnson Space Center Oral History Project, JSC.

happening in the public space of the United States. The gender discourse seemed to have moved from the outside into NASA's offices; however, some people did not see it entirely that way. Nancy Fee, for example, thought that NASA was at the forefront of social change. According to Fee

> [i]f you were female back in the '60s, even early '70s, and before that, there was not much unless you wanted to sit there and battle and battle and battle. Now the floodgate has been opened, and they [women] are even and equal to everything. I think NASA even started that before some of the outside world. As far as diversity, NASA was number one, starting out of the gate, versus the outside industry.[215]

4.12 Office space and the space community

This chapter looks into NASA's offices in the 1960s and 1970s and particular attention is paid to those women who worked for the R&D organization but not in the scientific field. The findings do not really confirm Nancy Fee's assertion that NASA was a trail blazer for gender equality. It rather appears that the traceable change in attitudes and treatment of women in the space community offices over the decades went in accordance with the changing cultural gender norms in U.S. society and NASA not only had to adjust its public image accordingly – eventually it also had to change up its gendered hiring, promotion and social office system in a drawn-out transculturation process which was objected by various male but also female space community members. Women of color's experiences were considerably worse and they faced tremendous backlash when they entered the workforce.

However, one important aspect that provoked an early gender discourse in the media space was largely missing in the social space of the NASA offices – the Cold War. The Cold War and the rivalry with the Soviet Union had been an important factor in the discourse surrounding the possibility of American women flying to space in NASA's nationalism phase.[216] Although the Cold War did not immediately help potential female astronauts in the early 1960s to speed ahead with gender equality or lead American society in implementing social changes, it was a significant factor giving them a competitive edge in the

215 Nancy L. Fee, interview by Rebecca Wright, November 29, 2011, p. 4, Oral History Transcript, NASA Johnson Space Center Oral History Project, JSC.
216 See the periodization of NASA's public images: Byrnes.

public discourse. This particular edge did not exist for the women in administration, although one could argue that the American Cold War cultural and social order was very much part of NASA office culture in the image of the nuclear family, the very distinct gender roles, and the clear ideas of women not depending on having a career. However, another group of women who tried to make it in another space at NASA, the laboratories, were most likely more touched by the impact of the Cold War on the workspace.

5 Lab space in the late 1950s and 1960s

Unlike the women who wanted to become astronauts but were not given a chance, or the women who diligently worked for American success in space in the administrative field but were not given recognition, female scientists and engineers seemed to have enjoyed unanimous support from the media and high-ranking government and NASA officials already in the early 1960s. NASA actually authored various essays about its notable female scientists and engineers as well as touched on the topic of outstanding women scientists in speeches and space symposia. The R&D agency tried to battle the negative press it received, after Jerrie Cobb and her female pilot colleagues were not considered to be astronauts, by highlighting the importance of women in the space effort through their work as scientists and engineers.[1] NASA officials often emphasized that "[o]nly a handful of people, male or female, will fly space ships in the next decades. But it will take many thousands of experts on the ground – in laboratories, universities and factories – to get those few into orbit."[2] Women were freely given a place in space within the ground forces on Earth and a lot of women embraced this point of view. Especially, women who already worked for NASA, like NASA's chief of astronomy Nancy Roman, backed the agency's position up:

> I believe that there will be women astronauts some time just as there are women airplane pilots, but there are so many other ways that women can contribute importantly to the space program that the fact that there are no women astronauts as yet should not worry us. After all there are thousands of men in the space program who are not astronauts either, all of whom are contributing importantly to the success of the manned flight program.[3]

Space community officials, such as NASA administrator James E. Webb, repeatedly emphasized this sentiment. In a speech addressed to a room full of women, he told his all-female audience that the people at NASA "really are convinced

---

1 For more information see chapter three. See also: Jennifer Ross-Nazal, *Women Scientists at NASA*, paper presented at Women, Gender, and Science, a Conference at Suny New Paltz, October 22, 2005, p. 3, 008988, HRC.

2 Erik Bergaust, 'Is Space a Place for the Ladies?', *This Week*, September 2, 1962; William McGaffin, 'Female Space Barrier', *New York Post*, March 19, 1962; Helen Waterhouse, "It is just a matter of time'. Space-minded Distaff Speak Out', *Christian Science Monitor*, October 5, 1962.

3 Nancy Roman, *The Role of Women Scientists in the Space Program*, speech at Marymount College, Tarrytown-on-the-Hudson, January, 31, 1963, MC 575, Box 160, Folder 2036, Betty Friedan Collection, Schlesinger. Also see: Nancy Roman, *Space Age Exploration and the Role of Women*, ca. early 1960s, unknown newspaper, p. 37, MC 575, Box 160, Folder 2036, Betty Friedan Collection, Schlesinger. See Fig. 5.1.

https://doi.org/10.1515/9783110629828-005

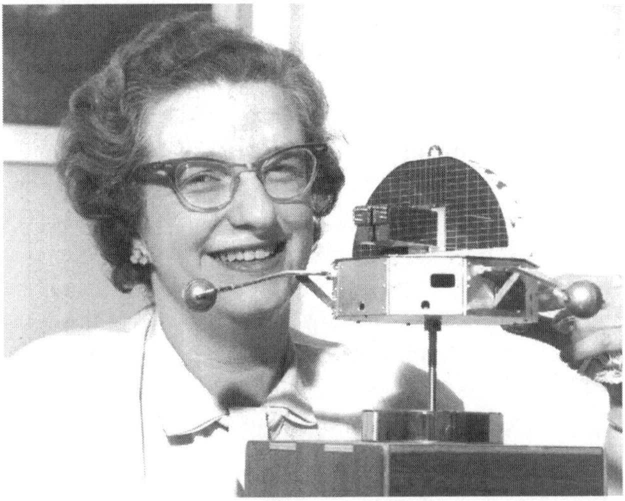

Fig. 5.1: Nancy Roman, 1962, NASA.

that women are here to stay, and that they have an important place in our Nation's space activity."[4] Robert C. Truax, a retiree from the United States Navy and a former president of the American Rocket Society, went a step further and called upon all women to be scientists as well as wives and mothers. For him pursuing a career in science and being a devoted wife and mother were not contrasting but compatible components in a woman's life in the space age. In his keynote speech at the First International Women's Space Symposium in Los Angeles on February 22, 1962 he told the mainly female audience:

> Maybe they call you eggheads, but you are the fortunate ones who have the chance to become pioneers, too, pioneers in the space age, and pioneers in the coming-of-age of women. [...] It seems that you girls would move mountains to achieve your scientific goals, you'd swim rivers, cross oceans! But you can't walk past that old wedding chapel! [...] The statistics show that tomorrow's woman can, and will, be both a wife and a career woman.[5]

So while spacefaring women were not wanted just yet and the women working in the office spaces in the space community were expected to leave their jobs once they had children, women with degrees in the sciences and engineering were seemingly urged to work for the American space effort in the labs and combine motherhood with their patriotic pioneering duty. This is quite

4 James E. Webb, *Address before the General Federation of Women's Clubs*, June 27, 1962, p. 11, Harris/MSRC.
5 Dee Dickson, 'Girl Students urged to be Space Pioneers', *Los Angeles Herald-Examiner*, February 23, 1962.

interesting because the established postwar narrative of American society usually expected its women to stay at home after being married despite the education they had received in the late 1950s and early to mid-1960s.[6] As seen in the kitchen debate between Vice-President Richard Nixon and Soviet Premier Nikita Khrushchev at the opening of the American National Exhibition at Moscow's Sokolniki Park on July 24, 1959, the U.S. made a connection between the domestic fulfilment of their homemakers and the success of the American economic and cultural way of life.[7] In order to fit the cultural and economic expectations, American women were supposed to be happy housewives and be the antonym to Soviet women, just like the wives and secretaries of the male NASA employees.[8] The sociocultural coding of femininity in U.S. society thus demanded that American female bodies to be pretty, fragile, graceful, shapely and childbearing ready, showing off American society's productive capacity without having to sacrifice femininity.

The main stereotype of Soviet women in the U.S., on the other hand, inscribed the pitfalls of communism onto female Soviet bodies. Most Soviet women were characterized as unfashionable, "graceless, shapeless and sexless" because of the manual labor they were forced to do in their communistic homeland.[9] According to American propaganda, Soviet women had laboring bodies, and therefore masculine physiques, while due to communistic politics and economic primitiveness, they also had no way of getting their hands on supposedly desired western fashions and cosmetics.[10] Although this view of Soviet women was not challenged, there were two additional stereotypes about Soviet women added to the canon; one that portrayed the Soviet women as smart and beautiful and one that depicted them as a grandmotherly type. The three stereotypes were contradictory but were needed for cultural and political reasons. However, in order to counterbalance those stereotypes on the American side of the ambitious competition between the Soviet Union and the United States, the sociocultural and economic opportunities of American women had to be expanded as well. While Americans looked down upon the unskilled,

6 Rosen, pp. 41–42.

7 Laville, 'Gender and Women's Rights in the Cold War'; William Safire, 'The Cold War's Hot Kitchen', *New York Times*, 2009 <http://www.nytimes.com/2009/07/24/opinion/24safire.html>; Greg Castillo, *Cold War on the Home Front. The Soft Power of Midcentury Design* (Minneapolis: University of Minnesota Press, 2010), pp. x–xii.

8 Douglas, p. 45. For more information see chapter three.

9 Griswold, p. 881.

10 Vera Bacal, 'The Latest Fashion News from Moscow', *New York Times Magazine*, November 20, 1960, p. 125; Kathleen Canning, 'The Body as Method? Reflections on the Place of the Body in Gender History', *Gender and History*, 11 (1999), 499–513; Robert L. Griswold, pp. 883–886.

unfeminine Soviet women in the Soviet workforce, they were willing to make exceptions for women like Nina Khrushcheva who was described as "serene, kindly, gentle, considerate, friendly, observant, adaptable, [and] sweet" in the *Washington Post* in 1959 despite her unshapely physique.[11]

Several Soviet women were even applauded by the American public. Sophisticated and educated women like Valentina Tereshkova and her female colleagues pursuing engineering and natural sciences were admired for their commitment to the nation and were regarded as valuable brain power the Soviet Union harnessed while the United States ignored the cunning and intellectual prowess of its women.[12] Therefore, a lot of people in the space community felt that the use of women in the sciences by the Soviet Union was to the advantage of the USSR because they were using all of their existing brain power to win the Cold War, and particularly the space race. While the US produced 90,000 scientists a year with one percent being women, the Soviet Union produced 190,000 a year, of which 26 percent were women.[13] The American public started to get concerned that the Soviet Union generally educated and employed more scientists of both sexes in all scientific disciplines while the scientific workforce in the U.S. seemed to decline.

A lot of Americans were concerned that the engineering personnel shortage would get more severe in the United States in the 1970s because a study by the National Science Foundation had estimated that the demand for engineers and scientists would double by 1970 while the enrolment of college freshman in technical courses had dropped from 38,000 in 1959 to 36,000 in 1961. Most Americans were shocked and disturbed that it was the opposite case in the Soviet Union, which had expanded its engineering education and where 126,000 engineers finished university in 1962. Thus, the perceived shortage of skilled and trained personnel in the natural sciences and engineering was regarded as one of the United States' most critical problems and a threat to the American way of life.[14] Therefore,

---

11 Griswold, pp. 886–889.

12 Griswold, pp. 889–896.

13 Dee Dickson, 'Girl Students urged to be Space Pioneers', *Los Angeles Herald-Examiner*, February 23, 1962; Ty Jurras, press release, February 22, 1962, Cobb/Schlesinger; James E. Webb, *Address before the General Federation of Women's Clubs*, June 27, 1962, p. 14, Harris/MSRC.

14 Gregory Donald H., *Designing the Space Age*, speech at Freeport Texas, Texas Society of Profession Engineers and Colorado Engineering Society in Denver, February 19 and 21, 1963, GH 51, General Reference Series, JSC; Scully Blotnick, *Otherwise Engaged* (Oxford: Facts on File Publications, 1985), p. 21. Nancy Roman was especially concerned that the Soviet Union employed more astronomers than the U.S. See: Lillian Levy, 'Career Girl 1959. A Space Career. She doesn't want to be launched', *Sunday Star*, May 31, 1959; *Statement of Senator James Abourezk before the Senate Aeronautical and Space Science Committee*, January 24, 1974, Harris/MSRC; Amy Sue Bix, 'A History of Women's U.S. Engineering Education', *NWSA Journal*, 16 (2004), 27–49 (p. 36).

Captain Truax and others warned the American public that "[n]ever before has America, in terms of its own survival, been so in need of utilizing all of its intellectual capabilities as we must now in this great space race."[15]

Cold War demands thus started to push the U.S. government and the public to bend the established gender order and encouraged both sexes of the next generation as well as women in science careers to join in the national military effort in order to compete with the Soviets in the Cold War to "make America first on earth and first in space."[16] This step seemed necessary in order to match the sheer numbers of Soviet engineers, including their "secret resources" – the women.[17] The Soviets also believed that their utilization of women gave them a scientific advantage, or at least this is what they contributed to the American discourse. The Russian space scientist Alla G. Massevitch, for example, stated: "In a sense, women represent one of Russia's secret resources as far as the space program is concerned. The work of some of our top theoretical woman scientists has great value to our space program. There are also many woman technicians in the space program."[18]

Although the contributions of the female Soviet scientists were recognized in the discourse within the American space community and the public, care was also taken to emphasize the shortcomings of the communist system regarding science education and employment for women. Nancy Roman, for example, remarked that Soviet women were given much more encouragement to go into the sciences thus far, but she also pointed out that she had the impression that Soviet women were decidedly discriminated against in all other levels of Soviet life, both economically and socially as well as in their workspace.[19]

When American women were encouraged in the early 1960s to pursue the sciences and engineering to compete with the Soviet Union in the space race, it was stressed that they could still keep all their feminine advantages and should still conform to their traditional gender roles.[20] James Webb, for example, pointed out that "most of the engineering work done in our agency is performed at desks in air-conditioned offices. The requirement is brains and skills, not brawn."[21] Most

15 Ty Jurras, *press release*, February 22, 1962, Cobb/Schlesinger.

16 See Fig. 5.2; See also: Douglas, pp. 22–23; Sheffield, p. 5.

17 Erik Bergaust, 'Is Space a Place for the Ladies?', *This Week*, September 2, 1962.

18 Erik Bergaust, 'Is Space a Place for the Ladies?', *This Week*, September 2, 1962.

19 Nancy Roman, *letter to Dan Lerner*, February 3, 1958, Series 2 Correspondence, Box 4, Folder 4, Nancy Grace Roman Papers, Niels Bohr Library, American Institute of Physics, College Park (hereafter NBL).

20 Douglas, pp. 22–23; Sheffield, p. 5.

21 James E. Webb, *Address before the General Federation of Women's Clubs*, June 27, 1962, p. 11, Harris/MSRC.

Fig. 5.2: TIMES-HERALD, Newport News, February 2, 1963.

professional women wholeheartedly agreed with this message at the beginning of the space age, because they had already been advocating that women were just as capable as men without needing the physical strength of a man or having to behave like a man, since "brains have no sex" and "[l]ipsticks and slipsticks are entirely compatible, and a pretty hat does not mean the head under it is empty."[22] It seems that the Sputnik moment changed the spirit of the times and not only led

22 Ethaline Cortelyou, 'The Status of the American Woman Scientist', *Sigma Delta Epsilon News*, November 1959, p. 3, 008998, HRC.

to the establishment of hundreds of new schools and the National Defense Education Act in 1958, but also established a new set of Cold War demands inviting, even urging, women and young students, including the girls, to carry out their patriotic duties by pursuing a career in the sciences or engineering in order to beat the Soviet Union to dominance in space. At the same time, the U.S. never discarded its ideal of the American woman as a happy, trim homemaker fulfilling her patriotic duty in a private role on the home front.[23] The women of the early 1960s were thus confronted with these two conflicting sets of Cold War demands and they had to find a life philosophy for themselves, which could somehow make these contradictory standards for females in U.S. society work together.

American men, for their part, had to find a way to encounter the women who had to include these inconsistent Cold War demands into their gender performance. So while American public rhetoric definitely invited women into the men's space of engineering and science, there were to the same time signals from employers and grad schools that seemed not that encouraging to women after all.[24] The statistics certainly reveal that female scientists must have faced serious barriers to their employment despite the inclusive rhetoric. Although the number of American scientists nearly tripled between 1950 and 1970, with the majority working in teaching and management, women scientists were overrepresented in low-paying teaching positions while only 3.1 percent were working in higher paying management positions in the late 1960s and early 1970s.[25] Furthermore, women scientists were typically hired at a lower grade and a lower salary than male scientists doing comparable work, and a lot of job offers for women did not even require graduate level education despite record numbers of women attending grad schools.[26] Teaching, the area in which

23 Laville, 'Gender and Women's Rights in the Cold War'; Rossiter, *Women Scientists in America. Vol. 2. Before Affirmative Action 1940–1972*, p. 61.

24 Rossiter, *Women Scientists in America. Vol. 2. Before Affirmative Action 1940–1972*, p. 68; Foster, *Integrating Women into the Astronaut Corps. Politics and Logistics at NASA, 1972–2004*, pp. 16–17.

25 NSF (National Science Foundation) survey: The median starting pay for all scientists is $10,000 and for women scientists $8,500 (or 84.2% of the median): after five years the typical scientist is making $13,100, the typical woman scientist $11,000 (83.9%). After ten years she is getting 79.8%, after 20 years 76.5% and after 25 years only 74.8% of the median $14,300 vs. $19,000. See: William Hines, 'Salary Unequal for Women in Science', *Los Angeles Times*, May 9, 1972, HRC; Rossiter, *Women Scientists in America. Vol. 2. Before Affirmative Action 1940–1972*, pp. 95, 108–109.

26 Committee on the Education and Employment of Women in Science and Engineering of the Commission on Human Resources at the National Research Council, *Women Scientists in Industry and Government: How much Progress in the 1970's? An Interim Report to the Office of Science*

female scientists were mainly employed, was not greatly important in the space community. NASA and the space industry were mainly in need of researchers and managers, and the statistics do not reveal a deviation from the U.S.-wide female employment trend in the sciences. In the early 1960s, NASA employed around three percent women engineers and scientists.[27] In 1973, the numbers looked pretty similar: 2.21 percent female engineers and scientists (government wide 2.7 percent), 2.22 percent female life scientists, and 2.23 percent women technicians working at NASA.[28] In 1982, the proportion of women scientists and engineers stood at 5.8 percent and that of female technical support staff at 3.7 percent.[29] By 1987, women represented 10.8 percent of the science and engineering positions at NASA, and in 2004, twenty percent of NASA's engineers and scientists were women.[30] The jump in the numbers in the 1980s can be ascribed to the affirmative action and equal employment opportunity programs starting in the mid-1970s.[31]

So was the Cold War demand for women joining the engineering and science workforce just lip service to keep up, at least rhetorically, with the Soviet Union in the 1960s? While the statistics seem to suggest just that, a considerable number of female NASA scientists and engineers emphasized that the agency never discriminated against women joining their ranks.[32] Women like

---

and Technology Policy, 1979, 008987, HRC; Rossiter, *Women Scientists in America. Vol. 2. Before Affirmative Action 1940–1972*, pp. 69–94.

27 Sylvia D. Fries, *The History of Women in NASA*, paper presented at Women's History Week, March 28, 1988, p. 7, 008995, HRC; Catherine Dryden Hock, *Space, the next Goal*, speech presented at the Student-Parent Night, Society of Women Engineers, February 24, 1965, p. 1, 008998, HRC.

28 Personnel Analysis Division, Office of Personnel, *Minority and Female Employment. A Comparison of NASA vs. Total Federal Government Full Time Permanent Employees as of June 30, 1973*, March 27, 1974, 013685, Box 32, HRC.

29 *Women as Percent of Total in Occupational Group, Permanent Employees, FY 80- FY 82*, 1984, 18506, Box 3, HRC.

30 *Response to Questions posed by terry Cuddy on behalf of Javon R. Bea prepared by NASA Educational Affairs Division*, Code XEE, May 3, 1988, p. 2, 008988, HRC; Jennifer Ross-Nazal, *Women Scientists at NASA*, October 22, 2005, p. 21, 008988, HRC.

31 For more information about EEO and affirmative action at NASA see chapter *Equal space in the 1970s*. For more general information about women scientists and affirmative action see: Rossiter, *Women Scientists in America. Vol. 3. Forging a New World since 1972*.

32 Beverly Swanson Cothren, interview by Sandra Johnson and Rebecca Wright, June 15, 2001, p. 23, Oral History Transcript, NASA "Herstory" Oral History Project,JSC; Eilene Galloway, interview by Sandra Johnson and Rebecca Wright, September 14, 2000, p. 10, Oral History Transcript, NASA Johnson Space Center Oral History Project, JSC; Beverly E. Golemba, 'Human Computers: The Women in Aeronautical Research', *Manuscript*, 1995, p. 93 <http://crgis.ndc.nasa.gov/crgis/images/c/c7/Golemba.pdf> [accessed 8 September 2016].

NASA's chief of astronomy and solar physics, Nancy Roman, for example, supported the agency's hiring practices. She believed that there was no discrimination, but a lack of qualified female candidates because not enough women had entered the fields of science and engineering. According to her, NASA's hiring and promotion policy was to treat "both men and women as individuals selecting the best candidate for the job rather than using sex as a discriminating factor either pro or con. To my mind, this is a far healthier attitude than promoting or not promoting women just because they are women."[33] In her mind that was the only way to proceed because as a young agency in a highly competitive field, the agency's practice of hiring and promoting only competent women was merely an act of self-preservation.[34]

Roman was correct that NASA was a new agency in a new scientific field which still had to prove itself; it did not have a corporate memory to resort to when in need of strategy in the early 1960s. This greenhorn status might have caused the agency's practices to revert to traditional values generally accepted in American society, including gender performance. Alternatively, its freshman status and the pressing need for well-educated employees might have made it easier for NASA to include new ideas as well as new people, including women scientists and engineers. Considering that it is the goal of this book to challenge the established postwar narrative of gender relations in the history of American society through the lens of the space community, this section will now take a closer look at the workspace of NASA's scientists and engineers – the labs. Keeping in mind the conflicting Cold War demands that women serve as homemakers as well as direly needed brain power in the space race, this chapter will analyze the sociocultural coding of masculinity and femininity in the lab space and the influence of the gender performance on women's ability to exercise more agency compared to their female colleagues in other workspaces at NASA or the space industry, in order to possibly reveal another side of the American mentality and attitude during the Cold War.

There are several points one has to keep in mind when analyzing the topic of female scientists and gender in the lab spaces of the American space community in the 1960s. The workspace 'lab' is also a socially produced space which is retained and reproduced on a daily basis through the work performance of the engineers and scientists and their execution and interpretation of administrative policies, technological systems, and social conventions. Furthermore, the concept of 'lab space' in this book describes the experiences of various people who

---

33 Nancy Roman, *letter to Penelope H. Thumberg*, December 28, 1966, 1821, HRC.
34 Nancy Roman, *letter to Penelope H. Thumberg*, December 28, 1966, 1821, HRC.

worked in the field of science or engineering at NASA in different branches and different NASA centers. The 'lab' is thus an umbrella term for the workspace of scientists in the space community, and was chosen because a lot of the scientists and engineers called their offices 'the lab,' even though their workspaces did not necessarily resemble a lab in the traditional sense.[35] The term was also needed in this book to differentiate the work place of the scientists from the office space of the administrative and clerical workers. Another important fact to keep in mind is that the term 'scientist' itself is already an umbrella term for people who have been educated in different disciplines and who have worked in different areas in the space community. It is very likely that engineers, chemists, physicists, astronomers, doctors, and mathematicians, to name just a few, had different experiences and prospects because they worked in different disciplines which may have been valued unequally within the space community and American society. The varying levels of esteem gave rise to competition and rivalry between the different disciplines, which must be considered because quite a few remarks of former NASA employees indicate that their sometimes was a status battle between, for instance, engineers and scientists.[36] The reputations of the disciplines have to be looked at as well, because all of these factors might have had an effect on the ease or difficulty of women's access to the lab space in the space community. For example, the majority of NASA's mathematicians were women, while engineering was dominated by men; distributions like these undoubtedly had an influence on the gender dynamics and performances in the lab spaces. Furthermore, the space community's lab spaces were not detached from the policies and structures of Cold War science.

Historians of science are in agreement that the Cold War was a political and cultural phenomenon which was very influential in defining, forming, practicing, and financing science in the United States. Considering that the space race was an instrumental part of the Cold War and that space exploration was directly linked to the Department of Defense, the scientists of the space community were very much influenced in their work by the politics of the Cold War through funding and research specifications from the government as well as internal and outside pressure to conform to the mainstream American political point of view. The policies and

35 For example: Annie J. Easley, interview by Sandra Johnson, August 21, 2001, p. 22, Oral History Transcript, NASA "Herstory" Oral History Project, JSC.

36 Jeanne L. Crews, interview by Rebecca Wright, August 6, 2007, p. 21, Oral History Transcript, NASA Johnson Space Center Oral History Project, JSC.

structures of Cold War science applied to male and female scientists equally.[37] However, there is the possibility that the sociocultural coding of masculinity and femininity in American society left wiggle room in the practice of Cold War science, giving one sex more free space to implement ideas and research, especially in the lab spaces of the space community.

While chapter seven will mainly be concerned with gender and the workspace in the 1970s, this section will take a closer look at the lab workspace in the 1960s before affirmative action laws. This is especially interesting with regard to the enthusiasm surrounding technical progress and the popularity of modernization theory in the early 1960s, and the social criticism and loss of trust in modernization and science during the war in Vietnam and the Civil Rights era during the late 1960s in American society.[38]

5.1 The significance of a support system – family

The careers of all the female scientists and engineers started long before they entered the lab spaces at NASA. Like a lot of the male NASA scientists, quite a few of the women also dreamed about their career in space long before they entered NASA's workforce and long before the possibility of space exploration and the space race even existed. An aero technologist and aerospace engineer at the NASA Space Task Group at Langley Research Center and Johnson Space Center, Dorothy B. Lee, remembered that she had always been interested in astrophysics, even before the term was coined. She liked to read books penned by George Gamow and other astrophysicists as a child and dreamed of space exploration. In 1937 Lee already knew that humankind would eventually go to the

37 Leslie; Jessica Wang, 'Scientists and the Problem of the Public in Cold War America, 1945–1960', in *Science and Civil Society*, ed. by Lynn K. Nyhart and Thomas H. Broman (Chicago: The University of Chicago Press, 2002), pp. 323–47; Ronald E. Doel, 'Scientists, Secrecy, and Scientific Intelligence: The Challenges of International Science in Cold War America', in *Cold War Science and the Transatlantic Circulation of Knowledge*, ed. by Jeroen van Dongen (Leiden: Brill, 2015), pp. 11–35; Unger; Audra J. Wolfe.

38 See for example: Nils Gilman, *Mandarins of the Future: Modernization Theory in Cold War America* (Baltimore: Johns Hopkins University Press, 2003); David C. Engerman and Corinna R. Unger, 'Towards a Global History of Modernization', *Diplomatic History*, 33 (2009), 375–85; David Ekbladh, *The Great American Mission: Modernization and the Construction of an American World Order* (Princeton: Princeton University Press, 2010); Marc Frey, *Geschichte Des Vietnamkriegs. Die Tragödie in Asien Und Das Ende Des Amerikanischen Traums*, 10th edn (München: C.H. Beck, 2016).

moon and that she wanted to be part of it.[39] Bertha M. Ryan, an engineer at Dryden Flight Research Center, grew up wanting to explore air and space from a very early age. She had always liked airplanes and started flying in a Piper J-3 Cub while still in high school. In the back of her mind, she always knew that she wanted to do something in aviation. It was in pursuit of this goal that she went to Emmanuel College and the Massachusetts Institute of Technology to study math and aeronautical engineering.[40]

These individual examples show that women did not only end up in the space community because of the Cold War demand for more brain power in the space race, but that quite a few women who worked in the labs of the space community had their own space agenda long before the public discourse of the 1960s invited them to join the national space effort. However, not all of the women who ended up working in the NASA labs had pursued the objective of space exploration from childhood on. One of the reasons for that lack of enthusiasm at an early age was that young women, unlike young men, did not have immediate role models to turn to, neither in science nor in fiction.[41] Sheryll Goecke Powers, a co-op student and later an engineer at Dryden Flight Research Center, remembered that she had difficulties deciding what to do after high school until she read an article in the Des Moines Register Sunday paper about a woman engineer. The article was significant for her because she had always liked planes and engineering but had never considered a career in engineering beforehand, probably because it was such an unusual career choice for a woman before the affirmative action programs of the 1970s.[42] Powers' example demonstrates that role models of one's own sex, even when they are only encountered in the media, are significant to young people, in this case girls.

A role model can show a young and impressionable mind that something can be done because someone has done it before. Strong female role models were therefore important for women, especially in a time when there were not a whole lot of fierce female role models available; most women had to start their careers without a woman in whose path they could follow. The majority of the women who worked in the labs in the late 1950s and 1960s had to find their way on their own. If they were lucky, they could find support and guidance at home from their

39 Dorothy B. Lee, interview by Rebecca Wright, November 10, 1999, p. 1, Oral History Transcript, NASA Johnson Space Center Oral History Project, JSC.

40 Bertha M. Ryan, interview by Sandra Johnson and Rebecca Wright, June 13, 2001, pp. 2–4, Oral History Transcript, NASA "Herstory" Oral History Project, JSC.

41 Rosen, p. 47.

42 Sheryll Goecke Powers, interview by Rebecca Wright and Sandra Johnson, June 13, 2001, p. 2, Oral History Transcript, NASA "Herstory" Oral History Project, JSC.

parents, who often instilled a sense of pride and work ethic in them which helped them navigate obstacles in their later science careers. Quite a few of the female NASA scientists of the 1960s were incredibly grateful to their parents for actively supporting them in their interest in the sciences from an early age. Shirley Hunt Hinson, who worked as a mathematician and engineer at Langley Research Center and Johnson Space Center, for example, fondly recalled that her father had been so proud of her that "he couldn't see straight" because she was the first girl in her family to graduate from college in math in the late 1950s.[43] Jeanne L. Crews, an engineer and astronomer at Johnson Space Center, considered her parents to be wonderful facilitators of her sense of curiosity because they provided all the materials she needed to explore her interests. They got her a telescope when she showed interest in astronomy or a microscope when she wanted to see what was inside of water. They not only encouraged her with gifts, but more importantly, they never discouraged her with words. They never told her "You can't do this. You can't do that."[44] Annie J. Easley, a mathematician at Lewis Research Center, considers her own mother to be her greatest role model because she always encouraged her to (literally) reach for the stars. She repeatedly told her daughter that she could be anything she wanted to if she work hard for it. Her mother's boost of confidence was especially important for the African American Easley, who not only had to struggle with American society's grasp of gender propriety, but also with racial discrimination and segregation in the 1950s and 1960s. Her mother always told her that "[i]t doesn't matter what you look like, what your size is, what your color is. You can be anything you want to, but you do have to work at it."[45]

However, not all women who became scientists in the space community had the same experience of growing up in a household that allowed its daughter to pursue all of their interests, especially when their propensities led them down a path into an area untypical for women in the 1950s and 1960s. Sheryll Goecke Powers, for example, recalled that her mother did not want her to go into engineering. She urged her daughter to take another subject, one that was less contentious for a woman to take. Her mother most likely anticipated how hard it was going to be for her daughter to be a trailblazer; she wanted an easier life for Powers.[46]

43 Shirley Hunt Hinson, interview by Rebecca Wright and Kevin Rusnak, May 2, 2000, p. 2, Oral History Transcript, NASA Johnson Space Center Oral History Project, JSC.

44 Jeanne L. Crews, interview by Rebecca Wright, August 6, 2007, p. 1, Oral History Transcript, NASA Johnson Space Center Oral History Project, JSC.

45 Annie J. Easley, interview by Sandra Johnson, August 21, 2001, pp. 4–5, Oral History Transcript, NASA "Herstory" Oral History Project, JSC.

46 Sheryll Goecke Powers, interview by Rebecca Wright and Sandra Johnson, June 13, 2001, p. 44–45, Oral History Transcript, NASA "Herstory" Oral History Project, JSC.

Beverly Swanson Cothren, who was a mathematician at Langley and Dryden Flight Research Center, also remembered that her mother had tried to push her into studying art. She therefore decided initially to major in art and minor in math, only to learn that this was not possible. Despite being conflicted, she ultimately decided to take math as a major at the Georgia State Woman's College (now part of the University of Georgia) after all.[47] Although Cothren did not feel that anybody ever discouraged her from becoming a mathematician, her example also shows that encouragement and the lack of discouragement are not the same thing, because at first she did follow her mother's recommendation to take art as a major in college.[48] As it turns out, the pioneering women of NASA's lab spaces were often held back by their own mothers, who did not necessarily discourage their daughters' interests and plans so much as influence them through the demonstration of concern and steer them into another occupational perspective. Bertha M. Ryan, for example, recalled that her mother, on the one hand, encouraged her to be brave enough to be different and that she could do anything she set her mind to if she just worked hard enough, but on the other hand she discouraged her from her interests in planes and flying. However, she did not stop her either. Ryan assumed that planes scared her mother, who was born in 1884, into a world where the Cold War, atomic weapons and rockets did not exist.[49] Although the cosmochemist Lauel L. Wilkening was always cheered on by her mother, her mom's occupational destiny was an experience she carried with her as well. Her mother had been a high school teacher at the beginning of the twentieth century, when only single women were allowed to teach, and had had to quit her career when she got married. Her mother's experience was shocking to Wilkening, but it certainly did not hold her back in the pursuit of her professional dreams.[50]

The experiences of parents as well as their discouragement or, often, lack of encouragement did, however, very much influence a lot of young girls when they had to decide what kind of career they should seek after high school. The findings of a survey by the National Science Foundation, quoted by a technical assistant to the director of the Manned Spacecraft Center, in a speech at Freeport Texas in 1963, showed that 90 percent of people who opted for an engineering career had already made the decision to go into engineering by the

47 Beverly Swanson Cothren, interview by Sandra Johnson and Rebecca Wright, June 15, 2001, p. 2, Oral History Transcript, NASA "Herstory" Oral History Project, JSC.

48 Beverly Swanson Cothren, p. 31.

49 Bertha M. Ryan, interview by Sandra Johnson and Rebecca Wright, June 13, 2001, p. 5, Oral History Transcript, NASA "Herstory" Oral History Project, JSC.

50 Lauel L. Wilkening, interview by Carol Butler, November 15, 2001, p. 14, Oral History Transcript, NASA "Herstory" Oral History Project, JSC.

time they entered the ninth grade. It also revealed the importance of encouragement at a young age for getting girls into the sciences.[51] Who knows how many girls might have chosen a career in the sciences if their parents had not withheld their support and motivation. However, the parents were not the only possible source of encouragement and assistance in the lives of the future space scientists.

5.2 The significance of a support system – school

Johnson Space Center engineer Irene M. Piatek, for example, recalled only positive reinforcement from her high school science teachers in the mid-1970s. Their encouragement most likely helped her overcome her mother's irritation with her career plans.[52] Despite affirmative action programs across the entire nation, her mother was still puzzled when Piatek decided to go into engineering. Yet twenty years earlier, Mary (Tut) Hedgepeth, a mathematician at Langley and Dryden Flight Research Center, had already been able to share her love of math with her father as well as an outstanding high school teacher who had made her even more interested in the subject.[53] These two examples demonstrate that the support of parents and teachers did not necessarily grow stronger with the passage of time into the EEO era, although it did become more likely. Some NASA scientists were able to find support and encouragement already in their early years, but the majority of the women did not find this type of help and motivation in their high schools.

Some women, like Nancy Roman, were outright discouraged in the pursuit of their career goals. Roman's high school counselor, for instance, told her that ladies should be studying Latin instead of math. According to Roman, only her firm belief in her astronomical career and the realization that the natural sciences were much more interesting and easier for her than the social sciences and humanities, kept her from leaving astronomy due to

51 Gregory Donald H., *Designing the Space Age*, speech at Freeport Texas, Texas Society of Profession Engineers and Colorado Engineering Society in Denver, February 19 and 21, 1963, GH 51, General Reference Series, JSC.

52 Irene M. Piatek, interview by Sandra Johnson, August 12, 2015, p. 2, Oral History Transcript, NASA Johnson Space Center Oral History Project, JSC.

53 Mary (Tut) Hedgepeth, interview by Rebecca Wright and Sandra Johnson, June 12, 2001, p. 17, Oral History Transcript, NASA "Herstory" Oral History Project, JSC.

outside pressures.[54] The implication that Roman was not demonstrating proper behavior for a woman because of her choice of subject was a type of allegation with which young girls who aspired to a career in the sciences were often faced in the first half of the twentieth century. It was also a rather effective deterrent for a girl who still wanted to be known as virtuous and ladylike despite her interests in science. Most girls, however, were not outright discouraged in high school, but were instead not even considered. Most teachers and counselors apparently did not even think about the possibility of a girl wanting to pursue a career in science or engineering. NASA Deputy Administrator Lori Garver, for example, remembered that because her high school did not offer calculus in the late 1960s and early 1970s, the students were allowed to go to calculus classes at the nearby university. However, the school only asked the male students if they wanted to take the opportunity, because the teachers had assumed that the girls were not interested in pursuing math.[55] Ames Research Center's chief of the Life Science Division Bonnie P. Dalton had a similar experience a decade earlier. She had never taken calculus in high school because no one expected a woman to need calculus in her future job. When she finally did take calculus for her MBA, it was so easy for her that she could not quite wrap her head around the fact that it was not recommended as a subject to the girls in her high school.[56] Harriett J. DeVries Smith's counselor also missed the opportunity to give her fitting advice. The girl who ended up as an engineer at Dryden Flight Research Center was fond of physics in high school. It was her favorite subject and she was most interested in the mechanical parts of physics – the basis for engineering. Her counselor should have suggested to her that she go into engineering at that time, but in 1950 it was unusual for women to enter that field, and this career choice for a girl had apparently not occurred to her counselor. Because of the lack of proper career preparation, DeVries Smith got a degree in physics and math at the University of California Berkeley.[57]

54 Nancy Roman, *The Role of Women Scientists in the Space Program*, speech at Marymount College, Tarrytown-on-the-Hudson, January, 31, 1963, MC 575, Box 160, Folder 2036, Betty Friedan Collection, Schlesinger.

55 Garver. See also: Sally Richmond, mathematician in the advanced orbital programming branch at GSFC, was not able to take mechanical drawing in high school because she was a girl. 'A Woman's Space in Science', *Goddard News*, July, 1963, pp. 3–5.

56 Bonnie P. Dalton, interview by Rebecca Wright, April 22, 2003, p. 55, Oral History Transcript, NASA "Herstory" Oral History Project, JSC.

57 Harriett J. DeVries Smith, interview by Rebecca Wright, November 7, 2001, pp. 2 and 7, Oral History Transcript, NASA "Herstory" Oral History Project, JSC.

Not just teachers and counselors but also male classmates were potential hurdles for science-minded high school girls. Dalton, for example, recalled that she was the only girl in her high school's physics class. The boys in her class were often mad at her because she always got the highest grades and she felt that she had to apologize for that.[58] It must have been an extremely uncomfortable and discouraging feeling to be the target of harsh judgement just because one was a girl who was better at something than her male classmates. In hindsight, Dalton was able to laugh about it, but it is very likely that quite a few girls did not bear up against the negative feelings they were faced with and quit their science dreams in order to get positive male attention. Girls of color were faced with the same barriers as white girls in high school, in addition to which they had to overcome racial discrimination. Langley Research Center's mathematician Katherine Johnson, for instance, went to high school in White Sulfur Springs, W.Va. in the 1930s, a time when African Americans were only educated until the eighth grade. In order for his bright daughter to continue her education, her father had to drive her 120 miles to Institute, W. Va., where she was able to finish high school and go to West Virginia State College.[59] While Johnson was able to push her way through gender and racial discrimination, most African American girls were forced to give up their aspirations of becoming a scientist. The double discrimination of the Jim Crow laws and contrary gender expectations were often too great of an obstacle to pursuing their interests in science into university and beyond.

Even without being discriminated against for one's race, being a woman at an American university in the first half of the twentieth century was a challenge of its own.[60] Female scientists who worked in the space community labs in the 1960s had scads of stories to tell about their university experiences – good and bad. Most of the women were often the only female student in their classes, like, for example, Jet Propulsion Laboratory engineer Donna L. Shirley at the

---

58 Bonnie P. Dalton, interview by Rebecca Wright, April 22, 2003, p. 55, Oral History Transcript, NASA "Herstory" Oral History Project, JSC.

59 Jim Hodges, 'She Was a Computer When Computers Wore Skirts', *NASA*, 2008 <http://www.nasa.gov/centers/langley/news/researchernews/rn_kjohnson.html> [accessed 13 September 2016]; Shetterly, pp. 72–76.

60 Paula S. Fass, *Outside In: Minorities and the Transformation of American Education* (Oxford: Oxford University Press, 1989), pp. 156–188; Patsy Parker, 'The Historical Role of Women in Higher Education', *Administrative Issues Journal*, 5 (2015), 3–14; Patricia Albjerg Graham, 'Expansion and Exclusion: A History of Women in American Higher Education', *Signs*, 3 (1978), 759–73; Judith McIlwee and Gregg Robinson, *Women in Engineering: Gender, Power, and Workplace Culture* (Albany: State University of New York Press, 1992), pp. 46–78.

University of Oklahoma.[61] Being the only woman in a room full of men had its downsides. The women often became social piranhas who had to spend a lot of time by themselves. Dorothy M. Simon of Lewis Research Center, for instance, remembered that nobody wanted to sit next to her at first in the chemistry classes at Southwest Missouri State College in the 1940s. After her fellow male students got used to her presence that changed little by little, but she still had to go through a phase of being an outsider for a while.[62] Sheryll Goecke Powers also recalled her time as an aerospace engineering student as a time of solitude in her dorm. She always had to study by herself because she was the only woman in her program, while all the men were able to study together in the evening because they were also living together. They were also able to make use of notes left behind by previous students of the program which were stored in the male dorm, an opportunity Powers did not have.[63]

Another difficulty a lot of the female lab workers had during their time at university was an awkward relationship with their advisors. Powers' advisor, for example, was not quite sure how to behave around a female student. He was relatively young and Powers assumed that he would have been much more at ease with male students. Although he did not discourage her from studying aerospace engineering, Powers was able to feel his uncomfortableness around her, which made her uneasy as well.[64] This was obviously not an ideal, encouraging learning environment. Other advisors went a step further in their treatment of female students. The astronomer Jocelyn Gill, who was NASA's chief of in-flight sciences for the manned space program, was forced to accept that she had no say in her own Ph.D. topic at Yale in the late 1950s – it was chosen by her male adviser.[65] Joanne Simpson, who was head of the Severe Storms Branch at Goddard Space Flight Center, had a similar problem. She started her Ph.D. in 1947 at the University of Chicago after struggling to find an advisor who would take on a female Ph.D. student. In her search for a willing professor, she was told by several potential advisors that "it was totally inappropriate for a woman to be a meteorologist" because she would have to work night shifts, leave the airport in the middle of

61 Donna L. Shirley, interview by Carol Butler, July 17, 2001, p. 107, Edited Oral History Transcript, NASA Headquarters Oral History Project, JSC; Sheryll Goecke Powers, interview by Rebecca Wright and Sandra Johnson, June 13, 2001, pp. 44–45, Oral History Transcript, NASA "Herstory" Oral History Project, JSC.

62 Dorothy M. Simon, interview by Rebecca Wright, August 23, 2012, p. 2, Edited Oral History Transcript, NASA Headquarters Oral History Project, JSC.

63 Sheryll Goecke Powers, pp. 44–45.

64 Sheryll Goecke Powers, pp. 44–45.

65 Ruth Dean, 'Watching Skies with Memories', *Washington Star-News*, August 23, 1973.

the night, fly in airplanes to do research, and generally "do all kinds of things women can't do."[66] Carl-Gustav Rossby, who was an important meteorologist in the twentieth century, apparently was at least not opposed to the idea of female meteorologists, but only because, as he told Simpson (then Malkus), studying clouds would be an "excellent problem for a little girl to work on because it is not very important and few people are interested in [them.]"[67] Lauel L. Wilkening also had difficulty finding an advisor for her Ph.D. at the University of California in San Diego. Initially, she wanted to work with Professor James R. Arnold, but he told her that he was not interested in a female Ph.D. student because he had already had two female Ph.D. students and both of them had gotten married and quit science. Prof. Arnold's rebuff was not Wilkening's only shocking and demoralizing university experience. She also had to deal with an extremely chauvinistic postdoc in the university's lab. He told her that "women don't belong in the lab. Women and dogs, I mean, they just don't belong in science."[68] Being classified alongside canines by a colleague who performed similar tasks in the lab to oneself or facing uninvited questions and assumptions about one's body and fertility gnawed at the will, nerves and determination of the women who had decided to go into science or engineering.[69] Quite a few women left science because they did not care for the social pressure. Those who stayed developed thick skin in order to not be hurt too much by being compared to dogs and made the subject of other questionable analogies.

It is important to note that the same woman could have both bad and good experiences in their university career. While Wilkening had to deal with several put downs during grad school, she was able to enjoy inspirational professors in her classes at Reed College in Portland. At Reed College, Wilkening experienced "nothing but good advice, goodwill, [and] mentorship."[70] There were even excellent female role models, because there were a couple of women in the Reed faculty in the chemistry and math departments. Seeing female professors was really inspiring for Wilkening because they were her first exposure to women in public scientific roles. The professors at Reed were also significant to Wilkening because they fostered her interest in science on a personal basis, too. One chemistry

66 Jack Williams, 'The Woman in the Eye of the Storms', *USA Today*, November 3, 1998, pp. 15–17.

67 Jack Williams, 'The Woman in the Eye of the Storms', *USA Today*, November 3, 1998, pp. 15–17.

68 Lauel L. Wikening, interview by Carol Butler, Novemer 15, 2001, pp. 6–7, Oral History Transcript, NASA "Herstory" Oral History Project, JSC.

69 Bix, 'A History of Women's U.S. Engineering Education', p. 35.

70 Lauel L. Wikening, pp. 6–7.

professor in particular encouraged her to work hard for her success and gave her extra projects to do in the lab during the holidays when she was not able to go home, and she loved it.[71] Like Wilkening, a lot of the women were especially able to blossom as scientists at university when there were other women present. Bertha Ryan, for example, was especially grateful to two very talented women who had preceded her at MIT. She was convinced that their talent and professional manners had made the path of engineering easier for her.[72]

The women often went to women's colleges because there they would not become social outcasts because of their interest in the sciences. The women's colleges also pushed their students as hard as, or even twice as hard, as other colleges. Their university mottos often reflected their aspirations. Mary (Tut) Hedgepeth's university motto at the Woman's College at the University of North Carolina (now: University of North Carolina at Greensboro), for example, proclaimed: "Educate a man and you've educated a man, but educate a woman and you've educated a family."[73] Even though it was easier to connect with role models of one's own gender because they proved to the aspiring female scientists that it could be done, a lot of the women also found support and guidance from some of their male professors. Dr. William W. Schiefflin Claytor, for instance, took Katherine Johnson under his wing after he recognized how exceptionally smart Johnson was. Claytor, who was the third African American in the country to receive a Ph.D. in math, helped Johnson to become a research mathematician by creating special classes, like analytic geometry of space, just for her.[74] Although Johnson was incredibly bright and qualified, she had to wait nearly 15 years to get a shot at fulfilling her dream of becoming a research mathematician. In the meantime she had to resort to teaching math and French in a small grade school in Marion, Virginia.[75]

---

71 Lauel L. Wikening, p. 4.

72 Bertha M. Ryan, interview by Sandra Johnson and Rebecca Wright, June 13, 2001, p. 23, Oral History Transcript, NASA "Herstory" Oral History Project, JSC.

73 Mary (Tut) Hedgepeth, interview by Rebecca Wright and Sandra Johnson, June 12, 2001, p. 25, Oral History Transcript, NASA "Herstory" Oral History Project, JSC.

74 'Katherine Johnson: A Lifetime of STEM', *NASA*, 3013 <http://www.nasa.gov/audience/foreducators/a-lifetime-of-stem.html> [accessed 10 September 2016]. See also: African American Elva Glover, mathematician at GSFC, was encouraged by her high school physics teacher. 'A Woman's Space in Science', *Goddard News*, July, 1963, pp. 3–5.

75 David Gutman, 'West Virginian of the Year: Katherine G. Johnson', *Charleston Gazette-Mail*, 2015 <http://www.wvgazettemail.com/news/20151226/west-virginian-of-the-year-katherine-g-johnson> [accessed 23 September 2016]; Shetterly, pp. 24, 72–76, 119; MediaWorldWatch, *Hidden Figures: Katherine Johnson: One of the Black Woman Mathematicians Who Worked with NASA on the Space Race* (North Charleston: CreateSpace, 2017).

Johnson was not the only space community scientist who had to work in the 'typical female occupation' before joining the space community labs.[76] Nancy Roman and Dorothy M. Simon, for example, taught at the University of Chicago and the University of Illinois respectively. Like Johnson, Roman, and Simon, a lot of the women who ended up in the space community labs did not go there straight from university. Quite a few of them had previous work engagements outside of the space community which left a lasting impression on them. As an assistant to a professor at the University of Illinois in Chicago in the 1940s, Simon had to deal with male students who expected female instructors to be easily manipulated with the promise of romance and love. One of her students asked her in earnest: "If I took you out for dinner would you give me a better grade?"[77] Roman also had her fair share of unwanted male attention. In a letter to her parents, Roman complained about unfair treatment by a colleague who behaved antagonistically towards her and had tried to block any discussion of her work. Roman was shocked to see that this male colleague constantly tried to undermine her work, not only by ignoring her personally but also by publicly discussing her topic of strong- and weak-line stars, only to mention her work with a statement "so blatantly wrong that he must have known that it was a lie."[78]

Roman and all the other women with similar college and work experiences were hurt, but they also learned how to rise above and fight for their dream of being a scientist. This is why Donna L. Shirley was convinced that "[a]nybody who makes it through college in one of these technical fields, I think, has the tools to be okay because getting through college in these fields is not easy."[79] Her statement shows that the women who worked in the space community's lab spaces were not empty canvases when they arrived there, that they started their jobs with a professional and social tool kit which they had gained through their previous work and study experiences. Most of them had experienced gender-based discrimination and because of it most of them had developed an enhanced resilience regarding sexism, patronizing, machismo and misogyny before entering NASA's lab spaces. It stands to reason that their previous experiences affected their work and gender performance in the labs.

---

76 Shetterly, pp. 10–14.

77 Dorothy M. Simon, interview by Rebecca Wright, August 23, 2012, p. 4, Edited Oral History Transcript, NASA Headquarters Oral History Project, JSC.

78 Nancy Roman, *letter to Roman's parents*, July 12, 1953, Series 2 Correspondence, Box 4, Folder 2, Nancy Grace Roman Papers, NBL.

79 Donna L. Shirley, interview by Carol Butler, July 17, 2001, p. 107, Edited Oral History Transcript, NASA Headquarters Oral History Project, JSC.

5.3 A gendered hierarchy of the scientific disciplines

Depending on what type of discipline the women were in, they were even used to different types of machismo. W. Patrick McCray rightly remarked that there is nothing inherently masculine about technology, science, and engineering; nevertheless, the technical endeavors of the United States are a space where the active participation of men has always been welcomed while women were relegated to being their support system.[80] The concept of masculine technology was especially fostered in the beginning of the twentieth century by American magazines and novels which starred the new American hero of the industrial age – the engineer with his superhuman laboring powers. Technology, especially engineering, acquired an aura of masculinity, of machismo. This impression was reinforced by federal policies in the 1940s and 1950s, such as draft deferments for men in engineering school, the military's Reserve Officers' Training Corps programs, and most importantly the Servicemen's Readjustment Act of 1944 – the G.I. Bill – which gave returning World War II veterans the opportunity to attend university, high school or vocational school at the expense of the American government. A lot of the young men chose a career in engineering and thus assured that the occupation retained its aura of machismo.[81]

Although this exclusive space for men was disrupted for a brief period of time during World War II, when a lot of women actually entered the engineering workforce and female college students were encouraged to pursue engineering, this changed yet again after the men returned from the war. During World War II half of the students enrolled in aeronautical engineering at Iowa State in 1943, for example, were women.[82] The university experiences of the female NASA scientists, who were usually the only women in classes full of men, show that this had not been the case before or after the men served on the battlefields in Europe and the Pacific. The machismo returned to engineering in full force after the war; engineers were perceived as very masculine, vigorous, and creative men and personified the future of the United States. It actually

---

80 McCray, pp. 272–73.

81 Sylvia D. Fries, *The History of Women in NASA*, paper presented at Women's History Week, March 28, 1988, p. 3, 008995, HRC; Glenn Altschuler and Stuart Blumin, *The GI Bill: The New Deal for Veterans* (Oxford: Oxford University Press, 2009).

82 Sheryll Goecke Powers, *Women in Flight Research at NASA Dryden Flight Research Center from 1946 to 1995* (Washington, D.C.: NASA History Office, 1997), p. 16; *Women in Engineering: Pioneers and Trailblazers*, ed. by Margaret E. Layne (Reston: ASCE Press, 2009); Bix, 'A History of Women's U.S. Engineering Education'.

seemed as though men were uniquely gifted to be engineers. According to former NASA chief historian Sylvia D. Fries, however, "the only thing that was unique about them was that never before and never since has the federal government done so much to guarantee that one profession would be dominated by a single group in American society."[83] Nevertheless, the men who had pursued engineering in the first half of the twentieth century were the men who shaped the space community. Along with their skills, they brought their sense of masculinity with them. The engineers were by far not the only group to do so – the test pilots definitely matched their level of machismo – but other disciplines' reputations were not as 'macho' as that of engineering. This is why a lot of the women did not go straight into engineering in the first half of the twentieth century, but started out in 'softer sciences' which were more associated with life and nurturing, such as medicine, or other space career pipeline subjects like physics, biology, astronomy, chemistry, and above all, mathematics.[84] Many women who ended up working in engineering had studied math or physics instead, just like Harriett J. DeVries Smith. Contemplating her career, she emphasized that the "flight-test business is very macho" and furthermore assumed that she "had a harder time in that respect at Dryden" than "any of the other NASA centers."[85]

Unlike DeVries Smith, most of the women who studied math worked as computers for NACA and NASA.[86] These were the women who did the calculations for the engineers and scientists in the space community. They were given data by the engineers and then plotted the data, created simulations and trajectory computations and assisted them in analyzing tracking and communications data. The computers initially worked with slide rules and illuminated light boxes with which they had to measure line traces from grainy strips of oscillograph film, and later with huge calculators the size of cash registeres built by Friden and Marchant which could add, subtract, multiply,

83 Sylvia D. Fries, *The History of Women in NASA*, March 28, 1988, pp. 3–4, 008995, HRC.
84 Sylvia D. Fries, *The History of Women in NASA*, March 28, 1988, pp. 5–6, 008995, HRC; Foster, *Integrating Women into the Astronaut Corps. Politics and Logistics at NASA, 1972–2004*, pp. 17, 19.
85 Harriett J. DeVries Smith, interview by Rebecca Wright, November 7, 2001, p. 12, Oral History Transcript, NASA "Herstory" Oral History Project, JSC.
86 Catherine T. Osgood, interview with Rebecca Wright, November 15, 1999, p. 17, Oral History Transcript, NASA Johnson Space Center Oral History Project, JSC; Annie J. Easley, interview by Sandra Johnson, August 21, 2001, pp. 39–40, Oral History Transcript, NASA "Herstory" Oral History Project, JSC; Beverly Swanson Cothren, interview by Sandra Johnson and Rebecca Wright, June 15, 2001, p. 11, Oral History Transcript, NASA "Herstory" Oral History Project, JSC. See Fig. 5.3.

Fig. 5.3: Computer Melba Roy, 1964, NASA.

and divide.[87] The female computers worked in a separate lab, or at least in a separated part of the room. Their workspace often felt like a library because no one spoke; the only noise came from the hum of the Friden calculators. Although the computers worked for the engineers, they usually did not have any contact with them. They were removed from the rest of the project and contact was usually neither wanted nor initiated. During their rare exchanges, though, the men often referred to them as 'computresses.'[88] The treatment of

87 Mary (Tut) Hedgepeth, interview by Rebecca Wright and Sandra Johnson, June 12, 2001, p. 6, Oral History Transcript, NASA "Herstory" Oral History Project,JSC; Annie J. Easley, interview by Sandra Johnson, August 21, 2001, p. 8, Oral History Transcript, NASA "Herstory" Oral History Project, JSC; Richard P. Hallion, *On the Frontier: Flight Research at Dryden, 1946–1981* (Washington, D.C.: NASA, 1984), p. 11; Holt, *Rise of the Rocket Girls: The Women Who Propelled Us, from Missiles to the Moon to Mars*; Shetterly.

88 Betty S. Love, interview with Rebecca Wright, May 6 2002, p. 6, Oral History Transcript, NASA Johnson Space Center Oral History Project, JSC; Shirley Hunt Hinson, interview by

the computers already shows the hierarchy of the disciplines and its attendant concepts of masculinity and femininity at NASA. While the female computers were treated as junior engineers during World War II, their status changed in the early 1950s because there was no male engineering shortage anymore.[89] In the 1950s and 1960s, the women who did the calculations were only considered computers, not junior engineers anymore, because engineers were men and the computers were women, according to former NASA computer Beverly Swanson Cothren.[90] Even women who actually were engineers had to fight to get the men in the labs to consider them as such, because the men did not even want to give them the title of 'engineer.' Women were supposed to be computers and computing was considered a female occupation because it was a tedious, time-consuming, and repetitive activity which required a lot of patience – a quality not associated with men and masculinity. Because women were supposedly more patient than men, they were in charge of recording the numbers.[91] Once the women were able to use electronic computers for their calculations and started programming code in the 1960s, however, their job description changed from 'computer' to 'mathematician.' With that change more and more men came into the field, probably because computing with 'machines' had been promoted and received a considerable bigger budget and represented a more attractive career path for men. Considering that most female computers and their female supervisors, such as Marcie Roberts of the Jet Propulsion Laboratory, did not want men to work as computers in the first place because they feared men would likely behave like their bosses instead of a colleague, maybe that particular fear also dispersed a bit once women started to work in the male domain of engineering. Thus, men could work as computers/mathematicians, too.[92]

Rebecca Wright and Kevin Rusnak, May 2, 2000, p. 5, Oral History Transcript, NASA Johnson Space Center Oral History Project, JSC; Holt, *Rise of the Rocket Girls: The Women Who Propelled Us, from Missiles to the Moon to Mars*, p. 164. It might have been different from center to center. Holt describes a different working atmosphere in the computer office at JPL: Holt, *Rise of the Rocket Girls: The Women Who Propelled Us, from Missiles to the Moon to Mars*, pp. 64–65, 140–141.

89 Powers, pp. 3, 18–19.

90 Beverly Swanson Cothren, interview by Sandra Johnson and Rebecca Wright, June 15, 2001, p. 11, Oral History Transcript, NASA "Herstory" Oral History Project, JSC. See also: Holt, *Rise of the Rocket Girls: The Women Who Propelled Us, from Missiles to the Moon to Mars*, p. 64.

91 Annie J. Easley, interview by Sandra Johnson, August 21, 2001, pp. 39–40, Oral History Transcript, NASA "Herstory" Oral History Project, JSC; Powers, p. 14.

92 Beverly Swanson Cothren, interview by Sandra Johnson and Rebecca Wright, June 15, 2001, p. 11, Oral History Transcript, NASA "Herstory" Oral History Project, JSC; Annie J. Easley,

The example of the engineers and computers already shows that there was some type of hierarchy in the lab spaces of the space community. It was a hierarchy of scientific disciplines coherent with the concepts of gender and masculinity, or at least comments from female scientists and engineers suggest as much. What is interesting, though, is that they do not necessarily perceive the hierarchy the same way. Bonnie P. Dalton, for instance, was convinced that she received a lower salary at Ames Research Center in 1963 not just because she was a woman but because she 'only' had a degree in biology instead of in engineering. According to her "at that time, biologists were at a lower rating than engineers, and of course, this is supposed to be an engineering center [Ames]. NASA is an engineering organization."[93] Thus, Dalton felt that the engineers were ranked higher in the lab space than biologists, a fact that was later on acknowledged and changed by the EEO office in the 1970s.[94] Although Dorothy B. Lee worked at another NASA center, she also felt that the engineers at Johnson Space Center, especially the ones her age "had great difficulty" letting her "be an equal," particularly since she was a math major and not an engineer.[95] Jeanne L. Crews, on the other hand, had the feeling that scientists were ranked higher at Johnson Space Center than engineers because when she was supposed to work together with a couple of scientists, they just treated her like an assistant. She was not happy about the treatment: "Of course, dumb me. What they wanted, they needed an engineer to help their scientists. Well, at the time I was a senior engineer with experience, and that angered me. So once again the battle raged with the scientists. [Laughs]"[96] Even within one field there was rivalry between the disciplines. Connie Eaton Harney, for example, remembered that analog programmers looked down on digital programmers, and that women thus had an easier time becoming digital programmers than getting into analog programming.[97]

---

interview by Sandra Johnson, August 21, 2001, pp. 39–40, Oral History Transcript, NASA "Herstory" Oral History Project, JSC; Holt, *Rise of the Rocket Girls: The Women Who Propelled Us, from Missiles to the Moon to Mars*, pp. 64–65; Shetterly, p. 205.

93 Bonnie P. Dalton, interview by Rebecca Wright, April 22, 2003, pp. 3–4, Oral History Transcript, NASA "Herstory" Oral History Project, JSC.

94 Harriett G. Jenkins, interview by Jennifer Ross-Nazzal, August 5, 2011, p. 21, edited Oral History Transcript, NASA Headquarters Oral History Project, JSC.

95 Dorothy B. Lee, interview by Rebecca Wright, November 10, 1999, p. 17, Oral History Transcript, NASA Johnson Space Center Oral History Project, JSC.

96 Jeanne L. Crews, interview by Rebecca Wright, August 6, 2007, p. 21, Oral History Transcript, NASA Johnson Space Center Oral History Project, JSC.

97 Connie Eaton Harney, interview with Rebecca Wright and Sandra Johnson, June 12, 2001, pp. 6–7, Oral History Transcript, NASA "Herstory" Oral History Project, JSC.

The competition between and within the different scientific disciplines did not necessarily start in the space community. Connie Eaton Harney, for example, remembered that certain rivalries had already been in full swing at university. She recalled that one of her mathematics professors despised applied mathematics and engineering so much that he did not even want to teach a class in an engineering building. His attitude obviously rubbed off on his young and impressionable students. Harney herself admitted that she did not want anything to do with that "applied stuff" because of her professor's dismissal of engineering, an attitude she came to regret later on.[98] All of these remarks by female scientists and engineers demonstrate that there were power relations at play between the different scientific disciplines in the space community labs.

These power relations were intrinsically tied to gender relations, which becomes evident when looking at the relationship between 'masculine engineering' and 'feminine mathematics.' However, gender is not just powerfully present when women and men share the same space, in this case the lab, it is also always attendant among a 'men only' group. Thus, women often came into established lab spaces with hierarchies structured around masculine relationships. Multiple masculinities had been constructed in this gendered workspace power relationship. Some forms of masculinity were hegemonic whereas others were subordinate. Hegemonic masculinity is the dominant form of masculinity, and also the more accepted form in a social, cultural, and work context. The concepts of hegemonic and subordinate masculinities establish an understanding of the way in which the scientists and engineers tried to 'do' their masculinity in the lab space.[99] Due to the rival masculinities, women had different opportunities to present themselves as gendered members of the particular lab space. It must thus be assumed that women had to use different strategies when joining a lab space where masculinities were more dominant than when joining lab spaces where the sociocultural coding of masculinity was less vigorous.

Considering that the female scientists and engineers were certainly not in a political or gendered state of tabula rasa when they joined the space community's labs in the 1960s, due to their previous experiences at home, in high school, in university, and in other workspaces, their various assertions that there was no gender-based discrimination at NASA receives a new poignant quality.[100] The women who felt that there was no discrimination in the space

98 Connie Eaton Harney, interview with Rebecca Wright and Sandra Johnson, June 12, 2001, p. 33, Oral History Transcript, NASA "Herstory" Oral History Project, JSC.

99 R.W. Connell and James W. Messerschmidt, 'Hegemonic Masculinity: Rethinking the Concept', *Gender and Society*, 19 (2005), 829–59; Maier and Messerschmidt; Llinares, pp. 12–13.

100 See for example: McIlwee and Robinson, pp. 23–143.

community, knowingly or unknowingly, already put their statements into the greater social context of the gender expectations of American society in the early 1960s, because they were more or less aware that they were "part of that culture" in "those kind of times."[101] Furthermore, they sometimes were not even aware of discrimination, because what would constitute chauvinism and sexism in the twenty-first century was not necessarily viewed the same way in the 1950s and 1960s, when these women worked in the labs. So when Goddard Space Flight Center's Eleanor Pressley stated that "[t]here's no discrimination against women at NASA because we need scientists and engineers badly," she was also aware that there were complaints from some women that they might have received a higher government rating if they were men.[102]

In the same article about female scientists, it was also revealed that the lab women only received fair treatment from the space community "for the most part."[103] While Nancy Roman repeatedly declared that "[s]pace as a career depends on the individual who's considering it," she was also aware that "women meet appreciable discrimination, particularly in the universities" and that women "are discriminated against to some extent in all the sciences" although the discrimination is "rapidly lessening" because the demand for scientists is so great.[104] Johnson Space Center's Ivy Schneider also did not feel that there had been gender-based discrimination, because from an "intellectual standpoint, there is no reason why more women shouldn't be in science or math."[105] However, she also remarked that the field is "obviously" dominated by men and that there "may be some die-hards among men in the field who are prejudiced against women," but she was convinced that these men were an exception.[106]

These statements make clear that there was gender-based discrimination in the lab spaces of the space community, but that compared to what these women

---

101 Harriett J. DeVries Smith, interview by Rebecca Wright, November 7, 2001, pp. 5–6, Oral History Transcript, NASA "Herstory" Oral History Project, JSC; Hunt Hinson and J. Kirby Hinson, interview by Rebecca Wright and Kevin Rusnak, May 2, 2000, pp. 43, 2, Oral History Transcript, NASA Johnson Space Center Oral History Project, JSC.
102 Judith Viorst, 'Women find Place in Space', *New York World-Telegram and Sun*, September 1964.
103 Judith Viorst, 'Women find Place in Space', *New York World-Telegram and Sun*, September 1964.
104 Zoe von Ende, 'NASA Astronomer: Women not barred from Space Program', *Denver Post*, March 18, 1966; Sue Cronk, 'Orbitable Gals: U.S. can send them', *Washington Post*, April 21, 1963.
105 Blair Justice, 'Her Career in in Science. They have a common Trait: 'Stick-to-Itiveness'', *Houston Post*, February 7, 1965.
106 Blair Justice, 'Her Career in in Science. They have a common Trait: 'Stick-to-Itiveness'', *Houston Post*, February 7, 1965.

had already experienced before joining the space community, it did not seem like a big deal anymore because they had gotten used to a certain amount of sexism or even misogyny and a hyperbolic emphasis on the fact that they were not only scientists but women scientists. Physiologist Carolyn Huntoon, who started working for the National Research Council at the Manned Space Center in 1968, explained that she and the other women had to accept

> that because that was the community at the time, and I really didn't have a choice. You either accepted the ground rules, or you didn't play in the game. So I accepted that. The NASA program in general did not discriminate, and the NASA program in general was very good to me as an individual, as it was to many people. There were individuals in the program that did discriminate, did make life hard for me and other women. Luckily, they moved on. I outlasted most of them.[107]

In her explanation, Dr. Huntoon actually revealed the crux of the matter. There was no legal discrimination against women at NASA and other companies in the space community, not even in the 1960s. Nor was there any written law that prohibited women from doing anything in the space community. However, there was individual and structural discrimination the women in the labs had to deal with.[108]

5.4 Discriminatory informal structures

Because gender-based discrimination was not decreed on an institutional or legal level, but enacted on an individual and structural level, it is considerably harder to identify and trace, even for people who were right there in the same lab space when it happened.[109] Engineer Walter W. Guy, who joined NASA in 1959, for example, was convinced that he had never seen gender-based discrimination at NASA because women were present in his lab at his first day of work: "I'm never a good champion for the women's underdog position at NASA. I have never seen that. People say that's because I didn't look; I don't know. I've never seen it. When I hired in, the day I hired in, and the group – I think there were two or three of us,

107 Carolyn L. Huntoon, interview by Rebecca Wright, June 5, 2002, pp. 40–41, Oral History Transcript, NASA "Herstory" Oral History Project, JSC.

108 Fred L. Pincus, 'Discrimination Comes in Many Forms: Individual, Institutional, and Structural', in *Readings for Diversity and Social Justice*, ed. by Maurianne Adams and others (New York: Routledge, 2000), pp. 31–34.

109 Toni House, 'Woman Space Launcher Comes Home', *The Evening Star*, December 24, 1970.

there was a female engineer."[110] The fact that NASA had hired a female engineer was for him synonymous with the fact that there was no discrimination; her existence in the lab was proof enough for him. He also witnessed "lots of women in lots of different positions" get a promotion. Even the first person to become a GS-15 expert in his division was female. Following his line of argument, if a person was deserving and had gained enough experience over the years, they got promoted, no matter their gender. According to him, there just were not enough female scientists and engineers around in the beginning of NASA and therefore not that many women could receive a promotion. That is why he felt that the allegations of discriminatory practices were without substance and invidious: "I don't think it's fair to look back and say none of the Apollo managers were women. The pool was not that large at that point."[111] This particular explanation for the relatively rare granting of promotions to female scientists in the space community was often repeated by various men and women in the community. They felt that only the individual merit of a person influenced NASA's promotion decisions and the considerably smaller pool of female engineers and scientists tipped the scale toward men most of the times.[112]

There were even some men in the space community, like thermal systems mechanical engineer James R. Jaax, who insinuated that women were often promoted too early, especially in the late 1970s, to gender-balance NASA's promotion record. According to Jaax, these women had produced only mixed results because they had been promoted to positions of responsibility too soon.[113] Taking a closer look at Jaax's statement reveals a subliminal expectation that women in the space community had to live up to. A mistake made by one man did not necessarily reflect on other men just because they belonged to the same gender, but it was only men who were permitted to make human errors. Women who were promoted were not allowed to misjudge, make mistakes, or least of all fail in their abilities and decision making. They had to have an entirely clean record, in order for male colleagues not to judge them and not to ruin another woman's chance of getting hired or promoted. Anne L. Accola, for

---

110 Walter W. Guy, interview by Rebecca Wright, January 29, 2007, p. 36, Oral History Transcript, NASA Johnson Space Center Oral History Project, JSC.
111 Walter W. Guy, p. 37.
112 Alan B. Shepard, *Women in the Space Age*, May 11, 1965, pp. 6–9, 008998, HRC; Nancy Roman, *letter to Penelope H. Thumberg*, December 28, 1966, 1821, HRC; James R. Jaax, interview by Rebecca Wright, December 14, 2010, p. 175, Edited Oral History Transcript, NASA Johnson Space Center Oral History Project, JSC.
113 James R. Jaax, p. 176.

example, remembered being warned by a female co-worker that she had "better succeed" and that if she did not she delivered "an excuse never to hire another woman" to her superiors.[114] Furthermore, women were not only not allowed to make mistakes but they were always under pressure to perform better than their male colleagues in their jobs. Sheryll Goecke Powers explained that there was no true equality in NASA's hiring practices. For her,

> true equality is when the people of equal ability get to have the same chance for the job. And for women and I suspect for a lot of minorities, it's not equal ability. You have to be vastly superior before you can be considered for the job, and there's still a lot of that. It's not equal ability, so you have to be more capable than your competition. If you're equally capable, then chances are you're not going to get it.[115]

Women not only had to have superior abilities to get a job or receive a promotion in the lab but they also had to work and delegate in exactly the same fashion men would do. Women who had developed their own management style were often regarded as incapable because they did not handle things like a man. The fact that some people in the space community could not get over the "idea of women not being just like men" was ridiculous to Huntoon. She complained that women were often hired and promoted to "do jobs, and the fact that they didn't behave just like another guy would do, meant they couldn't do the job. [Some managers] weren't willing to give them a change to develop their own way of doing things."[116] However, women in the lab spaces sometimes did get the opportunity to develop their own management style after being promoted, though not usually before the 1970s and 1980s. Exceptions prove the rule, though. Aeronautical engineer Shirley Hunt Hinson, for example, received rave reviews from the people in her division. The people under her had to fill out forms and rate her performance as an instructor, and her feedback was so good that other instructors at a management seminar told her: "I don't think anybody has ever gotten any better comments than you have. I don't know what you do to these people. You must have brainwashed them or something."[117]

114 Anne L. Accola, interview by Rebecca Wright, March 16, 2005, p. 18, Oral History Transcript, NASA Johnson Space Center Oral History Project, JSC. Women of color felt double pressure because of their gender and race: Shetterly, p. 49.
115 Sheryll Goecke Powers, interview by Rebecca Wright and Sandra Johnson, June 13, 2001, p. 32, Oral History Transcript, NASA "Herstory" Oral History Project, JSC.
116 Carolyn L. Huntoon, interview by Rebecca Wright, June 5, 2002, pp. 40–41, Oral History Transcript, NASA "Herstory" Oral History Project, JSC.
117 Shirley Hunt Hinson, interview by Rebecca Wright and Kevin Rusnak, May 2, 2000, p. 34, Oral History Transcript, NASA Johnson Space Center Oral History Project, JSC.

Most of the women, however, met trouble when they wanted to get promoted or receive the same paycheck as men working in the same position in the lab. The list of women who felt that the structures and working attitudes in the labs of the space community worked against them because of their gender is quite long. Jet Propulsion Laboratory computer Barbara Paulson, for example, faced criticism when she became the supervisor of the computer pool. A male engineer colleague felt that a man should be in charge because, according to him, she would eventually "get married, get pregnant, and then quit."[118] Kirby Hinson, husband of Shirley Hunt Hinson, for instance, was convinced that she was treated unfairly in regard to raises. So unfairly, in fact, that it "would bring anybody to tears."[119] Harriett J. DeVries Smith also felt passed over when she worked at Dryden. According to her she was not even considered for a position although she had "an experience which went beyond what any of them [her competitors for the job] had."[120] Women like the mechanical engineer Kitty Joyner were capable but did not get an opportunity to prove themselves further because "there was a lot of feeling that women shouldn't be engineers," much less instructors.[121] Harriett J. DeVries Smith was also aware that her male colleagues got better promotions than her and that therefore she did not make the same kind of money.[122] When Katharine H. Armistead was up for a promotion but did not receive it in the 1950s, her supervisor explained her why she had not gotten it. According to DeVries Smith, Armistead's supervisor told her that there were only limited funds available for promotions, and it was more important that the men who had families to support got the promotions.[123] So the female lab workers often did not get a promotion or received a smaller paycheck because they were not expected to support a family during the 1960s. Furthermore, female scientist rarely objected to receiving a smaller wage than their male co-workers, and recruiters were very aware that women could be hired for less.

118 Quote of Bill Hoover in: Holt, *Rise of the Rocket Girls: The Women Who Propelled Us, from Missiles to the Moon to Mars*, pp. 153–154. She was later fired for being pregnant despite wanting to go to work. Holt, p. 163.

119 Shirley Hunt Hinson and J. Kirby Hinson, interview by Rebecca Wright and Kevin Rusnak, May 2, 2000, p. 43, 2. Oral History Transcript, NASA Johnson Space Center Oral History Project, JSC.

120 Harriett J. DeVries Smith, interview by Rebecca Wright, November 7, 2001, pp. 15–16, Oral History Transcript, NASA "Herstory" Oral History Project, JSC.

121 Edwin C. Kilgore, interview by Sandra Johnson, May 3, 2008, p. 28, Oral History Transcript, National Advisory Committee for Aeronautics (NACA) Oral History Project, JSC. See Fig. 5.4.

122 Harriett J. DeVries Smith, interview by Rebecca Wright, November 7, 2001, pp. 5–6, Oral History Transcript, NASA "Herstory" Oral History Project, JSC.

123 Harriett J. DeVries Smith, pp. 6–7.

Fig. 5.4: Kitty Joyner at Langley Research Center, 1952, NASA.

Nancy Roman, for example, was unofficially told that "we don't discriminate against women, we just get them for less."[124] This statement shows that there were no official regulations in the space community which discriminated against women in the labs in the 1960s, but that there were informal standards which did. These discriminatory informal structures within the space community were often acknowledged to some extent. Marjorie Townsend, who was the first woman to supervise the launching of a U.S. spacecraft was, for example, told by a male colleague: "[i]f you were a man, you'd find it easier to sell your ideas."[125]

124 Cody Calamaio, 'Friendship Heights Woman, NASA's first Director of Astronomy, reflects on Career and Future', *gazette.net*, November 22, 2010, 1821, HRC.

125 Ruth Dean, 'Woman Heads Space Probe Launches', *The Evening Star*, January 12, 1972.

Dorothy B. Lee heard something similar from her assistant division chief, who said, "Dottie, if you were a man, you would have been a boss."[126] Although Lee never wanted to become the boss and Townsend eventually found people she could interest in her ideas, their examples clearly show that their gender was somewhat of a handicap when it came to promotions and credibility.

5.5 Sharing a workspace – male and female responses

If a woman did actually receive a promotion and was in charge of other employees in the lab, her status as a leader was often dismissed and she had problems being taken serious because some of her subordinates of the other sex were not willing to accept a female supervisor at first. Time and competent leadership on the part of the women in the labs usually changed people's minds, however. Jeanne L. Crews, for instance, recalled that she had issues with contractors who did not want a female monitor. An engineer from Martin Marietta apparently had been very skeptical about her leadership abilities, but within six months he called her "boss," which she felt was a huge compliment in light of his initial skepticism.[127] In 1970, when Marjorie Townsend was the first female project manager for the first orbital x-ray mission, which was the first U.S. spacecraft launched outside the U.S. by a binational crew, the 100 Italian and 40 American men Townsend oversaw were at first irritated at having a female boss but they, too, got over it.[128] Their experiences show that although women had to prove themselves, they were able to convince their male colleagues of their abilities and break through the structure of discrimination.

Not only did women have to prove themselves in leadership positions, but they also had to prove themselves as co-workers, especially when they entered a lab where they were the first woman to join the team. Their male supervisors and co-workers were often so uncertain about how to handle themselves and their new colleague that they just ignored the new arrival. Jeanne L. Crews, for example, remembered that she had to work harder in order to be accepted by her male engineering colleagues in the Flight Crew Support Division at Johnson Space Center.[129]

126 Dorothy B. Lee, interview by Rebecca Wright, November 10, 1999, p. 44, Oral History Transcript, NASA Johnson Space Center Oral History Project, JSC.
127 Jeanne L. Crews, interview by Rebecca Wright, August 6, 2007, p. 16, Oral History Transcript, NASA Johnson Space Center Oral History Project, JSC.
128 'Spacecraft Launch Boss is a Mother', *The Huntsville Times*, December 15, 1970.
129 Jeanne L. Crews, interview by Rebecca Wright, August 6, 2007, p. 4, Oral History Transcript, NASA Johnson Space Center Oral History Project, JSC.

The men in her lab just did not know what to do with her. In the beginning, they had someone following her around with a little ash tray because they were overwhelmed and did not know how to accommodate a woman of equal work status. According to Crews, the men were disconcerted: "There's a woman. What are we doing?"[130] Anne L. Accola had a similar experience when she started to work in the Simulation and Training Branch of the Flight Control Division. The head of the Training Branch, Richard A. Hoover, did not know what to do with her or what kind of work assignment he should give her. It took him two weeks to decide which job he should assign her. In the time he took to make his decision, he "parked" her in the secretary's office outside his office, which left Accola looking like a secretary without a typewriter, a situation someone with a B.S. with Honors in Mathematics from Colorado State University did not appreciate. Accola also recalled that one of her supervisors, Jones W. Roach, had asked her how she wanted to be treated because they had never have a woman in their workspace before. The question seemed strange to Accola at first because she wanted to be treated decently with respect like everybody else. However, she soon enough had to realize that the question was not necessarily rude but essential because a lot of her male co-workers treated her "like a curiosity. They would come by to take a look at me. Others were just friendly, as nice as could be. And a lot of men had daughters and they were thinking how they would want me to be around their daughters. Then there were a couple that went out of their way to be a problem."[131]

Acola was not the only person to remember that women were a curiosity at the workplace to a lot of men. Aerospace engineer Larry D. Davis also recalled that there were only few women at Johnson Space Center in the Gemini days and so "[i]t was like an event if a woman walked down the hall or was out on the duck pond. People would beat on the window and say, 'Hey, somebody's walking by,' or something. It was an event. It sounds like when guys are in prison or something. [Laughs]"[132] Even though the men's gawking and exaggerated behavior has a funny touch nowadays, and thus a lot of the interviewees had to laugh when remembering the first times men and women shared their workspace in the space community, their behavior certainly did not put the college educated women in a laughing mood in the 1960s. The male attitudes toward their new

---

130 Jeanne L. Crews, p. 3.

131 Anne L. Accola, interview by Rebecca Wright, March 16, 2005, pp. 16–18, Oral History Transcript, NASA Johnson Space Center Oral History Project, JSC.

132 Larry D. Davis, interview by Sandra Johnson, February 15, 2007, p. 36, Oral History Transcript, NASA Johnson Space Center Oral History Project, JSC.

female colleagues not only made some of the women uncomfortable, but it additionally constricted them in their work performance. Accola was "parked" during a time when the United States was in a Cold War race to the moon and supposedly was in desperate need of skilled personnel. Other women's time was wasted with avocation work. Harriett J. DeVries Smith, for example, had told her supervisor that she wanted something more challenging to do, so he gave her a calculus problem with integral equations. Although the equations had already been solved in a report from Langley, a couple of her co-workers had gone through them and they had all ended up with different answers. Her colleagues were therefore convinced that the report was erroneous and they had almost written back to Langley to claim as much. Instead, they gave the calculus problem to DeVries Smith first and she not only solved the problem but also got the correct answer. The men were stunned and she had to explain to them how she got the same answer as the Langley paper. Their response was just: "Oh." Understandably so, because as it turned out they had not given the paper to DeVries Smith to check the equations, but for amusement. They had already been convinced that the equations in the paper were wrong and they had the right answer. According to DeVries Smith, "[t]hey just gave it to me just to amuse me. [Laughs]"[133]

Not only did women in the labs have to serve as punchlines for ignorant jokes from their male co-workers, but they were furthermore interrupted and restricted in their daily work routines by unwritten behavioral rules. At Dryden, for example, women were not allowed in the hangars because Walt Williams and Paul F. Bickle frowned on it. It did not matter if the woman was a secretary or an engineer, they did not like to see the other sex in the hangars because they felt that "women belonged where they were working and they should stay there" and that "women out there [in the hangars] would distract the guys, and the guys wouldn't get their work done, or their minds would wander."[134] This rule, which was never an official NACA or NASA rule, was solely based on the personal convictions of men in power in the space community who individually discriminated against female engineers and scientists at NASA and in the space industry.

---

133 Harriett J. DeVries Smith, interview by Rebecca Wright, November 7, 2001, p. 46, Oral History Transcript, NASA "Herstory" Oral History Project, JSC.

134 Betty S. Love, interview with Rebecca Wright, May 6 2002, pp. 21–22, Oral History Transcript, NASA Johnson Space Center Oral History Project, JSC.

Informal rules like these obviously had consequences for the professional women who wanted to excel at their jobs. DeVries Smith, for instance, had to pass on her instructions for the technicians and mechanics through one of her male fellow engineers. She was not allowed to do so personally because the hangars had been declared no-go-zones for women. Not surprisingly, this practice led to several problems and disruptions in the work flow: information got lost on the way, was transmitted improperly or inaccurately and, of course, the women felt out of the loop.[135] As the Center Director, Bickle was in a position of authority and he was in no way willing to give women the opportunity to get into positions of authority as well. Only when his daughter entered the workforce, shortly before he retired, did he stop his sexist ways.[136]

Before that, however, he found several opportunities to humiliate his female engineering staff. At a big X-15 conference, Bickle had ordered that the entire female staff should wait tables during a lunch served in the hangar. All the women had to do it, including the engineers. Considering that none of the men, no matter their work status, was asked and that all of the women were initially hired to perform office and lab work, engineers like Bertha M. Ryan were especially furious that she was supposed to serve food to their male colleagues of equal work status. So she said to Bickle: "You know, I'm kind of clumsy. [. . .] I'm liable to spill soup all over this guy and then later have to work with him in a professional capacity, and he won't take me seriously."[137] In the end, the women engineers did not have to wait tables but the secretaries still had to perform this task, despite being hired for an entirely different purpose. The case of Paul F. Bickle's personal discrimination at Dryden, however, also shows that there are always two point of views. Mary (Tut) Hedgepeth, for example, was convinced that Bickle "was really ahead of his time on women's rights."[138] She appreciated that he let her work up until a week before her child was born and that she was allowed to return to work two weeks after giving birth. She was really happy that she was not discriminated against at her

135 Harriett J. DeVries Smith, interview by Rebecca Wright, November 7, 2001, pp. 9–10, Oral History Transcript, NASA "Herstory" Oral History Project, JSC.
136 Sheryll Goecke Powers, interview by Rebecca Wright and Sandra Johnson, June 13, 2001, p. 31, Oral History Transcript, NASA "Herstory" Oral History Project, JSC.
137 Harriett J. DeVries Smith, interview by Rebecca Wright, November 7, 2001, pp. 10–11, Oral History Transcript, NASA "Herstory" Oral History Project, JSC.
138 Mary (Tut) Hedgepeth, interview by Rebecca Wright and Sandra Johnson, June 12, 2001, p. 11, Oral History Transcript, NASA "Herstory" Oral History Project, JSC.

workplace due to her pregnancy.[139] From a modern point of view Bickle's behavior does not seem supportive at all because she did not receive a long paid maternity leave, but Hedgepeth was satisfied that it was not expected of her to stay at home and that her status as a working mom was accepted.

There were various cases of individual discrimination in the lab space. The case of Anne L. Accola's example, however, demonstrates that individual discrimination did not always prove effective because even men in positions of authority, such as flight directors, were not above the law. Accola claimed that Eugene F. Kranz did not want women in his team, but he was forced to hire her because she would have blocked the federal register for the entire Houston area. This means that he would not have been able to just hire someone he liked better right away, but would instead have had to wait to hire his preferred candidate until somebody hired Accola. Considering that Accola had no intention of working anyplace else and that Kranz was the only hiring authority at Johnson Space Center at the time, he had to take her. Even though he waited for a month, in the end he had to employ her.[140] Although Accola wanted to be in the Flight Control Division, she was scheduled to start in the Simulation and Training Branch because for most of the 1960s women were not allowed to work in flight control.[141]

Again, there was no official rule prohibiting women from the floor of Mission Control, but the head flight controller and the head doctor did not want to see women in the Mission Operations Control Room (MOCR) because, just like the hangar ban at Dryden, it was again assumed that women were a distraction for the men. Those women who worked in any capacity on the current mission, however, still had to be available in case there was any issue with their contribution to the mission. They hence had to stay in the back rooms and were forced to overhear flight controllers they met in the elevator make comments like: "Well, it's certainly good we keep women out of the Mission Control."[142] Women started to work in flight control at the tail end of Apollo, and by the end of the Apollo-Soyuz Test Project women worked in all the different flight control disciplines. A systems engineer and former Deputy Director of Johnson Space Center recalled that most of the flight controllers in the late 1970s

---

139 Mary (Tut) Hedgepeth, p. 11.

140 Anne L. Accola, interview by Rebecca Wright, March 16, 2005, p. 14, Oral History Transcript, NASA Johnson Space Center Oral History Project, JSC.

141 Anne L. Accola, pp. 16–18.

142 Jeanne L. Crews, interview by Rebecca Wright, August 6, 2007, p. 17, Oral History Transcript, NASA Johnson Space Center Oral History Project, JSC.

thought it was great to have women in Mission Control and all the other areas of the community.[143]

Women did not, however, easily break into this stronghold of machismo. Considering the time women were able to start working in Mission Control, the 1970s, it is likely that they were only able to do so because of affirmative action legislation. Before, they had not been able to break into that powerhouse of hegemonic masculinity, despite various available and skilled candidates. Only the astronaut corps took longer to include women. Therefore, it cannot be surprising that getting into the Flight Control Division was not without its barriers. Accola, for example, had to stick it out for two months in 1981 after her branch chief had selected her, and the division chief had agreed to make her, head of the Orbit/Computer Section, because Kranz did not want her to be section head. So she had to wait, yet again, until he gave in.[144] It is possible that Kranz discriminated against Accola not because she was a woman but 'only' because he did not like her as a person. Nevertheless, her case shows that men in positions of authority were able to discriminate against women even though there was no official legislation forbidding women in the workspace. Furthermore, it also demonstrates that not all men were alike and that there were men who supported women despite men like Kranz treating women in flight control as a laughing matter.[145] It has to be emphasized, though, that Kranz was not necessarily against women generally, but maybe just against women as flight controllers, because Shirley Hunt Hinson referred to him as the "[g]reatest man to work for. He inspired me. He asked me to do a job for him, I had never done that job before in my life, and I learned it in two and half days and started doing it for him. He could make anybody produce 140 percent. Everybody wanted to work for him."[146]

Female scientists and engineers were often discriminated against already during their job interviews. Harriett J. DeVries Smith, for example, recalled that during the interview process for a job as an engineering aide in the late 1950s, she had to perform an additional typing test. None of the men had to do this

143 Brock R. Stone, interview by Sandra Johnson, October 31, 2006, p. 31, Oral History Transcript, NASA Johnson Space Center Oral History Project, JSC.
144 Anne L. Accola, p. 51.
145 See anecdote of: Robert L. Carlton, interview by Kevin Rusnak, April 10, 2001, p. 4, Oral History Transcript, NASA Johnson Space Center Oral History Project, JSC.
146 Shirley Hunt Hinson, interview by Rebecca Wright and Kevin Rusnak, May 2, 2000, p. 42, Oral History Transcript, NASA Johnson Space Center Oral History Project, JSC.

even though they were all applying for the same job.[147] Annie J. Easley had a similar experience in the 1970s. Before she got her math degree from Cleveland State she was considered a sub-professional at NASA. After earning her degree, she was told that she had to take more classes in order to become a professional at NASA, which was unreasonable because new hires with a degree but no previous experience in the space community were hired as a professionals right away. Considering Easley was African American, she might have been discriminated against due to her race. Either way, she was very aware of the subtle gender and/or racial discrimination.[148] Easley's and DeVries Smith's cases demonstrate that women often had to jump through more hoops than equally qualified white men before they could start or continue their careers. This was a phenomenon not limited to NASA, but rather a general attitude women in technical disciplines had to deal with in the 1960s.[149] Minority males often had similar experiences to those of the women. Several African American male co-op students at Marshall Space Flight Center, for example, were also given exams that none of the white students had to take in order to get a job in 1963.[150]

5.6 Lab dress code

Sometimes the barriers were also of a psychological nature. Female engineers and scientists often had to hear stories about their predecessors crying. Jeanne L. Crews and Anne L. Accola were, for example, told that there had already been a woman in their positions, but she crumbled under the pressure and ended up crying in the bathroom, which inevitably led to her quitting the job.[151] The women heard stories like that several times but never personally encountered these 'crying women' themselves, and eventually Crews told her co-

147 Harriett J. DeVries Smith, interview by Rebecca Wright, November 7, 2001, p. 2, Oral History Transcript, NASA "Herstory" Oral History Project, JSC.

148 Annie J. Easley, interview by Sandra Johnson, August 21, 2001, pp. 28–29, Oral History Transcript, NASA "Herstory" Oral History Project, JSC.

149 Rossiter, *Women Scientists in America. Vol. 2. Before Affirmative Action 1940–1972*, p. 106; Julie des Jardins, *The Madame Curie Complex: The Hidden History of Women in Science* (New York: Feminist Press, 2010), pp. 117–284; Bix, 'From "Engineeresses" to "Girl Engineers" to "Good Engineers": A History of Women's U.S. Engineering Education'.

150 Paul. The men were: Walter Applewhite, Wesley Carter, George Bourda, Tommy Dubone, William Winfield, Frank C. Williams Jr., and Morgan Watson. Their employment increased the number of black professionals at Marshall by five-fold.

151 Anne L. Accola, interview by Rebecca Wright, March 16, 2005, pp. 16–18, Oral History Transcript, NASA Johnson Space Center Oral History Project, JSC.

workers to let her know if they heard the story or encountered a woman like that because she wanted to talk to them.[152] Why did Crews want to talk to the people spreading this type of story? Probably because stories like these hurt the credibility of the women who wanted to be recognized as professionals of equal status. This type of story made the full-grown engineers and scientists sound like little girls and this perception in turn hurt their authenticity as well as their power of argument. Not only were female engineers and scientists subtly compared to girls, but, just like the female secretarial and clerical staff, they were often called 'girls' by their male co-workers in the 1960s. However, looking at the Johnson Space Center oral histories, it quickly becomes evident that the women, too, at that time often referred to themselves as the 'girls.' The habit of referring to women as girls was still practiced in the 1980s. One Kennedy Space Center engineer for mechanical systems on the Shuttle, Sue Dickinson, was very aware of the subtle put down and complained that "[i]f you're doing a job that is criticality-1 hardware on the Space Shuttle, you're a woman, not a girl."[153] The lab women of the 1960s were not as aware of the language discrimination just yet, something they had in common with their female colleagues in the offices.

While the scenarios described thus far were specific to professional women in engineering and science who tried to establish themselves in the lab spaces of the space community, these women also faced the same gender-based stereotypes that confronted female office workers. However, some of these stereotypes were especially obstructive for the women in the labs. Like the secretaries, the lab women faced concerns regarding traveling. Jeanne L. Crews remembered that her male supervisors did not want her to travel all by herself as a woman because they deemed it unsafe. She also recalled that the skepticism regarding women travelling on their own was a recurring theme up until the 1990s, thirty years later, when she needed to fly to the Island of Guadalupe for a life science test.[154] She faced even greater skepticism when she needed permission to fly in a zero-G plane to test and evaluate if the astronauts were able to use a specific manual she had prepared for them.[155] Like Crews, Shirley Hunt Hinson also remembered her first experience traveling for NASA, in her case as the first woman who got to travel for the Space Task Group, and she emphasized that when she "started flying for NASA like that, I wore high heels

152 Jeanne L. Crews, interview by Rebecca Wright, August 6, 2007, p. 39, Oral History Transcript, NASA Johnson Space Center Oral History Project, JSC.

153 Sarah Oates, 'Sexism on decline in space industry', *Orlando Sentinel*, September 17, 1989.

154 Jeanne L. Crews, interview by Rebecca Wright, August 6, 2007, pp. 4 and 5, Oral History Transcript, NASA Johnson Space Center Oral History Project, JSC.

155 Jeanne L. Crews, p. 4.

and hats and gloves."[156] Her emphasis on her wardrobe reveals that the dress code issue was as important to the women in the labs as it was to those in the offices. However, because their workspaces were utterly different in their composition and infrastructure, female engineers and scientists were in need of different clothing in order to perform their work satisfactorily, despite which they were expected to wear the same type of clothes as women in the offices, which made a lot of their tasks far more difficult than necessary.[157]

Although there was no official dress code, all the women in the space community in any job were expected to wear dresses, nylons, and high heels until the early 1970s.[158] Like the secretaries, the women in the labs also banded together the first time they wore pantsuits to work because they knew it would cause a stir. They decided and agreed on wearing pants to the lab together beforehand and showed up, dressed in pants, as a group.[159] Before American women in general and the women's movement in particular began to push boundaries and shift gender norms, including matters of appropriate attire, in the late 1960s, the lab women had never really thought about challenging the unwritten dress code, even though it would have been sensible for them to do so.[160] Instead, they often just reproduced the social conventions and gender code of American society in their labs like the secretaries did in the offices. The examples of Anne L. Accola and Sheryll Goecke Powers demonstrate as much. Accola considered herself "fortunate" and was genuinely happy that she only started training in the command module simulator and the lunar module simulator in 1971, because she was able to wear pantsuits by the time she had to lay on her back in them.[161] The idea of wearing pants to work in the 1960s did not even occur to her, which is why she felt "lucky" she did not have to work in a skirt in the simulator. Powers also did not consider the possibility of wearing

156 Shirley Hunt Hinson, interview by Rebecca Wright and Kevin Rusnak, May 2, 2000, pp. 36 and 6, Oral History Transcript, NASA Johnson Space Center Oral History Project, JSC.

157 Powers, pp. i–ii.

158 Beverly Swanson Cothren, interview by Sandra Johnson and Rebecca Wright, June 15, 2001, p. 39, Oral History Transcript, NASA "Herstory" Oral History Project, JSC; Harriett J. DeVries Smith, interview by Rebecca Wright, November 7, 2001, p. 43, Oral History Transcript, NASA "Herstory" Oral History Project, JSC; Sheryll Goecke Powers, interview by Rebecca Wright and Sandra Johnson, June 13, 2001, pp. 43–44, Oral History Transcript, NASA "Herstory" Oral History Project, JSC; Anne L. Accola, interview by Rebecca Wright, March 16, 2005, pp. 14–15, Oral History Transcript, NASA Johnson Space Center Oral History Project, JSC.

159 Harriett J. DeVries Smith, p. 43.

160 Brandon Marie Miller, pp. 89–90; Holt, *Rise of the Rocket Girls: The Women Who Propelled Us, from Missiles to the Moon to Mars*, pp. 208–209.

161 Anne L. Accola, p. 17.

pants to work in the 1960s. When Powers started to work on the XB-70, she had to climb on top of it. Climbing on top of any object was not a very ladylike thing to do to begin with according to the social conventions in the United States in the late 1950s and 1960s. Climbing on top of a plane in a skirt was even more inappropriate. However, in order to do her work, Powers needed to do just that, so whenever she needed to check her experiment on the plane, she had to change into coveralls. The men in her workspace did not have concerns like that. They were able to just climb onto anything they liked because their society dictated wardrobe choices did not restrict them in their work.[162]

Ironically, the lab women's compliance to society's gendered wardrobe expectations also sometimes backfired. The Kennedy Space Center engineer Ann Montgomery, for example, was denied entrance to the launch pad by a male security guard who had the order: 'no women allowed.' According to the security guard, anybody with skirts and heels would just "hurt themselves and everything else."[163] Although Montgomery actually wore jeans and flat shoes that day, her example demonstrates that the dress code issue was used, intentionally or not, to discriminate against the women in the labs. The security guard had received the general instruction that women were not allowed on the launch pads because of their attire. The fact that Montgomery was not even wearing the imperilling clothes did not even matter because she was a woman and was therefore supposed to be wearing a dress and heels. The guards' orders were eventually changed in the mid-1960s, but the case still shows the rocky road female scientists and engineers had to walk on in the space community. A similar incident happened to Jeanne L. Crews, who was forced to wear a man's size-eight boots because of a general regulation on zero-G-flights. When she expressed her wardrobe concerns, she was immediately shut down: "If I had to jump out, they would come off." – "That doesn't matter. It's regulation boots."[164] It was accepted that Crews might have to fly without boots before allowing her to wear something that actually fit her body and did not limit her in her work performance. Her case shows that there was no real concern for female bodies in the space community in the 1960s – regulations and societal standards were more important than the women's work performance. On the other hand, Anne L. Accola, was once rebuked for complying with the informal

162 Sheryll Goecke Powers, interview by Rebecca Wright and Sandra Johnson, June 13, 2001, pp. 43–44, Oral History Transcript, NASA "Herstory" Oral History Project, JSC.

163 Ernie Reyes, interview by Dr. Aaron Gillette, July 28, 1992, pp. 25–28, Tape Transcript, HRC.

164 Jeanne L. Crews, interview by Rebecca Wright, August 6, 2007, p. 4, Oral History Transcript, NASA Johnson Space Center Oral History Project, JSC.

dress code and wearing a dress. She was told by one of the contractors: "Young lady, you'd better invest in some pants if you're going to stay around here."[165] Although she thought his remark was funny, his comment on her wardrobe choices also illustrates that it was almost impossible for the women in the labs to dress their female bodies in a way that male colleagues and supervisors did not take offense to in one way or another.

5.7 The matter of sexual harassment

Female bodies were as much a topic in the lab space as they were in the other spaces of the community. However, it was only an issue that lingered underneath the surface. Jeanne L. Crews, for instance, very clearly recalled that her male co-workers were way more interested in "my styles, my fashion, what I was wearing, than they did what I could do [. . .] and if you were cute, they really didn't want you to have to do anything. [Laughter]."[166] Even though she was convinced that the patronization of her male co-workers was meant well, it did not help her establish herself as a professional engineer. Powers also reported that women were more restricted because their co-workers and supervisors even worried about what they did after work-hours. Allegedly it was because of their security clearance, but she assumed that the men's after-hours activities were not watched the same way. Powers even remembered a case where the supervisors had a talk with a woman about her after-hours activities.[167] If she was not a Russian spy or a criminal, there was really no reason to talk to her except for the fact that most men in the 1950s and 1960s thought it was alright for them to question a female colleague's choices in her private life. Women like Crews and Powers found it patronizing, but there were also women who appreciated the men's concerns. Dorothy B. Lee, for example, liked that the men at NASA had created "a very sheltered world." She also denied that there were cases of sexual harassment in the lab space:

> Never in my entire career was I ever embarrassed by a man. [. . .] Never. And today when I read about women that – what is it they call it? Harassment? Not that I want to say that they ask for it, and I'm not a beauty like my mother, but I was never embarrassed. If they

165 Anne L. Accola, interview by Rebecca Wright, March 16, 2005, p. 17, Oral History Transcript, NASA Johnson Space Center Oral History Project, JSC.

166 Jeanne L. Crews, interview by Rebecca Wright, August 6, 2007, p. 40, Oral History Transcript, NASA Johnson Space Center Oral History Project, JSC.

167 Sheryll Goecke Powers, interview by Rebecca Wright and Sandra Johnson, June 13, 2001, pp. 43–44, Oral History Transcript, NASA "Herstory" Oral History Project, JSC.

> embarrassed me, I didn't know it. I was too naive and I brushed it off and dismissed it as if it was a joke for the day. Never would I have any occasion to have to – of course, I worked with professional men and in an environment, a very sheltered world. NASA really provided a very sheltered world. I never had a problem.[168]

Lee's concept of sexual harassment is interesting. She indicated that women have to be both willing and in possession of certain physical features, to be regarded as 'pretty,' in order to experience harassment. Misguided assumptions which intrinsically dismissed the experiences of numerous women. While she claimed that she was never harassed herself, it sounds like she was but had suppressed it. After all, if there was no harassment, what did she have to "brush off" as a "joke for the day?" Harassment can be obvious, like Donna L. Shirley's experience with her "horny old boss at McDonnell Aircraft" who chased her around, or more subtle, like Barbara Paulson's or Jeanne L. Crews' experiences.[169] Two of Crews' supervisors had tried to make her uncomfortable on purpose; Johnson Space Center Director Christopher C. Kraft was one of them. The two men discussed the astronauts' underwear in her presence for no scientific reason but with the sole intention of embarrassing her.[170] Barbara Paulson, a mathematician at the Jet Propulsion Laboratory, and her female colleagues did not go to a specific male colleague's office because of all the 'naked ladies' he had decorated his office walls with.[171] Despite the women's experiences, it must be acknowledged that there is no indication that (sexual) harassment was an everyday occurrence in the lab spaces of the space community. Furthermore, the lab women also did not report a sexual atmosphere like that the women in the offices experienced in the early days of NASA but there were flirtations from time to time.[172]

Although sexual harassment was not a daily subject for the women in the labs in the 1960s, they did have to deal with it on their own when it did occur because most employees were not even familiar with the concept and did not

168 Dorothy B. Lee, interview by Rebecca Wright, November 10, 1999, p. 47, Oral History Transcript, NASA Johnson Space Center Oral History Project, JSC.

169 Donna L. Shirley, interview by Carol Butler, July 17, 2001, p. 107, Edited Oral History Transcript, NASA Headquarters Oral History Project, JSC.

170 Jeanne L. Crews, interview by Rebecca Wright, August 6, 2007, pp. 40–41, Oral History Transcript, NASA Johnson Space Center Oral History Project, JSC.

171 Holt, *Rise of the Rocket Girls: The Women Who Propelled Us, from Missiles to the Moon to Mars*, pp. 63–64.

172 Jerri Bell, interview by Dr. Aaron Gillette, July 24, 1992, pp. 31 and 33, Tape Transcript, HRC; Emogene Greville, interview by Dr. Aaron Gillette, July 27, 1992, p. 20, Tape Transcript, HRC; Holt, *Rise of the Rocket Girls: The Women Who Propelled Us, from Missiles to the Moon to Mars*, p. 233.

quite grasp which areas the term 'sexual harassment' encompassed. Furthermore, anti-sexual harassment laws were only passed in the 1980s, during which decade Johnson Space Center started to repeatedly educate its employees about sexual harassment and held seminars on the topic for its management and supervisors. In a 1981 newsletter, the EEO office at Johnson Space Center, for example, explained what sexual harassment is and that it needed to be taken seriously:

> What is sexual harassment? Is it undesirable sexual behaviors – both overt and subtle? Is it physical or verbal? Is it a conscious act toward someone to gain power? Does it happen all the time or just occasionally to people? Is it a continuance of old sex stereotyping? All of these question plus more can be answered yes regarding the definition of sexual harassment. [...] Supervisors and managers must take the subject seriously. When an employee comes to a supervisor with a complaint, the supervisor should listen and solicit what the employee thinks is the solution to the problem and most importantly, follow through.[173]

In order for supervisors to act adequately, they had to attend seminars on sexual harassment. Anne L. Accola was actually one of the people who attended the very first seminar. Before she went, she asked her boss: "Why me? Has somebody complained about me, some of the guys that worked for me?" To which he just replied: "Well, everybody is going to have to go."[174] While she went to the seminar, a lot of her male colleagues claimed that they had scheduling conflicts and did not go; demonstrating that even in the 1980s sexual harassment was not taken seriously by a lot of male supervisors in the space community. The men that did attend the seminar further reinforced this supposition because, according to Accola, "that seminar could have been the subject of sexual harassment charges because of the behavior of one or two of the guys in it. They just made all kinds of rude and crude remarks about women and why couldn't they take it and what was the matter with saying this and that."[175] Although the seminar instructor was excellent, some of the men were just not interested in what he had to teach. The most shocking part about Accola's anecdote, however, was that the introduction of this particular sexual harassment

173 *Equal Opportunity News JSC*, March 25, 1981, Box 2, Equal Opportunity 1981, Center Files, JSC. The management often repeated their announcements about important topics such as sexual harassment. For example: Joseph D. Atkinson, *Policy Statement and Definition on Sexual Harassment*, March 2, 1983, Box GR 1018, Center Series, JSC.

174 Anne L. Accola, interview by Rebecca Wright, March 17, 2005, p. 2, Oral History Transcript, NASA Johnson Space Center Oral History Project, JSC.

175 Anne L. Accola, pp. 2–3.

seminar led to her being being singled out for her sex. She was asked a second and a third time to attend the seminar, probably because a lot of the men ducked out of going. On the third occasion, Accola, the only female supervisor in the directorate, replied: "If you call me to take this again before everybody else at the Center has taken it, I'm going to consider my own charges," and with that she was never called again.[176]

Neither the term nor the concept of 'sexual harassment' existed in the 1960s. The women of the space community, like all the other women in the United States, therefore had to figure out a way to deal with this kind of unwanted male attention and their expectations of what is proper for a woman both in general and in the lab.[177] Women actually developed several strategies to deal with gender-based treatment, which will be discussed shortly. Beforehand, however, it is necessary to look at more sources of discrimination, or rather omission, to understand women's reactions. Probably because female engineers and scientist were, percentage wise, rather rare, their needs and demands were often not considered. In a letter to her parents, Nancy G. Roman revealed one of these down-sides of being a pioneering female scientist in the man's world of science – missing out of important aspects of professional socialization. When Roman attended a meeting at the State University of Iowa she also experienced "the only drawback" of the conference: she had to stay in the women's housing a mile away from the other people – all men – attending the meeting, because she was a woman. Thus, she had no one to spend time with in the evening and, more importantly, she missed all the informal deals and discussions the men had after-hours.[178] Bertha M. Ryan faced a similar but more severe problem when she had to travel to Langley on official business. Her assigned workspace was not in the same building with her fellow engineers, but in the same space with the female computers. Although she was fond of the women, she was not able to share work discussions with them. After calling her supervisors at Edwards, her workspace issue was resolved and she was allowed to share a workspace with her male engineering colleagues. Ryan also realized that the supervisors at Langley handled several

176 Anne L. Accola, pp. 2–3.

177 Baxandall and Gordon, pp. 175–210; Susan Brownmiller, *In Our Time: Memoir of a Revolution* (New York: Dell, 1999), pp. 279–294; Margaret A. Crouch, *Thinking About Sexual Harassment: A Guide for the Perplexed* (Oxford: Oxford University Press, 2001).

178 Nancy Roman, *letter to Roman's parents*, June 27, 1962, Box 1, Folder 15, Nancy Grace Roman Addition to Papers, NBL.

situations differently from her supervisors at Edwards. After going to the wind tunnel tests a couple of times in the evenings to see if there was anything to do or prepare for her, she was told that: "No female works at night."[179] Ryan's example demonstrates that even within one agency, like NASA, the different centers and directors dealt very differently with issues regarding access and gender. Procedures and activities that were possible in one lab space were not necessarily possible in another.

Yet another issue women in the labs had to deal with was the puffed up egos of their male colleagues. Women like Jeanne L. Crews frequently had to operate as intermediaries and guinea pigs in order to appease and motivate their co-workers. For example, in the early 1970s, she had to act as an agent between astronauts and principal investigators (PI) or between astronauts who came from the Navy, Air Force, and Army, competing branches of the military, to get things done for Skylab. Crews was convinced that "having a woman in there really helped them, actually. I don't think they could have gotten it done very readily at first with a man, because the reason it worked is they didn't want to kill me. They were still trying to patronize me, and yet they still wanted to fight, but I had a little leverage there."[180] So being patronized could sometimes have positive effects on work progress, or maybe she was just an excellent mediator. Either way, she not only had to put up with colliding male egos and masculinity performances, she was thrown into the middle of the conflict. In some cases, the fact that she was a woman was also used in another form. Crews was occasionally asked to perform certain tasks just to motivate the men in her team. When they saw her doing something difficult, the men's reaction was, "Well, a girl can do it," so obviously they must also be able to do it. Although Crews thought that having been used as bait to harness the masculinity performances into something productive was funny decades after the fact, she did not appreciate being used like that at the time. Reflecting on her work relationship with her male colleagues, she came to following conclusion:

> the funny thing is with men, and no offense men, if you're reading this, but once you show you can do it, it's almost like, "Oh, well, sure." But just doing it that first time when they're – it's more like they're half protecting you, and they're half protecting themselves,

179 Bertha M. Ryan, interview by Sandra Johnson and Rebecca Wright, June 13, 2001, pp. 11–12, Oral History Transcript, NASA "Herstory" Oral History Project, JSC.
180 Jeanne L. Crews, interview by Rebecca Wright, August 6, 2007, p. 12, Oral History Transcript, NASA Johnson Space Center Oral History Project, JSC.

> because of their ego. But in the end, most of them will come around; they accept if you just show them.[181]

Thus, the rival masculinities, the gender performance of the men in the lab, ironically, sometimes gave the female scientist an opportunity to demonstrate that the actual work they were doing in the lab had nothing to do with gender but only with skill and education.

5.8 One sex – different opinions

Female scientists and engineers not only had to convince their male colleagues of their skill and work status, they frequently also had to show other women in the community, especially the secretaries, that they were scientists and engineers and needed to be treated just like their male colleagues. Nancy G. Roman, for instance, recalled that she had to use the prefix 'Dr.' with her name when she started to work at NASA because "[o]therwise, I could not get past the secretaries."[182] Most women as well as men had to learn to come to terms with the unusual, for the 1960s, situation of having female engineers and scientists, female bosses. However, some of the secretaries did not appreciate that women were pushing into the male workspace of science and engineering. Just like some men, they wanted women to stay homemakers and men to be the providers for their families. Therefore, some of the women engineers and scientists in the space community had uncomfortable confrontations with female coworkers from other workspaces. Jeanne L. Crews, for instance, was chastised by a secretary who told her: "Jeanne, did you know you're keeping a man with a family from a job?"[183] Her comment reveals that just because they both were of the same sex and both worked for NASA, they were not necessarily both on the same page regarding gender performance and values. Just like the men, the women in the different workspaces of the space community therefore cannot be lumped together; they were all individuals with different opinions and did not necessarily band together just because they were all of the same sex. Thus, more women in the workspace did not necessarily alleviate the situation of the female engineers and scientists, because members of their own sex, but of a

181 Jeanne L. Crews, pp. 12–13.

182 Nancy Roman, chat with Nancy Roman @ Goddard, *Female Frontier QuestChat*, November 4, 1997, 1821, HRC.

183 Jeanne L. Crews, interview by Rebecca Wright, August 6, 2007, p. 40, Oral History Transcript, NASA Johnson Space Center Oral History Project, JSC.

different occupation, sometimes also discriminated against them and their push into a workspace denoted as male. The Deputy Director of Marshall Space Flight Center, Carolyn Greiner, however, assumed that by the twenty-first century it was easier for women in the workspace because the number of women in the workspace have increased.[184] This is probably true because the discrepancies regarding women's place and gender performance in the workspace are no longer as great as they were before the women's movement.

In the 1960s, though, not even the pioneering women necessarily backed each other up. Nancy G. Roman, for example, was very outspoken against the pioneering women who wanted to become astronauts in the 1960s. She was convinced that women should contribute in another way to the American space effort, and furthermore believed that female astronauts would create a logistical problem at this early point in the space program because there were no "ladies' rooms" on the spaceships, no female quarters on the recovery ships and the living conditions in Hangar S at Cape Canaveral were not suitable for a woman. She believed that female astronauts would fly eventually but that including women in the space program during the race to the moon would cause the United States to lose "more ground than we'd have gained because of the duplication of effort and the complication. And the space program is already complicated enough."[185] Her reaction not only demonstrates that the women in the space community did not necessarily agree on everything, but also shows that pioneering women were just as stuck with the current gender norms as the men. The idea that female astronauts could use the same quarters or toilets as the men did not even occur to her because it was not proper for a woman to come this close to a man they were not married to.[186]

It has to be emphasized, though, that some of the female scientists and engineers were very much in favor of female astronauts, which again demonstrates that the spectrum of opinions was wide, even between the lab women. Dr. Jocelyn R. Gill, who was the Chief of In-Flight Sciences working in the Office of Space Science and Applications at NASA Headquarters, for example, was

184 Sherrie Super, 'Carolyn Griner, one of NASA's female pioneers, to retire from Marshall Center at year's end', *Marshall Star*, December 14, 2000.
185 Sue Cronk, 'Orbitable Gals: U.S. can send them', *Washington Post*, April 21, 1963; Ruth Wagner, 'Astronomer Down to Earth', *Washington Post*, July 26, 1962.
186 Women were actually not able to use the urine collection device on Gemini and Apollo because it was designed for the male anatomy. However, on the very first Mercury flight, the device had not been designed yet. If women had joined the space program during the early days, it would have been possible to design a device that fit both anatomies. Foster, *Integrating Women into the Astronaut Corps. Politics and Logistics at NASA, 1972–2004*, pp. 124–126.

convinced that women would have made excellent astronauts because she knew the women and they were of the "same breed" as the astronauts and would have "stood right along with Alan Shepard and Gordon Cooper and Scott Carpenter."[187] Her positive assessment of the pioneering female pilots who wanted to become astronauts is interesting and valuable because Gill knew all the astronauts personally due to her job of teaching and explaining the science experiments which needed to be performed on their missions to them.[188] She furthermore knew some of the women who wanted to become astronauts as "very tidy little packages" with "wonderful skill and wonderful patience."[189] Although Gill did not see an immediate future for them in space exploration, she was convinced that women would definitely join the "red-blooded American" men on "the longer voyages because the men won't be able to stand it."[190] So, interestingly enough, even an accomplished female NASA scientists considered sexuality the only option these accomplished female pilots had to join the astronaut corps. Her point of view shows on the one hand how masculine and potent she perceived the astronauts to be, and on the other hand that the lab women, who had already had to overcome several gender-based barriers during their education and daily work life, also occasionally resorted to using sex to overcome obstacles and advance women.

The way they were portrayed was yet another obstacle for women in the lab spaces. If one takes a closer look at the picture of Marjorie Townsend and her colleague next to the Small Astronomy Satellite (SAS) at Goddard, you can see a middle-aged man who is working on the satellite while literally standing on a pedestal. Meanwhile the small women in a light dress and kitten heels is looking up to the man adoringly. If one had to guess who the manager is in this picture, no one would assume that the young woman, Marjorie Townsend, was the project manager of SAS.[191] She is not doing anything in the picture. She is just a decorative element, positioned similarly to the former NASA beauty queens when they were photographed with NASA equipment.

Not only Townsend but most of the female lab workers were in a similar predicament. They were not portrayed and recognized according to their status at work because American society, including the space community, still continued to portray women like decorative accessories as it had in the past. The fact that they were at eye level with the men did not change the women's portrayal

187 Jocelyn Gill, interview, December 26, 1966, p. 13, 30/22, HRC.
188 'Woman Scientist to make Flight', *The Evening Star*, June 29, 1963.
189 Jocelyn Gill, interview, December 26, 1966, p. 13, 30/22, HRC.
190 Jocelyn Gill, interview, December 26, 1966, pp. 13–14, 30/22, HRC.
191 See Fig. 5.5.

Fig. 5.5: Marjorie Townsend and colleague, December 2, 1970, NASA.

as admiring worshippers of ostentatious displays of masculinity. Not only in pictures, but also in their daily work routine, the women were often not identified as engineers, scientists or supervisors. Mary (Tut) Hedgepeth, for example, had several experiences of people mistaking her for the secretary of the group of men she instructed.[192] So what did the women do in order to be recognized for their status and skills? How did they handle discrimination? Although they often were in similar situations, they developed very different strategies to deal with discrimination of any kind.

192 Mary (Tut) Hedgepeth, interview by Rebecca Wright and Sandra Johnson, June 12, 2001, pp. 11–12, Oral History Transcript, NASA “Herstory” Oral History Project, JSC.

5.9 A variety of coping strategies

Not all the women who started working in the labs of the space community experienced discrimination. Doris Chandler, who started working as an engineer at Marshall Space Flight Center in 1953, for example, was glad that she always "was in the right place at the right time. I didn't run into any of the discrimination that I know exists for other people."[193] The women who, in fact, did experience sex discrimination and who were not willing to put up with the sexist, condescending, and patronizing attitudes of their male, and sometimes also female, colleagues often just quit. Some women left right away after realizing what they were dealing with, while others left at mid- or high-level careers because they were no longer willing to accept the gender barriers they had to overcome. Some women, like Harriett J. DeVries Smith, made the decision to leave when they recognized that they had no future in the space community because they had reached the glass ceiling and were not able to advance further in their careers. Other women, like Jet Propulsion Laboratory computer Janez Lawson, did not even attempt to apply for certain jobs in the first place in anticipation of being overlooked.[194]

The women that did apply and stay had to develop a thick skin. Although most of the women had already developed defensive mechanisms in high school and university, they needed to develop attitudes that did not let chauvinistic behavior and discriminatory practices get them down. Annie J. Easley, for example, was very aware that she was often judged by her looks instead of her skills. This is why she always told herself that, "You may control my purse strings, but you don't control my life."[195] Easley was not the only woman who was very aware of what was happening but who chose to not do anything against it. Before she quit and "before the equal rights and all that," Harriett J. DeVries Smith also did not complain that she did not get the kind of promo-

193 'Paula Bernstein, It's no longer a Science Fiction Fantasy', *Daily News*, December 8, 1975.
194 Sheryll Goecke Powers, interview by Rebecca Wright and Sandra Johnson, June 13, 2001, p. 33, Oral History Transcript, NASA "Herstory" Oral History Project, JSC; Harriett J. DeVries Smith, interview by Rebecca Wright, November 7, 2001, pp. 15–16, Oral History Transcript, NASA "Herstory" Oral History Project, JSC; Carolyn L. Huntoon, interview by Rebecca Wright, June 5, 2002, pp. 40–41, Oral History Transcript, NASA "Herstory" Oral History Project, JSC; Betty S. Love, interview with Rebecca Wright, May 6, 2002, pp. 20–21, Oral History Transcript, NASA Johnson Space Center Oral History Project, JSC; Holt, *Rise of the Rocket Girls: The Women Who Propelled Us, from Missiles to the Moon to Mars*, p. 81.
195 Annie J. Easley, interview by Sandra Johnson, August 21, 2001, p. 15, Oral History Transcript, NASA "Herstory" Oral History Project, JSC.

tions and assignments the men did. DeVries Smith did not put up a fight, did not even consider doing so or blaming the men for their behavior toward her, because she was as much "part of that same culture" as the men were.[196]

Accepting and even helping to cultivate the sexist culture in the labs was not a strategy for every female engineer and scientist in the space community. Jeanne L. Crews, for example, pursued a much more active approach to dealing with her colleagues' and supervisors' patronizing and condescending behavior. Although she had never sworn before and thought that it was a terrible thing to do, she started to swear and quit being sweet to men because otherwise she would not "have survived at least the first ten, fifteen years."[197] She turned herself into "Jeanne Lee Crews, the – excuse me – the bitch that runs the lab, the witch."[198] According to her "[y]ou always knew you did good when you left a meeting of 200 men, and you heard, 'Bitch.' 'Yes!' [Laughs]."[199] Crews started to become more assertive in her choice of words and behavior to be on the same eye level with the men, a strategy that worked well not just for her but also for other women. Sheryll Goecke Powers confirmed her strategy when she observed that a lot of the women on "the senior management level, are definitely more aggressive than a lot of women are. It's not to say there's anything bad about it, their personalities are just much more aggressive than most of the other women are."[200]

Not only men but also quite a few women lacked understanding for this more dominant approach. Dorothy B. Lee, for instance, missed the charm in the 'tough gal' attitude. She had been taught by her family that "sugar will get you farther than vinegar any day."[201] That motto stuck with her and this was also how she dealt with the men at work, an approach Betty S. Love wholeheartedly agreed with. She

---

196 Harriett J. DeVries Smith, interview by Rebecca Wright, November 7, 2001, pp. 6–7, Oral History Transcript, NASA "Herstory" Oral History Project, JSC.

197 Jeanne L. Crews, interview by Rebecca Wright, August 6, 2007, pp. 5 and 41, Oral History Transcript, NASA Johnson Space Center Oral History Project, JSC.

198 Jeanne L. Crews, pp. 43–44.

199 Jeanne L. Crews, pp. 43–44.

200 Sheryll Goecke Powers, interview by Rebecca Wright and Sandra Johnson, June 13, 2001, p. 32, Oral History Transcript, NASA "Herstory" Oral History Project, JSC.

201 Dorothy B. Lee, interview by Rebecca Wright, November 10, 1999, p. 17, Oral History Transcript, NASA Johnson Space Center Oral History Project, JSC.

> always felt that, hey, you treat them right, they're going to treat you right. If you're going to be snippy with them and want to pull your female [attitude], what do I want to say? Well, being a gal, you're just supposed to have privileges over the male, and that didn't work. You have to be one of them, and you have to have fun with them, and you have to work hard with them.[202]

It was thus important to a lot of women like Love and Mary (Tut) Hedgepeth to integrate well, to be a part of the team, to be 'one of the guys.' Hedgepeth, for example, recalled a situation in which a man had told her that he did not think of her as a woman and that she still considered that statement to be a huge compliment.[203] To just be a plain member of the team, to be recognized for one's skill and not for one's sex, was probably the most common strategy for the lab women. They tried to blend in. If they were a supervisor, however, they were not able to just be one of the guys because they were the instructor. Some women were able to develop their own management style. For example, Dorothy B. Lee's modus operandi when she became a manager over men was to never tell anyone how to do the job but to ask specific questions which turned the men's thinking around and steered them in the desired direction, a strategy which she had learned from the Manned Space Center's first director, Robert Gilruth.[204] Other women were forced to more or less take over their predecessors' way of managing things because it was expected that she should behave just like another guy would.[205] According to the predominant gender performance of men in charge, men had to be competitive and assertive, both admired qualities of hegemonic masculinity in the space community. When women performed their jobs in the same way, something that was actually often expected of them, they were not respected for their assertiveness but were deemed 'aggressive' and 'bitchy.'

Some women like Jeanne L. Crews embraced the stamp successful, assertive women were given, but most of women did not want to be known as a 'bitch' and thus held back in various situations and let the discrimination wash over them.[206] Connie Eaton Harney, for example, decided to simply not

202 Betty S. Love, interview with Rebecca Wright, May 6 2002, pp. 20–21, Oral History Transcript, NASA Johnson Space Center Oral History Project, JSC.
203 Mary (Tut) Hedgepeth, interview by Rebecca Wright and Sandra Johnson, June 12, 2001, p. 12, Oral History Transcript, NASA "Herstory" Oral History Project, JSC.
204 Dorothy B. Lee, interview by Rebecca Wright, November 10, 1999, pp. 6–7, Oral History Transcript, NASA Johnson Space Center Oral History Project, JSC.
205 Catherine T. Osgood, interview with Rebecca Wright, November 15, 1999, p. 42, Oral History Transcript, NASA Johnson Space Center Oral History Project, JSC.
206 Jeanne L. Crews, interview by Rebecca Wright, August 6, 2007, pp. 43–44, Oral History Transcript, NASA Johnson Space Center Oral History Project, JSC.

react when a simulation engineer refused to work for her because of her sex. Incidents like that were just dismissed. According to Harney, "[i]t just wasn't prevalent as far as I was concerned. I didn't see it. I wasn't looking for it, either. I really didn't."[207] Physiologist Rita Rapp had a similar strategy when she had to deal with remarks and questions she was not comfortable with. In an interview with *Today*, she chose not to react to the condescending questions of the reporter, something which the latter even acknowledged in her article: "There are times when Miss Rapp just doesn't talk and nobody could be silent more graciously" or "[s]he smiles, and is just as reticent about herself as her work."[208]

However, remaining silent was not always an appropriate option. This is why some of the women sought male allies when they tried to get the recognition they deserved. Roger Burke, an engineer at the Jet Propulsion Laboratory, for instance, added the name of his female colleague from the computing section, Kathryn L. Thuleen, to an academic paper she contributed to. The women's contributions had always been overlooked previous to Burke's measure.[209] Harriett J. DeVries Smith, for example, received backup from Isaac Gillam, Dryden Flight Research Center Director, when Milt Thompson, the technical director of the project for which she was chief engineer, tried to force her to defer the technical responsibility for the program to someone else, even though she was officially in charge. In response, she proceeded to highlight all her responsibilities in her work contract and put it on Gillam's desk. She assumed that Gillam must have spoken to Thompson, because the issue was quickly resolved afterwards. The only clue that Thompson had tried to exclude her was the way Thompson would „sheepishly" gaze at her afterwards.[210] A lot of the women, however, tried to avoid head on confrontations and attempted to work around the men if it was not possible to work with them.[211]

Some of the strategies for being on eye level with the men at work were sometimes quite literal according to Bonnie P. Dalton. She always wore high heels to work to literally be on the same eye level as the men. She found that the

207 Connie Eaton Harney, interview with Rebecca Wright and Sandra Johnson, June 12, 2001, p. 7, Oral History Transcript, NASA "Herstory" Oral History Project, JSC.
208 Juli Daves, 'She feeds Astronauts', *Today*, October 10, 1968, p. 1E-2E.
209 Holt, *Rise of the Rocket Girls: The Women Who Propelled Us, from Missiles to the Moon to Mars*, pp. 203–204.
210 Harriett J. DeVries Smith, interview by Rebecca Wright, November 7, 2001, p. 13, Oral History Transcript, NASA "Herstory" Oral History Project, JSC.
211 Annie J. Easley, interview by Sandra Johnson, August 21, 2001, pp. 30–31, Oral History Transcript, NASA "Herstory" Oral History Project, JSC.

men exhibited a very different attitude when she was wearing high heels because she was able to look them straight in the face, instead of having to look up.[212]

5.10 The experience of women of color

The women in the labs had to overcome several barriers due to their gender, and also developed various strategies to deal with their colleagues' or their supervisors' discriminatory, chauvinistic, patronizing and condescending behaviour. There were only a few women of color working in the labs but they, too, had to find ways of handling not only sexism but also racism. Because there are almost no statistics that have tracked the participation of minority women in the labs in the early 1950s and 1960s, it is difficult to nowadays reconstruct the number of women of a particular ethnic background who worked at NASA and in the space industry. Margot Lee Shetterly, for instance, estimates that there were around one thousand human computers working at NASA over the years.[213] They were probably the most prominent women of color in the space community.

These African American mathematicians and scientists were all well-educated and started their careers similarly to the white women. Despite prevailing stereotypes, they usually grew up in households that had stressed education and schooling. Annie J. Easley especially emphasized that this had been the case for black girls in the Southern states. She herself and her friends were always encouraged by their relatives to educate themselves even though state legislation in the South did not require African American girls to be educated past eighth grade.[214] Like their white colleagues, most of them had not intentionally set out to be space pioneers; they had just wanted a job in the field of science or math.[215] Like their white colleagues, or even more so, they were also very aware of their treatment in the labs of the space community. However, unlike their white colleagues it was not always quite clear why they were discriminated against. Did they face problems in their workspace because of their gender, their color, both or because of individual reasons? Considering that Jim

212 Bonnie P. Dalton, interview by Rebecca Wright, April 22, 2003, pp. 30–31, Oral History Transcript, NASA "Herstory" Oral History Project, JSC.

213 Shetterly, pp. xvi–xvii.

214 Annie J. Easley, interview by Sandra Johnson, August 21, 2001, p. 43, Oral History Transcript, NASA "Herstory" Oral History Project, JSC; Hodges.

215 Annie J. Easley, p. 14.

Crow did not stop at the gates of NACA/NASA, most of the African American women's grievances were due to their race.

While Janez Lawson and Annie J. Easley were able to work in integrated teams of computers at the Jet Propulsion Laboratory in California and Lewis Research Center in Ohio from the start in the 1950s, the African American computers at Langley in Virginia had to do the same job as the white computers, but in a different room.[216] Kathryn Peddrew, Dorothy Vaughan, Miriam Mann, and Mary Jackson, for example, started working in the new West Section at Langley, which was a segregated lab space for computers of color. They not only had to work in a segregated space but they also had segregated dining and restroom facilities as well. In small acts of defiance Mann sometimes removed the humiliating 'colored' signs on their dining table in the cafeteria and Johnson used all bathrooms, not only the ones marked 'colored.'[217]

The West Section was a workspace a lot of the white computers were not even aware of, even though there were limited and brief periods of integration in special cases. Although the African American and white computers had to work together when the workload of one section became too overwhelming and they were not able to process it in time, most of the time the segregated West Section had to perform the less interesting and more tedious assignments, while the white women were able to pick the more challenging jobs.[218] In the early and mid-1950s, there were apparently no female computer supervisors of color, even in the segregated West Section at Langley. The women also reported that there were no African American male engineers in the early years at Langley. Langley was eventually integrated in the late 1950s and its African American employees gained the opportunity to become supervisors. Vaughan, for example, became head of a group of computers in 1958 and Jackson, who started at Langley in 1951, became the first African American computer to be integrated in a mixed-race team when she was sent to the 4 ft. tunnel.[219] Katherine Johnson was able to integrate best because of her extraordinary mind. Due to her inquisitive nature, she was not content with just doing the calculations for the engineers; she wanted to know more about the context of the problems she worked on. This curiosity led her to start asking questions, a habit which was not at all welcome in the beginning because the female computers, black or white, were not supposed to care about anything other than their specific assignments. Because of this mentality, the computers

216 Annie J. Easley, p. 14; Holt, *Rise of the Rocket Girls: The Women Who Propelled Us, from Missiles to the Moon to Mars*, pp. 81–82.

217 Golemba, pp. 42–43; Shetterly, pp. 39–48, 108–109, 129.

218 Golemba, pp. 43–44.

219 Golemba, pp. 43–44; Shetterly, pp. 114, 172.

were not invited to participate in briefings or meetings. When Johnson found out that there was no rule against it, she started to attend the meetings anyway. Because of her pointed and valuable questions in the meetings, she began to work more and more with the Space Task Force and ultimately became the very first woman, black or white, to ever be pulled from the computer section to work together with the engineers on trajectory calculations for the very first space flights.[220] Johnson was in good company, other women of color in the space community also managed to get recognized for their talent, skill, and work ethic. JPL's Janez Lawson, for example, was one of only two computers who was picked to attend the IBM training school. Therewith, she became one of the very first computer programmers at the Jet Propulsion Laboratory.[221]

Although the integration of African Americans into white workspaces had been effected according to the guidelines, their social integration did not proceed with the same speed. Jackson, for example, remembered that African Americans still had to know "which restrooms were for which race" even though segregation of the races had officially been terminated.[222] Annie J. Easley was also very aware of racial issues. She recalled one instance in which she was cut out of an official picture for a display as part of an open house. She had been one of six people in the picture, but was the only one to not make into the display. She was very embarrassed about it and went to her room supervisor, who was nice to her but did not do anything about the situation. She just told her: "Oh, I don't blame you. I'd be upset, too."[223]

That was just one of many occasions on which Easley was discriminated against because of her race. In another instance, she did not receive reimbursement for work-related math courses at a local community college although the costs had been covered by NASA for other employees. Her supervisor, Bert Henry, told her that NASA only covered the costs of training for professionals. He did not listen to her when she explained to him that it was a junior college and that there had been a flyer regarding those courses from management. Easley went anyway and paid for the courses herself. When an engineer inquired of Henry why Easley had not received financial help for her work-related courses, he just replied: "Oh, well, I didn't think you did that." Easley had also asked for paid leave to go to university and finish a degree, just like other

220 'Katherine Johnson: A Lifetime of STEM'; Shetterly, pp. 178–182.

221 Holt, *Rise of the Rocket Girls: The Women Who Propelled Us, from Missiles to the Moon to Mars*, pp. 89–91.

222 Golemba, pp. 43–44.

223 Annie J. Easley, interview by Sandra Johnson, August 21, 2001, p. 15, Oral History Transcript, NASA "Herstory" Oral History Project, JSC.

employees, but she ended up paying for herself, yet again. It is not quite clear why Easley did not receive funding for work-related education and other employees, including women, did receive NASA's support.[224] There are several possible reasons. Maybe she was actually not eligible for funding or her supervisor was just generally ignorant and unaware of advanced training opportunities for his employees. Apparently, her supervisor was very apologetic afterwards, claiming that he just did not know, but he also never offered to pay for the course retrospectively. The most likely remaining explanation is that her supervisor did not try to help her because she was a woman or because she was African American. Considering that other women, like Connie Eaton Harney, were allowed to go to university courtesy of NASA, it is very likely that Easley was discriminated against due to her race.[225] On the other hand, Harney and Easley had different backgrounds, worked at different NASA centers, and asked for funding in different years, so it is also possible that it was just pure capriciousness on the part of her supervisor. Like sexism, it is not easy to always uncover racist attitudes because they were usually masked with other explanations.

Racial discrimination was only on the surface at the very beginning of NACA, during the official existence of segregation. Kathryne Peddrew, who had received a degree in chemistry at Storer College, was, for example, not allowed to work as a chemist at NACA when she started at Langley in 1943. Although she was hired for a position in the chemistry division, she was transferred to the segregated West Section when her supervisors realized that she was African American. Despite her education, she had to work as a computer because there was no segregated chemistry laboratory, and the existing laboratory did not want to integrate African Americans.[226] Janez Lawson, a chemical engineer trained at the University of California, Los Angeles, also started her career as a computer at the Jet Propulsion Laboratory in Pasadena in 1952 because there were no engineering positions open to someone of her race and sex.[227]

224 Annie J. Easley, pp. 30–31.

225 Connie Eaton Harney, interview with Rebecca Wright and Sandra Johnson, June 12, 2001, p. 3, Oral History Transcript, NASA "Herstory" Oral History Project, JSC.

226 Golemba, p. 18.

227 Nathalia Holt, 'How Sexism Held Back Space Exploration. America's Early Aerospace Engineers Ignored Computers Because They Considered Programming to Be Women's Work', *The Atlantic*, 2016 <http://www.theatlantic.com/science/archive/2016/06/how-sexism-held-back-space-exploration/486644/> [accessed 20 October 2016].

By the 1960s there were no longer any segregated sections at NASA. However, that did not mean that there was no racial discrimination anymore. Just like Annie J. Easley's inability to get her training paid for demonstrated, discrimination was still prevalent but hidden beneath the surface. Yvonne Clark, an African American associate professor for mechanical engineering at Tennessee A&I State University and a NASA contractor for the Saturn 5, also remembered that "[s]ome of the men resent it at first, having a woman around," however, "it usually wears off."[228] As one of the fewer than 8,000 female mechanical engineers in the United States, she was used to discrimination and thus became relatively indifferent toward the attitudes of her white, male colleagues.[229] What is interesting, though, is that she assumed that her colleagues resented her because she was a woman, not because she was African American, or because she was a woman of color. Female engineers were rare in the early 1960s, and African American women in engineering were even more unusual, but Clark nevertheless identified the source of discrimination as gender-based and not race-based. Apparently, her status as a minority did not define her in her workspace. Easley also recalled not feeling like a minority in her lab, although she was still very aware of the possibility of people perceiving her differently.[230] Katherine Johnson remembered that she did not even feel the segregation at NASA because everyone was so determined and worked continuously on the space mission for their country.[231] However, Johnson's colleagues from the segregated computer pool in the West Section did not quite feel the same way. They have actually expressed some bitterness about the segregation, although they were not upset with their immediate colleagues because of it.[232]

Like their white colleagues but even more so, these women did not have a choice and they had nowhere to turn to when they did have a complaint regarding race or gender; therefore, they had to develop an indifference regarding race- and gender-based issues.[233] Dorothy Vaughan, for example, remembered:

228 'Tennessee State's Lady Engineer. Professor Splits Time between Teaching, Space Project Work', p. 75.

229 'Tennessee State's Lady Engineer. Professor Splits Time between Teaching, Space Project Work', pp. 75, 76.

230 Annie J. Easley, interview by Sandra Johnson, August 21, 2001, p. 14, Oral History Transcript, NASA "Herstory" Oral History Project, JSC.

231 Yvette Smith, 'Katherine Johnson: The Girl Who Loved to Count', *NASA*, 2015 <http://www.nasa.gov/feature/katherine-johnson-the-girl-who-loved-to-count> [accessed 13 September 2016].

232 Golemba, pp. 43–44.

233 Annie J. Easley, interview by Sandra Johnson, August 21, 2001, p. 16, Oral History Transcript, NASA "Herstory" Oral History Project, JSC.

"I changed what I could and what I couldn't, I endured."[234] Annie J. Easley had a similar attitude. She continued her career despite various obstacles, biases, and prejudices without ignoring the issues at hand:

> My head is not in the sand. But my thing is, if I can't work with you, I will work around you. I was not about to be so discouraged that I'd walk away. That may be a solution for some people, but it's not mine. So yes, I'm sure, I, like many others, have been judged not on what I can do, but on what I look like.[235]

5.11 Managing cultural expectations

African American and other minority women faced the same problems their white counterparts faced, and in addition they were also discriminated against because of their race. However, it was often not quite clear if their discrimination was rooted in gender- or race-biases. When it came to combining family and professional work, white and black women more or less faced the same difficulties and had similar concerns. First and foremost, they all had to deal with the prevalent expectation that women were supposed to stay at home and take care of the children or look after house and husband if there were no children yet. The unmarried women often felt the pressure to find a husband and the married women had to defend their decision to keep working despite their marital status. Mary (Tut) Hedgepeth remembered that she was frequently asked if her husband did not mind that she went to work because a lot of spouses would not have appreciated a working wife. Especially women who did not pursue a career outside the home often looked down upon those women who did work, and Hedgepeth was never quite sure how to respond to these condescending questions – at least until a co-worker told her how she always replied to such queries. According to her colleague one needed to respond with "Well, he would kill me if I tried to stop" and sure enough these women were at a loss of words.[236] Even women who worked in the space community in the 1980s, such as Jane Campbell, a propellant and life support design engineer at EG&G Florida Inc., a contractor of Kennedy Space Center, recalled that "[p]eople tell me right to my face that I should be home having babies."[237]

---

234 Golemba, pp. 43–44.

235 Annie J. Easley, p. 15.

236 Mary (Tut) Hedgepeth, interview by Rebecca Wright and Sandra Johnson, June 12, 2001, pp. 14, 22–23, Oral History Transcript, NASA "Herstory" Oral History Project, JSC; Holt, *Rise of the Rocket Girls: The Women Who Propelled Us, from Missiles to the Moon to Mars*, pp. 70, 88.

237 Sarah Oates, 'Sexism on decline in space industry', *Orlando Sentinal*, September 17, 1989.

The expectation that women had to become mothers, and thus stay-at-home moms eventually was so prevalent in the 1950s and 1960s that space community upper management as well as the women themselves struggled with the implications. Before the birth control pill was approved as a contraceptive in 1960 and slowly started to become popular because it gave women the option to plan their pregnancies, women could become pregnant all of a sudden. Some lab women wanted to continue to work after giving birth. However, there were no regulations, much less maternity leave, in place in the space community as well as the rest of the country. After Barbara Paulson was fired for being pregnant, her female colleagues at JPL had to find ways to keep their jobs after becoming pregnant. While Janet Davis tried to hide her pregnancy as long as possible, Helen Ling saved up her vacation and sick days to deliver the baby and come right back to work.[238] Most of the lab workers, though, did not consider going back to work after the birth of their children right away. Marshall Space Flight Center engineer Doris Chandler, for instance, wanted to quit after the birth of her first child in the late 1950s, but her husband actually encouraged her to continue working because he was convinced that she "would go nuts" if she stayed at home.[239] JPL's computer Sue Finley, in fact, did develop mental health issues because she stayed at home with her children, missing the challenges at work.[240] Beverly Swanson Cothren, on the other hand, remembered that "[i]t never entered my mind not to stay home with my children [in the late 50s], and I don't know where I would have found a babysitter in the desert in the first place."[241] While the men never had to worry about childcare, because finding a babysitter was not in accordance with

238 Holt, *Rise of the Rocket Girls: The Women Who Propelled Us, from Missiles to the Moon to Mars*, pp. 163, 170, 183, 187; Lamphier and Welch, p. 211; Elizabeth Siegel Watkins, *On the Pill: A Social History of Oral Contraceptives, 1950–1970* (Baltimore: Johns Hopkins University Press, 1998); Nancy MacLean, *Freedom Is Not Enough. The Opening of the American Workplace* (Cambridge: Harvard University Press, 2008), pp. 135–136. It was only in 1978 that the Pregnancy Discrimination Act granted protection to pregnant women in the workforce. Martha May, *Women's Roles in Twentieth-Century America* (Westport: Greenwood Press, 2009), p. 25.

239 Paula Bernstein, 'It's no longer a Science Fiction Fantasy', *Daily News*, December 8, 1975.

240 Holt, *Rise of the Rocket Girls: The Women Who Propelled Us, from Missiles to the Moon to Mars*, pp. 216–217.

241 Beverly Swanson Cothren, interview by Sandra Johnson and Rebecca Wright, June 15, 2001, p. 11, Oral History Transcript, NASA "Herstory" Oral History Project, JSC.

their gender role, children were the sole problem of the space employee who was also a mother.[242]

While Soviet women could rely on their country's public daycare centers, not only Cothren but all the women in the American space community from the secretary to the engineer needed to find care for their children in the relatively rare case of them going back to work, and that was not always an easy thing to accomplish.[243] Hedgepeth and Dorothy B. Lee were lucky enough to find full-time help when they went back to work; a nanny was an absolute necessity because there were no childcare facilities at their NASA centers.[244] No one even considered childcare facilities at NASA before the 1970s despite the public call for young women to join the space community due to the Cold War threat of the Soviet Union surpassing the United States in space. Childcare facilities came to NASA as late as May 1990, when Johnson Space Center opened a daycare center for 20 infants, twelve toddlers and 30 preschoolers.[245] However, there were other NASA centers which offered childcare for its employees as early as June 1973. Goddard's daycare was staffed full-time and looked after kids aged 32 months to six years.[246] Mothers who wanted to work at NASA in the 1950s and 1960s, though, were not able to benefit from the agency's childcare facilities just yet because NASA only started to consider childcare with the start of EEO in the early 1970s.[247] NASA's non-existent childcare facilities thus indicate that the Cold War induced, public invitation for women to join the space effort as engineers or scientists in order to match the Soviet Union's use of brainpower, was not followed by deeds which would have helped the women to actually do so. This is actually quite interesting, because space community officials, such as Catherine Dryden

---

242 Paula Bernstein, 'They yearn for out of this World Discoveries', *Daily News*, December 9, 1975; Holt, *Rise of the Rocket Girls: The Women Who Propelled Us, from Missiles to the Moon to Mars*, pp. 172–173, 187. See also: McIlwee and Robinson, pp. 161–165.

243 Irene M. Piatek, interview by Sandra Johnson, August 12, 2015, p. 65, Oral History Transcript, NASA Johnson Space Center Oral History Project, JSC.

244 Mary (Tut) Hedgepeth, p. 22; Dorothy B. Lee, interview by Rebecca Wright, November 10, 1999, p. 5, Oral History Transcript, NASA Johnson Space Center Oral History Project, JSC.

245 'Grand Opening. JSC Child Care Center opens for Business, Play and Learning', *Roundup*, May 4, 1990; Aaron Cohen, *JSC provides Building and Utilities, the Day Care is operated by a Non-Profit Cooperation*, May 19, 1989, Box GR 1019A, Center Series, JSC.

246 Ruth Bates Harris, Memorandum to George Low, *GSFC's Proposal for Day Care Center*, January 11, 1973, 008987, HRC; *Status of NASA Activities on the Employment of Women*, 1973, 008994, HRC.

247 Ruth Bates Harris, *letter to Millicent Allewelt*, June 16, 1971, 18508, Box 1, HRC; MacLean, pp. 136–137.

Hock of NASA's Office of Manned Space Flight, especially emphasized in public speeches that marriage and a career in science could successfully be combined. However, while they stated that it could and should be done, they gave no indication of how to do it.[248]

Even in the early 1970s, when some NASA centers and space companies already provided daycare, it was complicated to combine motherhood with a lab space career. According to Goddard's Joanne Simpson, "trying to combine top level achievement with marriage and motherhood" was "close to prohibitive" because the "[g]reatest productive years in science" usually coincided with a woman's "childbearing years."[249] The women were, thus, not able to keep up with their male counterparts because they had to take breaks from work to give birth and possibly care for their children. Hedgepeth and Marjorie Townsend, on the other hand, had a different outlook on combining motherhood, marriage, and career. Townsend wanted young women to know that "you can have it all. It is possible to give enough time to each aspect of one's life. My message is not to turn young women off all the things they want to do. I don't think anyone has ever been without problems. But it's what you do with the problems that matters."[250] Hedgepeth and Townsend were able to 'have it all' because they had full-time childcare on the one hand, and on the other, they had proud and very supportive husbands.[251] Townsend and her husband, for example, did not have classical roles in their household. They divided the chores however they saw fit and their four sons also had to help out at home, including performing chores considered 'female.'[252]

Other women who worked in the labs had different arrangements with their spouses. Shirley Hunt Hinson's husband, who also worked at NASA "was taken care of in the old-fashioned way" according to Hinson.[253] Although they

248 Catherine Dryden Hock, *Space, the next Goal*, February 24, 1965, p. 7, 008998, HRC; Robert C. Truax in: Dee Dickson, 'Girl Students urged to be Space Pioneers', *Los Angeles Herald-Examiner*, February 23, 1962.

249 Jack Williams, 'The Woman in the Eye of the Storms', *USA Today*, November 3, 1998, pp. 15–17.

250 'One Giant Step for Womankind', interview with Marjorie Townsend, *Working Woman*, May 1980, pp. 53–54.

251 Mary (Tut) Hedgepeth, interview by Rebecca Wright and Sandra Johnson, June 12, 2001, p. 25, Oral History Transcript, NASA "Herstory" Oral History Project, JSC.

252 'One Giant Step for Womankind', interview with Marjorie Townsend, *Working Woman*, May 1980, pp. 53–54.

253 Shirley Hunt Hinson and J. Kirby Hinson, interview by Rebecca Wright and Kevin Rusnak, May 2, 2000, pp. 26–27, 2. Oral History Transcript, NASA Johnson Space Center Oral History Project, JSC.

both had a full-time job, she took care of the household all by herself in the evening and on the weekends. Like the astronauts, he did not even have to mow the lawn, a job which is usually considered masculine in U.S. society.[254] Her husband admitted that he "never would have thought of cooking. I didn't think very much of taking her out to eat so that she didn't have to cook."[255] Shirley Hunt Hinson was by far not the only woman who served double duty, in the lab space and at home, because most of the women themselves had traditional gendered role allocations in their minds.[256] Emily Good, a physicist with the Data Reduction Section of New Mexico State University's Physical Science Laboratory, which fed data into NASA's Goddard Space Flight Center, also did not expect her husband to help her with the housework. She did, however, expect him to be responsible for the financial security of the family by himself. Even though housework and financial responsibility could have been equally distributed between wife and husband, by mutual consent the finances were her husband's job while she had to take care of the household.[257] Connie Eaton Harney also realized that she had a traditional, gendered conception of husband and wife when she started to earn more money than her husband. Shortly before her promotion which offered her more money, she told her colleague Mary Little: "Well, I don't ever want to make more money than my husband. That might cause a problem."[258] Although she quickly dismissed this notion, and considered it "kind of crazy" and "funny times" in hindsight, she was forced to confront her own ideas about the gender roles of men and women.[259]

Women earning more money than their husbands, women who chose their careers instead of children or who struggled with 'doing it all,' often also struggled in their relationships with partners, especially when the spouse had a traditional conception of family. Keeping the balance between family and work as well as coming to a mutual understanding of the gendered concepts of the roles of husband and wife was not easy, and some couples were unable to come to an agreement and had to separate because of it. Bonnie P. Dalton

---

254 Shirley Hunt Hinson and J. Kirby Hinson, pp. 26–28.

255 Shirley Hunt Hinson and J. Kirby Hinson, pp. 26–27.

256 Holt, *Rise of the Rocket Girls: The Women Who Propelled Us, from Missiles to the Moon to Mars*, p. 210. See also: McIlwee and Robinson, pp. 158–161.

257 Lillian Levy, 'Good is the Name! Tracking Satellites is the Game!', *NASA News*, February 1975, 008994, HRC.

258 Connie Eaton Harney, interview with Rebecca Wright and Sandra Johnson, June 12, 2001, p. 4, Oral History Transcript, NASA "Herstory" Oral History Project, JSC.

259 Connie Eaton Harney, p. 4.

was separated from her husband for fifteen years [1978 to 1993], because their ideas about marriage, family and career completely clashed. While Dalton wanted to advance in her career, her husband embraced the traditional concept of a wife and they had to separate because of it.[260] However, it is quite interesting to note that his concept of a wife was not so traditional after all when her job as a scientist at Ames Research Center put him through college.[261]

Dalton was not the only female scientists or engineer who financially supported her husband and family. Mary (Tut) Hedgepeth's wage, for instance, helped to pay for the family's house in Lancaster and Marjorie Townsend supported her family with her income as an engineer because her husband was not able to earn enough money to support a family at the beginning of his medical career.[262] According to her, it was all just a "big juggling act. Keeping a husband thinking he is the most important thing in your life, raising four kids and doing the best you can on the job."[263] The Hinsons confirmed that working for NASA put a lot of pressure on a lot of relationships, but they also agreed that it only happened if one allowed it to. According to them, "[i]t was a personality thing more than anything else."[264] Considering that the Hinsons both worked in professional capacities at NASA and dated for eight years before they got married, even though they never dated at work, their case might be special because they both had the same employer and were aware of the kind of expectations the community put on its workers.[265]

One of these expectations that a spouse might not understand was that lab workers were expected to move to another location if necessary or to advance one's career. Women were expected to follow their husbands when they had to move to another city, perhaps in another part of the country, for work. However, husbands in the 1960s were not expected to do the same for their wives. Shirley Hunt Hinson, for example, turned down the opportunity to do a Ph.D.

---

260 Bonnie P. Dalton, interview by Rebecca Wright, April 22, 2003, pp. 29–30, Oral History Transcript, NASA "Herstory" Oral History Project, JSC.

261 Bonnie P. Dalton, pp. 29–30.

262 Mary (Tut) Hedgepeth, interview by Rebecca Wright and Sandra Johnson, June 12, 2001, pp. 7–8, Oral History Transcript, NASA "Herstory" Oral History Project, JSC.

263 'One Giant Step for Womankind', interview with Marjorie Townsend, *Working Woman*, May 1980, pp. 53–54.

264 Shirley Hunt Hinson and J. Kirby Hinson, interview by Rebecca Wright and Kevin Rusnak, May 2, 2000, p. 25, 2. Oral History Transcript, NASA Johnson Space Center Oral History Project, JSC. See also: Holt, *Rise of the Rocket Girls: The Women Who Propelled Us, from Missiles to the Moon to Mars*, pp. 114, 134, 216, 247.

265 Shirley Hunt Hinson and J. Kirby Hinson, pp. 2 and 4.

in Colorado in order to stay with her husband in Houston and Katherine Johnson did not relocate to Houston because of her husband.[266] They were far from being the only female career scientists who declined an opportunity or who put their husband's career and needs above their own. Hedgepeth, for instance, transferred from Langley to MUROC (Edwards) because her husband got a job there in the late 1940s, and Harriett J. DeVries Smith had a deal with her husband John, who also worked for NASA, that they would only move to Washington, D.C. on a permanent basis if he found a position at headquarters he liked as much as she liked her new job in Congress, otherwise they would have both gone back because he had a good job back at Dryden.[267]

Considering that the Smith's deal was in the 1980s, it becomes obvious that the conception of 'traditional' gendered roles for husband and wife, in their case who follows who for which job, persisted even after the second women's movement. The space community employers often operated along the same lines. NASA's Johnson Space Center, for example, offered various activities for the families of its employees, such as monthly family friendly, semi-technical symposia. The goal of the event was to include the families of the employees in order for them to better understand what their spouses were working on and to establish a better rapport between the different divisions. The invitation, however, did not call on all spouses, but only targeted the 'wives,' demonstrating that the female employees and their partners were not on the radar of the agency because of prevalent gender roles in the 1960s.[268]

The lab women and their families faced challenging times in the space community in the 1960s. The women had to deal with contradictory and sometimes confusing gender expectations and gender-based discrimination, and their partners and co-workers had to find a way of accepting and integrating women into the field of science and engineering. One way of showing support was to honor women with prestigious awards such as the Federal Women's Award. Eleanor Pressly, Nancy Roman, Jocelyn Gill, and Marjorie Townsend,

---

266 Shirley Hunt Hinson and J. Kirby Hinson, p. 28; Holt, *Rise of the Rocket Girls: The Women Who Propelled Us, from Missiles to the Moon to Mars*, p. 210.

267 Harriett J. DeVries Smith, interview by Rebecca Wright, November 7, 2001, pp. 16–17, Oral History Transcript, NASA "Herstory" Oral History Project, JSC; Mary (Tut) Hedgepeth, interview by Rebecca Wright and Sandra Johnson, June 12, 2001, p. 6, Oral History Transcript, NASA "Herstory" Oral History Project, JSC. Also see the case of Kathryn Thuleen: Holt, *Rise of the Rocket Girls: The Women Who Propelled Us, from Missiles to the Moon to Mars*, pp. 243–244.

268 Robert R. Gilruth, *MSC Family Night Symposium*, March 23, 1966, GR 1013, Center Series, JSC.

for example, all received the Federal Women's Award in 1962, 1963, 1966, and 1973, respectively.[269] Although Jocelyn Gill was happy and proud to receive the award, she was also convinced that only "when there no longer is a need for the award. Then we'll know women have made it."[270] Her statement shows that she was very much aware of the fact that women still needed their own award instead of being integrated into the same categories as men. However, the award not only revealed that gender equality had not been reached just yet but also demonstrated that there was at least already appreciation for women's contribution to science and engineering in the space community as early as the 1960s. What is interesting, though, is that there was no specific award targeted toward female scientists and engineers at NASA; these women were eligible for the same awards as their male colleagues. Marjorie Townsend, for example, received the NASA Exceptional Service Medal in October 1971.[271] This is a significant difference from their female co-workers in the NASA offices who were not eligible and were only able to receive awards in their own categories, such as the Outstanding Secretary Award.[272]

The women in the lab and office spaces had a lot of things in common. Like the secretaries, the female scientist and engineers also reported that the gendered work culture had changed significantly since the early 1970s, since Apollo, and since some of the old, chauvinistic managers and supervisors retired or passed away and were replaced with a younger management which tried to encourage women and minorities.[273] Before Apollo, NASA and its employees were "in a race to space. They did work long hours. There was a lot of divorce. There was a lot of domestic issues outside of work because of the incredible number of hours."[274] The agency culture was able to change after Apollo in the early 1970s because, on the one hand, the space race was *de facto* over due to the successful American moon landing and the agency's new capacity to rearrange its priorities, and, on the other hand, the culture and attitudes had to change because of the prevalent social consciousness in the 1970s and the federal legislation that actually began to tip the scale in favor of women.

269 *Three Women honored for achievements in Space Age*, news release, May 7, 1963, 1821, HRC; Ruth Dean, 'Watching Skies with Memories', *Washington Star-News*, August 23, 1973.

270 Ruth Dean, 'Watching Skies with Memories', *Washington Star-News*, August 23, 1973.

271 Ruth Dean, 'Woman heads Space Probe Launches', *Evening Star*, January 12, 1972.

272 'Outstanding Secretary Award to be Given', *Roundup*, August 3, 1973.

273 Harriett J. DeVries Smith, interview by Rebecca Wright, November 7, 2001, pp. 10 and 12, Oral History Transcript, NASA "Herstory" Oral History Project, JSC.

274 Irene M. Piatek, interview by Sandra Johnson, August 12, 2015, p. 65, Oral History Transcript, NASA Johnson Space Center Oral History Project, JSC.

5.12 Lab space and the space community

Women were welcomed into the laboratories of the space community during the height of the space race to match the Soviet Union's use of female scientists and engineers. However, in its nationalism and romantic period in the 1960s and early 1970s, NASA and other companies in the space community had no special programs to bring women in, or to change their employees' attitudes regarding gender in general and working women in particular. The women who actually did overcome gender-based hurdles and discrimination during high school and university and who became engineers and scientists in the space community labs in the 1950s and 1960s also never mentioned Cold War policies that might have helped them open doors into the labs. It seems like the Cold War-induced openness to women in the lab spaces of the United States, including the space community, was mainly just lip service in the media and public space. NASA, this new agency in an up and coming scientific field, did not use its novelty status to transfer its innovative disposition regarding science and technology to its composition of employees. This greenhorn agency without a corporate memory of its own did not include new types of employees, including women and minority scientists and engineers, as a distinctive corporate strategy to tackle the pressing need for well-educated employees. It did not comprehensively challenge the prevalent cultural and social norms and values in American society, including gender performance. Therefore, the women who did enter the space community workspaces had to stand their ground in the labs by themselves and they also had to deal with gender-based discrimination by themselves; there was no one to complain to. Thus, the women in the space community developed various individual strategies to access the space community and deal with sexism and racism at their workspace. A lot of women, for example, opted to work in the labs in scientific disciplines that were regarded as less macho, such as mathematics.

As there were no written rules which discriminated against women based on their gender, it was furthermore difficult to detect discrimination in the first place. It was only with the implementation of EEO and affirmative action in NASA's pragmatic period that women's issues and grievances in the office and lab spaces were truly considered in the space community. This shows that the lab spaces, just like the office spaces, in the various NASA centers and space companies were social spaces which were mainly reproducing the prevalent concepts of gender performance and social conventions that both male and female employees were familiar with from other spheres of their lives, despite the needs and urgency of the Cold War that might have suggested otherwise.

6 Work space in the 1960s and 1970s

Office and lab spaces at NASA were preassigned social spaces, perpetuated and reproduced on a daily basis. Men and women within these spaces performed their gender and portrayed administrative policies and social conventions in their performance. The Cold War factored into the performance and thus opened up possibilities for the women in these social spaces within the space community according to the prevalent lip service of government officials. However, the social structures of the lab and office spaces were set and rigid due to societal standards. Even though these social spaces were also subject to manipulation and change, transformations usually occurred neither frequently nor expeditiously. Thus, it was often a tedious challenge to facilitate change in the NASA offices and labs in order to provide more possibilities for women to participate in the well-established social spaces of the space community. Considering that NASA was a fairly new organization in the 1960s, it is reasonable to assume that there were also several workspaces that were not as defined and predetermined as the mainstream workspaces which also existed in older communities that were not as new and innovative as the space community. As an experimental agency, NASA had to create hard- and software that had not existed before. Accordingly, NASA also had to create jobs that had not existed prior to itself. It is now the goal for this section to look into workspaces that were novel or not that well-established in order to find out if the gender relations and performances there varied from those in the more traditional workspaces in the space community, and if women were able to exercise more agency or were given more opportunities in them. Deviations in the sociocultural coding of masculinity and femininity in the more or less nouveau workspaces of the space community can show us another aspect of American society during the Cold War, especially if the free workspace equaled free mind space. Thus, this section intents to analyze the jobs of several women in the space community to find out if a free workspace also produced free minds regarding female agency and gender equality. Accordingly, the jobs of Cece Bibby, Nichelle Nichols, Ursula Neill and Eilene Galloway will take center stage in this section.

6.1 Art space – Cece Bibby

First off is Cece Bibby, a NASA contract space artist who worked for Chrysler Aerospace at the Cape. Cecilia – Cece – Bibby was a high school graduate who enrolled in art school while working as an operator for a telephone company. After

https://doi.org/10.1515/9783110629828-006

receiving her art degree she was looking for a job. Although she was acquainted with Dr. Theodore von Karman, one of JPL's founders, Bibby applied to the Jet Propulsion Laboratory because of an ad in the newspaper. She was hired and had to move to Florida, where, as a draftsman for the Radio Corporation of America's Missile Division at Patrick Air Force Base, she mainly created as-built drawings for the downrange missile tracking site. Bibby soon transferred to the company's publication department to do space drawings for the engineers in the think tank. In 1959 Chrysler Aerospace Corporation, a NASA sub-contractor, hired Bibby. As a contract artist, she did not work in the Chrysler building, but directly at Kennedy Space Center in the graphics department, where Bibby was under the direct supervision of a NASA employee even though Chrysler was her employer. She was the only female artist in her workspace, but that did not hold her back. She performed the same job as her male colleagues and harnessed her creativity and skills to every project she was assigned to, from instruction manuals to the artistic implementation of engineers' and scientists' ideas for presentations. Her workspace was located in immediate proximity to the astronaut office at Cape Canaveral, meaning that she worked in an office space but in an artistic occupation, which most likely gave her more free space. Her work for the astronauts indicates as much.[1]

For his space mission, John Glenn wanted his insignia to be a personal design in script, hand painted on his capsule. Alan Shepard's and Gus Grissom's spacecraft had only been marked with block lettering which was spray painted with a stencil, and Glenn wanted a more distinctive and artistic representation of his mission. Bibby received the honor of designing Glenn's mission insignia because her boss figured she, as a woman, had better hand writing than her male colleagues. Her boss' line of thinking shows that gender stereotypes can also be advantageous for a woman's career at times. Bibby created three designs which her boss presented to Glenn. He picked the one he liked best and asked for the artist who had created his mission insignia to paint it on the capsule personally. However, her boss declined Glenn's request because of her sex.[2] The trained surrealist painter recalled the conversation between her boss and John Glenn, as told by John Glenn, as follows:

---

1 Robert Pearlman, 'Cece Bibby, Artist behind NASA Space Capsule Logos, Dies', 2012 <http://www.foxnews.com/science/2012/11/16/cece-bibby-artist-behind-nasa-space-capsule-logos-dies.html> [accessed 19 July 2016]; Lawrence McGlynn, 'Cece Bibby: The Mercury Artist', 2013 <http://www.spaceartarchive.com/2013/10/cece-bibby-the-mercury-artist-.html> [accessed 18 July 2016]; Lawrence McGlynn, 'Breaking through the Glass Gantry', 2005 <http://www.collectspace.com/news/news-080705a.html> [accessed 18 July 2016].

2 Colin Fries, 'Will It Go 'Round in Circles? A Colorful Mosaic of Art Flown in Space', *Quest*, 11 (2004), 12–21 (pp. 13–14).

John: "I want the artist who designed that to put it on by hand."

Boss: "Well, that's a woman."

John: "So?"

Boss: "She'd have to go out to the launch pad and up to the top of the gantry."

John: "Is she handicapped in some way?"

Boss: "Well, she's a woman." (To boss being a woman was a handicap.)

John: "Is she afraid of heights?"

Boss: "I don't know . . . but she's a woman."

John: "Why don't you find out from her whether she has some objection to going up to the top of the gantry to paint this for me. Let me know what she says."[3]

Bibby did not have any objections and was allowed to recreate her own artwork on Glenn's spacecraft Freedom 7. Usually, women were not allowed on the launch pad and a lot of the pad workers had a problem with her being there. Pad Leader Günter Wendt even told her to leave immediately. However, she insisted that they would have to take it up with Glenn and started working on the capsule. Glenn's backing made her into the first woman to work on a rocket gantry.[4] This incident shows that women also faced objections due to their gender in the less established workspaces in the space community, but also that there were apparently also less bureaucratic ways to resolve the gender scruples.

One possibility to overcome the gender-based prejudices and taunts was to ensure the help of important men in the space community. In this case, Bibby even received support from the very same men who would see to it that the Mercury 13 women would stay grounded. According to Bibby, Glenn and his Mercury buddies did not have any problem with her being a woman and working on their spacecraft. This shows that the astronauts were very much capable of working with and accepting women in non-traditional female workspaces – they just did not want them to take their jobs. The same probably goes for most of the male workers in the space community. It was easier to accept and respect women's contribution to the American space cause if they contributed in another area and did not represent any direct competition. Because Bibby worked in a rather small and particular workspace, she did not face that much gender-

3 Cece Bibby, 'Friendship 7 with Cece Bibby' <http://freepages.genealogy.rootsweb.ancestry.com/~cecebibby/friendship-7/friendship_7.htm> [accessed 18 July 2016].
4 McGlynn, 'Cece Bibby: The Mercury Artist'. See Fig. 6.1.

Fig. 6.1: Cece Bibby and John Glenn, 1962, NASA.

based backlash at NASA. The men at Chrysler were jealous of her, but their feelings were ignored by the astronauts because "she was the best design and layout artist in the organization," according to Gordon Cooper.[5]

In the end, Bibby needed a week to apply the paint job on the capsule because of the taunts of the pad workers and the constant movement and testing of the capsule. She also had to be very careful about the transferal of the designs to the surface of the capsule because she could not use graphite type transfer paper, which might have caused cracks in the capsule shell; she had to chalk the back of her design and gently trace it onto the uneven surface of the capsule. Although Bibby could not remember how many coats of paint she applied to the spacecraft, she remembered that she did not solely paint for art's sake, as her insignia was used to test a new type of heat-resistant paint. Her artistry sadly almost entirely disappeared during the spacecraft's reentry.[6]

Bibby also designed twelve 'Friendship 7' gold medallions for John Glenn. These were mementos he took with him on his flight to give to his wife and children as well as to his secretary and his nurse, Dee O'Hara. Bibby believed that

5 Gordon Cooper, *letter of recommendation*, July 29, 1965, in: McGlynn, 'Cece Bibby: The Mercury Artist'.
6 Colin Fries, p. 14; Pearlman.

her mission charms "might have been the precursors of the [mission] patches. People knew about the charms and wanted some memento of their own. Some enterprising salesman probably approached someone at NASA and the patch idea was born."[7] Bibby also created the artwork and mission mementos for Wally Schirra and Scott Carpenter. Gus Grissom told her later that he wished his Liberty Bell could have been painted by an artist and that it really bugged him that Glenn had thought of it and he had not.[8]

6.1.1 Naked ladies gotchas

When Bibby met the astronauts for the first time, Cooper told her that she was not what they had expected. Bibby did not know if they had expected a male artist, which would have been very likely. However, Cooper told her exactly what was unusual about her. He said: "You don't giggle when we talk to you."[9] Cooper's statement gives insight into one of the problems the career women in all workspaces frequently faced in the space community – gender fellows who discredited the career women's actions through their fangirl behavior.[10] The so-called "cape cookies" were not only a thorn in the astrowives' sides but also an

---

7 Bibby, 'Friendship 7 with Cece Bibby'. Officially, the idea of mission patches is ascribed to Gemini 5 astronauts Gordon Cooper and Charles Conrad, who convinced NASA leadership of their idea of an official mission patch in 1965. Considering that Cooper did have contact with Bibby before, it is possible that her mission charms inspired Cooper. However, this has never been confirmed by Cooper. More likely, he was inspired by the tradition of military shoulder patches. Tamara Chuang, 'Denver Artist Kacena Tapped by Sister to Design NASA Patch for Rodent Research IV Mission. The History of How Patches for Space Mission Are Created Is a Mystery', *Denver Post*, 2017 <http://www.denverpost.com/2017/02/17/nasa-patches-cu-boulder-artist-kacena-spacex/> [accessed 20 February 2017]; Travis K. Kircher, 'More Than Just a Merit Badge', *collectSpace*, 2000 <http://www.collectspace.com/resources/patches_astronauts.html> [accessed 20 February 2017]. For more general information about the history of the NASA logo see: Chambers and Chambers.

8 Burgess, *Liberty Bell 7: The Suborbital Mercury Flight of Virgil I. Grissom*, p. 65.

9 Burgess, *Sigma 7: The Six Mercury Orbits of Walter M. Schirra, Jr.*, p. 91.

10 For more information about fangirls hitting on the astronauts in the 1960s, see: Burgess, *Sigma 7: The Six Mercury Orbits of Walter M. Schirra, Jr.*, p. 103. For more information about the space groupies in the Space Shuttle era, see: Mike Mullane, *Riding Rockets. The Outrageous Tales of a Space Shuttle Astronaut* (New York: Scribner, 2006), p. 52. Mike Mullane recalls that "there was an even more powerful pheromone than jet-jockey wings and the SEAL insignia: the title 'astronaut.' The fact that none of us had been any closer to space than an airline flight attendant didn't seem to matter. To the space groupies the title was good enough. We males found ourselves surrounded by quivering cupcakes. Some were blatantly on the make, wearing spray-on clothes revealing high-beam nipples and smiles that screamed, 'take me.' [...]

obstacle for women seeking a career in the space community, because they encouraged the men to expect fangirl treatment.[11] Thus, career women in all workspaces had to convince their male colleagues of their serious career intentions. At the same time, they had to demonstrate their 'buddy qualities.'

In order to become a part of the space community workspaces, the women had to go along with the tradition of workspace pranks. The astronauts in particular loved to pull practical jokes on one another. Jamye Coplin, the secretary in the astronaut office, is one of many people to confirm the astronauts' prankster behavior:

> Most of them [the astronauts] had just great senses of humor and appreciated that a good laugh and a lot of levity helped to ease the stress in the situation. So we did have a good time. We had some fun times. [. . .] You had to [have humor], because they would play jokes on us just for the sport of it. So you had to learn to give as good as you got. That was the way they let off a little bit of steam and got out from underneath a lot of the stress. They worked horrendous hours, absolutely horrendous hours. We were good, easy targets. So were the training guys and the team leaders.[12]

Cece Bibby became aware of the love of pranks in the NASA workspaces pretty fast as well. The astronauts and Bibby became sort of friends and she was roped into their game of pranking, famously dubbed 'gotchas' by Wally Schirra.[13] During a day on which she was working on Glenn's capsule design, Bibby met Grissom on the stairs next to the astronaut office. He inquired about the paint job on the "boy scout's" capsule.[14] Then he came up with the idea of painting "naked ladies" instead of Glenn's mission insignia on the capsule. Grissom thought that this would be hilarious because Glenn was considered a "goody-two-shoes" by the other astronauts.[15] Bibby told Grissom that she could not do that because her job would be on the line. She was fully aware that NASA would fire the artist, not an astronaut, over improper pictures on the spacecraft. Grissom apparently just laughed and told her she was a "chicken." Not wanting to be a spoilsport, Bibby thought of another way to pull a gotcha over on Glenn. So she did paint a

---

Even the gold bands on the fingers of us married TFNGs were no deterrent to many of these women. They were equal opportunity groupies."

11 Koppel, p. 58.

12 Jamye Flowers Coplin, interview by Rebecca Wright, November 12, 2008, pp. 28–29, Oral History Transcript, NASA Johnson Space Center Oral History Project, JSC.

13 Burgess, *Sigma 7: The Six Mercury Orbits of Walter M. Schirra, Jr.*, p. 45.

14 Cece Bibby, 'Story Time: A Naked Lady' <http://freepages.genealogy.rootsweb.ancestry.com/~cecebibby/nasa-stories/naked-lady.htm> [accessed 18 July 2016].

15 Milt Salamon, '"Goody-Two-Shoes" Glenn Got Eyeful', *Florida Today*, 17 November 1998, p. 12A.

suggestive naked lady with the caption “It’s you and me John-Baby, against the world;” however, she painted the woman and slogan on paper and not the capsule. She then gave her drawing to a friend in the photo lab who produced a positive print of it. Finally, she gave the print to Sam Beddingfield, an engineer in Glenn’s launch crew, who installed the picture in the periscope view of the capsule, the idea being that the naked lady cover located on the exterior of the lens would be removed shortly before the launch.[16]

Although the flight was scrubbed due to technical problems, Glenn still got “a big kick” out of the drawing. He had immediately recognized Bibby as the prankster because of the word “Baby” in the caption. There had been a mix-up and a lot of people at NASA thought Bibby’s last name was actually ‘Baby,’ thus she was often called Cece Baby instead of Cece Bibby at her workspace. Glenn told Bibby in a note that he was not upset about her gotcha and that he planned to frame the picture and hang it in his den at home. He also asked her if he should expect another drawing for the rescheduled launch. Glenn’s amusement aside, it turned out that Bibby’s prediction that she would get fired if she were to paint a naked lady on the capsule was not too far from the truth. Rocco Petrone, the launch director for Glenn’s flight, wanted to bar Bibby from the launch pad and even asked her boss to fire her because she allegedly upset Glenn with the drawing and endangered the success of the flight. In the end, Glenn and Grissom had to intervene in order to protect Bibby so she could keep her job. This incident did not hold her back from painting another picture for Glenn, though. For his next launch, on February 20, 1962, which was also Bibby’s birthday, she painted an old lady in a house dress with a mop and a Friendship 7 bucket in hand, despite a warning from Petrone. This time the inscription said: “You were maybe expecting someone else. . .John Baby??”[17]

When it was Scott Carpenter’s turn to fly, he immediately asked Bibby whether he should also be expecting a lovely “nekkid” lady for his periscope view. He wanted her to paint one for him, too. Bibby actually did create a naked lady for Carpenter; however, the real gotcha was not the naked lady, who sat next to two pillows and an Aurora guitar, but the caption of the drawing, “. . .but Scott-Sweetie – ‘Yellow Bird’. . .again!??,” because it made fun of Carpenter’s love for the song, which he played on repeat and annoyed his

16 Bibby, ‘Story Time: A Naked Lady’; Burgess, *Sigma 7: The Six Mercury Orbits of Walter M. Schirra, Jr.*, pp. 94–96; McGlynn, ‘Cece Bibby: The Mercury Artist’. See Fig. 6.2.

17 Bibby, ‘Story Time: A Naked Lady’; McGlynn, ‘Breaking through the Glass Gantry’. See Fig. 6.3.

colleagues with.[18] Undeterred, Carpenter ignored the taunting of his favorite song *Yellow Bird* and still kept playing his signature tune after his stint in orbit.[19] Although Bibby had planned to paint a naked lady for Schirra's flight, she decided against it because she did not want to push her luck. Around this time, Bibby had been suspected of sneaking one of her naked ladies into an illustration in a NASA manual on gantry safety. Only after several hundred copies of said manual had been printed and distributed to members of the gantry crews did NASA detect the topless blond lady within a group of serious looking men in a gantry elevator. The picture was supposed to demonstrate safety regulations, but people only talked about the naked lady for obvious reasons. Naturally, the bosses suspected Bibby to be the creator of the artwork; however, it turned out that the creation of the manual had been contracted out to yet another company. Thus, Bibby was acquitted of any wrongdoing, but the incident also acquitted her of the desire to paint naked ladies for the astronauts.[20] Bibby briefly left Chrysler in 1962 because she had wanted to move back to California, but the thrill of being part of the space community lured her back to her work at NASA shortly after Cooper's flight.[21] As a contract employee, she stayed with NASA until 1970 and, although she had been cured of naked ladies, she still very much participated in other workspace gotchas.[22]

Gotchas were not popular only with the astronauts; rather, the pranking culture was actually widespread in the space community, in all social spaces, not only the workspaces. On the one hand, the practical jokes were a way to have fun and let of some steam, and on the other hand, they also worked as a tool to connect the different people of the community, making gender or social status irrelevant for at least a brief period of time. However, sex and sexism still played an important role and Bibby's naked ladies were far from the only naked ladies at NASA. Some naked ladies even made it to the moon.

All astronauts who walked on the moon had 'cheat sheets' with them, so-called wrist checklists, which outlined the activities and tasks for the astronauts during an extra vehicular activity. When Pete Conrad and Alan Bean went about their astronaut business on the moon during Apollo 12 in 1969, they were surprised to find their very own naked ladies in their cuff checklists after

18 Walter M. Schirra, Jr., interview by Roy Neal, December 1, 1998, p. 9, Oral History Transcript, NASA Johnson Space Center Oral History Project, JSC; McGlynn, 'Cece Bibby: The Mercury Artist'. See Fig. 6.4.

19 'Nation celebrates as Carpenter gets Fast Tour, Hero's Welcome', *Roundup*, June 13, 1962, p. 1.

20 Bibby, 'Story Time: A Naked Lady'.

21 McGlynn, 'Breaking through the Glass Gantry'.

22 For example: Burgess, *Sigma 7: The Six Mercury Orbits of Walter M. Schirra, Jr.*, pp. 99–100.

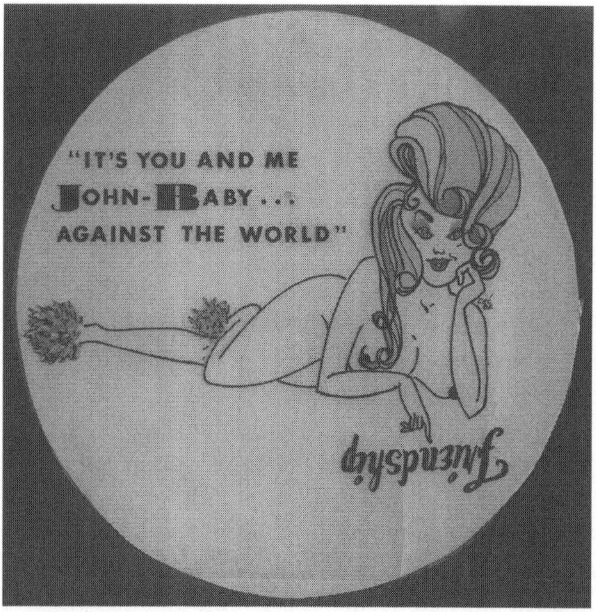

Fig. 6.2: Cece Bibby, 1962, NASA.

Fig. 6.3: Cece Bibby, 1962, NASA.

Fig. 6.4: Cece Bibby, 1962, NASA.

already having spent two and a half hours on the lunar surface. Their checklists included pictures from *Playboy* magazine furnished with sassy captions. Bean found a naked Cynthia Meyers, Miss December 1969, with the caption "Don't forget – Describe the protuberances" and Leslie Bianchini, Miss January 1969, with the instruction to "Survey her activity." Conrad was also surprised to find himself looking at the bare breasts of Angela Dorian, Miss September 1967, along with the question "Seen any interesting hills and valleys?" and Reagan Wilson, Miss October 1967, who was named his "Preferred tether partner."[23] Although Bean and Conrad were very excited about their naked travel companions a quarter of a million miles away from Earth, they did not talk about it. They just showed each other their naked companions in silence and giggled about it. They actually laughed so hard that some people thought they were drunk or suffering from 'space rapture.' Bean later explained that they did not say anything on air because the American people were listening to them during

23 D.C. Agle, 'Playmates on the Moon', *Playboy*, 41 (1994), 138, 213 (p. 138); Koppel, p. 337. Playboy is very proud about their role in the space community. They even issued Playboy/ Apollo 12 merchandise. 'Playboy on the Moon?', *Playboy*, 2016 <https://www.playboy.com/articles/playboy-on-the-moon> [accessed 27 July 2016]. See Fig. 6.5.

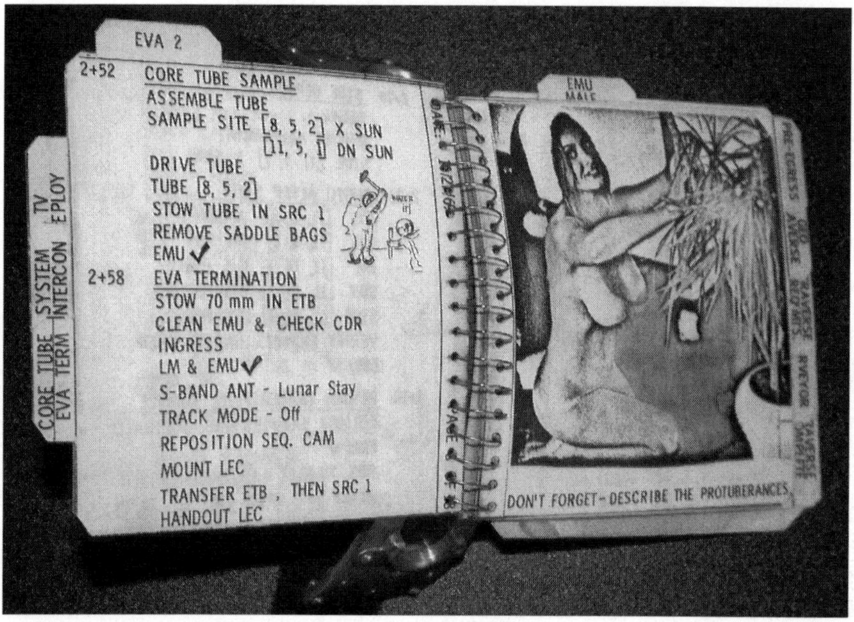

Fig. 6.5: Alan Bean's Apollo 12 Cuff Checklist, Playmate Cynthia Meyers, 1969, NASA.

their time on the moon: "We thought some people back on earth might become upset if they found out we had Playboy Playmates in out checklists. They would have said, 'This is where our tax money is going?'"[24] According to Bean, Conrad even put the photos on restricted access after their return to Earth, unlike the other pictures from the mission, because he did not want the press to get wind of the naked ladies on the moon. Thus, their pranksters did not get a public gotcha, though they certainly got a good laugh out of it.

The practical jokers were none other than their fellow astronauts – the Apollo 12 back-up crew Dave Scott, Jim Irwin, and Al Worden. They had been in charge of assembling the checklists and did not want to miss the opportunity to pull a gotcha on the Apollo 12 crew. Richard Gordon, who had to orbit the moon by himself during Conrad's and Bean's EVA, also received a naked lady: Miss August 1967, DeDe Lind, was hidden in his locker. It was not an easy ordeal to smuggle the ladies from *Playboy* onto the mission.[25] The naked ladies'

24 Agle, p. 138.
25 See pictures: Eric M. Jones, 'Apollo 12 CDR Cuff Checklist', 2013 <http://history.nasa.gov/alsj/a12/cuff12.html> [accessed 27 July 2016].

pictures had to be shrunk and printed on fireproof plastic-coated paper in order to fit them into the tiny wrist checklists without endangering the crew. The chief prankster and backup commander for the Apollo 12 mission, Scott, admitted that gotchas were "a family thing" at NASA and that they had actually "spent a lot of time going through the checklist to see where we could insert something humorous."[26] The back-up crew and other people from the ground personnel also annotated the in-flight documentation with other messages, jokes, their signatures, funny misspellings, and Snoopy cartoons.[27]

Gotchas were a fun activity pursued not only in the workspaces of the space community but also in the private space. For example, the astronauts' wives were just as good humoured as their husbands. The wives of the Skylab II astronauts, Helen-Mary Garriott, Sue Bean, and Gratia Lousma, worked together with French journalist Jacques Tiziou and Houston-based artist Ardis Shanks Settle to create a special wives' mission patch. The patch for the wives was based on the mission patch of their husbands' Skylab II insignia, which displayed Da Vinci's Vitruvian Man. On the women's patch the male figure was substituted by a female figure and the men's names were replaced with the women's first names.[28] Tiziou later recalled that "[t]he fact that their patch displayed a nude man and that women's lib is fashionable now dictated the idea."[29] The wives loved their personal patches and they displayed them everywhere, on their blouses, on their cars, and on their purses. The men on board the 1973 Skylab mission found out about the 320 patches when they opened the storage lockers in the command module to get out their gear. They found their gear within, as well as decals of the wives' mission patch pasted to the inside of the locker doors. Jack Lousma thought that this gotcha "was a great idea and consistent with one of our mottos: 'Never lose your sense of humor.'"[30]

Although Dave Scott considered gotchas "a family thing," it is not quite clear who was part of the family according to him.[31] However, it is most likely that he

26 Agle, p. 138.

27 Matthew H. Hersch, 'Checklist: The Secret Life of Apollo's "fourth Crewmember"', *The Sociological Review*, 57 (2009), 6–24 (pp. 17–18). For more information about the special relationship of Snoopy and NASA in terms of 'gotchas' as well as in more serious matters see: *Historical and Cultural Readings of Peanuts*, ed. by Peter Lee (Jefferson: McFarland, 2018 forthcoming).

28 David Hitt, Owen K. Garriott and Joe Kerwin, *Homesteading Space: The Skylab Story* (Lincoln: University of Nebraska Press, 2008), pp. 60–61. See Fig. 6.6 and Fig. 6.7.

29 Amy Clark, 'Skylab wives play joke on husbands', *Today*, September 23, 1973.

30 *HSA Space Exploration and Aviation Auction Catalog #6000* (Dallas: Heritage Air and Space, 2008), p. 5.

31 Agle, p. 138.

Fig. 6.6: Women's Skylab II mission patch, 1973, Jacques Tiziou.

Fig. 6.7: Skylab II mission patch, 1973, NASA.

referred to the NASA family – a popular term that was often used to describe differing space community concepts. It seems that Scott included all NASA acquaintances, not just the astronauts, into his understanding of 'the family,' considering that the Mexican-American engineer Ernie Reyes played an important part in pranking the Apollo 12 crew. The gotchas described above show that the practical jokes were an important part of the space community's code of conduct, first and foremost to boost morale and have fun. Furthermore, the pranks unintentionally functioned as a connector between the different camps in the community. Race, gender, sex and status in the community did not matter. Everyone could just, or even had to, participate, including the wives of the space workers, who were mostly on the outskirts of the community, as well as the secretaries, whose contribution to the space community were often regarded as a matter of course, and women like Cece Bibby, who worked in rather unconventional jobs but still contributed to the successes of the space community.

Pranking was a communal activity and women were often a part of it, if not in person then through imagery. The naked female body was a recurring theme in the space community gotchas in the 1960s, even when women participated in the prank. In order to belong and to confirm one's community membership, the women had to be good sports and could not be offended by the objectification of female bodies because it supposedly was all just fun and games. The naked lady pranks noticeably disappeared in the 1970s; there are, at any rate, no documents or stories about gotchas of that kind anymore. Considering when the sexist gotchas were discontinued, it seems logical to assume that the naked ladies started to become inappropriate when the women and men in the community started to identify with the values of the women's movement and were no longer willing to accept the naked female body as the butt of the jokes.

6.2 Visitor space – guardettes

Along with the naked ladies, some specifically female space jobs have disappeared as well. In the 1960s, there was one job exclusively for women in the space community besides the occupation of secretary – the guardette. The guardettes were clerical workers as well as tour guides who worked for the Wackenhut Corporation, which was a subcontractor to Trans World Airlines (TWA), which itself was a prime contractor for Kennedy Space Center to run its base support operations.[32] The Wackenhut Corporation operated KSC's Visitor Record Center (VRC) and was in charge of badging guests with the adequate security clearance level required to gain access to Kennedy Space Center. The Visitor Record Center issued an average of 10,000 to 12,000 badges each month.[33] Starting in January 1965, Kennedy Space Center started to allow self-guided tours of the NASA site and the guardettes had to hand out information about NASA and Kennedy Space Center as well as making sure that the visitors did not leave their cars.[34] In 1966, TWA also started a guided bus tour through Kennedy Space Center and the guardettes had to serve as the tour guides.[35] All women who wanted to work as a Wackenhut guardette had to wear a work uniform consisting of French-blue skirts and jackets, white blouses, red pillbox hats and belts, red piping trim, and matching shoes.[36] They were Kennedy Space Center's lovely and pretty welcoming party, responsible for greeting all visitors from Vice President Hubert Humphrey to tourists from all over the world.[37]

Looking at pictures of the guardettes, one has to immediately think of the 1960s stewardesses, as flight attendants were originally called, whose uniforms almost exactly resembled those of the guardettes. Considering that the guardettes' employer was under contract from TWA – a major American airline which operated from 1925 until 2001 and was considered as one of the 'Big Four' domestic airlines in the United State – it cannot come as a surprise that the guardettes' work clothes at the one and only American spaceport were inspired by the clothes of a popular female flying occupation in the 1960s. Being a stewardess was an absolute dream job for a lot of women, and if one was young, white, single, slim, attractive, and female, she had a good

---

32 See pictures of the guardettes: Guardette Ursula Neill and James A. Lovell December 4, 1965; General Reference Series; GH 63; Astronaut Biographies; Lovell, James A., JSC; Guardette Ursula Neill in *Spaceport News*, January 7, 1965 and the *Orlando Sentinel*, January 5, 1965; Guardette Mary Klatt in *Spaceport News*, July 21, 1966; Caption: "Charming Mary Klatt of TWA has the welcome sign out for visitors who will board the first KSC-sponsored bus tours of the Spaceport."; Guardette Pat Blalock in *Spaceport News*, March 26, 1964; artwork by Chrystal Jackson entitled *Tour Agent and Guardette*, NASM.

33 'KSC Security Badging covers wide Spectrum', *Spaceport News*, November 21, 1968, p. 5.

34 'Hundreds take Bus Tour', *Orlando Sentinel*, January 5, 1965, p. 12; *Spaceport News*, January 7, 1965.

35 'Astronauts' Home away from Home', *Palm Beach Post*, June 4, 1966; 'Public Bus Tours Off to Resounding Start', *Spaceport News*, August 4, 1966, p. 1; 'Center Congratulated on Start of Bus Tours', August 4, 1966, p. 7.

36 *Spaceport News*, March 26, 1964.

37 Gay Debus, 'Wife of first Center Director, remembered for ease with People', *Spaceport News*, August 12, 1994, p. 2.

chance of getting the supposedly glamorous job. Because the experience of travelling and flying was historically unprecedented in the 1960s, the pretty, young women were able to experience rare career opportunities while meeting rich, famous, and influential people.[38] However, the air hostesses also faced a lot of discrimination from their predominantly male employers, pilots, and passengers.[39] The young women were often treated like mere sex objects, especially in the airlines' advertisements which promoted the sexual availability of the supposedly 'promiscuous,' 'naughty' stewardesses. The women also had to comply with the airlines' corporate look, a rigorous age, weight, and dress code.[40] The parallels to the job of the guardettes are uncanny. Mainly young, pretty and white women were employed and their job also entailed a certain amount of glamour,[41] since they were part of a new futuristic industry, a cutting edge endeavor with white, patriotic, handsome, masculine highly trained professionals at its center; virile military trained men who were family men one the one hand but also epitomized sexual prowess on the other. Thus, there were not only similarities between the guardettes and stewardesses but also between their co-workers/bosses, the plane captains and astronauts.[42] Just like their colleagues in the sky, the guardettes in fact did also have immediate access to the male heroes of their workspace, as the picture of guardette Ursula Neill and astronaut James Lovell shows. The guardettes also attended to the needs of other important and

38 Louisa Peacock, 'Being a Pan Am Air Stewardess Was Best Job for a Woman', *Telegraph*, 2013 <http://www.telegraph.co.uk/women/womens-life/9909671/Being-a-Pan-Am-air-stewardess-was-best-job-for-a-woman.html>; Bruce Handy, 'Glamour with Altitude', *Vanity Fair*, 2014 <http://www.vanityfair.com/news/2002/10/stewardesses-golden-era> [accessed 28 July 2016].

39 Paula Kane and Christopher Chzandler, *Sex Objects in the Sky: A Personal Account of the Stewardess Rebellion* (River Grove: Follett Pub. Co, 1974); Libbie Escolme-Schmidt, *Glamour in the Skies: The Golden Age of the Air Stewardess* (Stroud: History Press, 2009).

40 Handy; Anke Ortlepp, 'Flugreisen Und Männlichkeit Nach Dem Zweiten Weltkrieg', in *Väter, Soldaten, Liebhaber: Männer Und Männlichkeiten in Der Geschichte Nordamerikas. Ein Reader*, ed. by Jürgen Martschukat and Olaf Stieglitz (Bielefeld: transcript, 2007), pp. 293–312 (p. 8); Anke Ortlepp, 'Friendly Skies? A Cultural History of Airtravel in Postwar America', *GHI Bulletin*, 115–125 (2007), pp. 122–123; Gail Collins.

41 There were also a few women of color who worked as stewardesses. Christine Reiko Yano, *Airborne Dreams: 'Nisei' Stewardesses and Pan American World Airways* (Durham: Duke University Press, 2011). If that has also been the case for the job of the guardettes is not known due to a lack of data regarding the women's race and ethnicity.

42 Ortlepp, 'Flugreisen Und Männlichkeit Nach Dem Zweiten Weltkrieg', pp. 295–298.

Fig. 6.8: Guardette Ursula Neill and James A. Lovell December 4, 1965, NASA.

glamourous visitors to Kennedy Space Center, such as movie and pop stars or (vice) presidents who often came to KSC to watch the historic space launches.[43]

Like the stewardesses, the guardettes also wore uniforms. In comparison to other air hostess uniforms of the 1960s, such as the miniskirt uniforms of Pacific Southwest Airlines, the TWA air and space hostess uniforms were relatively modest with a skirt that reached the knees.[44] The guardettes' uniforms

---

43 See Fig. 6.8; 'Gay Debus, Wife of first Center Director, remembered for ease with People', *Spaceport News*, August 12, 1994, p. 2.

44 Michael Waldock and Johanna Omelia, *Come Fly With Us: A Global History of the Airline Hostess* (Portland: Collectors Press, 2003), p. 118; Ortlepp, 'Friendly Skies? A Cultural History of Airtravel in Postwar America', pp. 122–123. See Fig. 6.9.

Fig. 6.9: Kennedy Space Center headquarters concept, tour agent and guardette, watercolor by Chrystal Jackson, Smithsonian National Air and Space Museum.

thus did not hypersexualize the women wearing them, although they still showcased their womanly features and deliberately turned them into sexual objects. Therefore, it is safe to assume that the guardettes were faced with similar types of (sexual) discrimination as her stewardess colleagues and female space community co-workers in the office spaces.

The job of the space hostess eventually lost its glamourous appeal in the 1970s, leaving the sexist image of the sexually explicit 1960s stewardess working under male imposed conditions to nostalgia; as did the occupation of air hostess, along with the integration of the workforce and the intoxicating thrill of the space race after the moon landing.[45] The workspace of guardettes was not free from prejudice and expectations. In fact, guardettes' workspace gender relations and performances were predetermined by the well-established hostess workspace, meaning that the guardettes worked in what can, in fact, be considered a traditional, preset workspace, even if 'guardette' is not among the familiar, established occupations most people would generally expect to find in

45 Drew Whitelegg, *Working the Skies: The Fast-Paced, Disorienting World of the Flight Attendant* (New York: New York University Press, 2007), pp. 3–8.

workspaces in the space community. The women working as guardettes were thus not able to exercise more agency and were not given more opportunities than the women who, for example, worked in NASA's more traditional workspaces such as the office.

6.3 Science fiction space – Lt. Uhura

The television show *Star Trek* and NASA always had a special relationship. The franchise and the National Aeronautics and Space Administration constantly referenced each other and even switched employees on rare occasions, creating a free workspace at NASA for one of *Star Trek*'s most recognizable characters – Lt. Uhura of the *USS Enterprise*, or rather the actress who portrayed her in the television show, Nichelle Nichols. Considering the important influence science fiction had in the American space community in the 1930s as one of the few ways to circulate ideas about space flight and the conquest of the stars before the rise of the space program, it cannot come as a surprise that the very social and participatory genre of science fiction still played a significant part in the space community in the second half of the twentieth century.[46] In order to understand Nichols' contribution to NASA's workforce and gendered workspace dynamics, it is essential to know the background of the reciprocal relationship between NASA and *Star Trek*, as well as of NASA's image adjustments according to current political environments.

Building on Mark Byrnes hypothesis that NASA has to repeatedly adjust its public image according to the always changing political and social environment of the United States over the years, NASA's connection to Star Trek was very important to retain its public support.[47] In the romantic period in the late 1960s, NASA put an emphasis on the exciting aspects of space exploration, shifting the agency's image to romantic themes such as exploration, heroism, curiosity and the emotional reward of space exploration.[48] All of those romantic themes were also picked up in the then new *Star Trek* television series. Captain Kirk led an Americanized group of astronauts around the 'final frontier' to explore strange new worlds, to seek out new life and new civilizations, to boldly go where no man has gone before – a mission very similar to the American undertaking of landing a man on the moon and returning him safely

46 Cheng, pp. 1–7; Kilgore, p. 81.
47 Byrnes, p. 2.
48 Byrnes, pp. 8–87.

to the earth. Considering the two very similar missions, despite one being fictional, during the same time period, despite *Star Trek*'s setting in the 2260s, NASA's intermingling with *Star Trek* appears to be an obvious decision from a public relations point of view. The film studies professor Constance Penley noticed that NASA and *Star Trek* formed a reciprocal relationship because the agency and the TV show/movie franchise were able to push each other's popularity in the American public. *Star Trek*, just like NASA but unlike a lot of other American science fiction, has a utopian outlook on the future and space exploration. Due to their similar positive outlook regarding space scenarios, it was sensible to join forces and promote a positive space exploration in conjunction with one another. While *Star Trek* showed a beautiful and exciting vision of the twenty first century to twentieth-century humans in the midst of counter-culture and Cold War disillusionment, NASA provided the real-life tools to one day achieve *Star Trek*'s utopian future. *Star Trek*'s rosy future prognosis gave credit to NASA's human spaceflight missions and was even able to ease the damage to NASA's credibility caused by certain reports, such as media cover-ups about wild astronauts who partied a bit too hard. NASA, on the other hand, equipped Gene Roddenberry's creation with real-life cachet, which was possible because of *Star Trek*'s allusions to realism, hard science, and its setting in the Milky Way Galaxy, which also contains our solar system, and thus appeared within NASA's immediate reach. This bond between NASA and *Star Trek* worked so well because "[t]his new entity NASA/TREK shapes our popular and institutional imaginings about space exploration by humanizing our relation to science and technology. It also gives us a language to describe and explain the world and to express yearnings for a different and better condition; it is, then, a common language for utopia".[49]

Star Trek's and NASA's common language of Utopia was repeatedly expressed and portrayed in the *Star Trek* franchise as well as NASA's public relations efforts. There were a lot of NASA and *Star Trek* crossovers, and both sides profoundly benefited from the NASA/Trek collaboration.[50] The biggest

49 Penley, *NASA/Trek. Popular Science and Sex in America*, p. 16.

50 For more information about the merge of NASA and *Star Trek* as well as other franchises such as *Star Wars* see: Karin Hilck, 'The Space Community and the Princess: Reworking the American Space Program's Public Image from "Miss NASA" to Princess Leia', in *A Galaxy Here and Now: Historical and Cultural Readings of Star Wars*, ed. by Peter W. Lee (Jefferson: McFarland, 2016), pp. 33–61; Penley, *NASA/Trek. Popular Science and Sex in America*; Mae Jemison, *Find Where the Wind Goes* (New York: Scholastic Press, 2001), p. 177; Steve White, 'Star Trek Creator Gene Roddenberry's Ashes to Be Blasted into Deep Space', *Mirror*, 2013 <http://www.mirror.co.uk/news/world-news/star-trek-creator-gene-roddenberrys-1975563>; Jörg Hillebrand, Bernd Schneider and Mike Okuda, 'NASA References in Star Trek', *Ex Astris Scientia*, 2015

crossover, however, was the employment of Uhura actress Nichelle Nichols at NASA in 1977.[51] Starting in 1976 NASA was searching for new astronauts for the space transportation system – the Shuttle program – and for the first time in the agency's history they actually wanted to include women and minorities in their workforce because NASA needed to reconnect to the current issues of the American people and adjust its image to the pragmatic period of the 1970s and 1980s. In this era, NASA needed to prove that it understood the needs of the American people. The agency portrayed its space activities as reliable and routine activities with the potential to improve life on Earth through technological and economical stimuli such as satellite technology. NASA also needed an integrated astronaut corps to show that the agency was not only a forerunner in space exploration but also in social equality, improving the lives of Americans through technology as well as social integration and representation. The agency had to change its image strategy because in times of social upheaval and disillusionment with the Cold War and especially the failure in Vietnam, NASA's previous agency branding of nationalism and romanticism was no longer proving successful in the wake of the moon landing. Most Americans figured that the U.S. had won the space race and had explored enough; they were convinced that the money going to NASA could be used better on U.S. soil. Therefore, NASA had to emphasize the benefits of space exploration for every American citizen on Earth, including women and minorities who did not feel represented by the agency due to NASA's male, white astronaut corps.[52]

In order to show the agency's dedication to equal employment and social justice, NASA's Assistant Administrator for Public Affairs, John P. Donnelly, suggested in 1974 that:

<http://www.ex-astris-scientia.org/database/nasa_references.htm> [accessed 8 May 2015]; Jay Levine, 'Star Trek's Nichelle Nichols' Warp-Speed Visit to Dryden', *NASA*, 2012 <http://www.nasa.gov/centers/dryden/Features/nichols.html#.V6C1aKK2qE0> [accessed 4 August 2016]; 'NASA – Final Frontier Astronauts Land on Star Trek', *NASA*, 2005 <https://www.nasa.gov/vision/space/features/Astros_on_StarTrek.html> [accessed 20 March 2015]; Kevles, pp. 130–131; Samantha Cristoforetti, 'There's Coffee in That Nebula', 2015 <https://twitter.com/search?q=%23dragon %40astrosamantha&src=typd> [accessed 28 April 2015]; 'NASA Space Shuttle Missions Get Hollywood-Style Movie Posters', *The Telegraph*, 2009 <http://www.telegraph.co.uk/news/science/picture-galleries/7511902/NASA-space-shuttle-missions-get-Hollywood-style-movie-posters.html> [accessed 28 July 2016].

51 Shayler and Moule, pp. 151–156.

52 Byrnes, pp. 99–135; Kay, p. 114. NASA keeps spin-off records up to this day. For more information see: Dan Lockney, 'NASA Spinoff', *NASA*, 2015 <https://spinoff.nasa.gov/> [accessed 10 July 2015].

> it may be time for us to go out and get ourselves a black astronaut from the Air Force or Navy. [. . .] I think from a public affairs standpoint the announcement of a black astronaut would by far be the most visible means of demonstrating NASA's commitment to EEO [Equal Employment Opportunity].[53]

However, as in the 1960s, NASA opted not to choose a 'token' African American astronaut.[54] NASA's leadership wanted to include women and minorities into the astronaut corps *comme il faut*. One also has to note that there were no human spaceflight missions between 1975 and 1981, and the inclusion of a token minority astronaut would not have been very effective if the astronaut in question had no opportunity to fly to space. This gave NASA's management the time to develop a sound and systematic program for the integration of women and minorities. Thereupon, Johnson Space Center created a new gender and minority friendly astronaut program with new astronaut positions, the advertisement of which in 1976 was especially targeted at women and minorities.

Johnson Space Center even had a special plan to recruit from these two demographics. They informed and met special interest and minority groups such as the Tuskegee Airmen, each center and contractor had to screen their own employees for eligibility, and they informed minority media outlets as well as cooperating with the general media. However, only 1,500 people, including a handful of minorities and women, applied. After six months, only 75 women and approximately 30 minorities had actually sent an application. Another astronaut class without women and minorities was out of the question; NASA had to push its efforts to recruit from these demographics even further.[55] NASA's recruiters tackled the sluggish pace of applications on different channels, approaching hundreds of people, universities, aerospace companies, professional organizations, and the media.[56] NASA's recruitment efforts not only showcase the interconnectedness of different areas of the space community, but the networking endeavor also shows that NASA truly wanted women and minorities

53 Quote in: Atkinson and Shafritz, pp. 137–138.

54 For more information about African Americans in the space community see: Phelps; Paul and Moss; Von Hardesty, *Black Wings: Courageous Stories of African Americans in Aviation and Space History* (New York: HarperCollins, 2008); Shetterly.

55 James C. Fletcher, *letter to the Tuskegee Airmen, Inc.*, August 18, 1976, Harris/MSRC; Women in Motion Inc., *Introduction*, p. 1, NASA Astronaut Recruitment Final Report, August 10, 1977, HRC; Atkinson and Shafritz; Kevles, pp. 60–65; Phelps, pp. 62–66. Atkinson and Shafritz name a different figure. They write that after six months of recruiting only 93 women had applied. Atkinson and Shafritz, p. 151.

56 For a detailed list of NASA's recruitment efforts, see: Atkinson and Shafritz, pp. 153–154.

to join the thus far white, male astronaut corps. Aware of their white, male image problem, NASA reached out to Nichelle Nichols to ask for her counsel.

Nichols and NASA representatives had previously met at a *Star Trek* convention in the Chicago Hilton Hotel in 1975, where NASA had introduced its long range space program. Ever since the initial meeting, Nichols had taken an interest in America's space program, and after the convention, she had toured NASA's Ames Research Center in California. Although the NASA scientists were surprised to learn of Nichols' sincere interest in the space program, they were delighted about it and invited her to fly an eight-hour mission aboard the Kuiper C-141 Astronomy Observatory on an interferometric study of the atmospheres on Saturn and Mars. She was also a VIP guests at the Jet Propulsion Laboratory for the landing of Viking on Mars. Nichols even wrote an article for a National Space Institute publication for which she interviewed several NASA employees about Viking, and also about the topic of women and minorities at NASA.[57] One of the people she interviewed was Herb Rowe, who worked in NASA's External Affairs Office. She wanted to know why there was "not more public support for space exploration with all its amazing feats? Why had NASA earned a reputation for racism and sexism? Why were there no women astronauts and why could not an agency able to marshal its forces in so short a period of time to win the race to the moon be able to deal squarely and rapidly with the matter of Equal Opportunity?" She also told him that "[i]f my taxes helped pay for space exploration and research, then this was my space agency – my NASA."[58] Nichols felt very connected to NASA and a part of the space community through her trademark role as Lt. Uhura. Naturally, she also attended the roll-out ceremony for the Space Shuttle *Enterprise* with pride, because she genuinely enjoyed witnessing the positive influence of *Star Trek* on the American space program. Nichols also toured the Marshall Space Flight Center in Alabama, where she again interviewed NASA employees. This time around she spoke to female scientists who did experiments pertaining to astronaut crews in deep space.[59] Because of her interest in the space program and her status as a *Star Trek* cast member, she was also nominated as a member of the Board of Directors of the National Space Institute on January 21, 1977, a nomination she gladly accepted. As one of the few women on the board, Nichols gave a speech at their annual

57 *Profile: Nichelle Nichols, 1977*, NASA Astronaut Recruitment Final Report, Women in Motion Inc., August 10, 1977, HRC; Nichelle Nichols, *Trekking NASA from Mars or What's in It for Me?*, August 19, 1976, Harris/MSRC.

58 Nichelle Nichols, *Trekking NASA from Mars or What's in It for Me?*, August 19, 1976, Harris/MSRC.

59 For more information about these women see chapter five.

meeting in Washington, D. C. In her speech, Nichols criticized NASA for ignoring qualified women and minority applicants in their astronaut selection, even giving her audience pointed examples of qualified minorities and women who had applied to NASA but were repeatedly rejected. Nichols also told those in attendance that most women and minorities had stopped applying for and caring about the space program because they felt that they had no chance if they were not straight, white, and male. Her speech 'New Opportunities for the Humanization of Space' reached not only the National Space Institute members, but also NASA officials who were in attendance at the meeting.[60]

Nichols was invited to NASA Headquarters the very next day. NASA managers wanted her to assist them in recruiting women and ethnic minorities. Her advice was for the agency to work with prominent minority figures like Bill Cosby or Coretta Scott King to encourage women and minorities to fill out applications. However, they offered her a job as a consultant, an offer she could not believe at first. "I said you've got to be joking; I didn't take them seriously," she later recalled.[61] In the end, NASA was able to convince her to do it herself.[62] Thereupon, NASA awarded a contract totaling $49,900 to the Women in Motion Production Company (WIM) "for services associated with acquainting potential minority applicants with the opportunities available to them in the Shuttle Astronaut Recruitment Program."[63] The Executive Vice President of Women in Motion Inc. and project manager, Nichelle Nichols, and her female co-workers, Shirley Bryant Keith, Shannon O'Brien, and Janet M. Holbrook, thus received a free, previously nonexistent consulting workspace from NASA. Although NASA had employed women as consultants before, like Jerrie Cobb or Jacqueline Cochran, this consulting contract involved payment for the consultants' work and the agency was actually dependent on the vigor of Nichols' minority and women recruitment campaign. The contract NASW-3049 between NASA and WIM was signed in February 1977 and was limited to a half year. However, the influence and science (fiction) credibility NASA received from Nichols' relentless work still benefits the agency up to this day.[64]

---

60 *Profile: Nichelle Nichols, 1977,* NASA Astronaut Recruitment Final Report, Women in Motion Inc., August 10, 1977, HRC.

61 Jay Levine.

62 Phelps, p. 62.

63 *NASA News, February 24, 1977*, NASA Astronaut Recruitment Final Report, Women in Motion Inc., August 10, 1977, HRC.

64 Nichols still tours the NASA centers in the twenty-first century and does NASA/*Star Trek* outreach work, see: Monroe Conner, 'NASA's SOFIA Observatory Flies Star Trek Icon and Educators on Science Mission', *NASA*, 2015 <http://www.nasa.gov/centers/armstrong/features/sofia_flies_star_trek_icon.html> [accessed 8 August 2016]; Jay Levine.

Ms. Nichols' company, which had been formed in 1975, stopped its normal production work in order to fully concentrate on the task assigned to them by NASA.[65] The production company usually produced educational and motivational films as well as printed materials to reach young minority students and their parents. The goal of the company's work was to encourage students to seek out more technically oriented careers.[66] Through the NASA contract, WIM had the rare opportunity to provide strong, intelligent, and glamourous minority role models for the kids they usually targeted in their work. Keenly aware of the fact that minority and female kids were in desperate need of people they could look up to in real life, Nichols always made sure to let her audience know, too. She repeatedly emphasized in her recruiting tours that the "black woman and/or man who qualifies and cares to meet the challenge, carries a place in history, along with the other great names in history. They will, I assure you, not be made of ordinary stuff, and will serve as role models for generations of aspiring young black children who dare to dream of greatness."[67]

While *Lt.* Uhura provided a fictional role model already, Nichols understood the importance of Uhura's NASA equivalent and welcomed the unique opportunity to also be involved in finding NASA's very own Uhura. Taking the NASA job, Nichols was fully conscious of being the face of the strong, intelligent, female, black communications officer and the influence her *Star Trek* character had on so many female and minority people: "I have had women tell me they have sought a career they once thought impossible because of Uhura – the 'if she can do it, I can do it' idea – the possible dream that's there, and says that it's alright to be different, even if that difference means handicapped, poor, black, brown, red, yellow or white."[68] The most famous examples of Uhura's career influence are probably the first African American astronaut, Mae Jemison, and actress Whoopi Goldberg.[69] "Since I met/created her," Nichols repeatedly said of *Uhura*, "she simply won't let me alone. She prods me into action in my present world, and will fairly indict me unless I find ways in this

65 Women in Motion, *letter to the Congressional Black Caucus*, NASA Astronaut Recruitment Final Report, Women in Motion Inc., August 10, 1977, HRC; *Profile: Nichelle Nichols, 1977*, NASA Astronaut Recruitment Final Report, Women in Motion Inc., August 10, 1977, HRC.

66 Women in Motion Inc., *letter template*, NASA Astronaut Recruitment Final Report, Women in Motion Inc., August 10, 1977, HRC.

67 Nichelle Nichols, *sample presentation for professional organizations*, NASA Astronaut Recruitment Final Report, Women in Motion Inc., August 10, 1977, HRC.

68 *Profile: Nichelle Nichols, 1977*, NASA Astronaut Recruitment Final Report, Women in Motion Inc., August 10, 1977, HRC.

69 Jemison, p. 177; Nichelle Nichols, 'Influence on Whoopi Goldberg', *Makers* <http://www.makers.com/moments/influence-whoopi-goldberg> [accessed 4 August 2016].

century to help insure the existence of her century."[70] However, although Uhura was the impulse for Nichols' contract with NASA, she took on this role as a NASA spokesperson out of her own personal believes. Being a career woman, an African American, and a mother, she felt that the future of space exploration belonged to all Americans and that she had to do everything in her power to achieve a time in which a woman of color could be in a command position, in a position of respect and responsibility, without it having to be science fiction. According to Nichols, the Vulcan logic philosophy of crewmate Mr. Spock actually helped her to develop her belief in 'Project Equality' at NASA. It is basically just a science fiction-to-NASA translation of Mr. Spock's IDIC philosophy, which stands for 'Infinite Diversity and Infinite Combinations,' meaning that all people in their infinite diversity should be respected in their infinite combinations for their differences, not in spite of them. Thus, people should only be evaluated, accepted, or rejected, on a one on one basis.[71]

After taking the NASA job and reviewing NASA's recruitment efforts, Nichols and her colleagues found out pretty fast why there were no applications from women and minorities. Most of the targeted personnel were hostile or indifferent toward NASA because they had no trust in the agency's equal opportunity intentions.[72] The WIM campaign, thus, had to give credibility to NASA and make NASA's recruiting intentions known in the minority communities. WIM divided the workload into four main areas: Scheduling Mode, Information Dissemination Center, Recruitment Drives, and Media and Public Affairs Functions.[73] While WIM contacted various organizations and individual contacts and distributed the requested NASA information materials as part of the Scheduling Mode and the Information Dissemination Center, they actually ventured into the field during the Recruitment Drives. The Media and Public Affairs Function provided the media exposure for the recruitment of potential astronauts in the field. Of course, Nichols herself was the figurehead of WIM's recruitment campaign. She recorded several radio and TV commercials in which she was dressed like an astronaut,

70 *Profile: Nichelle Nichols, 1977*, NASA Astronaut Recruitment Final Report, Women in Motion Inc., August 10, 1977, HRC. See also: Nichelle Nichols, *sample presentation for professional organizations*, in: NASA Astronaut Recruitment Final Report, Women in Motion Inc., August 10, 1977, HRC.

71 *Appendix J, Likeness, Personal, Endorsement, Voice Over, Image*, NASA Astronaut Recruitment Final Report, Women in Motion Inc., August 10, 1977, HRC; Nichelle Nichols, *Trekking NASA from Mars or What's in It for Me?*, August 19, 1976, Harris/MSRC.

72 Women in Motion, *Introduction*, p. 13, NASA Astronaut Recruitment Final Report, Women in Motion Inc., August 10, 1977, HRC.

73 Women in Motion, *Introduction*, pp. 3–4, NASA Astronaut Recruitment Final Report, Women in Motion Inc., August 10, 1977, HRC.

each of which varied slightly.[74] There were six similar sounding public service announcements, four released radio spots involving voice overs, three television film spots of thirty second and sixty second duration, and two films of four and ten minute duration.[75] However, she usually addressed potential astronauts along the following lines:

> HI...THIS IS NICHELLE NICHOLS. PLAYING THE PART OF LT. UHURA ON *STAR TREK* WAS REWARDING AND FUN. THERE WERE NO SOCIAL BARRIERS ON THE STARSHIP ENTERPRISE ...MEN AND WOMEN ... ALL RACES AND IDEOLOGIES. THAT'S THE WAY NASA WANTS IT WITH OUR NEW SPACE SHUTTLE PROGRAM THAT BECOMES OPERATIONAL IN THE 1980'S. MINORITY AND WOMEN CANDIDATES ARE BEING SOUGHT FOR PILOT AND MISSION SPECIALIST ASTRONAUT POSITIONS.[76]

According to her contract, Nichols was only obliged to do ten NASA related TV and radio interviews. Committed to the mission however, Nichols appeared in 35 major TV and radio talk shows, which is even more impressive considering the time constraint WIM had to work within and that program scheduling for major talk shows is usually done three to four months in advance. She also gave myriad press interviews about NASA's new astronaut program and was the object of countless NASA/*Star Trek* photo series which resulted in an abundance of articles in all kinds of news outlets in the United States and worldwide.[77]

Even though the majority of the articles and telecasts were favorable, the media reports also reflected the questions and concerns of the minorities and women who were being approached.[78] The concerns expressed in the media coverage were the same as those WIM had to deal with during their recruitment drives, of which Nichols was also the main speaker. The WIM recruitment drives introduced NASA and the Shuttle and new astronaut program to potential candidates. According to WIM's final report, the Mr. Spock inspired "one on one interaction" had become "the backbone of recruitment success" because WIM had to use "a process of developing skills to handle the intrinsic mistrust

74 Phelps, p. 63.

75 Women in Motion, *Introduction*, pp. 12–13, NASA Astronaut Recruitment Final Report, Women in Motion Inc., August 10, 1977, HRC.

76 *NASA Public Service Announcement #2*, NASA Astronaut Recruitment Final Report, Women in Motion Inc., August 10, 1977, HRC.

77 Women in Motion, *Introduction*, p. 12, NASA Astronaut Recruitment Final Report, Women in Motion Inc., August 10, 1977, HRC. See also the various examples in the appendix of the Final Report.

78 Women in Motion, *Introduction*, p. 6, NASA Astronaut Recruitment Final Report, Women in Motion Inc., August 10, 1977, HRC.

and manifested hostility expressed by people who were contacted, toward participation in, and the sincerity of, the space program."[79] It is impossible to say exactly how many people the WIM campaign was able to reach and how many people applied because of WIM's recruitment efforts. Due to the media exposure, however, there is no doubt that WIM was able to reach millions of readers and viewers.[80] Because of the very public recruiting effort, Nichols received several awards and thank you notes for her tireless struggle to include women and minorities in the astronaut corps. More importantly, WIM and Nichols personally received myriad letters, from interested scientists to school girls, who had all heard the recruiting call and wanted to participate.[81]

After WIM's recruiting efforts, NASA received considerably more inquiries and applications, with 24,618 people officially inquiring about the open positions and 20,440 requesting an information packet. In the end, NASA received 8,037 applications, including 6,735 for the mission specialist positions and 1,302 for the pilot positions. It is hard to say how many male minorities applied because the applicants did not have to disclose their ethnic membership. It is known, however, that there were 1,142 female applicants – 1,067 more applications from women than there were before WIM's recruitment efforts.[82] The number of female applicants was by far above any predictions, so it is justified to assume that Nichols and her team were able to reach and dispel the concerns of American women and most likely, American (minority) men, too.[83] Two of Nichols most famous

---

79 Women in Motion, *Introduction*, pp. 5–6, NASA Astronaut Recruitment Final Report, Women in Motion Inc., August 10, 1977, HRC.

80 Women in Motion, *Introduction*, p. 15, NASA Astronaut Recruitment Final Report, Women in Motion Inc., August 10, 1977, HRC.

81 For example: Daniel W. Gilbertson, *letter to Nichelle Nichols*, June 10, 1977, NASA Astronaut Recruitment Final Report, Women in Motion Inc., August 10, 1977, HRC; Barbara Jordan, *letter to Nichelle Nichols*, March 14, 1977, NASA Astronaut Recruitment Final Report, Women in Motion Inc., August 10, 1977, HRC; Laura L. Kinser, *letter to Women in Motion Inc.*, NASA Astronaut Recruitment Final Report, Women in Motion Inc., August 10, 1977, HRC.

82 'Over 8,000 apply for Space Shuttle Astronaut Program at JSC', *NASA News*, July 15, 1977, NASA Astronaut Recruitment Final Report, Women in Motion Inc., August 10, 1977, HRC. John gives a different number. She says that 1,544 women applied to NASA. And that there were 8,079 applications in total. It is certain however, that NASA selected six women, three African American, and one Asian American male, along with twenty-five white males. According to John, women accounted for only 390 out of 2,937 applications for the next astronaut class in 1980. Mariel John, 'NASA: Making Space for Women', *MIT Undergraduate Research Journal*, 11 (2004), 19–25 (p. 21). See also: Kevles, pp. 60–65; Atkinson and Shafritz, pp. 150–157; Phelps, pp. 62–66; Foster, 'The Gendered Anniversary: The Story of America's Women Astronauts', p. 158.

83 Marv, *letter to Nichelle Nichols*, May 4 1977, NASA Astronaut Recruitment Final Report, Women in Motion Inc., August 10, 1977, HRC.

recruits were the NASA astronaut and administrator Charles Bolden and the first female African American astronaut, Mae Jemison.[84] In its final report WIM, itself assessed that their recruiting program and personal interaction with the targeted people "produced such emotional and behavioral changes as: tentative trust, measured anticipation, reduction in the expression or deep-seated feelings of hurt among women who genuinely wanted to be astronauts, and among minorities, who felt heretofore left out if not deliberately excluded."[85]

It was not just WIM and its lead project manager, Nichelle Nichols, who positively evaluated their recruiting efforts; most NASA officials were in favor of the *Star Trek*/NASA crossover because they wanted to make sure that the recruitment effort reached the targeted people in a positive way. An agency official explained that "we feel Ms. Nichols is uniquely qualified to do this."[86] Hans Mark, Director of NASA's Ames Research Center, felt similarly. He was convinced "that cooperative efforts such as this Symposium [with Nichelle Nichols as the main speaker] are a most effective means of making students, particularly minorities and women, aware of career opportunities in science and engineering to the mutual benefit of all participants."[87] NASA administrator James Fletcher was also a supporter of Nichols as the face of *Star Trek* and NASA: "Images meant a lot to people, and Blacks thought they weren't wanted so they didn't apply. It [hiring Nichols] was a necessary thing for NASA to do. I don't think there is any question about it."[88] However, not all NASA employees approved of her hiring. Deke Slayton, for example, the head of the Approach and Landing Tests (ALT) of the Space Shuttle at the time, thought her employment at NASA was unnecessary.[89]

Slayton was most likely not the only member of the space community who did not approve of Nichols' hiring and who could not understand why a person with no engineering, science or flying background, of all people, received an unprecedented workspace at NASA. However, Nichols' results speak for themselves and show how one woman was able to carve out a new type of workspace where (minority) women experienced a completely new and independent

84 Jay Levine.
85 Women in Motion, *Introduction*, pp. 13–14, NASA Astronaut Recruitment Final Report, Women in Motion Inc., August 10, 1977, HRC.
86 *NASA News*, February 24, 1977, NASA Astronaut Recruitment Final Report, Women in Motion Inc., August 10, 1977, HRC.
87 Hans Mark, *letter to Nichelle Nichols*, June 17, 1977, NASA Astronaut Recruitment Final Report, Women in Motion Inc., August 10, 1977, HRC.
88 Atkinson and Shafritz, pp. 155–156.
89 Atkinson and Shafritz, pp. 155–156.

way of expressing their agency in the space community. Considering that Nichols and her female colleagues had to travel extensively for their recruitment drives and meetings, they were not forced to permanently work in a preassigned space, perpetuated and reproduced on a daily basis, as in the regular NASA office space, and spearheaded by a white, male boss. Thus, the WIM women did not have to perform their gender as much and did not have to consider the male-female social conventions in their gender performance due to the female-led, all-female workspace. The social structures of Nichols' workspace were not as set and rigid as those in the mainstream workspaces in the space community. It was a new and innovative workspace, which makes sense when taking the assignment – bringing new types of people into the space community – into account.

The goal of WIM's contract was to spread NASA's gospel in communities that NASA itself had been thus far unable to reach. Had they used the same approach and methods as NASA had been using previously, their work would scarcely have been seen as necessary. NASA actually wanted and needed a fresh, new kind of recruiting effort. The utopian science fiction allure of *Star Trek* was a perfect way of polishing the NASA image and bringing new people on board, because the agency already knew that the cooperation would most likely work due to previous NASA/*Star Trek* crossover experiences, such as the *Star Trek* inspired naming of the first Space Shuttle.[90] Furthermore, Nichols' appearances implied the good old times of the adventurous and successful romantic period, but in the pragmatic period, as well as the good new times of the *Star Trek* future.

It is also important to note that Nichols, one of the few female actresses in *Star Trek*, and not one of her co-stars was asked to represent NASA. It seems a logical choice considering that Nichols embodied the two groups NASA targeted in their recruiting effort – women and minorities. As Mary Jo Deegan has pointed out, however, the women in the original *Star Trek* series are rather secondary figures who mainly received screen time to provide romance.[91] Given her limited agency on screen, it is perhaps somewhat surprising that NASA chose Lt. Uhura as a spokeswoman for the space community. Nevertheless, while it is true that Uhura was only a supporting character on the show and had less agency than Captain Kirk or Mr. Spock, she did have some agency and worked in the sort of non-menial occupation a lot of minority (women) did not

90 Weitekamp, 'Selling Education in the Shape of a Shuttle', p. 214.

91 Mary Jo Deegan, 'Sexism in Space: The Freudian Formula in *Star Trek*', in *Eros in the Minds's Eye. Sexuality and the Fantastic in Art and Film*, ed. by Donald Palumbo (New York: Greenwood Press, 1986), pp. 209–24 (p. 209).

even dare to dream of. Furthermore, there was not really another character on *Star Trek* who could inspire women and minorities to literally reach for the stars. The other woman on board the Enterprise, Nurse Chapel, was white and worked in a traditional female job, which made her a less compelling choice of spokeswoman. The other non-Caucasian on the *Enterprise* was Lt. Hikaru Sulu, the helmsman of the spaceship who was supposed to represent all of Asia. Although NASA was very much interested in highly qualified minorities of all backgrounds, NASA was especially interested in involving the African American community, which might not have reacted to Sulu as the spokesperson the same way they welcomed Uhura. Nor were there other options from other science fiction franchises, either because they were dystopian stories, had no female or minority characters, or only female or minority characters with less cultural power than Uhura.[92] It turns out that although Uhura did indeed have limited agency on *Star Trek*, she was still the minority woman with by far the most agency in popular American TV and movie science fiction. Of course NASA also could have given the job to a minority and/or female employee who already worked for NASA in another capacity, but none of them had the star power and influence of Nichelle Nichols, or the tools to capture the imagination of the targeted Americans to beam them to new, exciting times in space exploration.

6.4 Newsroom space – Lillian Levy

When thinking about typical jobs in the space community, science, engineering and piloting are usually the first to come to mind, followed by administrative and clerical jobs. Most people have also most likely heard about the unusual jobs, like *Star Trek* consultant, in the newspaper because they are such rare and curious occupations. However, there are also jobs and workspaces almost every venture needs, but which almost never take center stage because they lay the groundwork or build the framework for said venture but do not make headlines. The space community had its share of such workspaces and women also participated in them. NASA, for example, needed people, to compile, arrange, and abstract information about the agency's work and accomplishments for easy consumption by its own workforce as well as by the press and the American and international public. Thus, NASA needed its own journalists, editors,

92 For more information why Princess Leia was not a valuable option, see: Hilck, 'The Space Community and the Princess: Reworking the American Space Program's Public Image from "Miss NASA" to Princess Leia'.

writers, and publicists to work in their newsroom, the Office of Public Affairs. Like almost all workspaces at NASA, the newsroom was mainly staffed with men; however, there were also women like Lillian Levy, who started working as a senior science writer in the newsroom already in 1963. Although she became branch chief of the Office of Public Affairs in 1972 and Special Assistant to the Division Director for Women's Activities and Science Writing in 1973, she initially started out as a consultant to James E. Webb, then NASA Administrator. She also worked as an author, editor, journalist, lecturer, and producer of TV programs on science and international issues while being on the staff of the NASA Office of Public Affairs.[93] Levy, for example, edited the book *Space: Its impact on man and society*, in which she introduced readers to the questions, concerns, and complexities of the impact of space exploration on humanity and American society in particular.[94]

Women were mainly absent or relegated to 'female topics' such as food, fashion and 'who's who' stories in the newsroom workspaces of most American television news programs and print media. They also often had to work as researchers rather than writers in the 1960s.[95] Even throughout the early 1970s, employers like *Newsweek, ABC News*, or the *New York Times* did not hire female reporters in any position of importance.[96] Female underrepresentation was also an issue in the NASA newsroom, where most employees were white men. In contrast to her colleagues who worked for media outlets, however, Levy did not have to write about supposedly female issues, such as clothes or food, but supervised a small staff, planned special PR and ceremonial programs and events, and handled almost all of the Apollo briefings and protocols as well as the briefings for robotic space missions. She also wrote and edited scripts, briefing material and statements, and had her own weekly column on research developments for the *NASA News*, the press releases of NASA, which were created for

93 *National Federation of Press Women visit to the NASA National Space Technology Laboratories*, June 3, 1977, Bay St.Louis, Mississippi, NASA Astronaut Recruitment Final Report, Women in Motion Inc., August 10, 1977, HRC; O.B. Lloyd, *Memorandum to all Public Services Division Personnel*, April 17, 1973, 013690, Box 32, HRC; *Attachment 1 to Complaint of Discrimination*, June 12, 1973, 013690, Box 32, HRC.

94 Lillian Levy, *Space: Its Impact on Man and Society* (New York: Norton, 1965).

95 Rhode, p. 9; Davis, p. 110; *History of the Mass Media in the United States: An Encyclopedia*, ed. by Margaret A. Blanchard (New York: Routledge, 1998), pp. 700–703; Carolyn Kitch, 'Women in Journalism', in *American Journalism: History, Principles, Practices*, ed. by W. David Sloan and Lisa Mullikin Parcell (Jefferson: McFarland, 2002), pp. 87–96 (pp. 92–94).

96 McQuaid, '"Racism, Sexism, and Space Ventures": Civil Rights at NASA in the Nixon Era and Beyond', p. 424.

inclusion in magazines and newspapers.[97] Considering that there were no 'typical' female topics at NASA, her assignments never resembled the stories of the women's pages in regular news outlets.

Levy herself was nevertheless not free of gender-based sterotypes about men and women. In an interview with Nancy Roman, which she conducted for the *Sunday Star* in 1959 before she started working at NASA, she described Roman very much in terms typical of the women's pages. Levy called her an "attractive brownette" and reassured the readers that "Dr. Roman is evidence that 'looks and books' do go together. She is a well-built 5′5', with fine, regular features. Her flawless complexion is accented by eyes of a color that defies description."[98] Levy also described for the readers what "pretty Dr. Nancy Roman" liked to do in her free time and showed pictures of "the very feminine scientist" sitting at her desk and cooking.[99] Despite being a woman herself, Levy followed in the footsteps of the male discourse about the female members of the space community, in which the female body and household skills took center stage. She also adopted their language to describe these accomplished women; her article about Roman was intentionally feminized.[100] This shows that women were affected by gender stereotypes just as much as men, even when the stereotype was potentially offensive to one's own gender. Furthermore, as was the case for the first female pilots and the first women who tried to become astronauts in the early 1960s, the women themselves and most journalists felt that it was necessary to emphasize the femininity of these women working in a thus far male occupation in order to show that one can work like a man but still be a 'typical' feminine woman.

This attitude changed in the 1970s and 1980s, when men and women alike had to learn what gender equality truly meant and how it could be expressed in writing. Starting in the 1970s, most journalists, female and male, tried to portray women working in the technical field as competent workers irrespective of their

---

97 *Attachment 1 to Complaint of Discrimination*, June 12, 1973, 013690, Box 32, HRC; *Position Record*, April 4, 1972, 013690, Box 32, HRC; Lillian Levy, *Proposed Reorganization of Public Services Division and Effect on Public Activities Support Branch*, Memorandum to John Donnelly and O.B. Lloyd, April 10, 1973, 013690, Box 32, HRC.

98 Lillian Levy, 'Career Girl 1959. A Space Career. She doesn't want to be launched', *Sunday Star*, May 31, 1959.

99 Lillian Levy, 'Career Girl 1959. A Space Career. She doesn't want to be launched', *Sunday Star*, May 31, 1959.

100 For more information see chapter three.

gender.[101] However, news stories from this decade and the 1980s were very much interested in women who accomplished 'firsts' in areas previously reserved for men.[102] The same was true for NASA, which wanted to show off its female, technical workforce. Thus, in 1972 NASA started a weekly column, 'Down to Earth – A Woman's View of Space,' which was initiated and written by Lillian Levy.[103] In this newsletter, Levy often wrote portraits about female NASA employees and their work for the agency and American society. This focus is quite interesting because Levy, who would go on to become the Public Services Division Director at NASA, had begun working at the agency nine years earlier but only started writing about women and 'female' topics during the height of the women's movement in the 1970s. It has to be emphasized, though, that while her column now highlighted "a woman's view of space," she still mainly wrote about science and technical topics.[104] This new emphasis was most likely important to NASA as a way of showing that they had an integrated workforce in all areas. Ironically, Levy, too, faced sex-based discrimination and filed an official complaint in 1973 because she felt that she was being denied the opportunity to advance in her job to a higher grade, GS-15, despite being promised just that.[105] Her case shows that women in the space community, especially at NASA, had access to all kind of jobs, but it also indicates that the agency often paid more lip service to its female employees and the female public than it paid actual attention to the inclusion and advancement of women in the community. Like most of the women who worked in the lab spaces, Levy, too, had to face gender-based discrimination and had to find a way to turn the agency's lip service about women's opportunities into reality.

6.5 Legislation space – Eilene Marie Galloway

While gender equality and discrimination were a topic for writers at NASA, albeit not as big of a topic as it was to their colleagues outside of the space community, there actually were female writers in the space community who never

101 Some of the portrayals of the first female astronauts show that this new attitude of not centering the news story around the female body had not reached all journalists in the 1980s. For more information, see chapter three.

102 For more information see chapter *Civil Rights Ladies, Glamournauts and the Teachernaut in the Media in the 1970s and 1980s.*

103 John, p. 20. For example, see: Lillian Levy, *Down to Earth. A woman's View of Space*, Box 18, Center Series – Poindexter Reference Files, women in space, JSC.

104 Lillian Levy, *Down to Earth. A woman's View of Space*, Box 18, Center Series – Poindexter Reference Files, women in space, JSC.

105 *Complaint of Discrimination*, June 12, 1973, 013690, Box 32, HRC.

faced gender-based discrimination or doubts despite exerting a dominating influence in the space community. For example, the Library of Congress analyst Eilene Marie Galloway became known as the 'Grand Dame of Space' already in the late 1950s.[106] Amongst other accomplishments, Galloway helped to write the National Aeronautics and Space Act, assisted with Congressional hearings that led to the creation of NASA, served on the United Nations Committee on the Peaceful Uses of Outer Space for several decades and was also instrumental in creating the International Institute of Space Law. Her papers, speeches, research, reports, and suggestions had a deep and lasting impact on U.S. and international space activities and legislation.[107]

While women had been discriminated against in other workspaces in the space community or, at the very least, had to challenge gender stereotypes or to push their way into a male-dominated workspace, Galloway was always surprised when journalists asked her about gender discrimination. According to her, she "was never discriminated against, and the question always surprised" her.[108] Galloway was only confronted with the question of gender discrimination as late as 1987. Paul Dembling, former General Counsel of NACA and NASA, asked her about it, and her astonished reaction was that she "could not believe anyone was thinking along those lines" and she began to wonder about the reasons for her not having been subject to any discrimination.[109]

Galloway herself could think of six reasons why the topic of gender and discrimination, because of her being a women, were never an issue for her and her male and female co-workers. One reason, according to her, was that there was no competition between the employees because she was working in the Congressional Research Service (CRS), which was a small organization that was in a phase of expansion, meaning that there was enough work for everyone who actually wanted to work there. Furthermore, it did not matter if you were a man or a woman, if a person could write properly, research and analyze the

106 Billings, 'Eilene Galloway, the Woman Who Helped Create NASA, Dies at Age 102'; Cory V. Langley, 'Library Analyst Helped Launch NASA', *Library of Congress Magazine*, May/June (2014), 20–21.

107 *Eilene M. Galloway Papers 1940s-2000s*, NASM; Jonathan F. Galloway and Marcia S. Smith, 'Eilene M. Galloway (4.5.1906 – 2.5.2009)', in *Pioneers of Space Law: A Publication of the International Institute of Space Law*, ed. by Stephan Hobe (Leiden, Boston: Brill, 2013), pp. 126–48.

108 Eilene Galloway, interview by Sandra Johnson and Rebecca Wright, September 14, 2000, p. 9, Oral History Transcript, NASA Johnson Space Center Oral History Project, JSC.

109 Eilene Galloway, interview by J.D. Hunley, February 12 and July 1, 1993, p. 60, 011689, HRC.

facts and meet the deadlines "that person was free to operate on the Hill."[110] Another reason for the lack of gender discrimination in the Congressional Research Service, according to Galloway, was that the committees in the House and Senate needed assistance; they were dependent on information from the Congressional Research Service if they did not want to solely rely on information from lobbyists. However, in her function as a library analyst she was not a threat to anyone because she was not the one actually making the decisions. Furthermore, she was a married mother of two sons and thus did not pose a threat in sexual or career terms. Members of Congress were also used to women running their households and offices and thus were not surprised when a woman organized and analyzed the facts surrounding space exploration and space laws.[111] Galloway assumed that there was one more element she was adding that was in short supply during the time of upheaval in the United States caused by Sputnik. "That was assembling all relevant facts in a given situation and analyzing interactions and setting forth probable consequences. [...] I was in a position to put the pieces together with a final multidisciplinary result."[112]

The reasons Galloway gives for her non-discriminatory experience in working for the U.S. government are quite interesting, especially because she assumed that her experience working with space law was true for all women who worked "in all the space things."[113] However, the previous chapters have shown that this was not the case, and that most of the women who worked in the space community had to deal with expectations regarding their gender from their male and female co-workers, regardless if they worked in a 'traditional male' or 'traditional female' job. They had to defend their femininity, while at the same time proving that they had skills. Female engineers and scientists were often perceived as a threat by their male colleagues, even if they were mothers and despite the fact that the space community was in desperate need of engineers and scientists, so their experience was different from Galloway's. The secretaries, on the other hand, were not perceived as a threat, even if they were the ones to bring order to the office in a time of Sputnik-induced social upheaval. The secretarial staff in the space community did, however,

---

110 Eilene Galloway, interview by J.D. Hunley, February 12 and July 1, 1993, pp. 62–63, 011689, HRC.

111 Eilene Galloway, interview by J.D. Hunley, February 12 and July 1, 1993, pp. 62–64, 011689, HRC.

112 Eilene Galloway, interview by J.D. Hunley, February 12 and July 1, 1993, p. 64, 011689, HRC.

113 Eilene Galloway, interview by Sandra Johnson and Rebecca Wright, September 14, 2000, p. 9, JSC.

have grave difficulties getting promoted, unlike Galloway, who entered government service in September 1941 as a P-1 (Professional 1, comparable to GS-5), at a salary of $2,000 a year and became a Senior Specialist in International Relations/National Security in the late 1950s/early 1960s (comparable to GS-17).[114] She was constantly promoted because of her good work, especially her analytical reports on a great variety of topics, including defense and space law.[115] Her experience was thus very different from those of the other women of the space community, including the first female astronaut and the women who wanted to become astronauts, because they were also perceived as a threat to the masculinity of the male space workers and they often had to establish their claim on space – something Galloway never had to do. In order to find out why this was so, it is important to take a closer look at Galloway's life and career.

Galloway was born as Eilene Marie Slack in Kansas City, Missouri, in 1906 as an only child. She was awarded a four-year scholarship to Washington University in St. Louis in the fall of 1923, five years after the university admitted its first female medical students and three years after women received the right to vote in the United States. Galloway was therefore unusual in attending university in a time when it was not the norm for women to do so, much less receive a scholarship. During her first week at university, she met her future husband George Galloway, who was a graduate student there. Two years later they got married and moved to Washington, D.C. and then Philadelphia for her husband's new job. Galloway gave up two years of her scholarship in the process, but she finished her education at Swarthmore, studying political science, philosophy, economics and history, before staying on there for two more years to teach in the political science department. Then her husband got a job in Washington, D.C. and they moved back to the capital. She was just starting to work at the Federal Emergency Relief Administration (FERA) when her husband got another job at the National Economic and Social Planning Association. For this job they left Washington, D.C. and toured the United States together for one year, during which she did research for reports her husband wrote.[116]

When they returned to Washington, D.C., Galloway started working for the Legislative Reference Service, which later became the Congressional Research

114 Eilene Galloway, interview by Sandra Johnson and Rebecca Wright, September 14, 2000, p. 8, JSC; Eilene Galloway, interview by J.D. Hunley, February 12 and July 1, 1993, p. 60, 011689, HRC.

115 Eilene Galloway, interview by J.D. Hunley, February 12 and July 1, 1993, p. 61, 011689, HRC.

116 James W. Canan, 'Conversations with Eile Galloway', *Aerospace America*, January (2008), 12–15 (p. 12).

Service. The research arm for the House and Senate was housed in the Library of Congress, where Galloway started to work as the editor for postwar abstracts in 1941. According to Galloway, these abstracts were "like a congressman's Reader's Digest" of all the issues he should be knowledgeable about.[117] Although the United States was not in the war yet, the State Department anticipated that the United States would eventually join the battles of World War II. Her job was to conduct research about problems that might occur after the war was over. Congress just assumed that if the U.S. were in the war, they would eventually win and would have to solve several problems and crises.[118] Galloway wrote all the abstracts about international relations and national security, and when the position of defense analyst was created, she applied for it and got the job because except for her there was no one who had previously worked on the subject in terms of the library giving information.

When the atomic bomb went off, Galloway asked her boss, the director of Congressional Research Service, Dr. Ernest Griffith, if she could write a public affairs bulletin. No one in the CRS had then yet written about atomic energy, so her boss was delighted that she was yet again willing to take on an unexplored topic. Because she did not know anything about atomic energy or the atomic bomb, she first needed to get the science straight. Therefore, she read everything available about the scientific aspects of the subject before finally writing the public affairs bulletin 'Atomic Energy: Issues before Congress,' which was published in 1946. The bulletin, in which she identified the issues and problems of atomic energy, provided the background material for the Congressmen and Senators who were holding hearings to set up the Joint Committee on Atomic Energy. Thus, Galloway played a major part in reviewing a legislative concern which required the prompt organization of the executive branch and Congress, in order to cope with a current problem – in this case atomic energy, in another, space exploration.[119]

Her work on atomic energy was the same type of work that she had done before, with the difference that it was an emergency and had to be dealt with immediately. This experience helped her to gather information about space exploration and space law after Sputnik shocked the United States. Congressman McCormack, who was Majority Leader of the House, asked her to gather

117 Eilene Galloway, interview by Rebecca Wright, August 7, 2000, p. 8, Oral History Transcript, NASA Johnson Space Center Oral History Project, JSC.
118 Eilene Galloway, interview by Sandra Johnson and Rebecca Wright, September 14, 2000, pp. 1–4, JSC.
119 Eilene Galloway, interview by Sandra Johnson and Rebecca Wright, September 14, 2000, pp. 6–7, JSC.

all the relevant information and write a report about space exploration and the resulting issues before Congress, in order to bring the Congressmen and Senators up to speed. Senator Lyndon Johnson, who was Majority Leader in the Senate at the time, also worried about the use of outer space and that the Soviet Union was getting ahead. He, too, turned to Galloway and requested her assistance with the investigation. In 1957, there was no committee in the House or the Senate that was prepared to undertake the issue of space exploration. In the spring of 1958, Congress therefore set up two special legislative committees, the House Select Committee on Astronautics and Space Exploration and the Senate Special Committee on Space and Astronautics, to create NASA. McCormack became chairman of the committee in the House and Johnson became chairman of the Senate committee, while Galloway delivered all the relevant information. By January 23, 1958, Galloway and her co-workers had collected important information and written 1,377 pages concerning the testimonies of preparedness experts, scientists and engineers from many sectors. Their information about space exploration helped to change the perception of space as a military problem to one of space as an arena for peaceful uses of communications, navigation, meteorology and other purposes to benefit all humankind.[120]

Johnson and McCormack both asked Galloway, rather than one of her (male) co-workers, for help because time was of the essence: the Cold War had moved into space and the Americans were desperate to keep up. They needed information about the uses of outer space immediately and could not afford the time to think about the sex of their informant. Galloway was the logical choice because she was the only person in the Congressional Research Service that had worked in related topical fields and she had proven before that she was able to gather a lot of relevant information in a short amount of time. As the very first person in the CRS to start researching space exploration and space law she became the indisputable expert in the field and no one could afford to decline her expertise. Senators, Congressmen, NASA officials and international space researchers were frequent solicitors of her advice and opinions. In the course of responding to them, Galloway was able to expand her expertise about space legislation and thus consolidate her position as a space expert. The decision-makers even continued calling on Galloway long after her retirement. She was invited to write and present papers at conferences and in journals and to

120 Eilene Galloway, interview by Rebecca Wright, August 7, 2000, pp. 6–10, JSC; Eilene Galloway, 'Sputnik and the Creation of NASA: A Personal Perspective', *NASA*, 2007 <http://www.nasa.gov/50th/50th_magazine/gallowayEsaay.html> [accessed 22 August 2016].

review fiscal budgets for NASA and the ISS up to her death in 2009, two days shy of her 103rd birthday.[121]

So what was different for Galloway compared to other women in the space community? Galloway herself thinks that sex and gender was a non-issue in the space community because in this community "nobody cares anything about any of that, because we're studying a subject and trying to do something with it."[122] Considering the experience of other females in the space community, no matter the workspace, it has already become clear that this was not necessarily the case. Galloway also assumed that she was able to advance to such heights in the field of space legislation because she was not a threat to anyone.[123] This assumption seems very likely; although Galloway was very influential, she herself was not the one making the actual decisions. For example, while Galloway was the one to suggest that NASA should be an administration rather than an agency, she was not the one responsible for pressing the issue. McCormack was the one who had asked her what she thought about the NASA Act, and she told him her reasons why NASA should not become agency, namely that it would be too narrowly defined and too limited considering that there were a lot of other departments in the government that were involved with space exploration, such as the Department of State, the Department of the Interior, the Department of Commerce, and the Department of Defense. NASA as an agency, as opposed to an administration, would not have wielded enough authority to carry out international programs, which was important because space was an international area of interest.[124] Although it was Galloway who pointed out the importance of NASA becoming an administration, it was McCormack and other politicians and lawyers who elaborated the NASA Act and made it so, thereby strengthening international cooperation both within and outside the government.[125]

Galloway never wanted to become a decision maker or sought out the limelight, nor did she ever directly work in or for the space industry during or after her retirement. While she did volunteer at professional space organizations in the role of a government-oriented objective researcher, her job remained

121 Eilene Galloway, interview by Sandra Johnson and Rebecca Wright, September 14, 2000, p. 28, JSC.

122 Eilene Galloway, interview by Sandra Johnson and Rebecca Wright, September 14, 2000, p. 10, JSC.

123 Eilene Galloway, interview by J.D. Hunley, February 12 and July 1, 1993, pp. 62–64, 011689, HRC.

124 Eilene Galloway, interview by Rebecca Wright, August 7, 2000, p. 10, JSC.

125 Eilene Galloway, *letter to Steven J. Dick*, June 30, 2008, 000747, HRC.

limited to doing requested research for members and committees of Congress, and sometimes for the United Nations. She became an expert in all things space. Her assignments included research about the exploration of the Moon and Mars, communications, remote sensing, navigation, international cooperation, solar power, regulation of orbits, the relation between the military and civilian spheres, and appropriations.[126] Despite her expansive knowledge, Galloway was content with pulling the strings from the second row. She did not threaten anyone's position of power and she did not threaten any workspace to begin with because it was a brand new workspace and there was no rival, male or female, for the job. She had created her own space career by being pro-active and by not being afraid of new and difficult assignments.[127] It is also interesting to note that she never intended to have a career in space exploration to begin with; she was just the most qualified persons at the right time in the right place. There was no man at the time who could have legitimately claimed to do a better job than her and because of the Cold War emergency no one wanted to delay the founding of NASA. Thus, no one cared about the gender of the Congressional Research Service researcher.

We cannot know if gender would have been a more important issue had there not been such a sense of urgency, but the Cold War definitely helped to kick-start Galloway's space career. This was another reason why her experience was so different from that of other women in the space community. While the Cold War was also a factor for the female scientists and engineers, unlike those women, Galloway, working in the traditional female area of the library, did not have to break into a male dominated area. She was also never interested in breaking gender barriers like the first female astronauts. She was a wife and a mother, and her career came second to her husband's because it was his job that had dictated their living location and situation as well as their income. However, according to Galloway, "[a]s a working mother, incidentally, I did not have a 'nanny' problem. I had a full-time housekeeper who lived in, and full support of my family for the work I was doing. My husband was enthusiastic about my career and we shared the same interests."[128] Unlike most of the secretaries in the space community, Galloway was able to pursue a career with the full backing of her family and professional help without having the financial pressure of having to support one's family. Another plus for Galloway's space career was that, although she traveled a lot, she did not have to move for the

126 Eilene Galloway, interview by Rebecca Wright, August 7, 2000, p. 53, JSC.

127 Eilene Galloway, interview by Rebecca Wright, August 7, 2000, p. 47, JSC; Eilene Galloway, interview by Sandra Johnson and Rebecca Wright, September 14, 2000, pp. 6–7, JSC.

128 Eilene Galloway, interview by J.D. Hunley, February 12 and July 1, 1993, p. 58, 011689, HRC.

job. Working in the space industry, it was often necessary to move to a new city if one wanted to move up the career ladder, something which a lot of women were not able to do because of family obligations and the priority of the husband's career. In Galloway's case, her career did not interfere with her husband's career and so there was no competition within the family.

Despite her status as the 'Grand Dame of Space' and her influential position in the space community, Galloway did not see the need to facilitate other women in their space related workspaces with quotas. According to Galloway, it would be "demeaning for a professional career woman to have other persons assume she got a job because she was a woman and had no special qualifications for the work."[129] Considering Galloway's relatively unique work experience, her stance regarding EEO and affirmative action is understandable. Her experience, however, does not alter the fact that most of the men and foremost women working in the space community had to deal with gender expectations, performances, and discrimination in one way or the other, as we have seen in this and previous chapters.

6.6 Work space and the space community

The hypothesis that the space community and NASA, a fairly new organization in the 1960s, offered several workspaces, that were not as defined and predetermined as the mainstream workspaces which also existed in older communities that were not as new and technologically innovative as the space community, was correct. As an experimental agency, NASA had to create rockets and space capsules that had not existed before. Furthermore, the space agency needed to satisfy the international public interest in the national space endeavor in order to keep its political support and funding. Accordingly, the space community had to create workspaces that had not existed prior to the founding of NASA in order to accomplish its goals. However, the novel workspaces were not necessarily spaces for women to thrive outside of the set sociocultural norms and values regarding gender relations and performances. Often, gender behavioral codes of traditional workspaces, such as the newsroom or the service industry, were transferred to similar jobs in the space community without reflection. The guardettes of Kennedy Space Center, for example, had a similar code of dress and conduct to stewardesses. However, there were a couple of workspaces in

129 Eilene Galloway, interview by J.D. Hunley, February 12 and July 1, 1993, p. 65, 011689, HRC.

the space community where women were able to exercise more agency, irrespective of the prevalent American gender belief system and NASA's image period. Eilene Galloway, for example, was able to become a coveted space analyst with the Congressional Research Service in NASA's nationalism, romantic, and pragmatic phase. She was responsible for essential space legislation, without facing gender based barriers and discrimination because her job as a space legislation expert had not existed before the founding of NASA and there was no man available to do an equally good job. A change in the sociocultural coding of masculinity and femininity in most jobs, even in the unusual free workspaces, however, only started to change in NASA's pragmatic period. In line with the general spirit in American society, NASA and other space companies wanted to negate gender barriers in the space community and achieve racial and gender equality in the midst of the Cold War, but only after the height of the space race in the 1970s. Therefore, the next chapter will look into another space of the space community, the equal space. This was an area wherein existing administrative policies and social conventions were challenged, and which eventually changed the sociocultural norms and values regarding gender in the entire space community.

7 Equal space in the 1970s

Mae Jemison, who in 1992 had become the first African American woman in space, once remarked that "[i]t's a noble goal that science should be apolitical, acultural and asocial, but it can't be, because it's done by people who are all those things."[1] This insight is especially relevant to an R&D administration like NASA, which worked on engineering and scientific projects at various installations all over the U.S. Nevertheless, the relevance and influence of the prevalent American culture and politics on the agency, its employees and its missions was treated as a non-issue in the 1950s and 1960s. Even though NASA's managers and employees were probably not aware of, and thus did not acknowledge, the influence of their cultural and political environment and upbringing for their work, they nevertheless reproduced the social and cultural conventions of the Cold War American society of the 1950s and early 1960s, therewith turning the agency's labs and offices into spaces of white male privilege.[2]

Although women and minorities were allowed to participate from the get-go – there was no law outright prohibiting them from doing so – the agency and other space companies did not pay specific attention to their inclusion and promotion either in the 1950s or the 1960s.[3] The agency instead selected its Cold Warriors solely based on merit.[4] NASA only wanted Americans with the "right stuff" to participate in its human or robotic space exploration feats.[5] However, the merit system in place, as well as the definition of merit, generally privileged white men, because minorities and women were often barred or hindered from pursuing space pipeline careers in the sciences or the military due to their social status in U.S. society. Considering that NASA was still in its nationalism phase, focused on the Cold War competition of landing a man on the moon and returning him safely to the earth, it is not surprising that the agency's management and employees alike did not pay extensive attention to

1 Kevin Ryan and James Michael Cooper, *Those Who Can, Teach* (Boston: Wadsworth Cengage Learning, 2010), p. 133.

2 Sheffield, pp. 9, 13.

3 See chapters four and five.

4 Alan B. Shepard, *Women in the Space Age*, May 11, 1965, pp. 6–9, 008998, HRC; Nancy Roman, *letter to Penelope H. Thumberg*, December 28, 1966, 1821, HRC; James R. Jaax, interview by Rebecca Wright, December 14, 2010, p. 175, Edited Oral History Transcript, NASA Johnson Space Center Oral History Project, JSC.

5 Reference to: Tom Wolfe.

https://doi.org/10.1515/9783110629828-007

shifting political priorities in the United States in the mid-1960s that protested this particular culture of privilege.[6]

With many Americans already preoccupied with domestic socio-political, sociocultural, and economic issues such as the civil rights movement, the draft for the war in Vietnam and the start of the women's movement, the agency, which had been founded as a product of the all-pervasive Cold War Sputnik Shock, also did not turn its attention inward, but stayed focus on the Cold War challenge.[7] Several space community employees explained that they felt very much insulated during the time of social unrest in the country, because they were so focused on NASA's mission and did not pay attention to the discussions, demonstrations and sit-ins that swept the country.[8] The Johnson Space Center aerospace engineer Larry D. Davis, for example, explained that "[t]here wasn't Internet. There wasn't twenty-four-hour news. If you missed the news, you missed the news, and a lot of nights you missed the news because you were at work, or you were asleep getting ready to get up at two-thirty in the morning and go to work. So I think the down-and-in focus, we missed a lot of that."[9]

The NASA employees as well as the employees of the private space industry only had few points of contact with the current sociopolitical issues at their workspaces. Anne Accola, a mission planning engineer, for instance, remembered when she was faced with the topic of the Vietnam War at Johnson Space Center. At her first day of work in 1967, the colleague in the personnel department who handled her paperwork told her that he liked her because he did not have to deal with further paperwork for a draft deferment, which he always needed to fill out for the young men. Most Vietnam War comments, however, referred to the fact that she had taken the job in the first place – quite a few of her co-workers felt that she was keeping a man who could have gotten the job

6 Byrnes, pp. 1–47.

7 Jürgen Heideking, Jörg Helbig and Anke Ortlepp, *The Sixties Revisited: Culture, Society, Politics* (Heidelberg: Universitätsverlag Winter, 2001); Jürgen Heideking and Christof Mauch, *Geschichte Der USA* (Tübingen: A. Francke Verlag, 2008), pp. 285–365; Jürgen Martschukat, 'Gewalt in Den USA Der 1960er Und 1970er Jahre: Eine Einleitung', *Amerikastudien/American Studies*, 49 (2004), 321–28; Lamphier and Welch, pp. 83–272.

8 Jamye Flowers Coplin, interview by Rebecca Wright, November 12, 2008, p. 20, Oral History Transcript, NASA Johnson Space Center Oral History Project, JSC; Connie Eaton Harney, interview with Rebecca Wright and Sandra Johnson, June 12, 2001, p. 31, Oral History Transcript, NASA "Herstory" Oral History Project, JSC.

9 Larry D. Davis, interview by Sandra Johnson, February 15, 2007, pp. 35–36, Oral History Transcript, NASA Johnson Space Center Oral History Project, JSC.

from getting a draft deferment.[10] The war in Vietnam was therefore not directly discussed, but was only used to pick on a woman who supposedly 'stole' the job away from a 'more deserving' man. There were overall but few political discussions about the war or civil rights.[11] Lively political debates and activism were objectionable at NASA in the 1960s and had to be practiced outside of work.[12] Astronaut Buzz Aldrin, for example, participated in a peace march in Houston after Martin Luther King Jr.'s death, while the Marshall Space Flight Center technical artist Marshall Keith staged a sit-in with African American students in an Alabama Woolworth.[13]

Considering that President John F. Kennedy and Lyndon B. Johnson were convinced that the space industry might help to improve race relations in the South through innovation, economic advancement, and the merit system of 'equality according to ability,' it is quite astounding that the members of the space community displayed such indifference regarding domestic political, social, and cultural issues in the United States. Johnson had specifically lobbied for the new NASA facilities to be established in former Confederate states in order to help usher in better race relations.[14] He wanted to use NASA as a catalyst to advance social justice and equality in that particular region according to the popular belief in the 1950s and 1960s that "communities with advanced types of industry, with their people employed in research laboratories and in the development of new engineering techniques, should display a high level of social innovation."[15]

In 1962, the American Academy of Arts and Sciences sent sociologist Peter Dodd, on behalf of NASA, to the space facilities to find out to what extent this popular conception of the social impact of 'advanced communities' actually applied to this specific 'advanced community.' His study was thus in line with modernization theoretical thinking at the time, the idea that societies and communities 'advance' and become more 'modern.' Until the late 60s, NASA and

---

10 Anne L. Accola, interview by Rebecca Wright, March 16, 2005, pp. 14–15, Oral History Transcript, NASA Johnson Space Center Oral History Project, JSC.

11 Jamye Flowers Coplin, interview by Rebecca Wright, November 12, 2008, p. 20, Oral History Transcript, NASA Johnson Space Center Oral History Project, JSC; Connie Eaton Harney, interview with Rebecca Wright and Sandra Johnson, June 12, 2001, p. 31, Oral History Transcript, NASA "Herstory" Oral History Project, JSC.

12 There are at least no sources that suggest otherwise.

13 Paul; Koppel, pp. 309–310.

14 For more information see *Mapping the Space Community.*

15 *American Academy of Arts and Sciences Report*, 1962, in: Paul.

the space community were often used as an example for 'progress' and 'modernity,' markers that were only started to be challenged in the 1970s.[16]

In multiple visits to Johnson, Kennedy and Marshall Space Flight Centers, Dodd found out that there was "no evidence of strong pressure for Negro rights, nor of strong sympathy among technologists for civil rights."[17] Furthermore, he believed to have realized that NASA workers did not feel responsible in any way for changing the current social status quo. Dodd found that within the space community, "the Negroes appear to be an outside group presenting demands which would have to be dealt with in some way, but which are no concern of theirs."[18] These findings are especially interesting in regard to several studies undertaken at the time that had suggested otherwise. According to those studies, space workers were expected to demonstrate liberal political views and opinions because of their high levels of education as well as their youth and geographic mobility, "which are known to be correlated with liberal views."[19]

The space community members only dealt with social and political issues in the 1960s when they were literally standing at their gates. The Poor People's Campaign organized by Martin Luther King's Southern Christian Leadership Conference (SCLC) under the leadership of Reverend Ralph Abernathy came to protest the launch of Apollo 11 in 1969. The SCLC and a lot of other Americans questioned the space program's big budget in comparison to the federal budget for social issues and the existence of abyssal poverty in the U.S.[20] NASA officials responded by inviting some representatives of the SCLC to watch the launch of the space vehicle that would bring the first human beings to the moon and Abernathy, just like the majority of space community members, momentarily forgot about social justice because he was so captivated by the achievement. After the launch Abernathy admitted that "I really forgot we had so many hungry people" during the short period he sat in the front row of America's race to space.[21]

Abernathy's brief social justice memory loss might give an insight into the minds of the American people and the space community workers in particular. They were enamored with space, focused on the missions, and motivated to

16 Gilman; Thomas Mergel, 'Modernization', *European History Online*, 2012 <http://ieg-ego.eu/en/threads/models-and-stereotypes/modernization/thomas-mergel-modernization> [accessed 21 September 2016]; Natasha Zaretsky, *No Direction Home: The American Family and the Fear of National Decline, 1968–1980* (Chapel Hill: University of North Carolina Press, 2007); Jodi Dean, p. 67; Heinemann, p. 14.

17 *American Academy of Arts and Sciences Report*, 1962, in: Paul.

18 *American Academy of Arts and Sciences Report*, 1962, in: Paul.

19 *American Academy of Arts and Sciences Report*, 1962, in: Paul.

20 Makemson, pp. 182–185; Shetterly, pp. 240–242.

21 Makemson, p. 185.

beat the Soviet Union. This was important to all Americans; it was their patriotic duty, after all. However, after the success of Apollo 11 left most Americans feeling that they had won the space race, the perception of the space program and the space community changed. While the community's inactivity regarding social justice and equal opportunity was not questioned in the 1960s, the 1970s presented a different decade to NASA and the space industry, one in which the majority of citizens no longer felt that the space program was worth the cost, especially in comparison to other human needs.[22] According to several opinion polls in the 1960s, a majority of Americans already believed that the Apollo program was too expensive to begin with and that after the moon landing interest in the space program waned so much that "one has the image of the program as something akin to a limping marathoner straining with every muscle to reach the finish line before collapsing."[23] As the number of people who found NASA relevant shrank, so too did the agency's budget.

The smaller budget compelled NASA to have several reductions in force (RIF) . Its civil-service workforce, for example, was reduced by a third by 1974. Only foreign aid was more unpopular than the space program in terms of government spending in the early 1970s. Even the federal welfare budget was less contested. Considering that NASA, like all other agencies, needed political support to function, the National Aeronautics and Space Administration had to adjust its public image and begin relating to issues of importance to the American people.[24] Thus, the space community needed to face and include the long ignored gender and race debates of American society in their community. To do so, NASA and the space companies had to incorporate the inclusionary laws in their hiring policies and, more importantly, tackle exclusionary customs and individual and structural discrimination in their daily work routines.

Several federal laws in favour of gender and racial equality at the workplace were passed and changed in the United States in the 1960s and 1970s.[25]

---

22 Foster, *Integrating Women into the Astronaut Corps. Politics and Logistics at NASA, 1972–2004*, pp. 81–83; McQuaid, 'Race, Gender, and Space Exploration: A Chapter in the Social History of the Space Age', p. 408.

23 Roger D. Launius, 'Public Opinion Polls and Perceptions of US Human Spaceflight', *Space Policy*, 19 (2003), 163–75 (p. esp. 168).

24 Byrnes, pp. 1–3 and 115; McCurdy, *Inside NASA. High Technology and Organizational Change in the U.S. Space Program*, p. 124 ff; McQuaid, 'Race, Gender, and Space Exploration: A Chapter in the Social History of the Space Age', pp. 408–409.

25 Foster, *Integrating Women into the Astronaut Corps. Politics and Logistics at NASA, 1972–2004*, pp. 19–22 and 71–74; Lamphier and Welch, pp. 106, 124, 131, 133–136, 241; Hugh Davis Graham, *Civil Rights and the Presidency. Race and Gender in American Politics 1960–1972* (Oxford: Oxford University Press, 1992).

To apply those laws not merely on paper, but in the actual hiring and work practices in the space community, was far more challenging. As it turned out, it was considerably harder to change the mindsets and culture in the workspace regarding EEO and affirmative action than it was to change the law. During the 1960s, NASA had refused to even acknowledge the existence of race and gender issues in the space community because of its merit system of 'equality according to ability,' ignoring the fact that women and minorities often had no or only minimal access to NASA's pipeline careers such as engineering or military test pilot academy. As late as September of 1971, as Kim McQuaid rightfully pointed out, NASA did not have a systematic EEO program, even though a lot of its field centers were located in Southern states like Texas, Alabama, Mississippi, Florida, Louisiana, and Virginia.[26] Thus, NASA's first efforts to establish equal employment regulations in all its centers and with contractors was rushed, rather disorganized and not always very productive. This chapter now intends to take a closer look at the very workspace that tried to bring gender and racial equality to NASA in terms of employment: the EEO office. The EEO office was a workspace just like the labs or the offices, and like them it was also a socially produced space, perpetuated and reproduced on a daily basis. However, unlike the other workspaces, its sole purpose was to break through the established performance of social conventions and exclusionary customs in all the spaces of the space community and to implement new administrative policies and work routines. Its employees, programs, and work routines not only reveal the state of the American spaceflight program regarding gender and race, but also show the mindset and cultural and political structures of American society during the 1970s, a time of thawing Cold War marked by détente and arms control and a time of social transition.[27]

---

26 McQuaid, 'Race, Gender, and Space Exploration: A Chapter in the Social History of the Space Age', p. 409.

27 Neil A. Hamilton, *The 1970s* (New York: Facts on File, 2006); Hallvard Notaker, David J. Snyder and Giles Scott-Smith, *Reasserting America in the 1970s: U.S. Public Diplomacy and the Rebuilding of America's Image Abroad* (Manchester: Manchester University Press, 2016); Jan Eckel and Samuel Moyn, *The Breakthrough: Human Rights in the 1970s* (Philadelphia: University of Pennsylvania Press, 2014); Sven F. Kraemer, *Inside the Cold War From Marx to Reagan: An Unprecedented Guide to the Roots, History, Strategies, and Key Documents of the Cold War* (Lanham: UPA, 2015), pp. 97–146; David M. Walsh, *The Military Balance in the Cold War: US Perceptions and Policy, 1976–85* (New York: Routledge, 2008).

7.1 The formation of the EEO office – laws and policies

According to some researchers and NASA employees, “NASA decision makers were as courageous in their willingness to discuss workplace diversity, in newsletter and in the media, as they were in their efforts to walk on the moon.”[28] They believe that NASA sincerely committed itself to promoting and hiring women and minorities in accordance with equality and diversity standards from the get-go, before the agency was forced to do so by the federal government and the women’s movement. They were under the impression that the space community was at the forefront of social change and EEO activities because NASA publicly emphasized the work of its female employees early on.[29] However, as we have seen in the chapter *Lady engineers and astrowives in the 1960s*, NASA mainly did so to steer the press and the public away from the request to include women and minorities in the astronaut corps. The daily work routines of the women introduced in the previous chapters, as well as the exclusionary customs and individual and structural discrimination they were faced with in the workspaces of the space community in the 1960s, show that NASA was most certainly not an agency that pressed for equal employment and affirmative action for minorities and women from the outset. NASA always hired and worked according to the current law, but did not go the extra mile in the social innovation in its workspaces as suggested by some authors and employees. Only when the law changed, and it did eventually, did NASA and its contractors act accordingly.

In order to understand the work and obstacles of NASA’s EEO office in the 1970s, it is important to reconstruct the EEO legislative process that paved the way for women and minorities in the workspaces of the United States generally and at NASA in particular.[30] The first law to facilitate equal employment was passed as early as 1883, when the Pendleton Civil Service Reform Act established the merit system, thereby abolishing the spoils system. President Roosevelt’s Executive Order 8587 later outlawed race, color, and religious discrimination in public employment, while the Ramspeck Act provided a statutory ban on discrimination in U.S. Civil Service in 1940. Executive Order 9980, issued by President Truman in 1948, established a fair employment committee to serve as the government contract compliance agency within the Civil Service

28 Williams, p. 369.

29 Nancy L. Fee, interview by Rebecca Wright, November 29, 2011, p. 4, Oral History Transcript, NASA Johnson Space Center Oral History Project, JSC; Foster, *Integrating Women into the Astronaut Corps. Politics and Logistics at NASA, 1972–2004*, pp. 71–73 and 154; Williams, pp. 369, 370, 377, 379, 380.

30 MacLean, pp. 117–154.

Commission to ensure non-discrimination in employment and service. President Eisenhower's Executive Order 10950, issued in 1955, mandated that equal opportunity be afforded to all persons for employment in the federal government.[31] President Kennedy's Executive Order 10925 required government contractors to take affirmative action and established the President's Committee on Equal Employment Opportunity to oversee contract compliance in March of 1961. It was the first EEO law that was issued after the foundation of NASA. This executive order was shortly followed by the creation of the Presidential Commission on the Status of Women (PCSW), which was created by Kennedy's Executive Order 10980 in December 1961. The final report of the Presidential Commission on the Status of Women affirmed the roles of wives and mothers, on the one hand, but also stated, on the other hand, that there was female talent that could be tapped for the Cold War effort. It also disclosed the discrimination women had to deal with in the workforce. As a direct result of the report, Kennedy signed an order that instructed the Civil Service Commission to hire its employees solely on merit and not sex. John F. Kennedy also signed the Equal Pay Act, as part of his New Frontier Program, into law in 1963. This amendment to the Fair Labor Standards Act made it illegal to pay men and women different wages for doing the same job.[32] According to the African American commission lawyer and activist Pauli Murray, another result of the commission was that President Kennedy had unintentionally created the very "first high-level consciousness group" for women.[33]

So how did the space community handle the new EEO legislation in their hiring and working practices? NASA Administrator James E. Webb announced that NASA would follow suit on the President's orders about EEO:

> As Administrator of NASA I support fully the President's efforts. It is my intention to take positive steps to insure equal opportunity for all qualified persons. Consistent with this purpose there shall be no discrimination, direct or indirect, because of race, creed, color, or national origin of qualified persons employed or seeking employment in the National

31 *Brief E.O. History*, 1974, 18508, Box 1, HRC; Harris, pp. 299–300. See also: Article VI in the U.S. Constitution prohibits religious discrimination, issued in 1787; The Fair Labor Standards Act established minimum wage, overtime compensation, and other regulations of employment but did not cover federal employees, issued in 1938.

32 Harris, p. 298; Sara M. Evans, *Born for Liberty* (New York: Free Press Paperbacks, 1997), p. 275; Foster, *Integrating Women into the Astronaut Corps. Politics and Logistics at NASA, 1972–2004*, p. 19; Rosen, pp. 66–70; Gail Collins.

33 Susan M. Hartmann, *From Margin to Mainstream: American Women and Politics since 1960* (New York: Knopf, 1989), p. 53.

Aeronautics and Space Administration. I expect NASA employees at all organizational levels to give full support and cooperation to this program.[34]

While Webb did not mention women in his first EEO memorandum in 1961, he did cover the topic of sex-based discrimination at NASA in his Management Manual Instruction of February 1962, which directly referred to the President's Commission on the Status of Women. Similar to his first EEO memorandum, in which he had declared the prohibition of work place discrimination on the basis of race, creed, color, or national origin, he announced the very same for the discrimination of women – with almost the same exact wording.[35] Webb actually cancelled this particular Management Manual Instruction and turned it into NASA policy directive 3713.3 using the exact same words in March of 1966.[36] Although these new regulations technically protected women and minorities, and Webb's reaction to the new EEO legislation proved that the space community's management and administration definitely took notice of it, the enforcement of these regulations at the various workspaces was mostly neglected. There is only a very short list of measures NASA introduced to promote equal employment in the years before the 1964 Civil Rights Act became law. According to NMI 2-1-1.3A NASA declared its Director of Procurement to be responsible for also monitoring EEO compliance by contractors and sub-contractors.[37] However, there are few to no documents which might prove that the Director of Procurement actually implemented the new instruction and undertook the new duties.

The agency also initiated some direct measures to target EEO, such as recruiting events at high schools and colleges, and started a co-op program between NASA and several universities and technical/vocational schools. The work study program alternated between studying at one's college and working at NASA. The idea was that the students would gain practical experience and earn money.[38] Dryden, for example, started a co-op program with Iowa State University which included female engineering students such as Sheryll Goecke

34 James E. Webb, *Administator's Memorandum on Equal Employment Opportunity*, May 8, 1961, 18507, Box 4, HRC.

35 James E. Webb, *Administrator's Memorandum on Equal Employment Opportunity for Women*, Management Manual Instruction 3-1-5.3, February 20, 1962, 008994, HRC.

36 James E. Webb, *Administrator's Memorandum on Equal Employment Opportunity for Women*, Policy Directive 3713.3, March 31, 1966, Box GR 1006, NASA Management Instructions, Management Delegates/ Policy Directives, JSC.

37 *EEO Responsibility in NASA: A Chronology*, undated, EEO, General 1969–2001, 18507, Box 4, HRC.

38 Nathaniel Pilot, interview by Dr. Aaron Gillette, July 28, 1992, p. 5, Tape Transcript, HRC; Powers, p. i.

Powers, while the Marshall Space Flight Center invited students of historically black colleges to Huntsville in 1963. However, NASA discontinued the cooperative education program with Iowa State after a couple of years because the agency felt that the university did not provide enough minority students. NASA also worked together with a contractors' group in Alabama which made the effort to recruit African Americans for the space community. The agency, furthermore, hired Charlie Smoot, who was known as the "first Negro recruiter" and who travelled throughout the U.S. to convince black scientists and engineers to join NASA in its Southern field installations.[39]

NASA's early EEO activities were few and scattered. The was no systematic, large-scale EEO program at NASA but sporadic EEO efforts, which mainly concentrated on recruiting minorities and overseeing its contractors' contract compliance regarding EEO. Consciously bringing women into NASA's workspaces was not a priority despite the new EEO legislation and the shortage of skilled and trained personnel in the natural sciences and engineering alongside the public fear that the Soviet Union was using its women as national resources in the Cold War.[40] However, it has to be emphasized that NASA's early EEO efforts were in line with other federal agencies and private sector employment at the time and that Webb's memoranda on Equal Employment Opportunity for Women actually exceeded the federal law's requirement of NASA. Legislation that forced the agency to not discriminate against women came later with the Johnson administration. For Webb to already sensitize his managers to women's issues "showed remarkable open-mindedness and forward-thinking, a reflection of the fact that Jim Webb was an extraordinary man and career federal bureaucrat," according to historian Amy Foster.[41]

President Johnson continued what the Kennedy administration had started. After the March on Washington, the assassination of President Kennedy, 200

39 Sheryll Goecke Powers, interview by Rebecca Wright and Sandra Johnson, June 13, 2001, p. 11, Oral History Transcript, NASA "Herstory" Oral History Project, JSC; Paul.

40 Erik Bergaust, 'Is Space a Place for the Ladies?', *This Week*, September 2, 1962; Gregory Donald H., *Designing the Space Age*, February 19 and 21, 1963, GH 51, General Reference Series, JSC; Lillian Levy, 'Career Girl 1959. A Space Career. She doesn't want to be launched', *Sunday Star*, May 31, 1959; *Statement of Senator James Abourezk before the Senate Aeronautical and Space Science Committee*, January 24, 1974, Harris/MSRC; Blotnick, p. 21; Bix, 'A History of Women's U.S. Engineering Education', p. 36.

41 Foster, *Integrating Women into the Astronaut Corps. Politics and Logistics at NASA, 1972–2004*, p. 71; Kevin Stainback and Donald Tomaskovic-Devey, *Documenting Desegregation: Racial and Gender Segregation in Private Sector Employment Since the Civil Rights Act* (New York: Russel Sage Foundation, 2012), pp. 78–83; Polly Reed Myers, *Capitalist Family Values: Gender, Work, and Corporate Culture at Boeing* (Lincoln: University of Nebraska Press, 2015).

proposed civil rights bills, 500 proposed amendments and 534 hours of debate, Congress passed an important piece of legislation: the Civil Rights Act of 1964. Title VII, also known as 'the Bunny Law,' made any employment discrimination on the basis of race, sex, color, religion and national origin in the private sector illegal. It was the Southern Congressman 'Judge' Howard Smith and other conservative advocates of racial segregation who suggested adding 'sex' to the title as a way to kill it or to at least not make it another 'negro's hour.' While the unusual coalition of racists and feminists had some people worried that the amendment might risk African Americans' opportunity to gain civil rights, it was the support of the segregationist that made women a part of the civil rights legislation.[42] This fact is especially interesting in regard to the administration of the NASA EEO office, which always treated minorities' and women's issue as one and the same, a problem that will be discussed in the following sections.

Title IX of the 1964 bill created the Equal Employment Opportunity Commission (EEOC). The EEOC was established as a regulatory office to investigate individual complaints regarding discrimination, making sure that employers abided by the law. Although a third of the complaints received by the EEOC came from women charging their employers with sex discrimination, the commission did not really concern itself with sex discrimination complaints; its priority was race discrimination relating to African American men.[43] Executive Order 11246, issued by President Johnson in 1965, ordered the federal government to bar workplace discrimination based on race, color, religion, or national origin and placed the program under leadership of the Civil Service Commission.

The order did not, however, protect against sex based discrimination, which is most likely why NASA stated in its policy of non-discrimination that any "federal employee or qualified applicant for federal employment who believes he has been discriminated against because of race, color, religion, or national origin may file a written signed complaint."[44] 'Sex' as a reason to file a complaint was notably missing, and it was only in 1967 that Johnson outlawed sex discrimination by the federal government and its contractors with Executive

---

42 William Henry Chafe, *The Unfinished Journey: America Since World War II* (New York: Oxford University Press, 1984), p. 330; Rosen, pp. 71–74; Jo Freeman, *We Will Be Heard: Women's Struggles for Political Power in the United States* (Lanham: Rowman & Littlefield, 2004), pp. 171–190; Harris, pp. 298–299; Hugh Davis Graham, pp. 46–86.

43 Rosen, pp. 72–73; Terry H. Anderson, *The Sixties*, 2nd edn (New York: Pearson Longman, 2004), pp. 85–86; Herman Belz, *Equality. A Quarter-Century of Affirmative Action* (New Brunswick: Transaction Publishers, 1992), pp. 69–88; Hugh Davis Graham, pp. 102–116.

44 *Policy on Nondiscrimination*, July 12, 1965, GR 1013, MSC Announcements NOS. 65–89 through 65–170, July 6, 1965–December 28, 1965, Center, JSC.

Order 11375, which amended Executive Order 11246. The Civil Service Commission responded with the foundation of the Federal Women's Program, which was supposed to support women in advancing within the civil service. The Federal Women's Program was incorporated into the EEO program in 1969 through President Nixon's Executive Order 11478, which was, at the time, the most far-reaching policy exclusively targeting federal employment. It amended Title VII of the Civil Rights Act and laid the groundwork for the Equal Employment Opportunity Act which would follow three years later. Although the new order gave the CSC statutory authority to establish and oversee affirmative action policies and non-discrimination techniques and action plans, the orders had almost no impact on the work reality of NASA's female and minority workforce.[45]

Yet again, NASA's administrator published a policy directive affirming the importance of and compliance with EEO at NASA. However, NPD 3713.2A did not leave a profound mark on NASA's EEO activities, mainly because there was still no coordinated effort from headquarters to bring all field installations and contractors in line. NASA's Civil Service EEO responsibilities were pushed from one office to the next for a couple of years. NMI 3713.4 declared the Director of Personnel to be NASA's EEO officer in 1966. In 1969, NMI 3713.4A established that the Associate Administrator for Organization and Management would also become NASA's EEO Director and in January 1970, the Director of Personnel was elevated to become the EEO Deputy Director, according to NMI 3713.4B.[46]

This decision to combine EEO and personnel in one office was rather questionable due to the fact that most of the EEO complaints concerned personnel actions and thus both sides of this workspace clashed with each other.[47] Despite calls for "maximum contribution" and "extra steps" to "insure that those

45 Jennifer Ross-Nazal, *Women Scientists at NASA*, October 22, 2005, p. 4, 008988, HRC; *Brief E.O. History*, 1974, 18508, Box 1, HRC; Helen S. Markoff, 'The Federal Women's Program', *Public Administration Review*, 32 (1972), 144–51 (p. 145); Terry H. Anderson, p. 86; Harris, pp. 299–300; Blanche Linden-Ward and Carol Hurd Green, *American Women in the 1960s: Changing the Future* (New York: Twayne Publishers, 1993), pp. 12–13. The 1966 Age Discrimination in Employment Act, which amended the Fair Labor Standards Act, was also an important EEO law. It prohibited discrimination against employees 40 years of age or older because of their age. However, it did not cover federal, state and local governments.

46 EEO Responsibility in NASA: A Chronology, undated, EEO, General 1969–2001, 18507, Box 4, EO Programs, Office of, NASA Historical Reference Collection, Washington, D.C.

47 Ruth Bates Harris, Joseph M. Hogan, and Samuel Lynn, *Statement before the House Judiciary Sub-Committee on Civil Rights and Constitutional Rights*, March 13, 1974, Judiciary Committee 93rd Congress, Subcommittee on Civil and Constitutional Rights, Box 37 of HR. BX. 33, NASA Hearing Correspondence etc., NASA 1974 Background Information for Members for Hearings, RG 233, NARA I.

who have been disadvantaged or handicapped in any way are, in fact, given opportunity to develop their potential," NASA's EEO decisions in the late 1960s demonstrate that EEO was neither taken seriously, nor regarded as a major part of the agency's employment and workforce strategy.[48] That this was the case is made even clearer by the fact that the EEO responsibilities were only performed on a part-time basis by the joint office. The agency also did not establish a central control organ, instead splitting oversight of EEO between different offices and turning control of EEO matters over to individual field centers.

It was thus up to the directors' discretion how they implemented the administrator's EEO policies. The director of Johnson Space Center, Robert Gilruth, for example, took Webb's new policies seriously and named several EEO officers and regularly distributed management instructions starting in 1969.[49] However, just like at Kennedy Space Center, the EEO office at Johnson Space Center was not effectively and prominently positioned within the center. At neither location did it report directly to the director or deputy director, which would have been important in effecting the assertive and swift implementation of EEO at the respective centers. Meanwhile, the director of Marshall Space Flight Center, Eberhard Rees, staffed his EEO office very inadequately and thus was not able to implement change due to competency and trust issues. EEO officers at Marshall, such as Arthur Sanderson, were not able to instantiate change for minorities or women and thus minority employees had no confidence in him or his colleagues.[50] Trust issues were actually a widespread problem in the workspaces of the space community, when it came to the implementation of EEO. Because trust is essential to building productive workspaces in a hierarchical organization that was run similarly to a military organization, which needs trust to function smoothly, EEO efforts frequently failed or were delayed due to trust issues on both sides.[51]

NASA only started its EEO program in 1969, shortly after the first moon landing, because the agency was pressed by the Department of Labor to assume a role in overseeing if its contractors provided equal employment opportunities. In response to this pressure, a freshly hired contract compliance specialist was asked to plan an appropriate contract compliance program in 1970. His original plan with proper staffing and favorable organizational placement was first

48 Charles G. Haynes, *Equal Employment Opportunity in NASA Headquarters*, memorandum to all Supervisors in NASA Headquarters, July 25, 1969, 18507, Box 4, HRC.

49 Foster, *Integrating Women into the Astronaut Corps. Politics and Logistics at NASA, 1972–2004*, p. 72.

50 Lloyd B. Gardner, *letter to Ruth Bates Harris*, April 7, 1972, 18508, Box 1, HRC.

51 Neville A. Stanton, *Trust in Military Teams* (Boca Raton: CRC Press, 2011).

approved in April 1970 and then obstructed in its execution. Instead of a centralized effort, each center was ordered to staff and operate an individual contract compliance program while the number of compliance specialist was arbitrarily cut by 25 percent, leaving 18 (instead of 24) contract compliance EEO officers for the centers and five (instead of six) officers at headquarters. Furthermore, the rank of the EEO officers was dropped by one GS grade and the program was placed under the control of the procurement office at each center, despite the repeated warnings of the contract compliance specialist that had created the original plan that contract compliance and procurement were incompatible. Although the program was eventually placed under the Director of Industrial Relations with a special announcement in February 1971, the warnings that "these decisions, if not reversed, would cause extremely serious problems and could cause a total failure of the program" were otherwise ignored.[52] Most of the field installations procrastinated and did not make it a priority to search for and select qualified EEO contract compliance officers, but they eventually had to choose officers because of an announced compliance training program at headquarters in January 1971. Out of the 16 officers, only one was a trained compliance specialist and only one was a woman. Six of eight teams were exclusively white men and four of eight field centers only used part-time personnel. Three centers actually did not even manage to appoint the required number of specialists in time.[53]

Not surprisingly, the selected officers at the different centers who had to work without a central control office were not able to bring about change because they were so severely limited in their effectiveness. Some of the officers even felt that NASA tried to sabotage EEO efforts from the get-go. Joseph M. Hogan, who started working at NASA headquarters as a contract compliance specialist in 1969, was convinced that the EEO program had been structured in a way calculated "to provide an appearance of compliance while real progress was thwarted." According to him, this kind of behavior "showed a contempt for the law. Equal Employment Opportunity Programs are intended to bring justice

52 Ruth Bates Harris, Joseph M. Hogan, and Samuel Lynn, *Statement before the House Judiciary Sub-Committee on Civil Rights and Constitutional Rights,* March 13, 1974, NARA I; See also: *NASA Hearings before the Subcommittee on Civil Rights and Constitutional Rights of the House Committee on the Judiciary: Background Information*, 1974, Judiciary Committee 93rd Congress, Subcommittee on Civil and Constitutional Rights, Box 37 of HR. BX. 33, NASA Hearing Correspondence etc., NASA 1974 Background Information for Members for Hearings, RG 233, NARA I; *EEO Responsibility in NASA: A Chronology*, undated, 18507, Box 4, HRC; *Background on Contractor Equal Employment Opportunity Program*, 1971, 18508, Box 1, HRC.

53 Ruth Bates Harris, Joseph M. Hogan, and Samuel Lynn, *Statement before the House Judiciary Sub-Committee on Civil Rights and Constitutional Rights,* March 13, 1974, NARA I.

to persons who have suffered discrimination. To frustrate these programs is an obstruction of justice."[54] It was not until the foundation of a central Equal Employment Opportunity Office in September of 1971 that NASA's EEO record was first properly assessed. However, it still took the EEO officers a couple of years to get all administrators, middle-management and employees on the same page regarding the importance of implementing equal employment strategies at NASA and its contractors.

NASA's management had constantly been urged to establish such an office by the headquarters contract compliance staff as well as by Associate Administrator for Organization and Management Richard C. McCurdy. They all saw the need for an EEO office because of the existing decentralized management practices as well as the burial of the three principle civil rights programs at different departments within NASA. The EEO program for civil service employees was in the Office of Personnel, the contractor's EEO program was administered by the Industrial Relations Office within the Office of Industry Affairs and Technology Utilization, and the minority small business program was in the Procurement Office. While the headquarters contract compliance staff lobbied for a new central EEO office that would report directly to top management, McCurdy suggested placing the office in his department, the Office of Organization and Management. Unlike the staff, he did not want to employ new personnel, but to transfer most of the existing personnel to the new EEO office while leaving the minority business function in the procurement division.[55] The EEO office was created as McCurdy had suggested, but was eventually elevated in April 1973 when it received an Assistant Administrator for Equal Opportunity Programs as its head.[56]

Some people in- and outside of the space community felt that the new EEO office was desperately needed in order to push "the agency to achieve the legal requirements of a federal agency, to abide by the law, to be able to achieve all of the aeronautical and space objectives that NASA was accomplishing with an integrated staff."[57] It had become obvious to them that it was about time for

54 Joseph M. Hogan, *testimony of Joseph M. Hogan in front of the Senate Appropriations Subcommittee on Housing and Urban Development, Space, Science and Veteran Affairs*, January 11, 1974,Harris/MSRC. See also: Joseph Hogan, *NASA Civil Rights Posture*, memo to Deputy associate Administrator for Organization and Management, March 9, 1971, 18508, Box 1, HRC.

55 Richard C. McCurdy, *memorandum to Administrator James C. Fletcher*, August 31, 1971, Harris/MSRC; Ruth Bates Harris, Joseph M. Hogan, and Samuel Lynn, *Statement before the House Judiciary Sub-Committee on Civil Rights and Constitutional Rights*, NARA I.

56 Harriett Jenkins, *NASA Equal Opportunity Program*, January 12, 1976, Box 1, Equal Opportunity 1976, Center Files, JSC.

57 Harriett G. Jenkins, interview by Jennifer Ross-Nazzal, August 5, 2011, pp. 2–3, edited Oral History Transcript, NASA Headquarters Oral History Project, JSC.

NASA to establish a central EEO office because NASA always compared itself, including its EEO status, with similar agencies in the federal government, such as the Federal Aviation Administration, the National Institute of Health, or the Department of Defense. At each of these agencies a single EEO office which reported to the Associate Administrator level had precedence and complied with Executive Order 11478.[58] Moreover, the Equal Employment Opportunity Act, issued shortly after the foundation of the EEO office in March 1972 by President Nixon, strengthened Title VII a good deal more because it placed public sector employees under the Civil Rights Act of 1964, giving federal employees access to the courts and ensuring that all federal agencies had affirmative action plans in place. The legislation validated equal opportunities for women that had previously been overlooked within various agencies such as NASA.

Even though the organizational culture for women at NASA, as at other federal agencies, still needed more coaxing over the next years and even decades, the 1972 amendment represented a watershed for women, securing job openings thus far denied them because of their sex. The 1972 act clearly demonstrated that sexual discrimination at the workplace would no longer be overlooked. Quite a few motivated female scientists in the U.S. actually used these new legal protections and filed class-action lawsuits against their agencies, including NASA.[59] The Department of Labor Office of Federal Contract Compliance's Revised Order No. 4, issued in December 1971, which had an effective date of April 1972, strengthened women's rights even further. The policy therewith expanded to encompass women employed by federal contractors. Federal employers and contractors were advised to develop goals and timetables to stop the 'under-utilization' of women in jobs they were overqualified for and to increase the number of women they employed, not only in sheer numbers but at the higher levels as well.[60]

NASA reacted to the change in legislation the same way as it had done in previous years. The new administrator, James C. Fletcher, distributed a policy directive which was in compliance with the new legislation. Again, it sounded

58 Harriett G. Jenkins, interview by Jennifer Ross-Nazzal, August 5, 2011, pp. 51–52.

59 Harris, p. 300; Markoff, pp. 146–147; Kessler-Harris, p. 315; Foster, *Integrating Women into the Astronaut Corps. Politics and Logistics at NASA, 1972–2004*, p. 20; Joyce Gelb, 'Equal Employment Opportunity Policy in the United States', *PS Online*, 2004; Rossiter, *Women Scientists in America. Vol. 3. Forging a New World since 1972*, pp. 189–191.

60 Ruth Bates Harris, *Dissemination of Initial NASA Guidelines on Contractor Equal Opportunity for Women*, June 13, 1972, 18508, Box 1, HRC; 'Regulations drawn to widen women's job opportunities', *Newsletter: NASA Activities*, January 15, 1972, GR 1701, NASA Information Summaries, General Reference Series, JSC; Diane Graham, *Revised Order No. 4*, April 4, 1972, 18508, Box 1, HRC.

similar to the previous NPDs, stating the purpose, applicability, authority, policy, implementation and cancellations:

> It is the policy of the National Aeronautics and Space Administration to provide equal employment opportunity for all persons, to prohibit discrimination in employment by NASA because of race, color, religion, sex, or natural origin, and to promote the full realization of equal employment opportunity on the basis of merit and fitness through a continuing affirmative program throughout the Agency. NASA's goal is the elimination of discrimination in employment and in the utilization and development of human skills in all activities, organizational segments, in all occupations, and at all levels. Every supervisor and manager is expected to take positive action toward achieving this goal.[61]

However, this time around the central EEO office already existed and the agency tried to put its policies into practice.

Still, the newly founded EEO office and its commitment was called into question by its own officers. The Deputy Administrator for the Equal Employment Office, Ruth Bates Harris, along with EEO and contract compliance officers Samuel Lynn, a retired Air Force Colonel and Tuskegee Airman, and Joseph M. Hogan, submitted a 40-page report on September 21, 1973 which accused NASA of having the worst statistical EEO record in the federal government. They assessed that "at the present rate of increase, NASA would reach only 9% minority by the year 2001!" The three EEO officers, furthermore, evaluated NASA's equal employment program as "a near total failure" and called out the Assistant Administrator of the EEO office, Dr. Dudley McConnell, as lacking integrity, experience, maturity, commitment, and sensitivity.[62] Shortly after, Ruth Bates Harris was fired and Hogan was transferred to a different office, while Lynn was allowed to stay in the EEO office "if he could get along with McConnell." He was able to stay because he supposedly was "the quiet one in the group" and was not expected to cause turmoil all by himself.[63]

Even though Congress was legally not allowed to directly investigate Harris' termination, several hearings regarding NASA's EEO and affirmative action efforts were held in Congress in January and March in 1974. Harris' case made waves in the media, especially because the members of Congress and the

61 James C. Fletcher, *NASA Equal Employment Opportunity*, NPD 3713.2B, October 5, 1972, Box 1, Equal Employment 1974, Center Files, JSC.

62 Ruth Bates Harris, Joseph M. Hogan, and Samuel Lynn, *letter and report to the administrator*, September 20, 1973, Harris/MSRC.

63 Ruth Bates Harris, Ruth Bates Harris vs. National Aeronautics and Space Administration Affidavit, June 19, 1974, pp. 15–16, Harris/MSRC.

public were in agreement that NASA had not committed to EEO wholeheartedly.[64] In reaction to the hearings, the agency was forced to submit quarterly reports to Congress.[65] These reporting requirements were only removed in 1978. Aware that the agency was watched by the public, they rehired Ruth Bates Harris and Deputy Administrator George M. Low ordered the major NASA centers and headquarters to hire at least one woman or a minority on the executive level within the next half year. The only condition for the appointment was that this person actually needed to have the appropriate qualifications. In order to accommodate the new executive level employees in the centers, Administrator James C. Fletcher had to allocate additional positions. Because Low knew that the public wanted to see more minorities and women in upper management but was also aware that the public would most likely not respond well to token appointees, he told his fellow NASA managers to keep quiet about it: "Everybody understands that these instructions are to be given verbally and will not be publicized."[66]

Damage control was not only necessary to appease the public, but also within the agency itself. Numerous employees had protested against Harris' termination in one form or another. Sixty-eight headquarter employees actually banded together and wrote a letter to the administrator, signed as "the concerned employees." In the letter, they expressed their concern about Harris' firing and how her loss undermined the agency's EEO program:

> We are deeply grieved at NASA's loss of Mrs. Ruth Bates Harris. We have known her as our staunchest supporter and advocate of equal opportunities for women and minorities at NASA. Her loss has shaken our faith in NASA's dedication to the Equal Opportunity Program and has stimulated serious concerns on the part of many employees.[67]

---

64 Various letters in support of Ruth Bates Harris written to Congress, in: Senate Aeronautical & Space Sciences Committee 85th-94th Congress, Decimal File 18.9–18.10, Box 108, RG 46, NARA I.

65 McQuaid, '"Racism, Sexism, and Space Ventures": Civil Rights at NASA in the Nixon Era and Beyond', p. 447.

66 George M. Low, *Improving Minority and Female Representation in NASA Executive Level Positions*, Memorandum to Deputy Associate Administrator for Organization and Management, January 21, 1974, Box 67, Folder 4, Low/RPI; Harriett G. Jenkins, interview by Jennifer Ross-Nazzal, August 5, 2011, p. 29, edited Oral History Transcript, NASA Headquarters Oral History Project, JSC; Correspondence regarding Ruth Bates Harris termination between NASA and Civil Rights/Women's Groups/Congressmen/Senators/private citizens, 1973–1974, 013703, Box 34, and 013704, HRC. Example for quarterly report: *NASA, Report on NASA's Equal Employment Opportunity Program, Report to the Subcommittee on HUD-Space-Science-Veterans Committee on Appropriations United States Senate*, April 1974, Senate Aeronautical & Space Sciences Committee 85th-94th Congress, Decimal File 18.9–18.10, Box 108, RG 46, NARA I.

67 *Loss of Mrs. Ruth Bates Harris*, memorandum to James C. Fletcher, October 1973, Harris/MSRC.

Fletcher met with ten of the letter's signatories on November 2, 1973.[68] In the meeting, they discussed the eleven concerns regarding EEO they had listed.[69] Administrator James C. Fletcher and Deputy Administrator George M. Low kept meeting the 'concerned employees' on a regular basis until the formation of the headquarters EEO Advisory Group (EEOAG) in spring of 1974. The EEO Advisory Group was supposed to advise NASA Headquarters upper management of the employees' concerns and views in EEO matters.[70] A replacement EEO Advisory Group was installed in 1976 with the same mandate as the first group, but with its members elected by the other employees this time around.[71]

One of the main critiques of the 'concerned employees' had been leveled at the Equal Employment Council, which they complained was exclusively represented through upper management and had not met since March 1973. Furthermore, they felt that most of the recommendations of the EO Council got lost in the system somewhere. Therefore, they suggested reestablishing the EO Council with more authority to require implementation of their recommendations and a membership that was more reflective of all levels of NASA's workforce.[72] NASA's management actually listened to its concerned employees on several accounts. The agency, for example, established the new EO Council in February 1974 according to the employees' suggestions. The elected council members from all the field centers and headquarters worked as an oversight committee

68 Jo Marie Di Maggio, Bill Wallace, Elizabeth Fox, Penny Kafka, Lynda Sampson, Annie Malone, Kay Voglewede, Sandra Scaffidi, Isaac Gillam, Charlotte Spann, Louis Fong.

69 They talked about confidence, commitment and results, affirmative action plan implementation, hiring goals, statistical program analysis, recruitment, NASA's Federal Women's Program Coordinator, Upward Mobility Program, Human Resources Council, NASA Equal Employment Opportunity Council, NASA's Incentive Awards Board and Personnel Management Review Committee and NASA's Equal Employment Opportunity Conferences. Jo Marie DiMaggio, *Memorandum to Concerned Employess 'Feed-Back on the Statement of Concern to Dr. James C. Fletcher,'* November 19, 1973, 013691, Box 32, HRC; *Points of Concern and Suggestions for Improving the Equal Opportunity Program at NASA*, 1973, Judiciary Committee 93rd Congress, Subcommittee on Civil and Constitutional Rights, Box 38 of HR. BX. 33, NASA Hearings Correspondence etc., RG 233, NASA 1974 Miscellaneous Documents, NARA I.

70 James C. Fletcher, *Headquarters Employees EEO Advisory Group*, May 1, 1974, Harris/MSRC. Members of the EEOAG: Hank Brady, Lou Fong, Elizabeth Gaboury, Paula Gillespie, Ruth Bates Harris (Chairperson), Delores Hart, Phil Kastelic, Bill Lucas, Mechthild Peterson, Jim Smith, Al String, Shirley Whitfield.

71 James C. Fletcher, *memorandum*, May 25, 1976, Harris/MSRC.

72 *Points of Concern and Suggestions for Improving the Equal Opportunity Program at NASA*, 1973, Judiciary Committee 93rd Congress, Subcommittee on Civil and Constitutional Rights, Box 38 of HR. BX. 33, NASA Hearings Correspondence etc., RG 233, NASA 1974 Miscellaneous Documents, NARA I.

that identified problem areas in EEO progress at the centers and at the agency in general. It also proposed solutions to the problems it identified.[73] Donald P. Hearth, the chairman of the EO Council, clarified that the council was "something that needs to be done because it is right! Our task is to search for and identify solutions; not just discuss problems. If the council is to be effective, those of us from the Field Centers must be willing to be open and candid with each other."[74] The council met four times a year for several years after its establishment and later on met twice a year. Its meetings rotated between the various centers in order for the members to meet different people and challenges from each installation, since each center was unique in its charter and the way it operated. This way, the council members had the chance to get to know the various NASA centers' missions and equal opportunity programs, and were able to learn from each other's successes and mistakes.[75] The meeting of the NASA leadership with the employees and the resultant EO Council is important to emphasize at this point because there not only was a direct reaction on the part of the NASA managers in the context of immediate crisis management but an actual turn toward a more participatory culture in an otherwise hierarchical organization.

Many outside factors continued to influence the development of EEO at NASA, among them the growing backlash against the women's movement and national debate over the Equal Rights Amendment, which saw the Federally Employed Women lobbying for ratification alongside other organizations such as the National Organization for Women, the Classroom Teachers Organization, Common Cause, and the ACLU.[76] Another law that influenced the work of NASA's EEO office was the Civil Service Reform Act (CSRA) sponsored by President Jimmy Carter in 1978, partly in response to the Watergate scandal. The Civil Service Reform Act reformed the civil service of the federal government by regulating that all agencies had to have a Federal Equal Opportunity Recruitment

73 Donald P. Hearth, *NASA Equal Opportunity Council/Establishment of the Revised Role and Membership*, memorandum, February 14, 1974, Box 1, Equal Employment 1974, Center Files, JSC; Harriett G. Jenkins, interview by Jennifer Ross-Nazzal, August 5, 2011, pp. 13–14, JSC.

74 Donald P. Hearth, *NASA Equal Opportunity Council/Establishment of the Revised Role and Membership*, memorandum, February 14, 1974, Box 1, Equal Employment 1974, Center Files, JSC.

75 Harriett G. Jenkins, interview by Jennifer Ross-Nazzal, August 5, 2011, pp. 13–14 and 24, JSC.

76 Linda Rogers, interview by Dr. Aaron Gillette, July 27, 1992, pp. 15–16, Tape Transcript, HRC; Rosen, pp. 331–340; Critchlow; Midge Decter, *The New Chastity: And Other Arguments against Women's Liberation* (New York: Coward, McCann & Geoghegan, 1972); Susan Faludi, *Backlash: The Undeclared War Against American Women* (New York: Crown Publishing Group, 1991); Rosalind Rosenberg, *Divided Lives: American Women in the Twentieth Century* (New York: Farrar, Straus and Giroux, 2008), pp. 225–226; Hugh Davis Graham, pp. 189–234; Davis, pp. 121–136; Rossiter, *Women Scientists in America. Vol. 3. Forging a New World since 1972*, pp. 13–14.

Program (FEORP) to recruit minorities and women. The FEORP was supposed to counteract and remedy the underrepresentation of these groups in agencies such as NASA. The Civil Service Reform Act also codified nine basic rules and procedures to govern all personnel practices and defined various prohibited practices in civil service. It was the first time that specific procedures and prohibitions were codified under the law. The CSRA also abolished the U.S. Civil Service Commission which had been NASA's main oversight agency in EEO matters. After the passage of the Civil Service Reform Act, this responsibility was transferred to the Equal Employment Opportunity Commission and NASA had to account for its affirmative action plans and annual reports at the Equal Employment Opportunity Commission instead of the Civil Service Commission.[77]

7.2 The EEO office

Although Cold War fears and speculations made the space community more accessible to women in the late 1950s and 1960s, according to statements of space community officials in the public space, it was the legislative process of the 1960s and 1970s that truly helped women and minorities to establish themselves in the workspaces at NASA and its contractors. While keeping the decentralized system, the creation of an EEO office at HQ in 1971 as an impulse generator for the centers was definitely an outcome of the social and cultural struggles of the long 1960s and the resultant legislation. The new EEO office was responsible for the agency-wide Equal Employment Opportunity Program for NASA employees, including the Federal Women's Program and the Contractor Equal Employment Opportunity Program.[78] Planning and implementing the agency's goals regarding EEO was the official job of the office. Informally, however, the office was also there to calm potential concerns of the employees and the public through communicating the agency's position and measures in regard to specific problems and concerns, always demonstrating that NASA was on top of it – whatever 'it' happened to be – and in the process of initiating a variety of strategies that were intended to solve the problem and to help the agency move forward. Thus, the EEO office needed to build trust within the 'NASA family.'[79] In order to accomplish NASA's EEO mission in concurrence with the changing legislation, the EEO

77 Harriett G. Jenkins, interview by Jennifer Ross-Nazzal, August 5, 2011, p. 26, JSC; Harris, p. 301.

78 James C. Fletcher, *Equal Employment Opportunity Office – Organization Realignment*, September 1, 1971, Harris/MSRC.

79 Harriett G. Jenkins, interview by Jennifer Ross-Nazzal, August 5, 2011, p. 7, JSC.

officers always needed to stay informed; they checked the papers for Supreme Court decisions and the varying interpretations of different regional judges. The officers also needed to see in which direction the law was developing and to assess to what extent the laws might change the agency's EEO plan and internal rules and regulations. Sometimes they needed to arrange for minimal adjustments, but once in a while they also needed to implement significant overall changes that impacted all areas of the community.[80]

Their jobs were made significantly easier after the EEO office was elevated to Assistant Administrator level in mid-April 1973 after several EEO officers had lobbied to position it more favorably and prominently in the agency, reporting directly to the director or deputy director and thus affording it more visibility and upper management support. Another common point of criticism was also eventually changed over the years. There were initially not enough full-time EEO officers at headquarters or in the centers, which slowed down the EEO implementation process.[81] Eventually, however, the 18 to 20 EEO officers on staff at headquarters during the first years increased in number to approximately 39 by 1992. Similar developments also happened at the field installations, where the number of officers practically doubled and they were therefore able to cover more ground in a shorter amount of time.[82] This was necessary because the officers had to oversee and initiate programs and training for around 25,000 employees and also had to monitor the implementation of EEO and affirmative action at NASA contractors with approximately 160,000 employees.[83] In order for the EEO officers and the employees to know who was responsible for specific concerns and programs regarding EEO, NASA used its system of management instructions in which everybody's role was explained. The agency clarified the role of the Administrator in relation to equal opportunity, the Assistant Administrator for Equal Opportunity's role in relation to the Administrator, and the Center Directors and their role in relationship to that function. The line of authority and the resulting expectations were thus already recorded in

80 Harriett G. Jenkins, interview by Jennifer Ross-Nazzal, August 5, 2011, pp. 74–75, JSC.

81 Joseph Hogan, *NASA Civil Rights Posture*, memo to Deputy associate Administrator for Organization and Management, March 9, 1971, 18508, Box 1, HRC; Ruth Bates Harris, *Mrs. Harris' remarks for the General Management Review*, June 12, 1972, 18508, Box 1, HRC; Ruth Bates Harris, Joseph M. Hogan, and Samuel Lynn, *letter and report to the administrator*, September 20, 1973, Harris/MSRC.

82 Harriett G. Jenkins, interview by Jennifer Ross-Nazzal, August 5, 2011, p. 17, JSC; *Ruth Bates Harris at NASA*, undated, Harris/MSRC.

83 Ruth Bates Harris, *Ruth Bates Harris. Some Areas of active Involvement in the Consumer and/or Public Interest Field in the Washington Metro Area*, ca. 1970s, Harris/MSRC. The amount of employees fluctuated depending on the specific year.

the early years of the EEO office and the management instructions were frequently updated over the years.[84]

Even though the majority of the EEO officers were Caucasian men before the establishment of the central EEO office in 1971, most of the appointed officers of the new EEO office were minorities and/or women because they were expected to know and understand the issues, sensitivities, and problems.[85] Ideally, an EEO officer had a multicultural experience and excellent management skills. According to Ruth Bates Harris' successor in the EEO office, Deputy Assistant Administrator Harriett Jenkins, it should be someone

> who knows the installation very very well, but also will know the various ethnic groups, and how to get along with them, and have a demonstrated track record, both of working with senior managers and working with the various groups that one has to persuade. The EO Officer would have to convey the exuberant message, 'Hey, believe NASA, they really do want talented outstanding people. You don't have to worry about getting your college degree and stopping off at NASA and having your career go south because you get into a sour work relationship.'[86]

Jenkins' description of the ideal EEO officer shows that he or she had to have a lot of technical but also soft skills. To make people accept change is a critical but also difficult endeavor which touched the social and cultural belief system of every employee involved. Considering that some employees were less willing than others to actually alter their belief system to implement the changes EEO brought about, EEO officers needed to be sensitive but also forceful. These were qualities which a lot of the EEO officers at first did not possess, because apparently the selection of EEO personnel at NASA did not, unlike the selection of personnel in the science departments, follow any standards. Samuel Lynn, for example, complained that the "contrast between NASA's selection of the most highly qualified scientific, engineering and technical personnel and NASA's settling for totally unqualified personnel for key slots in EEO is startling and unconceivable. Degradation of the requirements of the task, 'and any old one will do' attitude, or the assertion that there is 'no such thing as a qualified EEO

---

84 Harriett G. Jenkins, interview by Jennifer Ross-Nazzal, August 5, 2011, pp. 74–75, JSC. For example: James M. Beggs and Hans Mark, *Role and Responsibilities – Assistant Administrator for Equal Opportunity Programs*, NMI 1103.16A, May 27, 1983, 18507, Box 4, HRC.

85 Ruth Bates Harris, Joseph M. Hogan, and Samuel Lynn, *Statement before the House Judiciary Sub-Committee on Civil Rights and Constitutional Rights*, March 13, 1974, NARA I; Harriett G. Jenkins, interview by Jennifer Ross-Nazzal, August 5, 2011, pp. 27–28, JSC.

86 Harriett G. Jenkins, interview by Jennifer Ross-Nazzal, August 5, 2011, p. 28, JSC.

officer' sadly reflect NASA's attitude."[87] His statement shows that quite a few managers made the mistake of assuming that just because someone was female or not Caucasian, they had qualification enough to work as an EEO officer. Qualified minority scientists and women, such as the African American Dr. Dudley McConnell, were thus transferred from their field of expertise and placed within the EEO space.

Considering that EEO officers needed a specific education just like experts in other fields and that one of the main goals of the EEO program was to bring more minorities and women to the operating scientific and technical positions, it cannot be a surprise that some employees felt that the EEO office had become a 'ghetto' for minorities and women at NASA. The EEO office seemed to have become self-serving because in the beginning it appeared to be the only place where a minority or a woman could expect to find work and attain higher grades, and the highest ranking minority and female officials at NASA tended to all accumulate in the EEO office.[88] Witnessing Dr. Dudley McConnell becoming the first black Assistant Administrator at NASA and Ruth Bates Harris the first black Deputy Assistant Administrator and therewith the highest ranking woman at NASA, some employees saw their concerns regarding EEO officers confirmed.[89] The apprehension about an EEO 'ghetto' with unqualified personnel was eventually dispelled after the office's first few years due to personnel changes there and in all other NASA departments. The increasing financial investment in the EEO program most likely helped, too.[90]

---

87 Samuel Lynn, *Presentation by Samuel Lynn before the Senate Appropriations Subcommittee on NASA*, January 11, 1974, Harris/MSRC.

88 Samuel Lynn, *testimony*, 1974, p. 9,Harris/MSRC; Ruth Bates Harris, *Ruth Bates Harris vs. National Aeronautics and Space Administration Affidavit*, June 19, 1974, p. 11, Harris/MSRC; Panel of the National Academy of Public Administration, *Achieving Equal Opportunity in NASA: An Assessment of Needs and Recommendations for Action,* January 30, 1976, p. 17, Box 2, Center Files, JSC; Anonymous comments received from a cross section of NASA Headquarters Employees, November 1974, Harris/MSRC.

89 Ruth Bates Harris, Joseph M. Hogan, and Samuel Lynn, *letter and report to the administrator*, September 20, 1973, Harris/MSRC.

90 NASA's expenditures for EEO during FY 1975, for example, amounted to $6,076,000 EEO out-of-pocket costs; the agency also spent $13,875,000 for minority contractors and $2,867,000 for minority college programs. It continuously raised its EEO spending from the early 1970s through the mid/late-1970s. While the EEO office budget request for 1976, for example, showed $49,000 for Community Relation and Communication expenditures, the requested budget for 1977 in the same category was much higher at $125,000. Samuel W. Keller, *Equal Opportunity Program Expenditures*, Memorandum to Deputy Administrator George M. Low, December 22, 1975, Box 1, Equal Opportunity 1975, Center Files, JSC; Peter H. Chen, *Office of Equal Opportunity Programs Budget*, September 12, 1975, 18505, Box 2, HRC.

7.2.1 The Affirmative Action and Evaluation Division

The EEO officers worked in different divisions of the EEO office on various programs. There were the Minority University Research and Education Division, the Discrimination Complaint Division, and the Affirmative Action and Evaluation Division. The latter was in charge of formulating NASA's EEO plans and reviewing the agency's results in implementing them. One thing those EEO plans did was define the minority and female placement and promotion goals in numbers which might have given the impression of quotas having been set. This false conclusion is easy to leap to when looking at all the NASA EEO plans and reports, because they appear to emphasize numbers over content.[91] This was not, however, the case, and Jenkins especially emphasized that there were absolutely no quotas at NASA. The EEO goals were calculated based on the self-evaluation processes of each and every single NASA center starting in 1975, in which the center management had to illustrate how many women and minorities they employed.[92] They needed to specify in which grades these people worked and if the minority person was black, Spanish Surname, American Indian, or Oriental.[93] Furthermore, they needed to explain their hiring practices and estimate how many and what kind of job openings each center would be providing in the near future.

The feedback about the centers' and headquarter's very first EEO self-evaluation was rather mixed. At Ames Research Center some supervisors were concerned about the details requested in the self-evaluation because it singled out

---

91 For example: *FY 77 Goals and Progress through September 30, 1977*, October 18, 1977, Box 1, EO 1977, EEO 1973–1978, Center Files, JSC; *Contractor Status as of June 30, 1973*, Judiciary Committee 93rd Congress, Subcommittee on Civil and Constitutional Rights, Box 38 of HR. BX. 33, NASA Hearing Correspondence etc., NASA 1974 Contractor Status (June, December 1973), RG 233, NARA I.

92 Harriett G. Jenkins, interview by Jennifer Ross-Nazzal, August 5, 2011, p. 18, JSC.

93 These were the actual terms NASA used in the 1970s for African Americans, Hispanic Americans, American Indians or Alaska Natives, and Asian Americans and Pacific Islanders, following the prevalent categories used in federal laws and (in part) in the U.S. census. Some of the *nomenclature* of the categories were only changed recently. Johanna A. Brumberg, *Die Vermessung Einer Generation: Die Babyboomer Und Die Ordnung Der Gesellschaft Im US-Zensus Zwischen 1940 Und 1980* (Göttingen: Wallstein, 2015); 'What Census Calls Us: A Historical Timeline', *PewResearchCenter*, 2015 <http://www.pewsocialtrends.org/interactives/multiracial-timeline/> [accessed 11 November 2016]; Kamala Kelkar, 'Obama Signs Bill Eliminating "Negro," "Oriental" from Federal Laws', *PBS*, 2016 <http://www.pbs.org/newshour/rundown/obama-signs-bill-eliminating-negro-spanish-speaking-oriental-from-federal-laws/> [accessed 11 November 2016]; *Modernizing the U.S. Census*, ed. by Barry Edmonston and Charles Schultze (Washington, D.C.: National Academy Press, 1995), pp. 140–155.

individuals who appeared to get special treatment, a practice which they felt was discriminatory. Kennedy Space Center viewed the self-evaluation positively because it made the supervisors understand for the first time that the EO program was a reality and that they had a vital role in implementing the program. The Wallops Flight Facility felt that the central EEO office meddled too much in the centers' programs and EEO designs; they wanted the EEO office to restrict itself to developing guidelines. The self-evaluation at headquarters and JPL revealed that a lot of the employees did not quite grasp the meaning of EEO. While a lot of headquarters employees figured that EEO meant only blacks, employees at the Jet Propulsion laboratory equated the EEO program as complaints only.[94]

The self-evaluations of the centers and headquarters not only helped each facility reflect on its EEO programs, they also detected the existing problems of the employees and management with EEO. It furthermore delivered the numerical goals regarding EEO at NASA. Based on the centers' specified data, vacancies and rates of hiring, the EEO office assessed that the agency as a whole should strive for a 40 percent hiring goal for minority and female professionals, and a 35 percent hiring goal for minority non-professionals.[95] Although the hiring numbers of the EEO office were just meant as a general guideline, the recommended goals were regarded as unattainable both before and after the self-evaluation by the center managements. In 1977, Gerald D. Griffin, Deputy Director of Johnson Space Center, was still convinced that the prevailing method of setting EEO goals was arbitrary because they were 'top-down' goals which were assigned to each installation. He proposed to build a 'bottom up' system. However, he also emphasized that he was not "criticizing the somewhat arbitrary manner in which goals have been set in the past. They have served their purpose well in getting us started in the right direction. The fact is though that we all know more now than we did in 1973–74 and I think it's time we put that knowledge to use in establishing ambitious but meaningful EO goals."[96]

The Affirmative Action and Evaluation Division was not only in charge of designing the blueprint for NASA's EEO program, it was also responsible for the evaluation of the centers' Affirmative Action and EEO Plans which each center

94 *Minutes of NASA EO Council meeting held at Marshall Space Flight Center*, January 23, 1975, Box 2, Equal Opportunity Council – Meeting Minutes 1975–76, Center Files, JSC.

95 Harriett G. Jenkins, interview by Jennifer Ross-Nazzal, August 5, 2011, pp. 18 and 22, JSC.

96 Gerald D. Griffin, *letter to Bernard Lubarsky*, October 26, 1977, Box 1, Equal Opportunity 1977, Center Files, JSC. See also: Hatty H. Gorman, *Goals for Increasing Minority and Female Employee Representation in NASA/OMSF Institutional Goals*, memorandum to Dudley McConnell, January 11, 1974, Box 1, Equal Employment 1974, Center Files, JSC.

had to submit for every fiscal year.[97] Notwithstanding a few complaints, such as those mentioned above from Ames and Wallops, the EEO officers of the various NASA centers were usually glad to receive guidance regarding EEO measures from headquarters. In a letter to the central EEO office, EEO officer Jack R. Lister let the EEO Director know that: "it was a lot easier developing our Plan using the Headquarters draft as a starting point than if we had to start from scratch."[98] Still, the implementation of EEO goals at the different centers and at the contractors varied immensely, despite receiving the same guidance from HQ, depending on their location and management.

7.2.2 The Minority University Research and Education Division

NASA sponsored a lot of educational programs at schools and universities and for the public. The Minority University Research and Education Division was in charge of managing these programs. This agency started as early as 1963 with co-op programs, mainly with historically black colleges or other universities with a majority of enrolled minority students.[99] The work study program alternated between theoretical lessons at the colleges and gaining practical science or engineering experience at NASA. The idea was that the students would earn money while studying and come to work for NASA upon receiving their degree.[100] The co-op program was one of the few EEO measures the agency had already taken before the establishment of the EEO office. When the EEO office took over the "pale and lukewarm" program with minority universities in the 1970s, the Minority University Research and Education Division started to expand the program.[101] They, for example, launched the National Aerospace Fellowship which was awarded to 20 academically outstanding minority and female students graduating in fields related to space who already have completed at least two years of undergraduate education. The students received a $2,000 stipend, an allowance, and a summer internship at NASA as well as the

---

97 For example: *Affirmative Action Plan – Calendar Year 1973 (Effective Date – 1/1/73) National Aeronautics and Space Administration Manned Spacecraft Center,* January 29, 1973, MC 496, Carton 44, Folder 6, National Organization for Women Collection, Schlesinger.

98 Jack R. Lister, *Revised CY-73 Affirmative Action Plan,* Memorandum to Ruth Bates Harris, 1973, MC 496, Carton 44, Folder 6, National Organization for Women Collection, Schlesinger.

99 Paul.

100 Nathaniel Pilot, interview by Dr. Aaron Gillette, July 28, 1992, p. 5, Tape Transcript, HRC; Powers, p. i.

101 Harriett G. Jenkins, interview by Jennifer Ross-Nazzal, August 5, 2011, p. 53, JSC.

opportunity to later work for NASA after graduating from college.[102] The fellowship was just one of several new educational programs which were launched. While NASA only gave grants totaling $358,000 to seven minority colleges in 1969, the agency spent $2,000,000 in grants in 1974 for 45 minority colleges, and the investment in America's future scientists grew continuously from there.[103] A 1988 inventory of NASA's educational programs counted 162 in total: 59 elementary and secondary education programs, 37 university programs, 30 minority outreach programs, 19 employment programs, and 17 public education programs.[104] While the co-op programs and the disposition of research grants were relatively selective and the university programs only supported a limited number of individuals each year, the more public education programs reached millions of teachers, students, and space enthusiasts. Considering that only approximately six percent of the co-op participants have become NASA employees, the success rate pales in comparison to that of the public outreach.[105]

Because the Minority University Research and Education Division was tasked with changing the profile of the agency as it related to women and minorities, the EEO office created films, brochures, and books depicting these groups' contribution to NASA. Publications and projects like *Stepping Stones to the Future, Space Work for People – People Work for Space, Women at NASA, Seizing Opportunities to Strengthen America's Future,* and *Women at Work in NASA* were created to inform the public and demonstrate the variety of jobs that people work in at NASA. The brochures were supposed to tell people that NASA believed in diversity and that the space program was open to applications from minorities and women, as proven by the people portrayed in the brochures. The EEO officers usually distributed the materials themselves, but for

102 *NASA Policies and Procedures for Administering Fellowship Program*, undated, 013699, Box 33, HRC; 'NASA launches college program for minorities and women', *NASA News*, May 13, 1974, Judiciary Committee 93rd Congress, Subcommittee on Civil and Constitutional Rights, Box 38 of HR. BX. 33, NASA 1974 Correspondence with NASA, RG 233, NARA I; 'NASA wants Women', *Afro American*, July 12, 1975.

103 Dudley McConnell, *Statement of Dr. Dudley G. McConnell Assistant Administrator for Equal Opportunity Programs National Aeronautics and Space Administration before the Subcommittee on HUD-Space-Science-Veterans Committee*, 1974, p. 3, Harris/MSRC. He gave almost the identical statement in front of the Subcommittee on Civil Rights and Constitutional Rights of the Committee on the Judiciary, House of Representatives.

104 Asner, p. 403. See also: Harriett G. Jenkins, interview by Jennifer Ross-Nazzal, August 5, 2011,pp. 31–32, JSC.

105 Ruth Bates Harris, *testimony in front of the committee on the Judiciary U.S. House of Representatives Subcommittee on Civil Rights and Constitutional Rights*, March 14, 1974, Harris/MSRC; Asner, p. 403.

special occasions, such as the recruitment of a diverse astronaut corps in 1977, the EEO officers also contracted the job out to, for example, Nichelle Nichol's Women in Motion Inc. Normally, the brochures were sent to women and minority groups as well as to any other people who ordered them. They were also passed out at meetings, conferences, or conventions at which NASA had a booth, such as the *Star Trek* convention in the Chicago Hilton Hotel in 1975 or the National Women's Conference in Houston in 1977.[106] The EEO officers often went to conferences and meetings like these as a part of the division's special emphasis programs, such as the Hispanic Employment Program, Black Employment Program, and the Federal Women's Program.[107] For example, in only half a year between October 1978 and March 1979, NASA attended 53 conferences, visited 155 schools, and interviewed 1,300 students.[108]

7.2.3 The Federal Women's Program

The Federal Women's Program (FWP) was an integral part of the EEO office. It was integrated not only in the Minority University Research and Education Division but also in the other divisions. This is why the FWP Managers directly reported to the heads of the EEO office. The FWP Coordinators were involved in everything the EEO office was planning, to make sure that the representatives of all the NASA women were integrated in the decision making process. The main job of the FWP Coordinators was to advise the Assistant Administrator and the Deputy Assistant Administrator for Equal Opportunity on the employment and career challenges facing women and to ensure that their ideas for advancing women at NASA were given consideration. In order to remain up to date on the various topics, issues, and developments important to women in the workforce, the FWP Manager needed to constantly liaise with the female community in

106 Harriett G. Jenkins, interview by Jennifer Ross-Nazzal, August 5, 2011, p. 40, JSC; Ruth Bates Harris, *Equal Employment Opportunity Brochure and Film*, November 9, 1972, Harris/MSRC; *Profile: Nichelle Nichols, 1977*, NASA Astronaut Recruitment Final Report, Women in Motion Inc., August 10, 1977, HRC; 'Women at NASA –up, up, and away?', *NOW News Bay Area Chapter* 6/3 (1978), p. 1, PR-1 Carton 25, NOW Chapter Newsletter, Schlesinger.

107 The Hispanic Employment Program was originally called the Spanish Speaking Program. The three special emphasis programs which started in the 1970s were joined decades later by the Asian American and Pacific Islander Program (AAPIP) and the Native American Program which is based on President George W. Bush's Executive Order 13270 from 2002, which reaffirmed the Federal Government's commitment to tribal universities and colleges.

108 *NASA wide Recruiting Activity*, May 21, 1979, 1979–1981 EEO, Box 2, EO 1979, Center Files, JSC.

government, labor, industry, and educational fields through interfacing with organizations like the Federally Employed Women (FEW), the Society of Women Engineers (SWE) or Women in Science and Engineering (WISE).[109]

The Federal Women's Program was established at the national level in 1967 and was integrated into the overall EEO program in 1969 in response to Executive Order 11478. The program was supposed to work on two levels: awareness and action. It's goals were to recruit and hire qualified women, to open up dead-end jobs and give women the opportunity of advancement, to counsel women about the existing career opportunities and to encourage them to actually pursue a career instead of just doing a job, provide childcare for federal employees, and evaluate and communicate the progress and obstacles of women in the workforce.[110] However, the start of the FWP at NASA was rather bumpy. The Civil Service Commission assessed NASA's efforts in regard to the advancement of women as "rather phlegmatic," while the Federally Employed Women confirmed that the positive change in the status of women at NASA was "slow" in the early 1970s.[111] Their evaluation revealed that headquarters' and the centers' FWP "suffers a lack of program and management direction; in addition, some coordinators feel they do not have adequate time to dedicate to advancement of the program."[112] One of the major criticisms was that there was no Federal Women's Program Committee and that the FWP Coordinators all worked only part-time and had no track record in women's rights in the early years of the program.[113] The headquarters' FWP manager, for example, was

---

109 Harriett G. Jenkins, interview by Jennifer Ross-Nazzal, August 5, 2011, p. 30, JSC.

110 *NASA Headquarters Federal Women's Program Committee Report*, April 1973, MC 496, Carton 44, Folder 5, National Organization for Women Collection, Schlesinger.

111 U.S. Civil Service Commission, *A report on Equal Employment Opportunity National Aeronautics and Space Administration*, April 1974, p. 7, Judiciary Committee 93rd Congress, Subcommittee on Civil and Constitutional Rights, Box 37 of HR. BX. 33, NASA Hearings Correspondence etc., RG 233, NARA I; D.C. Chapter Federally Employed Women, *Report on Agency Visits: Federal Women's Program 1973*, Judiciary Committee 93rd Congress, Subcommittee on Civil and Constitutional Rights, Box 38 of HR. BX. 33, NASA 1974 Miscellaneous Documents, RG 233, NARA I.

112 U.S. Civil Service Commission, *A report on Equal Employment Opportunity National Aeronautics and Space Administration*, April 1974, p. 8, Judiciary Committee 93rd Congress, Subcommittee on Civil and Constitutional Rights, Box 37 of HR. BX. 33, NASA Hearings Correspondence etc., RG 233, NARA I.

113 *Points of Concern and Suggestions for Improving the Equal Opportunity Program at NASA*, 1973, Judiciary Committee 93rd Congress, Subcommittee on Civil and Constitutional Rights, Box 38 of HR. BX. 33, NASA Hearings Correspondence etc., RG 233, NASA 1974 Miscellaneous Documents, NARA I; D.C. Chapter Federally Employed Women, *Report on Agency Visits: Federal Women's Program 1973*, NARA I.

able to dedicate just one day a week to the Federal Women's Program and she was assisted by only one staff technician who also only spent one third of her time on the FWP.[114] The situation of the Federal Women's Program Coordinators in the field installations was similarly bleak. Johnson Space Center, for example, also had only a part-time FWP Manager who worked in the Administration and Program Support Directorate. Although the FWP Coordinator was transferred with the rest of the EEO office to the Johnson Space Center Director's staff and thus received a status upgrade in 1974, JSC proceeded to fill the job of the FWP Coordinator by lateral assignment instead of the statutory competition under regular Civil Service Merit Promotion procedures until 1978. Following several employee complaints to the NASA Administrator, Dr. Robert A. Frosch, and several inquiries from the Bay Area National Organization for Women chapter as well as Senator William Proxmire, Johnson Space Center opened up its job advertisement for the position.[115] Not only Johnson Space Center, but all the other installations, including headquarters, eventually started to enhance their Federal Women's Program in regard to skill, status, and scope. By 1974, NASA already had three full-time and eight part-time FWP Coordinators.[116]

Jo Marie DiMaggio and Oceola Hall in particular represented the driving force in launching the FWP as well as the Federal Women's Program Committee (FWPC). The FWP Committee was founded in 1973 to act as an intermediary between the women, management, and other employees at NASA. It was supposed to support and advise headquarters' FWP Coordinator and therewith help to implement the Federal Women's Program at NASA. Every major meeting within NASA HQ sent a representative to the Committee in order to have all opinions reflected in the FWPC.[117] One of the main jobs of the Committee was to educate

---

114 Ruth Bates Harris, *letter to Millicent Allewelt*, June 16, 1971, 18508, Box 1.

115 'Equal Opportunity Anyone??', *NOW News Bay Area Chapter* 6/1 (1978), p. 2, PR-1 Carton 25, NOW Chapter Newsletter, Schlesinger; *Equal Opportunity News JSC*, August 1981, Box 2, Equal Opportunity 1981, Center Files, JSC; *Affirmative Action Plan Addendum*, 1975, EEO, Box 2, Folder NASA JSC Affirmative Action Plan Addendum FY 1975, Center Files, 1979–81, JSC.

116 U.S. Civil Service Commission, *A report on Equal Employment Opportunity National Aeronautics and Space Administration*, April 1974, p. 5, NARA I.

117 Representatives of the FWPC at NASA in April 1973: Jo Marie DiMaggio; Mary Fling; Pat Jones Suzanne Garner; Charlinga Daniel; Liz Fox; Millie Hipsher; Charlotte Spann; Jean Robinson; Arlene Kahn; Peg Nelson; Barbara Goetz; Delores Hart; Phyllis Van Schaack; Saundra Smith; Sandy Scaffidi; Jane Sylvia; Doris Goodwin; Heidi Sanford; Nancy Roman; Marie Coleman; Betty Hand; Peggy Christian; Dale Eno; Jo Fessler; Diane Graham; Vivian Slack; Annie Malone; Nancy Pickering; Ruth Thomas; Jean Raymond; Carolyn Bronson; Penny Romine; Elma Green; Donna Geib; Jean Lecompte; Joyce Killam.

the employees, for which reason the FWPC published newsletter with important information. These newsletters briefed employees about the latest developments regarding the Federal Women's Program and informed them about why and how the FWPC representatives should be contacted. The newsletters were also a way to teach employees about the myths and realities of why women entered the workforce. According to a FWP newsletter, the myth that "[m]ost women don't need to work. They just want money for extra clothes and other luxuries" was not true because 70 percent of working women actually need their job to help support their families.[118] This example demonstrates that the Federal Women's Program not only had to initiate structural change in the agency, but also had to stimulate a cultural rethinking in regard to women in the workspace.

To implement structural and cultural change with the help of the Federal Women's Program at NASA, the FWP Coordinator needed to continuously educate themselves in order to properly pass on information and inspire the employees. The EEO office therefore started offering seminars for its EEO officers and FWP coordinators on topics such as 'Developing Managerial and Administrative Skills for the Professional Woman in a Technical Environment.'[119]

The FWP proceeded to educate and stimulate not just its coordinators, but all NASA employees as well as the public with events like the observance of the International Women's Year (IWY) at NASA in 1975. The International Women's Year and the first World Conference on Women, which was held in Mexico City in July 1975, were organized by the United Nations' Commission on the Status of Women and was a part of a larger United Nations program, the Decade of Women (1976–1985).[120] The IWY was created because women had, thus far, not

118 *NASA Headquarters Federal Women's Program Committee Report*, April 1973, MC 496, Carton 44, Folder 5, National Organization for Women Collection, Schlesinger; Harriett G. Jenkins, interview by Jennifer Ross-Nazzal, August 5, 2011, p. 73, JSC.

119 Other selected topics included: the functions of management, issues relating to power and management, women as effective managers, developing career objectives, sex/work-role stress such as institutional sexism (subtle and direct), role conflict (family vs. job) and image conflict ('being a woman' vs. 'being a manager', isolation, sexual harassment). Harriettt Jenkins, *Third Annual Agency-wide Federal Women's Program Seminar*, September 8, 1977, Box 1, Equal Opportunity 1977, Center Files, JSC.

120 'Resolution Adopted by the General Assembly. 3010 (XXVII). International Women's Year' <http://www.un-documents.net/a27r3010.htm> [accessed 16 November 2016]; 'World Conferences on Women' <http://www.unwomen.org/en/how-we-work/intergovernmental-support/world-conferences-on-women> [accessed 16 November 2016]; Peggy Antrobus, *The Global Women's Movement: Issues and Strategies for the New Century* (London, New York: Zed Books, 2004), pp. 38–62; Jocelyn Olcott, *International Women's Year: The Greatest Consciousness-Raising Event in History* (Oxford: Oxford University Press, 2017); Deborah Stienstra, *Women's*

been able to achieve equality in any country, despite decades' worth of progress in eliminating discrimination against women. Recognizing that NASA also had not attained gender equality in its workforce, yet, the agency wanted to put a spotlight on female issues in the community within the framework of the International Women's Year. NASA observed the IWY with a women's month during which all the field installations and headquarters were supposed to discuss women's issues and celebrate the achievements of women "in as many creative ways as possible" with a main kick-off event, several speakers, films, workshops, training courses, demonstrations and presentations.[121]

Even before, but especially after, the IWY events, various NASA centers regularly held FWP events, like the Federal Women's Week, which attracted big crowds of women and men.[122] The agency also recognized the women's history month two years after it was proclaimed by President Ronald Reagan in 1988 with Proclamation 5775 "to salute women's contributions to the American family and to society."[123] However, even at gender equality events like these, gender-based biases were alive and well – even within the people who set out to tear down gender-based biases and discrimination. Johnson Space Center's FWP Coordinator Viginia Hughes, for example, felt that it was necessary to emphasize that "the entire women's program [at the FWP week] was marked by professionalism in the presentation of facts, studies, and statistics and not by the kind of emotionalism that some people have come to associate with topics of a controversial nature."[124] 'Emotionalism,' was a quality often associated with social movements, but especially with women; a quality which was assessed negatively because it did not fit into an organization that was built on a self-understanding of objectivity, rationality, and masculinity. The women's rejection

Movements and International Organizations (Houndmills, Basingstoke, Hampshire: Macmillan, 1994), pp. 123–141.

121 *Federal Women's Program. Towards new Horizons*, booklet, November 1975, Harris/MSRC; *Minutes of the EO Council Meeting held at Langley Research Center*, February 19, 1975, Box 2, Equal Opportunity Council – Meeting Minutes 1975–76, Center Files, JSC.

122 *A Woman's Place is. . .Everywhere*, NASA Headquarters Federal Women's Week, November 4–8, 1974, 008995, HRC; Christopher C. Kraft, *Equal Opportunity Quarterly Report for the Quarter Ending June 30, 1977 for the JSC*, June 30, 1977, Box 1, Equal Opportunity 1977, Center Files, JSC.

123 *National Women's History Month Program*, March 5, 1990, GR 1019A, 1990, General Reference Files, JSC; Ronald Reagan, 'Proclamation 5775 of March 7, 1988. Women's History Month', 1988 <https://www.gpo.gov/fdsys/pkg/STATUTE-102/pdf/STATUTE-102-Pg4962.pdf> [accessed 16 November 2016].

124 Virginia Hughes, 'Federal Women's Program at JSC attracts unprecedented crowds', *Roundup*, February 28, 1978.

of 'emotionalism' as a supposedly negative quality, shows that they mostly tried to fit in the agency's prevalent ideals and cultural norms and code of conducts.

The agency's FWP also attended conferences outside of NASA such as the Women's Career Convention in Houston in 1981 or the National Women's Conference in Houston in November 1977. NASA representatives from different field installations participated at these events with a NASA exhibit which they claimed was "one of the best."[125] They handed out the brochures which showcased women's career opportunities at NASA and they had a photo booth where women were able to get their picture taken as an astronaut. In the very year NASA was looking for its first minority and female astronauts, which they also advertised at the National Women's Conference, women still needed to poke their head through a hole in a canvas painted with an astronaut in order to see an American woman in space.[126]

7.2.4 The Discrimination Complaint Division

The Discrimination Complaint Division had the responsibility for handling all allegations of discrimination. The EEO officers at NASA HQ, however, only handled the formal complaints and investigations. Before any employee was able to file a formal complaint about discrimination, the employee had to seek counsel at his or her installation first. Every NASA center had several counselors. As of December 31, 1974, NASA had 77 counselors in all its centers and at headquarters for its 24,881 civil servants. Because the recommended ratio of counselors to employees was 1 to 500, all the centers therefore had more counselors than recommended for a workforce that size. Johnson Space Center, for example, had 13 counselors for 3,798 permanent civil servants. The counselors worked in all areas and grades and could be located anywhere in the agency's organizational structure. Because they did not have to give an account for their actions and supervisees to the personnel office, they were also able to act relatively independently. They were male and female and of different races and ethnicities. The large number of counselors and their greatly varied backgrounds gave the employees a choice of who they wanted to confide in, and most were likely able to find a counselor they felt

125 *Minutes of the EO Council Meeting*, November 29–30, 1977, Box 1, Equal Opportunity 1977, Center Files, JSC; Harriettt Jenkins, *Third Annual Agency-wide Federal Women's Program Seminar*, September 8, 1977, Box 1, Equal Opportunity 1977, Center Files, JSC; *Equal Opportunity News JSC*, March 25, 1981, Box 2, Equal Opportunity 1981, Center Files, JSC.

126 See picture by Diana Mara Henry: http://www.dianamarahenry.com/img_detail.php?id=734&realname=NASA+Booth

comfortable with.[127] Depending on the NASA center, the employees were sometimes even able to nominate people to become counselors, though this was not always the case. At Johnson Space Center, for example, the employees were not allowed to get involved in the selection. Generally, it was the EEO officers, the Director of Administration and the Center Director who made the final selection.[128] All the counselors selected underwent training by the EEO officers, parallel to their jobs, for the responsibility of managing the discrimination process at their level. The very first step of the counseling process was for the counselor to sit down with the employee who wanted to file a complaint and discuss what had happened and why they felt discriminated against. The counselor then had to spell out all the options available to the employee, including an explanation about how to file a formal complaint and the possibilities of other options. Quite a few employees chose not to proceed to the next step but opted for an informal solution presented by the counselor.[129]

Like all newly established programs, the counseling program needed to become known before it could be effective. However, 28.8 percent of 733 NASA employees who responded to a questionnaire about the counseling program in 1974 were not yet aware of the counseling program, proving that ignorance was a problem the EEO office often had to counteract in the early 1970s.[130] The questionnaire furthermore revealed that 62.6 percent of the employees who were familiar with the counseling program, as well as 81 percent of the 42 counselors and 77.3 percent of the 178 managers who had responded to the questionnaire, felt that the counseling program was an effective tool for solving problems between management and employees.[131] Sixty-six employees had actually already used the counseling service and 97 percent of them were convinced that the counselors were accessible and it was not a problem to contact them.[132] However, 36.3 percent of the employees who had used the counseling program received the impression that the counselor had not presented them all their options, while 40.5 percent interviewed counselors believed that management was not fair and

127 *Report on Assessment of NASA's Equal Employment Opportunity Counseling Program Johnson Space Center*, 1975, pp. 6, 10, Box 1, EO 1975, Center Files, EEO 1973–1978, JSC.
128 *Report on Assessment of NASA's Equal Employment Opportunity Counseling Program Johnson Space Center*, 1975, pp. 6, 10, Box 1, EO 1975, Center Files, EEO 1973–1978, JSC.
129 Harriett G. Jenkins, interview by Jennifer Ross-Nazzal, August 5, 2011, pp. 50–51, JSC.
130 *Report on Assessment of NASA's Equal Employment Opportunity Counseling Program Johnson Space Center*, 1975, p. 2, Box 1, EO 1975, Center Files, EEO 1973–1978, JSC.
131 *Report on Assessment of NASA's Equal Employment Opportunity Counseling Program Johnson Space Center*, 1975, pp. 4, 5, 7, 17, Box 1, EO 1975, Center Files, EEO 1973–1978, JSC.
132 *Report on Assessment of NASA's Equal Employment Opportunity Counseling Program Johnson Space Center*, 1975, p. 12, Box 1, EO 1975, Center Files, EEO 1973–1978, JSC.

objective and lacked commitment to EEO and the counseling program.[133] Not only the employees, but management, too, used the counseling program if they felt discriminated against or if they had a problem with an employee who accused them of discrimination. The six managers out of the 178 who responded to the questionnaire and who had used the counseling service were content with the guidance they had received.[134]

Overall, the counseling service worked pretty well, but a lot of employees still decided to file their complaint formally. What is interesting, though, is that most discrimination complaints were filed because of age and race – not sex.[135] It is almost impossible to ascertain if women actually felt less discriminated against or if they did not opt for a formal complaint because they feared reprisal of some sort. This concern was expressed frequently by men as well as women.[136] Grace M. Ackerman, who had served as the Federal Women's Program Coordinator at Marshall Space Flight Center for ten years, for example, indicated that many women feared the "threat of intimidation and retaliation. When superiors openly made derogatory remarks about the 'women's give away program' and worse, there was little the women were in a position to do."[137] Considering that the women who worked in the labs in the 1960s and 1970s, including the first female astronauts, just wanted to be a genderless member of the team – one of the 'guys' – it seems likely that they might have sought more informal solutions or just kept ignoring some parts of the social exchanges they had with their male co-workers, like the women who participated in the sexist gotchas, because they did not want to be spoilsports.[138]

When an employee filed a formal complaint it went directly to the EEO office at headquarters. The EEO officer in Washington, D.C. then assigned an investigator, usually from outside the agency, to the case. Like a detective, the investigator looked into all the existing evidence and interviewed witnesses. The findings of the investigations were passed on to the central EEO office which reviewed the

---

133 *Report on Assessment of NASA's Equal Employment Opportunity Counseling Program Johnson Space Center*, 1975, pp. 5, 6, Box 1, EO 1975, Center Files, EEO 1973–1978, JSC.

134 *Report on Assessment of NASA's Equal Employment Opportunity Counseling Program Johnson Space Center*, 1975, p. 7, Box 1, EO 1975, Center Files, EEO 1973–1978, JSC.

135 Carl E. Grant and Harriettt G. Jenkins, *Report on NASA's Equal Opportunity Progress. Comparison of E.O. Progress for FY's 1974–1978*, Novemer 22, 1978, Box 1, Equal Opportunity 1978, Center Files, JSC.

136 *Report on Assessment of NASA's Equal Employment Opportunity Counseling Program Johnson Space Center*, 1975, p. 16, Box 1, EO 1975, Center Files, EEO 1973–1978, JSC.

137 Grace M. Ackerman, *letter to Wilma Scott Heide*, October 11, 1973, MC 495, Carton 20, Folder 5, Wilma Scott Heide Papers, Schlesinger.

138 See chapter six.

investigator's report to assess whether federal laws or internal regulations had been violated. The EEO officers then issued a recommended decision that was always reviewed by the legal counsel before the Administrator made the final decision regarding the complaint. Usually, the Administrator went along with the recommendation of the EEO office.[139] The agency stuck to this decision-making process for formal complaints even when the complaint was filed against the NASA Administrator personally. In such cases, the Administrator had to decide himself if he had discriminated against the person who had filed the complaint because of race, sex, age, national origin, or religious belief. In the case of Ruth Bates Harris' termination and her resulting formal complaint against NASA Administrator James C. Fletcher in 1973, Fletcher was able to decide his own fate. This procedure did not seem fair to Harris, for which reason she filed her complaint against Fletcher directly with the Civil Rights Commission. In her complaint Harris elaborated that while

> others share in acts of discrimination against me, such as the Deputy Administrator of NASA [George M. Low] and the Assistant Administrator for Equal Opportunity [Dudley McConnell], my complaint is being lodged against the top official of NASA. Therefore, it is necessary that I rile directly with the Civil Service Commission. I respectfully request a hearing or investigation.[140]

However, her request was denied by the Civil Service Commission and referred back to NASA because the Civil Service Commission had no provisions for the assumption of jurisdiction at the first stage of any complaint.[141] NASA's top management, because of the predicament the Administrator was in, actually shared Harris' desire to have the CSC take over her discrimination complaint. NASA's General Counsel, R. Tenney Johnson, therefore asked the Civil Service Commission to reconsider their decision to refer the Harris case back to NASA. The final decision about the complaint remained with the NASA Administrator, however, even though he did not want to be the one to make that decision about the complaint that was filed against him.[142]

Although Ruth Bates Harris ultimately did not have to go to court because the case was amicably resolved, Harris had actually planned to file a civil

139 Harriett G. Jenkins, interview by Jennifer Ross-Nazzal, August 5, 2011, pp. 50–51, JSC.

140 Ruth Bates Harris, *letter to Robert Hampton*, Chairman of Civil Rights Commission, November 3, 1973, Harris/MSRC.

141 Geral K. Hinch, Director Civil Service Commission, *letter to Ruth Bates Harris*, November 12, 1973, Harris/MSRC.

142 R. Tenney Johnson, *letter to Chairman of the Civil Service Commission Robert Hampton*, November 29, 1973, Harris/MSRC.

lawsuit and several lawyers had offered to take her case pro bono as a landmark case. She accepted the help of the lawyers for the National Association for the Advancement of Colored People (NAACP) who petitioned against her termination as an illegal reprisal.[143] At that point Harris had not been the only NASA employee who was seeking social justice outside of NASA's Discrimination Complaint Division; various employees had turned to the NAACP Legal Defense Fund, which had in turn filed class action lawsuits in U.S. District Courts against NASA for discrimination against African Americans and women. So numerous were the lawsuits that NASA management had actually launched an investigation into its own EEO personnel, accusing the EEO officers of giving information about complainants to the NAACP Legal Defense Fund.[144] However, it was the employees themselves who had turned to the NAACP, according to KSC's EEO officer Willie D. Gaither.[145] Jo Marie DiMaggio, Federal Women's Program Coordinator at headquarters, on the other hand, admitted that she, in fact, did hand over the names of some people who were being discriminated against by NASA to the NAACP Legal Defense Fund. When she refused to give up the names of people she had given to the NAACP, she was threatened with the remark: "Jo, if you can't work cooperatively with me in this office, your life is going to be dead as far as your job."[146] Instead of intimidating the EEO staff and the minority and female employees, it would have been more enlightening to take a look inward to investigate the effectiveness and problems of its EEO performance and discrimination complaint system, which the CSC had rated as "a very poor performer in processing employee complaints stemming from alleged discriminatory practices."[147] However, NASA management instead came to the conclusion that "[t]here are no systematic problems in the NASA complaint processing procedures which significantly impact handling time. Circumstances which led to three cases being poorly handled by NASA personnel have been corrected."[148] It appears that the discrimination complaint system was in fact not faulty, but that

143 Harris, pp. 272–273; McQuaid, 'Race, Gender, and Space Exploration: A Chapter in the Social History of the Space Age', p. 421.

144 Willie Gaither, Alvin Anderson, Jo Marie DiMaggio, and Joe Atkinson were investigated.

145 Willie D. Gaither, *testimony of in front of the Committee on the Judiciary U.S. House of Representatives Subcommittee on Civil Rights and Constitutional Rights*, March 14, 1974, Harris/MSRC.

146 Jo DiMaggio, *testimony in front of the committee on the Judiciary U.S. House of Representatives Subcommittee on Civil Rights and Constitutional Rights*, March 14, 1974, Harris/MSRC.

147 *Complaint Processing: Meeting Record between Low, McConnell and Gaither*, February 19, 1974, Box 32, 013690, Box 32, HRC.

148 Dudley McConnell, *Complaint Processing: Memo to Deputy Administrator*, February 18, 1974, 013690, Box 32, HRC.

the problem consisted of human errors on the part of various executives in NASA upper, but more so middle, management, long before as well as during the filing of the complaint.[149] Still, NASA's reaction to the lawsuits, which were discussed in the public space unlike the internal complaints, demonstrates that it often prioritized its public image instead of tackling the actual problem, in this case the effectiveness of its discrimination complaint system.

The disgruntled employees usually only turned to civil lawsuits after they had exhausted all of the administrative remedies. The NAACP, for example, filed a lawsuit in the name of two Johnson Space Center employees, Sylvester Barrett and Gloria Williams, who claimed that they did not receive promotions due to their race and sex. The plaintiffs, furthermore, accused NASA generally of steering all African Americans and women into dead-end jobs in the low grade levels GS-1 to GS-7 despite their eligibility for higher positions. They also alleged that African Americans and women were often told that there were no vacancies, when in truth the positions were kept open for white men, and that merit promotions were withheld from minority and female employees because the supervisors did not record their high potential the same way they did for white men and because the vacancies were not officially posted, only verbally discussed. This way, only the 'pre-selected' friends and favorites of the managers fulfilled the prerequisites for the jobs. Furthermore, women and African Americans often had to perform tasks and responsibilities that were not part of their job descriptions without additional compensation, reference in their record or an increase in grade, according to the plaintiffs. One of their main allegations was that NASA reviewed complaints only on an individual case basis and thus had "utterly failed to attack the pervasive and systematic employment discrimination against blacks and women prevailing in NASA-JSC and in the federal service."[150]

Taking a look at other lawsuits from female and minority NASA employees against the agency, the allegations were usually quite similar, irrespective of

149 Paul Delaney, 'NASA Ousts Top Black Woman in Dispute on Equal Employmemt', *New York Times*, October 28, 1973; George M. Low, *Miscellaneous Items of Interest*, July 8, 1974, Box 67, Folder 1, Low/RPI; Ruth Bates Harris, Samuel Lynn, Joseph M. Hogan and A. Diane Graham, *Statement prepared for Committee on Aeronautical, Space Sciences. Status of Equal Employment Opportunity in the National Aeronautics and Space Administration*, January 24, 1974, p. 15, Harris/MSRC.

150 Timothy S. Robinson, 'NASA Bias charged by NAACP', *Washington Post*, November 21, 1974; *Civil Action No. 74-1694 in the United States District Court for the District of Columbia*, 1974, 013690, Box 32, HRC; Dudley McConnell, *letter to Christopher C. Kraft*, September 28, 1973, Box 1, Equal Opportunity 1973, Center Files, JSC.

the NASA installation they worked for.[151] The agency's main defense for its poor EEO employment record was that there were not enough qualified applicants and that they had to reduce the workforce in the 1970s because of a lack of funding. Therefore, NASA was only able to improve slowly in EEO statistics due to the overall decrease of the workforce and their need of specialists such as scientists and engineers. Furthermore, the agency, by its own account, was often not able to employ the few minority and female specialists on the job market because NASA could not keep up with the salaries offered by private industry.[152] The NAACP, on the other hand, claimed that NASA just had "lame excuses" for having such a bad EEO employment record.[153] The agency, in return, alleged that NASA was being used by the lawyers who tried to persuade the courts to apply private sector law to the federal EEO allegations. This provision would have been favorable for the complainants because various cases had been decided primarily on the basis of statistics in private sector law, thus favoring remedies which were supposed to change the poor EEO statistics quickly.[154]

Private sector law was not applied to the NASA cases, much to the dismay of various potential future female and minority employees who would have benefitted from a legal obligation to improve the NASA EEO statistics. There were a great many disgruntled people who had applied to NASA but were rejected, allegedly due to their race or gender; however, these people did not have the opportunity to file a complaint with the Discrimination Complaint Division because they were not employees. Suing NASA was also problematic because it was considerably harder to prove discrimination in NASA's hiring procedures. Therefore, a lot of the declined applicants turned to the Civil Service Commission or their Senators and Congressmen and -women and described their stories in various

151 See, for example: *CA 75G0352 NE, in the United States District Court for the Northern District of Alabama North Eastern Division*, March 26, 1975, Harris/MSRC; Barry J. Casebolt, 'Job, Suits Filed against MSFC', *Huntsville Times*, March 27, 1975, p. 4; *CA 74-1832, United States District Court for the District of Columbia*, December 10, 1975, Harris/MSRC; 'Blacks, Women at NASA sue', *Federal Times*, January 8, 1975.

152 Frank Moss, *NASA's EEO Program*, May 14, 1974, MC 496 Carton 44, Folder 5, National Organization for Women Collection, Schlesinger; Dudley McConnell, *Statement of Dr. Dudley G. McConnell Assistant Administrator for Equal Opportunity Programs National Aeronautics and Space Administration before the Subcommittee on HUD-Space-Science-Veterans Committee*, 1974, p. 10, Harris/MSRC. He gave almost the identical statement in front of the Subcommittee on Civil Rights and Constitutional Rights of the Committee on the Judiciary, House of Representatives.

153 Timothy S. Robinson, 'NASA Bias charged by NAACP', *Washington Post*, November 21, 1974.

154 R. Tenney Johnson, *NAACP Legal Defense Fund Lawsuit against the Civil Service Commission (CSC) and NASA/JSC*, letter to James C. Fletcher, November 22, 1974, 013690, Box 32, HRC.

letters. Although it is now impossible to know why the agency had rejected these applications, whether because of a lack of skills, merit, available jobs, or due to more accomplished contenders, there is no denying that some of the stories give the impression of gender and race based discrimination. One of these people was Spann Watson, an African American Air Force officer. In excruciating detail, he laid out how he and other qualified minorities applied to NASA between 1964 and 1973 but were either ignored, rejected or faced with flat-out denial from NASA that the job they had applied for was actually vacant. When he had applied to several NASA centers for the first time in 1965, he did not hear back from most of them; only headquarters and Goddard Space Flight Center replied, and even they rated him as a GS-13 despite his CSC GS-14/15 rating, but did not pursue his application further. His next interaction with the agency was initiated by NASA. McConnell called him in 1969 to ask for his assistance in finding minority candidates, giving him the names of seven people, including himself, only one of whom, Margaret Webster, was ever actually invited for an interview. Webster was a minority female computer programmer and computer system analyst. Despite being rejected by NASA, she was employed by the Department of Health, Education, and Welfare (HEW) at a GS-15 level. In 1970, Watson was again contacted by NASA. Joseph Hogan asked him for a referral of a good EEO Compliance Officer. He named someone whom NASA actually accepted, but the applicant declined in favor of a better offer at HEW. He had also recommended Samuel Lynn, who was eventually accepted by NASA. In the summer of 1970, he was contacted by NASA yet again. This time an EEO officer needed a referral for an applicant in Huntsville. Watson recommended a suitable candidate and also contacted Alabama A&M College. His referrals all applied but never heard back from NASA. In September 1972, he was again contacted by the EEO office for recommendations for the position of Assistant Administrator for Scientific Affairs and Technical Utilization. Yet again, his referrals who had applied for the position were informed that NASA had no positions available. Finally, in December 1972/January 1973, Watson was begged to apply for an EEO position at NASA HQ. At first he was resistant because of his previous upsetting experiences with the agency, but they were eventually able to convince him to apply. However, when he inquired about the state of his application, he was informed by NASA that there were no vacancies. He then had to inform them that there were vacancies. Claiming an administrative error, Watson was eventually invited for an interview but was rejected in the end.[155] After his

155 Spann Watson, *Testimony – NASA Employment Practices*, March 14, 1974, Harris/MSRC. According to Ruth Bates Harris, the "no vacancy" response to minority and female applications occurred frequently. Ruth Bates Harris and David R. Cashdan, *amended Charge of Discrimination*, letter to James C. Fletcher, November 26, 1973, Harris/MSRC.

unsatisfactory application experiences and upsetting responses at all levels of NASA management, Watson, like many of NASA's female and minority employees, was convinced that EEO at NASA was nothing more than a public relations tactic and that people like him, who fell for the agency's lip service and "who were freely willing to assist NASA in mending its old ways lost widespread credibility in the minority community."[156]

The African American Deputy Personnel Director at the Drug Enforcement Administration in the U.S. Department of Justice, Vincent T. Oliver, had a similar experience with his four applications to NASA. He applied for the first time in 1968 for a GS-9 Personnel Staffing Specialist position. He was rejected. In 1972, Oliver applied for the position of Personnel Officer at Ames Research Center. However, in his interview, he felt that Ames' Glenn Goodwin was "inept, insulting, prejudiced and unkempt,"[157] which prompted him to withdraw his application and write a letter about his discriminatory interview experience as "an evaluative critique" of Goodwin's lacking management and people skills to Dr. Hans Mark, the Director of Ames Research Center.[158] Even though Mark asked him not to withdraw his application, Oliver did not comply.[159] Instead, he applied to NASA headquarters in September 1972 for the position of Director of Personnel. He was interviewed but rejected. Oliver actually applied a fourth time to the agency, for the position of Chief of Personnel Program Evaluation. He was again interviewed, this time twice, but never heard anything from NASA afterwards, not even a rejection letter. Although Goddard Space Flight Center tried to tap him on a fifth occasion for a position in Manpower Management, he did not apply, despite the agency's invitation to do so, because by this time he felt "it was patently obvious that NASA was seeking to insure that a minority would be 'considered' for such vacancies though there was no intent to select a minority for any of these positions."[160] Oliver was convinced that NASA was only playing the "minority consideration game" without actually wanting to improve their EEO statistics, which was also proven by NASA's appointment of women and minorities on only a token

156 Spann Watson, *Testimony – NASA Employment Practices*, March 14, 1974, pp. 7–8, Harris/MSRC

157 Vincent T. Oliver, *Testimony of Vincent T. Oliver before the Subcommittee on Congressional Rights and Constitutional Rights of the House*, March 14, 1974, p. 6, Harris/MSRC.

158 Vincent T. Oliver, *letter to Dr. Hans Mark*, June 16, 1972, Harris/MSRC.

159 Hans Mark, *letter to Vincent T. Oliver*, June 21, 1972, Harris/MSRC.

160 Vincent T. Oliver, *Testimony of Vincent T. Oliver before the Subcommittee on Congressional Rights and Constitutional Rights of the House*, March 14, 1974, p. 8, Harris/MSRC.

basis. He, too, evaluated NASA's early EEO efforts as lip service only – a problem that Congress had to rectify: "There is no doubt in my mind, however, that justice carried to its proper conclusion will ultimately bear out the assertion that NASA has and does discriminate both racially and sexually and is conducting its public business with public funds in blatant violation of the public law."[161]

It was not just rejected NASA applicants, but also current or former employees, who turned to the Civil Service Commission and Congress for support in EEO matters. They did this instead of or in addition to filing a complaint with the Discrimination Complaint Division in order to attract more attention to NASA's handling of EEO matters. Herbert M. Huie, for example, complained in 1974 that he was hired by NASA as a Contractor Relations Specialist at GS-13 with the promise of being elevated to GS-14 within the next six months, but he was never promoted by Kennedy Space Center and therefore lost money before his retirement.[162] Mary L. Sligh wrote to Congressman Don Edwards in the hope of getting his assistance. She worked in a dead-end position at NASA and she felt that her bosses discriminated against her because of her race and gender. Although she had good credentials and was initially promised a promotion, she did not receive one. Upon inquiring why she was denied a better position, her manager just told her that, "[i]f you don't like what's going on find yourself another job."[163] He also suggested that she would have to take an accounting course before she could be promoted, yet she was only allowed to attend the course after a white female friend wrote on her behalf to Senator Hubert H. Humphrey. Even though she was eventually promoted and her immediate supervisors appreciated her work, she was quickly replaced by a college student and had to go back to her old job – the entry-level position she had initially worked at when she had entered NASA six years ago.[164] Ellen C. Weaver, the acting Director of the Office of Research Development at San José State University and a former Ames employee, also turned to Senator Edwards

161 Vincent T. Oliver, *Testimony of Vincent T. Oliver before the Subcommittee on Congressional Rights and Constitutional Rights of the House*, March 14, 1974, pp. 4–5, Harris/MSRC.

162 Herbert M. Huie, *letter to the Chairman of the U.S. Civil Service Commission*, February 28, 1974, Harris/MSRC; Herbert M. Huie, *letter to Winton M. Blount* (Director, U.S. Postal Service), September 24, 1971, Harris/MSRC.

163 Mary L. Sligh, *letter to Don Edwards*, July 22, 1974, Judiciary Committee 93rd Congress, Subcommittee on Civil and Constitutional Rights, Box 38 of HR. BX. 33, NASA 1974 Correspondence with NASA, RG 233, NARA I.

164 Mary L. Sligh, *letter to Don Edwards*, July 22, 1974, NARA I.

in 1974. She sent him information material about women in the sciences and told him about the poor work conditions for women at NASA. She, too, felt that "NASA/Ames keeps making promises-promises, and yet its performance is reprehensible or worse."[165]

The allegations of rejected applicants and former and current employees all paint the same picture. They all felt that the agency paid more lip service to EEO than it put energy into making it a reality. They furthermore alleged that most managers only employed or promoted certain token minorities or women instead of instituting a complete turnover of employment practices in regard to EEO. Their experiences indicate that, although NASA's upper management had vowed to turn over a new leaf, quite a few individual managers did not want to or did not know how to comply with the EEO management instructions. To help the managers understand the EEO issues and to prevent sex- and race-based discrimination complaints, the EEO offices in all the centers and at headquarters offered sensitivity training and EEO seminars to its managers. The seminars were meant to explain the cultural change of race and gender inclusion that was going on in the country and at the NASA workspaces. At the seminars, the managers were, for example, encouraged to think about the social, economic and employment conditions which forced many women and minorities to mainly perform menial work. They were also asked to explore questions like: "Past role of women in society – How is it changing? Is there a difference between money for a woman and money for a man? If so, why? Was it lack of knowledge, intelligence, personality, emotional stability or ambition? Do these factors apply in today's world?"[166] The managers were also trained at the seminars to not fall into the trap of discriminating against someone or being accused of discrimination. They were told to treat all their employees the same way, to treat all professionals in the same professional capacity. Topics were explained to the seminar participants mainly with examples. For instance, if there was a special event and the staff needed to wait tables or serve coffee, the managers should not pick the female professionals for this tasks because they were just as overqualified as the male professional for that type of job.[167] They

165 Ellen C. Weaver, *letter to Don Edwards*, March 26, 1974, Judiciary Committee 93rd Congress, Subcommittee on Civil and Constitutional Rights, Box 38 of HR. BX. 33, NASA 1974 Correspondence with NASA, RG 233, NARA I.

166 NASA EEO Office, *NASA Human Rights Seminar Discussion Guide*, ca. 1970s, Harris/MSRC.

167 See for example the experiences of Harriett J. DeVries Smith and Bertha Ryan who were asked to wait tables at a big X-15 conference in the 1960s. That request would not have been possible any longer 15 years later. Harriett J. DeVries Smith, interview by Rebecca Wright, November 7, 2001, pp. 10–11, Oral History Transcript, NASA "Herstory" Oral History Project, JSC.

also clarified what sexual harassment is and how to avoid it. The managers were, for instance, instructed to refrain from physical contact of any kind, such as a friendly patting on the shoulders, because it might be misinterpreted.[168]

The EEO and sexual harassment seminars were needed, as evidenced by the experiences of the women and minority employees in the 1960s and the discrimination complaints filed. Several managers at different NASA centres seem to have recognized this and wanted more EEO training for management and the promotion of the EEO program through seminars for the employees.[169] The first attempted training workshops in April 1974 nevertheless had to be postponed due to resistance of some of NASA's managers.[170] There were also several reports that as late as the 1980s quite a few managers at different centers did not take the content of the EEO trainings that did happen seriously.[171] Johnson Space Center's Frank Hickey, for example, had to discontinue his EEO seminars in 1974 due to a lack of interest on the part of the managers. This outcome prompted Hickey to inform Deputy Assistant Administrator Low that a lot of the JSC managers did not care about EEO and sometimes even work against it:

> I have come to the unavoidable conclusion, however, that your frequent affirmances of this national policy on equal employment matters are generally falling on deaf ears. I could possibly live with that if I honestly believed that it was nothing more than human nature choosing to give self-serving interpretations, viewing your declaration in much the same manner as we shrug off periodic endorsements of the Savings Bond program. But it is much more than that, and something that will never appear in any reports forwarded to you through the usual management channels. I am personally observing a process of change take place here at JSC where each day of continuing ineffectiveness of the Equal Employment Opportunity Office is counterbalanced by a strengthening of the overt resistance to the law and to the agency policy. [...] Any effort to identify and establish goals to measure the progress of operating organization is fought, not merely delayed or neglected. We dare not fight our own Center leadership* [*I do NOT want to be reassigned][172]

Carl Praktish also sent a letter to Low in which he validated Hickey's frustrations with the response to EEO by Johnson Space Center management, which did not want to "start waves," but he also told him about Hickey's career set-

168 Harriett G. Jenkins, interview by Jennifer Ross-Nazzal, August 5, 2011, p. 49, JSC.
169 *Report on Assessment of NASA's Equal Employment Opportunity Counseling Program*, 1975, p. 16, Box 1, EO 1975, Center Files, EEO 1973–1978, JSC.
170 McQuaid, '"Racism, Sexism, and Space Ventures": Civil Rights at NASA in the Nixon Era and Beyond', p. 440.
171 Anne L. Accola, interview by Rebecca Wright, March 17, 2005, pp. 2–3, Oral History Transcript, NASA Johnson Space Center Oral History Project, JSC.
172 Frank Hickey, *letter to George M. Low*, June 20, 1974, 13700, Box 34, HRC.

backs and RIF experience, thus insinuating that Praktish was mainly frustrated with his personal work experience at NASA.[173] R. Tenney Johnson claimed that he had talked to Carlos Garza of the Johnson Space Center EEO office about Hickey's allegations and that Garza was neither aware of Hickey nor his allegations against the JSC managers.[174] It is therefore not quite clear to what extent Hickey's allegations against NASA management are genuine. Considering, however, how many people accused management of insincere lip service and a failure to act, the allegations most likely contain some truth.

7.3 Varying degrees of commitment to EEO

The work of the officers of the EEO spaces at headquarters and at the different centers was intended to change the performance of social conventions and exclusionary customs in the space community and to implement new administrative policies and work routines. While they always started to change their own social and intellectual space first, their goal was to bring the change to the other workspaces in the community. However, instilling the importance and justness of equality in people's minds was a difficult undertaking because quite a few space workers had trouble letting go of obsolete cultural and political structures that had granted them privileges which they now took for granted. Employees and management both, thus, needed to find a way to cope with the new EEO regulations. There were several strategies regarding the commitment to EEO which will be analyzed now. It is necessary to look at the space community's reaction to EEO in order to see how much the expedited re-coding of masculinity and femininity dictated by law, as well as the new concept of racial equality, changed the workspaces of the space community and if this change was in concurrence with the social transformation of American society at large.

The management's commitment to EEO was rather mixed for several reasons. One reason was that NASA's management consisted of a combination of individuals who all had differing approaches to EEO. Generally, Deputy Administrator George M. Low was regarded by his employees, including the EEO officers, as the principal facilitator of EEO within management. According to Kennedy Space EEO officer Nathaniel Pilot, Low was truly interested in EEO, unlike Administrator Fletcher and the other NASA managers who only did what

173 Carl Praktish, *letter to George M. Low*, July 1, 1974, 13700, Box 34, HRC.
174 R. Tenney Johnson, *memorandum to George M. Low*, July 3, 1974, 13700, Box 34, HRC.

was expected of them.[175] However, Harriett Jenkins considered both Low and Fletcher to be cordial, supportive, and very willing to listen.[176] Kennedy Space Center Director Dr. Kurt H. Debus, on the other hand, did not receive such positive reviews. Pilot felt that Debus was not really committed to EEO, but had only been following orders when he declared the implementation of EEO provisions at KSC.[177] EEO officer Willie D. Gaither had an even worse impression of the Kennedy Space Center management and called its management style "prerogative in an extremely arbitrary, discriminatory and conservative fashion."[178] NASA Director of Inspections, James Cummings, also received negative feedback. He was anonymously accused of outright rebelling against EEO at NASA by not hiring minorities and badmouthing Fletcher. Although he denied the allegations, Cummings supposedly told his staff that "Fletcher does not tell me who to hire in my office. He wouldn't like it if I told him who to hire in his office so why should I listen to him. Besides, I don't want any blacks working for me and I will not hire another one. I hired one, he left, thank God, and I won't hire another one. Who does Fletcher think he is anyway."[179] Johnson and Marshall Space Flight Center directors received a much better assessment. JSC Director Robert Gilruth was praised by historian Amy Foster for his serious commitment to EEO, while one of his successors, JSC Director Gerald Griffin, was lauded by the Federally Employed Women for "his efforts in furthering the cause of advancement for women and in advocating equal rights for both sexes while they climb the career ladder. He believes in the Equal Opportunity Program concepts and follows its criteria in his support of its goals."[180] Marshall directors Eberhard Rees and Wernher von Braun both received good EEO commitment evaluations from Ruth Bates Harris in her memoir. However, Harris also remembered that Rees lacked commitment at first, before he eventually turned into "a most ardent supporter of EEO and issued his affirmative action orders with courage and conviction."[181] The examples of these individuals within NASA

---

175 Nathaniel Pilot, interview by Dr. Aaron Gillette, July 28, 1992, pp. 37, 39, Tape Transcript, HRC.

176 Harriett G. Jenkins, interview by Jennifer Ross-Nazzal, August 5, 2011, p. 6, JSC.

177 Linda Rogers, interview by Dr. Aaron Gillette, July 27, 1992, p. 25, Tape Transcript, HRC.

178 Willie D. Gaither, *memo to Dudley McConnell*, December 4, 1973, 18508, Box 1, HRC.

179 Anonymous letter to James C. Fletcher, February 3, 1976, 13700, Box 34, HRC; James Cummings, *memorandum to George M. Low*, February 26, 1976, 13700, Box 34, EEO, HRC.

180 'Space Coast Award', *News & Views* 16/2 (1981), p. 6, F.E.W.'s News & Views, Schlesinger; Foster, *Integrating Women into the Astronaut Corps. Politics and Logistics at NASA, 1972–2004*, p. 72.

181 Harris, pp. 257–258.

management show that different managers felt differently about EEO, and that it was usually up to these individuals in charge to decide how rigorously they wanted to pursue EEO in their field of responsibility. Their different approaches resulted in sundry EEO performances – good, mediocre, and very poor.

Another reason for the very mixed EEO assessments was that high ranking managers, like Low, Gilruth or Rees, were not able to implement EEO by themselves; they needed the support of the managers who directly supervised the employees in the labs, hangars, or offices. However, various individuals and reports from the 1970s point out that there was a severe discrepancy between upper and middle management's willingness to thoroughly commit to EEO. Pilot believed that the reason for the discrepancy was that most senior managers were more exposed to EEO politics than middle managers, though he still did not believe that even the upper management did enough "to really drive home equal opportunity here."[182] A report from the Civil Service Commission contained a similar assessment. According to the 1974 report, NASA's top managers had voiced strong support for EEO but had not backed their words with actions, which gave middle managers and the employees the impression that EEO was exclusively handled by the EEO office. Middle management therefore had no incentives or supervision to integrate affirmative action in their own workspace, assuming that being non-discriminatory was enough.[183] An investigation by a panel of the National Academy of Public Administration in 1975 went a step further in their conclusion. The report revealed that while upper management undoubtedly believed in their commitment to EEO, their perception did not necessarily correspond with the impressions of the employees and middle management. The panel frequently found that

> NASA personnel at various levels were asking whether the statements endorsing equal opportunity by top administrators represent real commitment or a more or less pro forma attempt to demonstrate that something is being done. In the absence of a firmer conviction concerning commitment at the top, there remains a tendency to see equal opportunity as a side issue which can be fixed much as a problem in a hardware program can be fixed with some incremental addition to effort.[184]

182 Nathaniel Pilot, interview by Dr. Aaron Gillette, July 28, 1992, pp. 24–25, Tape Transcript, HRC.

183 U.S. Civil Service Commission, *A report on Equal Employment Opportunity National Aeronautics and Space Administration*, April 1974, p. 3, NARA I.

184 Panel of the National Academy of Public Administration, *Achieving Equal Opportunity in NASA: An Assessment of Needs and Recommendations for Action*, January 30, 1976, p. 11, Box 2, Center Files, JSC.

Considering that most middle managers assumed that NASA was only improving its equal opportunity profile as a side note to provisions seen as more important, it cannot be surprising that the internal social climate for women and minorities seeking equality at their workspaces was not ideal. Quite a few middle managers also perceived EEO provisions to be a diversion away from the agency's actual work or even considered EEO to be harmful to NASA's merit standards, to be reverse discrimination, or just another arduous political fad. Such attitudes led the panel to the conclusion that "from an agency-wide point of view, there still remains a considerable way to go before reaching the stage which can be clearly identified as commitment."[185] Yet another study, this one from management consultancy Arthur D. Little, came to the conclusion in 1977 that the "top managers were committed but the next lower level needed to be worked with."[186] Upper management, especially Low, was very aware of the problem because of the many complaints received about middle management not supporting EEO and a general lack of EEO leadership at the field installations.[187] Although not necessarily in public, he and Fletcher also frequently admitted to the agency's shortcomings in EEO implementation.[188] These shortcomings are why Low wanted his fellow managers to take more action and responsibility in EEO matters in their area of supervision and why he frequently had to tell the EEO officers that the responsibilities of the managers need to be spelled out specifically in the EEO plans, ideally linked to specific managers and deadlines for implementation[189]

However, most managers did not want to own up to the lack of commitment displayed by various NASA managers in middle as well as upper management. They believed in their own commitment to EEO but found various

---

185 Panel of the National Academy of Public Administration, *Achieving Equal Opportunity in NASA: An Assessment of Needs and Recommendations for Action*, January 30, 1976, pp. 12–13, Box 2, Center Files, JSC.

186 *Minutes of the EO Council Meeting*, November 29–30, 1977, Box 1, Equal Opportunity 1977, Center Files, JSC.

187 George M. Low, *Miscellaneous Items of Interest*, July 8, 1974, Box 67, Folder 1, Low/RPI.

188 George M. Low, *Women at NASA*, memorandum to Ruth Bates Harris and Dudley McConnell, August 14, 1973, Box 16, Chronological File, Folder 5,Low/RPI; James C. Fletcher, *Memorandum to all NASA employees*, November 2, 1973, Harris/MSRC; Gayla Salinas, 'The vocal majority', *Washington, D.C. Chapter* 5/1 (1974), p. 4, Pr-1 Carton 6, NOW Chapter Newsletter, Schlesinger.

189 George M. Low, *Comments on Affirmative Action Plan for Calendar Year 1974*, January 28, 1974, Chronological File, Folder 2,Low/RPI; George M. Low, *Comments on Draft Plan for Implementing Employee Suggestions Received November 5, 1973*, February 5, 1974, Chronological File, Folder 3, Low/RPI; George M. Low, *1974 Recruiting Plan for Minority and Female Candidates for Professional Positions*, February 5, 1974, Chronological File, Folder 2, Low/RPI.

excuses for their failure to integrate more women and minorities into their workspace. Diving into NASA's and the space industry's employment statistics was the most common way to prevaricate.[190] NASA underwent several reductions in force. Its civil-service workforce, for example, was by 1974 reduced by a third to 24,881 in-house employees, down from 36,200 in-house employees during NASA's peak employment in 1967.[191] So while the overall workforce was declining, NASA needed to employ more minorities and women, a confluence of factors which led managers like Johnson Space Center Director Christopher C. Kraft or JSC Chief of the Integration Division, John R. Sevier, to believe that there was no discrimination at Johnson Space Center because of the adverse conditions in the 1970s which made it considerably harder to change the employment status quo in regard to EEO. Sevier was convinced that JSC was "doing a good job to provide equal opportunities to minorities" within the existing limitations on the work force.[192] Kraft also liked to elaborate on the obstacles to EEO implementation at Johnson Space Center, such as limited job openings, average grade constraints, a shortage of qualified female and minority candidates for scientific and engineering positions in the higher grades, a lack of competitive salaries compared to private industry to attract the few minority and female engineering and science talents, a lack of direct-hire authority, veteran's preference and the rigid civil service classification system.[193] Although their explanations for NASA's slow progress in EEO are valid, NASA's RIFs and lack of competitive edge compared to the private sector were often abused as excuses for stagnancy. The panel of the National Academy of Public Administration came to the same conclusion in 1976. The

190 For example: *Statement of Dr. John F. Clark, Director, Goddard Space Flight Center, National Aeronautics and Space Administration before the Subcommittee on Congressional Rights and constitutional Rights of the House Committee on the Judiciary,* March 1974, Harris/MSRC; Dudley McConnell, *Statement of Dr. Dudley G. McConnell Assistant Administrator for Equal Opportunity Programs National Aeronautics and Space Administration before the Subcommittee on HUD-Space-Science-Veterans Committee,* 1974, pp. 9–18, Harris/MSRC.

191 *Report on Assessment of NASA's Equal Employment Opportunity Counseling Program*, 1975, p. 10, Box 1, EO 1975, Center Files, EEO 1973–1978, JSC; McQuaid, 'Race, Gender, and Space Exploration: A Chapter in the Social History of the Space Age', pp. 408–409; Nimmen, Bruno and Rosholt.

192 John R. Sevier, *NASA EO Self Evaluation Plan*, letter to WA/Manager, Program Operations, December 3, 1974, Box 1, Equal Opportunity 1973, Center Files, JSC.

193 Christopher C. Kraft, *Minority and Female Employment at JSC*, letter to NASA Headquarters J/Assistant Administrator for Institutional Management, January 18, 1974, Box 1, Equal Opportunity 1974, Center Files, JSC.

panel was very aware that NASA's employment data had to be put in the proper context in order to not be compared unfairly with other federal agencies, but they also emphasized that "there is no justification for magnifying them or using them as defensive devices to explain inadequate progress at NASA."[194]

The managers' approaches to overcoming a lack of commitment to EEO were manifold. While Kraft and other managers wanted to change structural problems with a modification of RIF procedures, the streamlining of examining and certification procedures or the provision of more hiring authority, Willis H. Shapley, Associate Aeronautics and Space Administrator, suggested transferring the successful direct accountability techniques used for handling the agency's space goals to the EEO program.[195] Donald D. Arabian, a manager in the Program Operations Office at Johnson Space Center, felt that directors and supervisors needed to apply more "loving care." He encouraged his fellow managers to take a more personal interest in the employees and to talk to them about their careers, problems and opportunities more often.[196] One of the most common lines of thought to improve commitment to EEO, or at the very least to give the impression of committing to EEO, was the idea of hiring a minority astronaut – a token astronaut. The idea was not new. President John F. Kennedy had already wanted to promote an African American into the astronaut corps and had therefore supported Edward J. Dwight's entry into test pilot school and the aerospace research pilot course at Edwards Air Force Base in June 1963. President Lyndon B. Johnson also wanted to see a minority astronaut. He, however, favoured Maj. Robert H. Lawrence Jr., who was accepted to the Air Force astronaut program in 1967.[197] After Lawrence's untimely death in a plane crash, there were no more minority astronaut appointees until the selection of the first astronaut class for the Space Shuttle in 1977. Several managers nevertheless generally supported the appointment of a minority astronaut on a token basis

194 Panel of the National Academy of Public Administration, *Achieving Equal Opportunity in NASA: An Assessment of Needs and Recommendations for Action*, January 30, 1976, Box 2, Center Files, JSC.

195 Christopher C. Kraft, *Minority and Female Employment at JSC*, letter to NASA Headquarters J/Assistant Administrator for Institutional Management, January 18, 1974, Box 1, Equal Opportunity 1974, Center Files, JSC; *Statement of Willis H. Shapley, Associate Aeronautics and Space Administrator NASA, before the Subcommittee on Civil Rights and Constitutional Rights Committee on the Judiciary House of Representatives*, 1974, p. 6, Harris/MSRC.

196 Donald D. Arabian, *Operating Level Affirmative Action Plan*, July 1, 1974, Box 1, Equal Opportunity 1974, Center Files, JSC.

197 Phelps, pp. 1–59.

in the early- and mid-1970s. Ruth Bates Harris, for example, tried to convince Christopher C. Kraft to push for women and minorities in the astronaut corps despite knowing that Apollo had just ended and that there was a surplus of astronauts in the corps in 1972.[198] She felt that a minority astronaut was exactly what children of color "needed to see to motivate them."[199] George Low also considered it in 1973, but found too many arguments against adding a token minority astronaut to the corps, including the fact that it would be unfair to existing personnel.[200] NASA Assistant Administrator for Public Affairs, John P. Donnelly, too, toyed with the idea of the "symbolic gesture" of employing an African American astronaut to publicly demonstrate NASA's commitment to EEO in 1974.[201] However, the minority communities would not have appreciated the tokenism, considering that they were already upset that tokenism was a common provision in EEO. They wanted to see a change in employment culture, not just tokens.[202] Considering that there was no major human space mission in the mid-1970s and that it would have violated the agency's code of merit, the idea of a token astronaut was dismissed in favor of eventually employing fully qualified persons later on.[203]

7.4 The accusation of EEO as reverse discrimination

It was not just the managers' commitment to EEO that varied; the employees, too, were often not on the same page regarding commitment, implementation, success, and the need for EEO and affirmative action in the space community. Generally there was an understanding that EEO was needed in the space community because "[i]f there was equality, we would not need an EEO office. We would not need any of that. We would not need any laws that say, this is what you need to do to bring things in compliance."[204] However, there were "die-hards who didn't see that there were problems that needed to be solved"

---

198 Ruth Bates Harris, *Minorities and Women into Space*, memorandum to Associate Administrator for Manned Space Flight, July 19, 1972, 18508, Box 1, HRC.
199 Harris, p. 267.
200 George M. Low, *personal notes*, January 23, 1973, Box 61, Folder 3, Low/RPI.
201 Atkinson and Shafritz, pp. 137–138.
202 'EEO Hypocrites', *Washington Afro American*, December 1, 1973.
203 Harriett Jenkins, *NASA Equal Opportunity Program*, January 12, 1976, Box 1, Equal Opportunity 1976, Center Files, JSC.
204 Annie J. Easley, interview by Sandra Johnson, August 21, 2001, p. 21, Oral History Transcript, NASA "Herstory" Oral History Project, JSC.

and bridled at the changing work culture and new administrative policies.[205] They flatly refused to acknowledge that society and NASA management called upon them to include people into their workspace previously considered not worthy. Welcoming minorities and women would have forced them to change their own behavior in regard to race and gender and overhaul their view about what is appropriate in social settings.

This attitude was especially prevalent in NASA installations and contractor workspaces in the Southern states, where a lot of the members of the space community were steeped in Southern culture. The people who had internalized this culture often had a hard time letting go of white privilege and embracing equality. According to contract specialist Linda Rogers, Kennedy Space Center in Florida, for example, had very much reflected the society it was in.[206] There were cases of several "hard core Southern" employees who were not willing to go along with EEO, such as a group of design engineers at KSC, who expressed through mockery their opposition to being, as they saw it, force-fed EEO issues.[207] Some non-minority employees were simply reluctant to accept EEO because they perceived their new African American colleagues as unqualified and "not as well equipped to do space oriented work" as their white counterparts who supposedly had to outlast a harder screening process for the job.[208]

There were also minority employees who felt that NASA was going too far in EEO matters. In May 1974, while Low was visiting Johnson Space Center, a Latino man confided to him that he had made it on his own in the agency's workspace and the other minorities should do the same, if they had what it takes.[209] Other minority and female employees, who made their feelings known anonymously, felt that NASA concentrated on the wrong people. Instead of bring new people to NASA, EEO efforts should mainly focus on upward mobility for the women and minorities already employed by NASA.[210] A sizeable number of non-minority employees also felt that the agency was going too far in implementing EEO. Non-minority employees usually accepted the EEO plans and deadlines as long as they were not personally affected; however, when NASA

205 Annie J. Easley, interview by Sandra Johnson, August 21, 2001, p. 21, JSC.
206 Linda Rogers, interview by Dr. Aaron Gillette, July 27, 1992, p. 30, Tape Transcript, HRC.
207 Ernie Reyes, interview by Dr. Aaron Gillette, July 28, 1992, pp. 33–35, Tape Transcript, HRC.
208 Nathaniel Pilot, interview by Dr. Aaron Gillette, July 28, 1992, pp. 11–12, Tape Transcript, HRC; Jerri Bell, interview by Dr. Aaron Gillette, July 24, 1992, pp. 19, 22, Tape Transcript, HRC.
209 George M. Low, *personal notes No. 120*, May 11, 1974, Box 67, Folder 2, Low/RPI.
210 Bob Strickland, *Eyewitness News Report on WTOP TV*, April 3, 1974, 18507, Box 4, HRC.

reduced its workforce in the 1970s, many felt that minorities and women had an undeserved edge competing for jobs in the space community.[211] Not only employees but also some managers were concerned by "individuals who appear to get special treatment."[212] They feared that this practice was discriminatory. In 1977, 76 percent of Lewis Research Center employees and 40 percent of Johnson Space Center employees felt that EEO at NASA had become reverse discrimination.[213] A letter to NASA's Weekly Bulletin gave an insight into the mind of a person disgruntled with EEO: "My question is simple. Would you explain why, on the flyer for the Fashion Show in *Weekly Bulletin* No. 24, the models were both black. Talk about discrimination. The only thing nondiscriminatory was that it was a black and white sketch. Us white folks have had it!"[214] The concept of reverse discrimination began to become popular with conservatives in the United States in the mid-1970s in the context of criticizing affirmative action efforts.[215] So it was not only at NASA, but also at other places of employment, in education, and in other areas of life in the United States, that white, heterosexual men started to complain about unfair treatment, among them the men who protested an annual summer program for college women on Nantucket Island in Massachusetts provided by the independent Maria Mitchell Association in 1977.[216]

A lot of the white men, and possibly also white women, at NASA were not able or willing to grasp the inherent idea of EEO and affirmative action. These efforts were supposed to gain women and minority groups access to workspaces previously denied or closed to them. It was basically a redistribution of access and power in order to establish social equality. However, equality is only possible when the people who have received preferential treatment due to the prevalent social customs and culture let go of their preferential treatment, their privileges. Considering that white men possessed more privileges than their male counterparts of color or women, it must have felt like discrimination when, in actuality, they only had to give up the privileges to which they had grown accustomed. Thus, EEO efforts, embodied by the people who were

211 Nathaniel Pilot, interview by Dr. Aaron Gillette, July 28, 1992, pp. 30–32, Tape Transcript, HRC.

212 *Minutes of NASA EO Council meeting held at Marshall Space Flight Center*, January 23, 1975, Box 2, Equal Opportunity Council – Meeting Minutes 1975–76, Center Files, JSC.

213 *Minutes of the EO Council Meeting*, November 29–30, 1977, Box 1, Equal Opportunity 1977, Center Files, JSC.

214 Larry Vogel, *Question for Weekly Bulletin*, Memo to Earl Ginyard, June 16, 1976, 18507, Box 4, HRC.

215 Fred L. Pincus, *Reverse Discrimination: Dismantling the Myth* (Boulder: Lynne Rienner, 2003), pp. 35–42.

216 Rossiter, *Women Scientists in America. Vol. 3. Forging a New World since 1972*, p. 74.

employed because of EEO in the 1970s, had to face not only the backlash against the women's movement, but also the allegations of reverse discrimination, in the United States at large and at NASA in particular. In this context, Emogene Greville's assessment that EEO had "no authority whatsoever" in the beginning makes sense. It was only in the 1980s that EEO came to be taken seriously and had a big influence at NASA, according to Greville.[217]

Most employees in the 1970s, like Greville, also felt that EEO at NASA was a 'sham,' hence why it was not only Ruth Bates Harris and her fellow EEO officers who felt that the EEO structure in place needed improvement.[218] At least a third of the employees wanted the EEO office at HQ to receive more authority. They were also in favour of more training for the EEO officers and counsellors and a stricter screening before their appointment. The Minority Employees at NASA (MEAN) even demanded that "NASA Headquarters EEO Counselors should be stripped of their titles, packed up, and returned to the corners of their offices to their former responsibilities on a full-time basis."[219] Quite a few employees felt that managers should be more involved with the enforcement of EEO and the counselling program, too. Furthermore, they wanted nonresponsive managers to be removed from their positions and a better system put in place to ensure that employees who were seeking counsel or filed a discrimination complaint be protected from reprisal.[220]

7.5 EEO: minorities vs. women

Other employees were in favor of EEO, just not for all the people targeted by it. Some white men were better able to relate to the needs and wishes of white women in regard to EEO, while others could only identify with the problems of minority males because they considered women's liberation to be a "big joke," much to the dismay of female employees who complained that "there's an economic problem that escapes them. They can't understand why women refuse sexist terms. They think of us in terms of radical lesbians not in terms of rational

217 Emogene Greville, interview by Dr. Aaron Gillette, July 27, 1992, pp. 42, 46, Tape Transcript, HRC.

218 Minority Employees at NASA (MEAN), *Third Party Complaint*, letter to Dr. Fletcher announcing MEAN, March 14, 1974, Judiciary Committee 93rd Congress, Subcommittee on Civil and Constitutional Rights, Box 38 of HR. BX. 33, NASA 1974 Miscellaneous Documents, RG 233, NARA I.

219 Minority Employees at NASA (MEAN), *Third Party Complaint*, letter to Dr. Fletcher announcing MEAN, March 14, 1974, p. 9, NARA I.

220 *Report on Assessment of NASA's Equal Employment Opportunity Counseling Program*, 1975, p. 16, Box 1, EO 1975, Center Files, EEO 1973–1978, JSC.

people wanting a decent chance to move upward."[221] Not only employees, but also managers and supervisors, sometimes had more sympathy for one group over the other. Female employees frequently complained that "[s]ome male supervisors seem to be threatened and some consider it a joke matter if women assume any role beyond that of passive individuals who know their 'places'; if they aspire to careers beyond that of secretaries; if they dare seek training for a promotion to management positions."[222] These remarks reveal one of the main problems of affirmative action and EEO efforts in general, and at NASA in particular: women and minority EEO issues were usually administered under the same EEO umbrella.

Considering the immense administrative effort and expenses needed to implement EEO and that both groups share a history of group discrimination, their pooling in one EEO office is logical. However, women and minority groups actually do have different histories and were excluded from the workforce for completely different reasons. While white women were not allowed to work in certain areas because they were expected to be homemakers and mothers, men of color were previously excluded from the workforce due to the belief that they were inferior to white people. Especially African Americans, who often are the descendants of enslaved Africans, have been discriminated against on an institutional, structural, and individual basis in all areas of life in the United States. Therefore, the civil rights and equality laws were mainly passed with this particular group in mind. This might explain why a majority of the employees during a self-evaluation at headquarters equated EEO with African American issues only.[223] Although the integration of African Americans into NASA's workforce was a primary goal, NASA's EEO efforts actually also targeted Hispanic Americans and, of course, women. The special emphasis programs of the EEO office, such as the Spanish Speaking Program, Black Employment Program, and the Federal Women's Program, attest to that fact. Although, Native and Asian Americans always appeared in NASA's EEO statistics in the 1970s, as American Indian and Oriental, their special emphasis programs would be established decades later.[224]

221 *Anonymous comments received from a cross section of NASA Headquarters Employees*, November 1974, Harris/MSRC.

222 Ruth Bates Harris, *Meeting with women from selected Headquarters and GSFC Offices*, letter to James Fletcher and George Low, August 9, 1973, Judiciary Committee 93rd Congress, Subcommittee on Civil and Constitutional Rights, Box 38 of HR. BX. 33, NASA 1974 Miscellaneous Documents, RG 233, NARA I.

223 *Minutes of NASA EO Council meeting held at Marshall Space Flight Center*, January 23, 1975, Box 2, Equal Opportunity Council – Meeting Minutes 1975–76, Center Files, JSC.

224 NASA, 'Special Emphasis Programs', 2010 <https://www.nasa.gov/offices/oeod/special_programs/Programs.html> [accessed 7 October 2016].

The few efforts within the special emphasis programs notwithstanding, women and the different minority groups were mostly pooled together for the EEO efforts. The EEO plans of the different centers can attest to this fact. For most of the 1970s, the plans were mainly concerned with EEO statistics and the numerical increase of EEO targeted people – be they minorities or women; there were almost no attempts to either acknowledge or find solutions to structural discrimination against the different groups. This type of discrimination was hard on all the groups because it was not as visible as institutional discrimination, which was amended with the change of the laws in the 1960s and 1970s. Even though the law had changed, structures holding back minorities and women remained in place. However, societal structures barring women from the workspaces were totally different from the social and cultural structures which kept people of color from joining the NASA workforce. Housing, for example, was a big issue for African Americans, who often lived in traditionally black, poor neighborhoods which were far away from the NASA installations. They often were not welcome in conservative white suburban neighborhoods, were not able to afford to move closer to a NASA center or pay for the expensive commute because public transportation was often lacking – for example at Johnson Space Center in Clear Lake or Goddard Space Flight Center in Greenbelt.[225] Although the agency was aware of this particular problem and observed the anniversaries of the National Fair Housing Law, Title VIII of the Civil Rights Act of 1968, NASA did not much more to solve the issue of either housing or transportation.[226]

Women, on the other hand, at least white women, had difficulties usually not with housing but with culturally ingrained expectations of their gender – women were equated with motherhood and marriage. How to balance working in the space community and taking care of husband and children was a problem the women were left to figure out alone. Topics such as daycare or part-time employment were rarely discussed and even more rarely implemented. Even though EEO officers Oceola Hall and Ruth Bates Harris started to consider childcare already in the 1970s, only a few NASA installations actually followed through with

225 *Report of KL-1/Management Intern, Contractor Equal Opportunity Programs to KL-1/Director, Contractor Equal Opportunity Programs, 'Minority Housing and Transportation at the Manned Spacecraft Center'*, August 2, 1971, 18508, HRC; *Statement of Dr. John F. Clark before the Subcommittee on Congressional Rights and constitutional Rights of the House Committee on the Judiciary*, March 1974, pp. 7–9, Harris/MSRC; Holt, *Rise of the Rocket Girls: The Women Who Propelled Us, from Missiles to the Moon to Mars*, pp. 98–99; Shetterly, p. 241.

226 For example: James C. Fletcher, *Ninth Anniversary of Fair Housing Law*, April 5, 1977, Harris/MSRC.

the suggestion, with holdouts lasting as late as 1990.[227] Furthermore, NASA policy or practices restricted part-time employment in the early and mid-1970s. This type of employment especially catered to the needs of women, who often wanted to split their time between children and work. Therefore, the Federal Women's Program Coordinator and the EO Council Women's Subcommittee recommended a change in regulations.[228]

The women were very aware of the structural discrimination against them. In a list of grievances, they told Fletcher to better inform the first line supervisors about their EEO responsibilities because "[m]any well-meaning, intelligent men do not necessarily intend to discriminate but rather, they, too, are victims of the traditional role imposed by society."[229] The women, furthermore, demanded that women should be counseled about their work opportunities at NASA "regardless of family status, number of children of school age, pre-school age, etc. Such data should not be used to discriminate against women."[230] They also called for a stop to the type of discrimination which required women to graduate from university and perform their jobs in a superior fashion in order to just be regarded as equal. They told Fletcher: "Ways are found to select men without proper qualifications. Why can't it be done for Women" – "Women employees reserve the right to be mediocre, too."[231]

So minority groups and women struggled with different issues, despite the many grievances they also had in common. Still, they were all looked after by the same EEO office and often felt like different peas in the same pod. They nevertheless had to compete for the same jobs, in the course of which the groups were frequently pitted against each other.[232] Minority Community Relations officer Roscoe Monroe, for example, recalled a complaint from

227 'Grand Opening. JSC Child Care Center opens for Business, Play and Learning', *Roundup*, May 4, 1990; Aaron Cohen, *JSC provides Building and Utilities, the Day Care is operated by a Non-Profit Cooperation,* May 19, 1989, Box GR 1019A, Center Series, JSC; Ruth Bates Harris, Memorandum to George Low, *GSFC's Proposal for Day Care Center,* January 11, 1973, 008987, HRC; *Status of NASA Activities on the Employment of Women,* 1973, 008994, HRC.

228 Ruth Klein Essnik, *EO Council Subcommittee*, October 19, 1977, Box 1, Equal Opportunity 1977, Center Files, JSC.

229 Ruth Bates Harris, *Meeting with women from selected Headquarters and GSFC Offices*, letter to James Fletcher and George Low, August 9, 1973, NARA I.

230 Ruth Bates Harris, *Meeting with women from selected Headquarters and GSFC Offices*, letter to James Fletcher and George Low, August 9, 1973, NARA I.

231 Ruth Bates Harris, *Meeting with women from selected Headquarters and GSFC Offices*, letter to James Fletcher and George Low, August 9, 1973, NARA I.

232 . Ruth Bates Harris, *Meeting with women from selected Headquarters and GSFC Offices*, letter to James Fletcher and George Low, August 9, 1973, NARA I.

black NASA employees who were upset that a white female was selected as the Upward Mobility Director, instead of a person of color, even though two qualified African Americans had applied.[233] This type of complaint shows that women and minorities did not necessarily regard each other as allies in their struggle to establish themselves in the workspaces of the space community, despite the belief of some feminists that "the war against sex discrimination should be linked to the principles and methods of the black civil rights movement."[234] Considering that a lot of the women active in the women's movement had left other movements, such as the Students for a Democratic Society (SDS) or the Student Nonviolent Coordinating Committee (SNCC), because they did not feel that these groups took women seriously or regarded them as full-fledged members apart from being a leading man's wife, girlfriend, or lover, it is understandable that the alliance between women and minority males also did not work out in the much smaller space of the space community.[235] Moreover, the double standards and burdens for women of color were not addressed in either the civil rights or (white) women's movements, nor was the topic properly discussed in NASA's EEO efforts. This is why women of color formed their own groups within the women's movement but they did nothing similar at NASA.[236] However, in the news, cases of "insidious racism – pitting a black man against a black woman," such as what happened in the case of Ruth Bates Harris and Dudley McConnell, were recognized and pointed out.[237] Like women of color, handicapped people were not specifically targeted in NASA's EEO efforts until the early 1980s, although they had expressed the wish to be part of it already in 1975.[238] In 1991, there were only 231 handicapped persons working at NASA, proving the lack of emphasis on this particular group in NASA's EEO efforts.[239]

Overall, the EEO umbrella concept was problematic. Linda Rogers, for example, was convinced that it had not worked very well socially at Kennedy Space Center in the beginning.[240] However, the EEO guidelines in the early years were rather chaotic in many regards, just one of which was the fact that

233 Roscoe Monroe, *assessment of personnel Policies and Practices*, memo to Dudley McConnell, March 11, 1974, 18508, Box 1, HRC.

234 Hugh Davis Graham, p. 232.

235 Rosen, pp. 94–105.

236 Breines; Davis; Blackwell; Rosen.

237 Ethel L. Payne, 'NASA under fire', *Courier*, December 29, 1073, p. 5.

238 *Minutes of NASA EO Council meeting held at Marshall Space Flight Center*, January 23, 1975, Box 2, Equal Opportunity Council – Meeting Minutes 1975–76, Center Files, JSC.

239 Harriett G. Jenkins, interview by Jennifer Ross-Nazzal, August 5, 2011, p. 58, JSC.

240 Linda Rogers, interview by Dr. Aaron Gillette, July 27, 1992, p. 29, Tape Transcript, HRC.

the hierarchical guidelines for EEO plans provided by HQ were not always in line with the centers' specific challenges and local problems.[241] This was one of the reasons for the centres' very different EEO success rates.

7.6 Mapping EEO performance

Despite all being part of NASA, the field installations were very different in size, function, projects, technical fields, budget, community outreach and location. One thing they had in common, though, was that all of them were unable to reach the EEO goals set for them by headquarters. While they all struggled with the underrepresentation of women and minorities, some centers did worse than others, with the Southern centers performing exceptionally poorly.[242] In June 1974, NASA employed 4,259 (17.1 percent) women and 497 (2 percent) minorities out of 24,854 permanent employees. HQ employed 593 (37.3 percent) women and 175 (11 percent) minorities. In comparison, Marshall Space Flight Center's female workforce numbered 672, but because Marshall was so big, the women only made up 15.3 percent of the Alabama based center's total workforce. Marshall had even fewer employees of color – only 31 minorities, who made up 0.7 percent of Marshall's workforce. Johnson Space Center's 626 women and 70 minority employees made up 17 percent and 1.9 percent, respectively, of the Texas based center's workforce. At the Florida based Kennedy Space Center, there were 407 (17.6 percent) women and only 21 (0.9 percent) minorities in the workforce.[243]

Even though Deputy Assistant Administrator of EEO Harriett Jenkins did not want to draw the conclusion that the Southern centers performed so poorly in regard to minority/female employment because of their location, she did admit that "culturally it would have had an impact."[244] Considering that the Jim Crow laws were enforced until the Civil Rights Act of 1964 in the Southern states and their abolition was often detested and rejected by white Southerners in private and public, it is not surprising that qualified minority employees were not flocking to NASA's Southern installations just a few years after racial segregation was ended in the South. Deterrent examples show the mentality of a lot of white

241 Hatty H. Gorman, *Goals for Increasing Minority and Female Employee Representation in NASA/OMSF Institutional Goals*, memorandum to Dudley McConnell, January 11, 1974, Box 1, Equal Employment 1974, Center Files, JSC.

242 Harriett G. Jenkins, interview by Jennifer Ross-Nazzal, August 5, 2011, pp. 15–16, JSC.

243 Harriett G. Jenkins, interview by Jennifer Ross-Nazzal, August 5, 2011, pp. 15–16, JSC.

244 Harriett G. Jenkins, interview by Jennifer Ross-Nazzal, August 5, 2011, pp. 16, JSC.

people in the South a couple of years after the Jim Crow era. For example, the 1969 Supreme Court case Alexander v. Holmes County Board of Education, in which deliberate delays in some school districts compelled the Supreme Court to order immediate desegregation of public schools in the South 15 years after having first ordered their integration in Brown v. Board of Education[245]

The cultural and social situation in NASA's Southern centers apparently did not differ too much from their surrounding environment, according to some NASA employees. At least in the 1960s, some NASA employees at Kennedy Space Center, for example, had a hard time embracing some of their new minority colleagues, likely also in part due to their upbringing in the South.[246] Keeping in mind that John F. Kennedy and Lyndon B. Johnson, had intended to help improve race relations in the South through the placement of the space centers, NASA's poor EEO performance in the Southern installations is especially interesting.[247] While it is understandable that minority employees would rather work in areas of the United States farther removed from Jim Crow, where they would possibly receive a warmer welcome from the local community, it is not quite clear why NASA, when it finally started its systematic EEO program in 1971, did so without paying special attention to the historical obstacles to EEO in the South.[248]

In contrast, NASA's centers in California performed exceptionally well in regard to EEO. The Jet Propulsion Laboratory especially regularly outperformed NASA's other field installations. In a 1978 comparison of JPL and NASA EEO statistics, the Jet Propulsion Laboratory substantially exceeded NASA in all categories of minority employment and also employed considerably more women as managers, officials, and technicians. While NASA only averaged three percent female managers and officials in its workforce, JPL had hired 11.7 percent. JPL also employed 10.8 percent female technicians, while the NASA average was only 2.8 percent.[249] In another comparison of the Jet Propulsion Laboratory,

---

245 Perry A. Zirkel, Sharon Nalbone Richardson and Steven S. Goldberg, *A Digest of Supreme Court Decisions Affecting Education*, 4th edn (Bloomington: Phi Delta Kappa Educational Foundation, 2001), pp. 144–145.

246 For example: Linda Rogers, interview by Dr. Aaron Gillette, July 27, 1992, p. 30, Tape Transcript, HRC; Ernie Reyes, interview by Dr. Aaron Gillette, July 28, 1992, pp. 31–35, Tape Transcript, HRC.

247 For more information, see chapter two: *Mapping the space community*. See also: Paul and Moss; Laney.

248 McQuaid, 'Race, Gender, and Space Exploration: A Chapter in the Social History of the Space Age', p. 409; Shetterly, p. 230.

249 *Comparison with JPL of NASA Workforce as of 9/3078*, May 21, 1979, 1979–1981 EEO, Box 2, EO 1979, Center Files, JSC.

Ames Research Center, and Dryden Flight Research Center with NASA, the California based centers clearly excelled in most of the 30 categories.[250] JPL had the highest percentage of minority (9.5 percent) and female (11.7 percent) managers, as well as the most female technicians (10.8 percent), while Dryden employed the most minority professionals (14.8 percent), and Ames counted the most female professionals (12.2 percent) and minority technicians (16.5 percent). Furthermore, Dryden had hired the most women (95.7 percent) and minorities (23.9 percent) in the clerical and administrative field. Overall, JPL had the most diverse workforce in regard to minorities (15.2 percent) and Ames had the most diverse workforce in regard to women (19.3 percent).[251]

One of the reasons, Dryden, Ames, and, of course, JPL were on top of their EEO employment was their favorable geographical location in California, a state with 19 million inhabitants in the mid-1960s. A lot of African and Native Americans and especially Asian and Hispanic Americans already lived in the state and kept on coming, turning California into the first majority-minority state in the U.S. by 2000. Unlike other centers such as Goddard Space Flight Center, the California centers did not have to rely much on workers from out of state. Goddard, for one, had difficulties attracting many minorities and women because of its location in the country as well as in the state. In addition, to not having to lure good personnel away from their home states, the fact that a lot of renown research universities and other companies related to the space industry had settled in California gave JPL and other West Coast centers an advantage, since the resource pool of minorities working in the desired fields of engineering and science was considerably deeper. Thus, the cultural diversity and the booming industry in California helped to attract the widely courted minority and female professionals.[252]

250 *Comparison with JPL of NASA Workforce as of 9/3078*, June 18, 1979, 1979–1981 EEO, Box 2, EO 1979, Center Files, JSC. DFRC is named the Armstrong Flight Research Center since 2014.

251 The NASA average only prevailed in three categories. NASA employed the most Native American professionals (1 percent) and had the most African American clerks (16.3 percent), because of which it also had the most African Americans in its total workforce (5.8 percent). *A further Comparison of JPL, with NASA, ARC, and DFRC, as of 9/30/78*, June 18, 1979, 1979–1981 EEO, Box 2, EO 1979, Center Files, JSC; *Comparison with JPL of NASA Workforce as of 9/3078*, June 18, 1979, 1979–1981 EEO, Box 2, EO 1979, Center Files, JSC.

252 *Statement of Dr. John F. Clark before the Subcommittee on Congressional Rights and Constitutional Rights of the House Committee on the Judiciary*, March 1974, pp. 7–9, Harris/MSRC; *Some Factors about JPL's EO Program which may account for their successful Accomplishments*, June 18, 1979, 1979–1981 EEO, Box 2, EO 1979, Center Files, JSC; Kevin Starr, *California: A History* (New York: Modern Library, 2005), pp. x, xiii, 245–270; Peter Schrag, *California: America's*

Location was not, however, the only factor in turning the Jet Propulsion Laboratory into a successful EEO employer – JPL also started its EEO efforts two to three years before the rest of NASA. This gave JPL a little head start in EEO provisions. All EEO efforts were carefully planned out and always accompanied by persistent management attention to the accomplishment of the set EEO goals. This included swift conflict resolutions and more flexible working hours.[253] JPL had actually defined four generations of EEO growth which were applicable to all NASA Centers. The first stage was to add and assimilate minorities in the work force. The second stage stipulated upward mobility and career training for minorities and women. The third stage paid attention to the retention and promotion of minorities and women in supervisory and management positions and the fourth stage was to further the sophistication of Office of Federal Contract Compliance (OFCC) definitions and enforcement. While the first two stages were ever present and had to be kept up continuously, the Jet Propulsion Laboratory positioned itself in stage three and four already in the mid-1970s, when the other NASA installations, especially in the South, were still mainly preoccupied with stage one.[254] So while other centers were still looking for minority and female employees, JPL had already created various special programs which found favor with women and minorities, such as academic part time employment or the summer employment plans.[255]

Even though it has to be emphasized that the Jet Propulsion Laboratory, as the only NASA field center that is not federally owned, had some benefits such as independence from some civil service restrictions, it also had a better, well-rounded EEO plan than the other centers from the get-go. Because the EEO performance of the field installations differed so much, the EEO personnel at headquarters always encouraged the centers to "share lessons learned" and "best practices" within the community, the "NASA family."[256] However, sometimes the EEO officers also used to compare the centers' EEO efforts as a means of applying pressure in hopes of encouraging an improvement by pitting the centers against each other in terms of their EEO performance, with remarks

High-Stakes Experiment (Berkeley: University of California Press, 2008), p. 2; Holt, *Rise of the Rocket Girls: The Women Who Propelled Us, from Missiles to the Moon to Mars*, p. 82.

253 *Some Factors about JPL's EO Program which may account for their successful Accomplishments*, June 18, 1979, 1979–1981 EEO, Box 2, EO 1979, Center Files, JSC; Holt, *Rise of the Rocket Girls: The Women Who Propelled Us, from Missiles to the Moon to Mars*, p. 204.

254 *Minutes of EO Council Meeting at Lewis Research Center*, September 22–23, 1976, Box 2, Equal Opportunity Council – Meeting Minutes 1975–76, Center Files, JSC.

255 *Some JPL Special Programs*, June 18, 1979, 1979–1981 EEO, Box 2, EO 1979, Center Files, JSC.

256 Harriett G. Jenkins, interview by Jennifer Ross-Nazzal, August 5, 2011, p. 44, JSC.

such as Johnson Space Center "has the weakest of the three manned space center EEO programs."[257] The pressure of the comparisons was sometimes necessary to demonstrate that it was within the realm of possibility to employ more minorities and women in all kinds of positions.

So while JPL was already reviewing its personnel records in 1974 to check if some women and minorities already on the payroll had been assigned improper jobs not in accordance with their education, Goddard Space Flight Center was still busy making excuses that they "were simply current, not pioneering" in matters of EEO.[258] John F. Clark, the Director of Goddard, for example, exculpated his center's EEO performance by emphasizing that they had only hired according to talent and experience because "in those hectic years nobody had time to practice discrimination."[259] The EEO pressure tactics certainly must have gotten to Clark at Goddard, because he overcorrected his previous omissions by requesting hiring explanations from his personnel in cases when they did not hire a woman or a minority.[260] Although the Civil Service Commission generally thought that this provision was "good affirmative action," it took issue with the fact that Clark only wanted an explanation when the successful candidate was a white male. The CSC thus demanded that hiring explanations be provided regardless of the race, ethnic origin, or sex of the person selected in order to make sure that the successful candidate fit the qualifications and merit needed for the position.[261]

In a letter to Goddard in 1974, Deputy Director of Marshall Space Flight Center William R. Lucas shared some of Marshall's EEO provisions with the team at Goddard, and thereby inadvertently revealed another problem. Marshall's outreach activities included, for example, assisting a children's dance group with giving a performance in Huntsville by providing transportation. The center also furnished and installed a generator for the City of Triana in their

257 Ruth Bates Harris, *Quarterly Reports from the Manned Space Centers in regard to the Center Affirmative Action Plan*, September 12, 1972, 18508, Box 1, HRC.

258 *Attachment to memorandum to NASA EO Council members, management actions at JPL which have contributed to the effectiveness of the JPL affirmative action program*, August 16, 1974, Box 1, Equal Employment 1974, Center Files, JSC; *Statement of Dr. John F. Clark before the Subcommittee on Congressional Rights and Constitutional Rights of the House Committee on the Judiciary*, March 1974, pp. 3–4, Harris/MSRC.

259 *Statement of Dr. John F. Clark before the Subcommittee on Congressional Rights and Constitutional Rights of the House Committee on the Judiciary*, March 1974, pp. 3–4, Harris/MSRC.

260 Jon Cramer, 'CSC Takes Issue with Goddard Hiring Policy', *Evening Star*, March 21, 1973, p. A2.

261 Jon Cramer, 'CSC Takes Issue with Goddard Hiring Policy', *Evening Star*, March 21, 1973, p. A2.

water system after a recent tornado and assisted the city in obtaining buildings for their fire station and daycare center. It also helped women and minorities find employment in the community and gave a local black college material and equipment for a TV studio, FM broadcasting, a computer and a cafeteria.[262] While Marshall's minority community outreach provisions were most certainly laudable, there is no obvious connection to the space community in Alabama. Marshall's efforts appear more like an act of charity than a step toward mending structural discrimination for minorities; instead of providing transportation or daycare for (potential) Marshall employees, the center assisted with these issues outside of Marshall. The Marshall Space Flight Center's EEO efforts in the early to mid-1970s thus demonstrate that some of the centers often had difficulties prioritizing, as well as a tendency to mistake community outreach for EEO provisions.

Because of their poor EEO performance, some centers decided to get outside help. Johnson Space Center, for example, hired the consultancy firm Arthur D. Little in 1977 to identify, investigate, and recommend how to improve EEO at JSC. The consultants found out that there was "a disparity in the degree of enthusiasm or optimism regarding possible success in this area, as well as in the amount of creativity and innovation explored by the managers to make Equal Opportunity (EO) efforts work despite the many obstacles that are clearly recognized by almost everyone."[263] Furthermore, the consultants revealed that EEO was mainly dictated from upper NASA management and most managers did not think that upper management was interested in their input, although they nevertheless felt properly informed about EEO and had a good relationship with the EEO office. Moreover, most managers had the impression that the minority and female employees were capable and only one in ten thought that there were discriminatory practices in place at Johnson Space Center. Most managers were also not able to detect a significant change in the EEO program over the years. It was evaluated "to be about the same as it was in the past, no better or no worse."[264] Arthur D. Little's findings and resulting recommendations for fixing JSC's poor EEO performance were not very surprising to most NASA employees.[265] They basically suggested more of the

262 W.R. Lucas, *EEO Actions and Programs at MSFC*, letter to Goddard Space Flight Center, May 22, 1974, Box 1, Equal Opportunity, Center Files, JSC.
263 Arthur D. Little, *Organization and Management of Equal Opportunity at the Johnson Space Center NASA*, August 1977, Box 1, Equal Opportunity 1977, Center Files, JSC.
264 Arthur D. Little, *Organization and Management of Equal Opportunity at the Johnson Space Center NASA*, August 1977, Box 1, Equal Opportunity 1977, Center Files, JSC.
265 *Minutes of the EO Council Meeting*, November 29–30, 1977, Box 1, Equal Opportunity 1977, Center Files, JSC.

same: redouble the search for minorities, more information gathering, meetings and training as well as an incentive system. Furthermore, they recommended employing more token minority and female employees in prominent positions who might have a trickle-down effect:

> We, therefore, recommend that JSC redouble its efforts to place minorities and women in positions of high responsibility at the Center. Our recommendation is not that JSC practice 'EO window dressing' but that these top priority new hires be looked upon as catalysts for changing the perceptions and attitudes throughout the Center regarding the availability and suitability of minorities and women for top-level jobs at JSC, and that such a step will not jeopardize the standards, the desire for quality, and out-standing performance at the Center.[266]

Considering that the consultants did not distinguish between minorities and women and their different backgrounds and needs, as well as the fact that they were not able to tell JSC anything they did not already know, it stands to reason that the consultants were able to help Johnson Space Center advance their EEO program. Following the Arthur D. Little report, Fletcher held several informal two and a half hour problem-solving discussions with small groups of managers to work on the issues brought up in the report, but not much more came of it.[267]

This had not been the first time that NASA had asked Arthur D. Little to assist in matters of EEO and affirmative action.[268] NASA had also used other consultants to improve its EEO program, including, for example, Madison Smith.[269] This shows that the agency as well as the individual field installations were open for suggestions and thus very willing to work on their specific EEO problems, at least in the mid- to late 1970s.

7.7 Discrimination within the EEO office

Just like all other workspaces in the space community, the EEO officers and managers were not exempt from experiencing gender- or race-based discrimination. Despite being the defenders and advocates of civil rights and equality, the EEO officers could thus also be the offenders – a problem unique to the EEO

---

266 Arthur D. Little, *Organization and Management of Equal Opportunity at the Johnson Space Center NASA*, August 1977, Box 1, Equal Opportunity 1977, Center Files, JSC.

267 *Letter from James C. Fletcher for Distribution*, July 16, 1977, 18507, Box 4, HRC.

268 McQuaid, 'Race, Gender, and Space Exploration: A Chapter in the Social History of the Space Age', p. 430; McQuaid, '"Racism, Sexism, and Space Ventures": Civil Rights at NASA in the Nixon Era and Beyond', p. 443.

269 Harriett G. Jenkins, interview by Jennifer Ross-Nazzal, August 5, 2011, p. 18, JSC.

office. Thus, the EEO office was not a discrimination free space by default, but the people working in the EEO office had to face discrimination or challenge their prejudices just like their colleagues in the other workspaces in the space community. The most prominent discrimination case, which also made waves in the public sphere and the media, was the termination of Deputy Assistant Administrator for Equal Opportunity Programs Ruth Bates Harris in 1973. However, there were several other instances of the EEO office becoming a space for discrimination. Women in particular had to deal with gender based expectations and negligence from their male co-workers and supervisors. For example, when Low visited Johnson Space Center in May 1974, he spoke to several employees, one of whom was Anita Davis, who inveighed against the location and leadership of the JSC EEO office as well as complained about the fact that there was no informal way of making oneself heard without filing complaints or protests. Although Low did believe her, he completely ignored her concerns due to the female stereotype of the 'nagging woman,' which was revealed in his personal notes: "My guess is that even though her complaint is probably right, I'm not sure whether she is the type of person who could ever be satisfied under any kind of management."[270]

Assistant Administrator for Equal Opportunity Programs Harriett Jenkins' feelings were also privately dismissed as an overreaction by Low in 1974.[271] The very incident Low had brushed off, was the very scenario Jenkins remembered when she was asked about challenges she had to face as a woman of color at NASA in 2011.[272] Jenkins broke out in tears at an EEO Council meeting in 1974 at which she requested that the chairman of the EO council, Don Hearth, be relieved of his position because of a derogatory comment he had made to her. Although she wanted Low to step in, the other people on the council felt that this would create a backlash against EEO.[273] Low consequently did not get involved, on top of which he had no regard for her being upset anyway. Jenkins later revealed that the tears had not welled up because of Hearth's rude remark but because of the anger she was trying to control about what had been said in the meeting. She remembered being treated "so beautifully" at NASA, but that

270 George M. Low, *personal notes No. 120*, May 11, 1974, Box 67, Folder 2, Low/RPI.
271 McQuaid, 'Race, Gender, and Space Exploration: A Chapter in the Social History of the Space Age', pp. 430–431.
272 Harriett G. Jenkins, interview by Jennifer Ross-Nazzal, August 5, 2011, p. 68, JSC.
273 George M. Low, *personal notes No. 137*, February 1, 1975, Box 66, Folder 4, Low/RPI; McQuaid, '"Racism, Sexism, and Space Ventures": Civil Rights at NASA in the Nixon Era and Beyond', p. 443.

incident was the “only time” she could recall “being emotionally upset about anything” in her job in the NASA EEO office.[274] Her treatment and her feelings were simply ignored and she was left to deal with it on her own. Considering that she remembered the incident 37 years later, it must have had a profound impact on her. Still, it was dismissed at the time, either because she did not start rattling cages or because it was just a singular occurrence. In any case, one reason could have been Low’s personal conviction that people should not bring their emotional baggage to work. After all, despite his resentment, Low, an Austrian Jewish refugee, had to work together with men like Wernher von Braun, who 30 years prior had been part of a regime which had classified him and his family as subhuman.[275]

The case of Assistant Administrator for EO Dudley McConnell, however, could not be dismissed that easily because of the frequency of the complaints from female co-workers. McConnell had received a Ph.D. in mechanical engineering from Case Western Reserve University in 1964 and did research for 12 years at Lewis Research Center. Before he became the new Assistant Administrator for Equal Opportunity Programs in April 1973, he was Director of the Scientific and Technical Information Office.[276] He was, overall, a well-educated man. The problem was that McConnell had absolutely no experience in the field of EEO. Even though he was African American, and could thus probably relate to the concerns and struggles of other minority employees, he had no idea how to deal with women on an equal basis. It was not just Ruth Bates Harris, the alleged recipient of the “silent treatment” from McConnell, who had issues with the Assistant Administrator.[277] There were various accusations against McConnell. The Federal Women’s Program Coordinator at Marshall Space Flight Center, Grace Ackerman, for example, referred to McConnell as a

274 Harriett G. Jenkins, interview by Jennifer Ross-Nazzal, August 5, 2011, pp. 68–69, JSC.

275 Bob Ward, pp. 202–203; McQuaid, ‘“Racism, Sexism, and Space Ventures”: Civil Rights at NASA in the Nixon Era and Beyond’, p. 429. According to Ward, von Braun was convinced that Low played a major part in Braun’s unhappy departure from NASA. This action would demonstrate that Low actually gave into his emotional hang-ups, although advising his employees against actions like these. However, this was later disputed by Michael J. Neufeld who emphasized that Low was not involved in von Braun’s resignation. Neufeld, *Von Braun: Dreamer of Space, Engineer of War*, p. 456.

276 George M. Low, *Key NASA Personnel Change: Assistant Administrator for Equal Opportunity Program*, April 16, 1973, Box 1, Equal Opportunity 1973, Center Files, JSC.

277 Grace M. Ackerman, *letter to Lynne Darcy*, November 4, 1973, MC 496 Carton 44, Folder 5, National Organization for Women Collection, Schlesinger. For more information about the friction between Ruth Bates Harris and Dudley McConnell, see chapter seven.

"sexist" and an "egoistical bastard."[278] An African American data scientist at Goddard Space Flight Center, Valerie Thomas, thought that McConnell was incapable of running a good EEO program and was insensitive toward women. Goddard's Hispanic American Personnel Director, Benita Sidwell, was also convinced that McConnell was insensitive toward women. She also figured that he had no leadership qualities, because when McConnell tried to recruit her for the job of being the Federally Employed Women representative, he told her that "she would only have to be concerned with black women because white women are satisfactorily accommodated within the system which now exists."[279] Even though women of color undoubtedly had to overcome more obstacles to become part of the workforce and establish themselves in the workspace, McConnell did not recognize that white women also still needed support because equality had not been achieved. In her goodbye letter to NASA, Diane Graham, a contract compliance officer at headquarters, expressed a similar sentiment. She regarded him as "not sensitive," and furthermore felt that he was even "close to hostile with regard to the concerns of women."[280] Considering that McConnell felt entitled to comment on a woman's breasts on a Skylab 2 poster or that he liked to summon his female employees with a dinner bell and refer to all women as 'girls,' her assessment was shared by a lot of female employees.[281] Graham was thus far from the only person to believe that McConnell was not capable of heading the EEO office because he lacked regard for women, the technical EEO knowledge, the commitment, the leadership abilities, and regard for the law. According to Graham, McConnell had "consistently stated his belief that NASA's previous transgressions of the law have not

---

278 Grace M. Ackerman, *letter to Lynne Darcy*, November 4, 1973, MC 496 Carton 44, Folder 5, National Organization for Women Collection, Schlesinger.

279 George M. Low, *notes*, January 14, 1974, Box 76, Folder 5, Low/RPI.

280 Diane Graham, *letter to George M. Low*, August 12, 1973, Harris/MSRC; Constance Holden, 'NASA: Sacking of Top Black Woman Stirs Concern for Equal Employment', *Science* 182 (1973), pp. 804–807.

281 Ethel L. Payne, news clipping, *Spectrum*, January 31, 1974, Harris/MSRC; Jo DiMaggio, *testimony in front of the committee on the Judiciary U.S. House of Representatives Subcommittee on Civil Rights and Constitutional Rights*, March 14, 1974, Harris/MSRC; F. George Drobka, *Memorandum for the Record*, June 29, 1973, Judiciary Committee, 93rd Congress, Subcommittee on Civil and Constitutional Rights, NASA Hearings Correspondence, Box 37 of HR. BX. 33, RG 233, NARA I; Gayla Salinas, *Testimony of the National Organization for Women before the House Judiciary Subcommittee on Civil Rights & Constitutional Rights*, March 13, 1974, p. 3, Harris/MSRC; Ruth Bates Harris, Samuel Lynn, Joseph M. Hogan and A. Diane Graham, *Statement prepared for Committee on Aeronautical, Space Sciences. Status of Equal Employment Opportunity in the National Aeronautics and Space Administration*, January 24, 1974, p. 8, Harris/MSRC.

hurt it" and that the EEO officers "therefore, cannot expect people in the agency to respond to the law in civil rights."[282]

Men, too, took offense with McConnell. Kennedy Space Center EEO officer Willie D. Gaither even quit because of McConnell's poor people skills and lack of commitment to the EEO program.[283] Robert L. Reecer, a former NASA employee who had worked under McConnell's supervision in the Scientific and Technical Office, also confirmed that McConnell was an "unbearable boss." He actually wrote to Congress "in hopes that somehow this will contribute to the preclusion of Dr. McConnell's harassment of those who have the misfortune to work with him."[284] However, not all men were offended by McConnell's behavior. Even though NASA Deputy Administrator George M. Low knew that McConnell still had "to learn how to communicate with people, and especially with women," he was loyal to McConnell and was convinced that McConnell was very capable and had the right ideas regarding EEO.[285] Most likely because McConnell was a "home grown product," some even considered him to be Low's protégé. Low had, after all, been the one to hire McConnell into NASA as well as into the very job he was now failing at.[286] The Deputy Administrator still wanted McConnell to be part of the NASA Personnel Management Review Committee after the Ruth Bates Harris incidence because it was his belief that McConnell was "eminently qualified." He did not further explain his assessment of his skills, though.[287] Low figured that McConnell just needed some help from him and Fletcher to improve his speed in implementing his EEO ideas: "I think Fletcher and I will have to 'lead him by the hand' for some time to come until he can really run the show on his own."[288] This is probably why Low personally wrote McConnell's statement

282 Diane Graham, *letter to George M. Low*, August 12, 1973, Harris/MSRC; Elizabeth R. Covert, *Resignation of Mrs. Ruth Bates Harris*, memo to James C. Fletcher, October 19, 1973, 13704, Box 34, HRC.

283 George M. Low, *personal notes*, February 19, 1974, Box 61, Folder 2, Low/RPI.

284 Robert L. Reecer, *letter to The Committee Investigating Ruth Bates Harris Case*, November 4, 1973, Judiciary Committee 93rd Congress, Subcommittee on Civil and Constitutional Rights, Box 38 of HR. BX. 33, NASA 1974 Miscellaneous Documents, RG 233, NARA I.

285 George M. Low, *personal notes No. 112*, January 20, 1974, Box 67, Folder 4, Low/RPI.

286 *Testimony of Andrea Diane Graham*, 1974, Harris/MSRC; McQuaid, '"Racism, Sexism, and Space Ventures": Civil Rights at NASA in the Nixon Era and Beyond', p. 432.

287 George M. Low, *letter to Dudley McConnell*, December 17, 1973, Chronological File, Folder 3, Low/RPI.

288 George M. Low, *personal notes No. 110*, December 23, 1973, Box 67, Folder 5, Low/RPI. See also: George M. Low, *personal notes*, October 1, 1973, Box 61, Folder 3, Low/RPI; George M. Low, *personal notes No. 112*, January 20, 1974, Box 67, Folder 4, Low/RPI.

for the Proxmire hearing. He had found McConnell's original statement to be "rather weak" and felt that he needed "to help him do his job" because McConnell was "overwhelmed and overworked."[289]

Eventually, McConnell was transferred to another position within NASA away from EEO and Harriett Jenkins took over from him in August 1974, but at no point did NASA's upper management admit to a transgression or incompetence in EEO matters on McConnell's part.[290] Considering that McConnell was the head of the EEO office, and thus the person primarily responsible for bringing gender equality and affirmative action to NASA and the space community, the agency's handling of him clearly demonstrates that the personal experiences of women in the agency's workspaces were not as important to NASA as their appearance in the statistics. The acceptance of the misogynistic behavior of NASA's head of EEO matters shows the degree of conviction in EEO efforts, especially because McConnell was never disciplined for his behavior toward women but was only transferred because of the public turmoil surrounding Harris' dismissal, even though the agency never admitted that this was the reason for his change in workspace.[291] It also illustrates that minority and female employees were not necessarily brothers and sisters in arms in the fight for equality. Furthermore, McConnell's behavior demonstrates that there was a gender gap at NASA – even at the heart of the very office whose job it was to fill this particular gap.

7.8 Equal space and the space community

Statements of space community officials in the media seemed to make the space community more accessible to women in the late 1950s and 1960s. However, it was not the Cold War fears and speculations that gave women an 'in' in the space community in NASA's nationalism and romantic period; it was the legislative process of the late 1960s and 1970s and the resulting nation-wide change of sociocultural norms that truly helped women and minorities to establish themselves in the workspaces at NASA and its contractors. Several federal laws in favor of gender and racial equality at the workplace were passed or changed in the United States in the 1960s and 1970s. NASA

289 George M. Low, *personal notes No. 112*, January 20, 1974, Box 67, Folder 4, Low/RPI.
290 McQuaid, '"Racism, Sexism, and Space Ventures": Civil Rights at NASA in the Nixon Era and Beyond', pp. 441–442.
291 George M. Low, *personal notes No. 126*, August 18, 1974, Box 67, Folder 1, Low/RPI.

had to apply those laws not merely on paper, but in the actual hiring and work practices in the space community. Although, it was a challenging feat for a community, whose inactivity regarding social justice and equal opportunity had not been questioned on a broad scale in the previous decade, it was a necessary step for the space agency in its pragmatic era in order to show the majority of Americans, who no longer felt that the space program was worth the cost, that space exploration was still a worthy endeavor and a trendsetting industry. Thus NASA established its own EEO office and program. While the program was not a centralized effort in the beginning of EEO at NASA and thus had little effect, the issue of EEO was elevated to a more prominent position within NASA with the establishment of an agency wide EEO office and program. Still, the EEO office faced several organizational and structural hurdles in the first years of its foundation. The main obstacles to equal employment opportunity, however, were the varying degrees of commitment from the different NASA centers and managers; especially centers in the Southern states had difficulties to comply with the EEO policies, while centers in California performed particularly well. There was also a discrepancy within management hierarchy. While upper management was relatively committed to the EEO goals set by the EEO office, middle management was often reluctant to accept that EEO goals should be as important as R&D goals. Furthermore, various employees feared that NASA's EEO policies were not compatible with the space community's merit standards. Some employees even felt EEO was reverse discrimination, while others figured the EEO office would become a ghetto for minority and female employees. Therefore, the EEO officers, management, as well as the employees who were hired because of the new legislation had to do a lot of convincing on behalf of EEO's necessity and importance. Especially minorities and women repeatedly had to prove their skills to their co-workers in a fashion other new hires did not have to, proving that discrimination and gender based scepticism did not disappear overnight. The top-down overhaul of the sociocultural norms and gender codes in the space community was a not so subtle, slow process because changing peoples' minds did not necessarily go hand in hand with changing institutional structures. Like the women who worked in the space community in the 1960s, women and minorities in the 1970s had to face individual gender and race based discrimination, even from within the EEO office. In their struggle for a spot in the space community, women and people of color sometimes even worked against each other. However, unlike their trailblazers in the 1960s, women and minorities now had a space to go to if they felt disrespected and poorly treated because of their race or gender: the EEO office.

8 Lady Astronauts, Lady Engineers, and Naked Ladies and the space community

Considering that not even the very office space which had the sole purpose of breaking through the performance of social conventions and exclusionary customs in the space community and implementing new administrative policies and work routines at NASA, the EEO office, was not necessarily always a space of equality, gives some insight into the struggles and obstacles all people in all of the space community workspaces had to go through in order to challenge and change the established gender norms, performances and relations in the community. Some managers in the EEO office, like Dudley McConnell and most managers and engineers in the space community in the 1970s, still reflected the mindsets and occupational patterns of the period they had been educated in, a segregated America in which women mainly existed to create a comfortable home for their male family members and rarely had the power to give orders to white males about anything. Thus, the implementation of the civil rights elements and affirmative action in the agency's employment program was stolid and lumbering in the early 1970s. NASA was not the only slow agency, though; it was in line with other government agencies and private businesses in its failure to apply EEO properly and in a timely fashion.[1] NASA's commitment to EEO in its early memoranda and management instructions remained mainly lip service and did not necessarily mean that the agency's managers were also willing to commit to EEO when it came to hiring, recruiting, and affirmative action activities.

Although the agency and its contractors produced thousands of EEO progress reports which became bigger and bigger as the years went on, the EEO problems remained almost the same at the end of the 1970s as they had been at the beginning of the decade. There were hiring limitations imposed upon all federal agencies by the Office of Management and Budget (OMB) and an overall decline in NASA's workforce. The disparity in average grades between minorities and non-minorities and between men and women did not become considerably smaller. There were still promotion constraints as a result of average grade ceilings and an underrepresentation of minorities and females in the resource pools to fill managerial and supervisory positions. The agency

---

1 *Statement of Clyde M. Webber, National President American Federation of Government Employees on the U.S. Equal Opportunity Program with particular reference to NASA submitted to the Subcommittee on Congressional Rights and Constitutional Rights of the House Committee on the Judiciary*, March 19, 1974, Harris/MSRC; Myers.

https://doi.org/10.1515/9783110629828-008

still could not compete with salary offers made by private industry to minority and female engineers graduating from college. The agency's managers still lacked the authority to make on-the-spot job offers during recruiting trips and they were also not able to convince most of the co-op students to stay with the agency after the end of their programs. The agency was also not able to reach large numbers of minorities on the civil service registers, particularly in the Southern regions. Furthermore, it was still not able to connect with female candidates on the registers because of the veteran preference and there was almost no progress in the removal of structural discrimination sources such as the creation of a flexible workspace.[2] Most of all, the process of changing people's mindsets regarding women and minorities was extremely slow. Depending on the location and leadership of the field center, some were even slower than others.

Notwithstanding that the managers' opinions about the necessity of EEO improved considerably with the generational change that happened in the community by the 1980s as the founding generation of NASA retired and thus made space for more equality, statistical progress still remained relatively slow.[3] By 1991, twelve percent of NASA employees working in science and engineering jobs were non-minority women, five percent were Asian Americans, four percent were African Americans, and four percent were Hispanic Americans.[4] By 2009, 15.5 percent of employees in these jobs were minority males, 16.7 percent were non-minority females, and 5.5 percent were minority females.[5] By 2012, over 20 percent of NASA engineers were female and about a third of NASA's 18,000 strong civil service employee workforce were women.[6] Considering that women and minorities comprise nearly two-thirds of the American population, their share of the science and engineering jobs remains comparatively low to this day.

2 Robert A. Frosch, Harriett Jenkins, *Equal Employment Opportunity Accomplishment Report for FY 1979*, October 11, 1978, Box 3, Center Files, JSC.

3 McQuaid, '"Racism, Sexism, and Space Ventures": Civil Rights at NASA in the Nixon Era and Beyond', p. 445.

4 McQuaid, '"Racism, Sexism, and Space Ventures": Civil Rights at NASA in the Nixon Era and Beyond', pp. 447–448.

5 Joan S. Burrelli and Jaquelina C. Falkenheim, 'Diversity in the Federal Science and Engineering Workforce', *National Science Foundation*, 2011 <https://www.nsf.gov/statistics/infbrief/nsf11303/> [accessed 6 December 2016]. In 2009, NASA employed 11,494 scientists and engineers, including 1,784 male minority and 1,921 non-minority female as well as 633 minority female employees.

6 Dina Spector, 'Meet The Beautiful Women Who Send Rockets Into Space', *Business Insider*, 2012 <http://www.businessinsider.com/the-women-of-nasa-2012-8?IR=T> [accessed 6 December 2016]; Garver.

However, it has to be pointed out that it does reflect the number of the enrolled female college graduates in the sciences and engineering at the time.[7]

Although still lacking their equal share of jobs in the space community, non-minority and Asian women in particular were able to find their space at NASA and in the space community. African American engineers and scientists, on the other hand, could only increase their numbers by 2.8 percent, from 3.9 percent in the late 1970s to 6.7 percent in 2009.[8] Considering that American society still struggles to pervasively own up to its history of racism and social and cultural suppression of African and Native Americans in the twenty-first century, it cannot come as a surprise that the space community did not do much better. It most certainly did not help that minorities and women were often pitted against each other when they were competing for a job due to their linkage within the EEO framework.

The statistical numbers are low and the obstacles were (and partially still are) immense for women and minorities. It has to be emphasized, though, that there indeed was also a lot of change within the space community between the late 1950s and late 1980s, change which was largely the result of women and minorities who did not accept the dominance of white men – people who took agency in their workspaces in the space community with intense personal effort, not necessarily with the goal to change the social order in- and outside of the space community but to realize their dream of working in a job that catered to their skills and interests. These women and people of color succeeded in establishing themselves in work environments which were dismissive in their initial attitude toward 'the other' – toward them. Even though there was no legislation at NASA and other space companies that outright excluded women and minorities, gender and race based hurdles were integral to the structures of the space community, by, for example, blocking pipeline careers, limiting educational opportunities, obstructing access to social gatherings, the use of excluding language, rituals, and dress codes, or the failure to provide structures that would enable women to combine a family and a career; just as they were an intrinsic component of all societal structures in the United States in the 1960s. Women were often not able to overcome this structural discrimination entirely, but they most certainly tried by using the agency they were able to generate in order to integrate or change the existing structures and attitudes. The women

7 McQuaid, '"Racism, Sexism, and Space Ventures": Civil Rights at NASA in the Nixon Era and Beyond', p. 448; Foster, *Integrating Women into the Astronaut Corps. Politics and Logistics at NASA, 1972–2004*, p. 154.

8 *Comparison with JPL of NASA Workforce as of 9/3078*, June 18, 1979, 1979–1981 EEO, Box 2, EO 1979, Center Files, JSC.

who wanted to become astronauts in the early 1960s demonstrated that much. They pushed their agenda into the limelight in the media space and managed to win over parts of the American public, therewith opening up the prevalent discourse about women in space and in the space community, during a time in American society when it was still quite unusual to see active, assertive women braving ridicule and gender stereotyping in the public sphere.

Structural discrimination was often accompanied by individual discrimination in the space community of the 1960s. While some men were already supportive of their female colleagues, recognizing their skills and positive work attitude, others actively or unknowingly discriminated against the pioneering women, who were perceived as disrupters of the prevalent social order in the space community, but more importantly in American society. Still, the women who worked in the space community offices, labs, and other workspaces were able to use their agency to prove themselves to their co-workers and also to pave the way for workspace equality. Especially the lab women were able to carve out pockets for themselves in the space community. These women, who usually excelled in their areas of expertise, demonstrated that women could be just as capable as their male co-workers, or even more so. Although the lip service of the American government and the NASA leadership in the public space of the early 1960s strongly indicated that the female lab workers received a considerable boost in agency in the space community due to the Cold War competition with the Soviet Union and the subsequent enabling of all possible national resources, like female 'mindpower,' the women working in the labs of the space community did not notice that implied impetus to their personal work experiences. The space agency upper management also already emphasized racial and gender equality in the early 1960s, sometimes even exceeding federal legislation's requirements of NASA, but the women in the agency's workspaces also did not directly register the leaderships' forward-thinking lip service in their daily work routines.

The women working for NASA in the 1960s mainly had to make their place in the space community themselves, without the backing of regulations that might have provided them with more agency. Thus, the women who did manage to exercise agency in their workspace were sometimes cautious of those women who entered the space community or were promoted in the EEO framework, established in the 1970s. NASA's commitment to diversity and equality in its workforce and the installment and enactment of a comprehensive EEO program mirrored the social transformation within American society and the American legislative process in the 1970s and 1980s. Civil Rights and women's groups had fought for these equality laws and women and minorities mostly welcomed them. Still, most women in the space community did not consider themselves feminists and did not acknowledge a possible boost they might have received in their careers due to EEO and affirmative action regulations. They referred only to their merit and skills when they were asked why they got the job or the promotion. Although their skills undoubtedly were the reason why the women received the tender for the job, EEO regulations made those women more visible to their bosses. Considering that the selection criteria for merit had been developed by white men, therewith favoring skills and qualities mainly exhibited by white men, the EEO policies that enabled women and minorities to advance into all areas in the workspaces of the space community were necessary. While the female scientists and astronauts usually tried to keep their distance from feminism and the women's movement at their workspace in order to be solely recognized for their skills, some of the women working in administration participated in women's groups, which assisted them in gaining more agency at the office space in which their work had often been treated as a matter of course.

The sociocultural coding of masculinity and femininity in the space community in the different decades, thus, mostly mirrors the changing social order, conventions, and policies as well as the complex, and sometimes even conflicting, cultural adaptions of American society between the 1960s and 1980s. Women did exercise agency in the discourse about women in space and in workspaces previously closed to them, not because the space community was especially accommodating toward these pioneering women from the get-go due to a Cold War boost, but because they had defied gender stereotypes and discrimination in their education, work and private lives all on their own. They had to do the same as their fellow women in workspaces in other areas of business and industry in the United States, persisting in isolation in male-dominated work environments. They had to fight the prevailing expectation that they, as representatives of the 'fair sex,' would eventually fail and drop out and they also had to ignore the fact that most people in American society would feel affirmed and more comfortable if they actually did. It was only in the late 1960s that women found their collective voice and started to press for political change in the public space. Female students, underemployed women, politically frustrated women, and economically exploited women from different occupations and origins started to publicly express their dissatisfaction with the status quo, which always linked their professional status to their sex. The momentum grew and the gender awareness started to unfold in the public sphere. The news media increasingly reported about feminist causes and sexist behaviors of people and organizations that would have passed unnoticed in the previous decade. In this context, the space community also became a target for criticism in the media space. With the nation's eyes all on the space agency, NASA was pressured to change its workspace composition and atmosphere in regard to race and gender, including its sociocultural coding of masculinity and femininity, in order to keep its political and societal support, which the space community needs to perform properly.

NASA has to repeatedly adjust its public image according to the always changing political, social and cultural norms and expectations in the United States over the decades, because all government agencies need political support to function. Considering the National Aeronautics and Space Administration is contingent upon the public and media for rallying political support, the agency and space exploration in general needs to be portrayed in a favourable light. Thus, NASA frequently had to change its approach regarding women and minorities according to the three distinct image periods NASA used in the twentieth century. While NASA's image was molded along the lines of the 'nationalism' and 'romantic' period in the 1950s and 1960s, there was not much room for women's issues in the space community because the public expected NASA to beat the Soviet Union at all costs before the decades of the 1960s ended. However, in NASA's 'pragmatic' era during the 1970s and 1980s, the space administration was forced by the public and federal legislation to recognize the shift in the sociocultural norms and values regarding race and gender in U.S. society. This is why NASA decreed a top down EEO program and marketed it in its public image campaign.

Even though most of NASA was not pioneering in bringing gender equality to its workspaces through special programs or an attempt to alter its social structures or individual attitudes from the get go, and mainly reflected the social transformation visible in all parts of American society, the women in the community nevertheless were instrumental in consolidating this very social transformation within the space community. These women used their agency to break through established social conventions and structures by wearing pants to work, refusing to perform assignments not recorded in their work contracts, earning more money than their husbands, demanding valuation, campaigning for promotions, pointing out gender based inequities, standing up to sexism, making themselves into indispensable experts, creating new workspaces, or by simply asking questions.

The complexity of social and workspace equality for women (and for minorities) in the space community was just as intricate and multilayered as it was in other communities in the United States. However, due to the space community's important role in Cold War politics, NASA's independence to develop the space program and its workspaces was undermined. The presidential directive to go to the moon put the agency on a tight schedule, which could have resulted in a more innovative, female friendly work culture at NASA, due to the utilization of female 'mindpower,' a decade before the rest of the United States. Instead, NASA mainly opted for the established cultural structures, which the space minded women had to breach, alongside the women from other communities and workspaces.

Today, most of the gender based obstacles have been removed in the space community; the space ceiling should be open. Although there has not been a female NASA Administrator so far (although there was an African American Administrator from 2009 to 2017, Maj. Gen Charles F. Bolden), there have been several female Deputy Administrators. The last four Deputy Administrators, including the current one, Lesa Roe, have been women. The first one to get the job was Shana Dale in 2005; however, her assumption of the office revealed that the social structures in the space community were not quite equal just yet. Her style of leadership and character were viewed differently from her predecessors'. According to Josie Soper,

> she's a definite indication that women have not really broken through what they call the 'glass ceiling' yet, because people treat her differently because she's a woman. She's exhibited a couple of idiosyncrasies, personal idiosyncrasies, that have drawn a great deal of conversation among the working troops. Those idiosyncrasies are no different from some of her male counterparts, okay? But because, I think, they're coming from her, I think they're more subject to scrutiny. I'm not talking in circles; I just don't want to get into the nitty-gritty of the details. It's no different than some of the men that have sat where she's sat have done, but it seems to be drawing a lot more conversation from the workforce, and I think that's because she's a woman.[9]

Seven years earlier, the National Organization for Women had campaigned for Jerrie Cobb to finally fly to space to take part in the study of the aging body Senator John Glenn took part in in 1998. NOW started collecting signatures on a petition, emphasizing that "[s]exism was the only thing that kept Jerrie Cobb out of space in the '60s, and it cannot be allowed to stand in her way."[10] The Cobb campaign in the 1990s was handled similarly to her campaign in the 1960s by NASA, which considered it a public relations problem, instead of actually contemplating the important contribution a study of a female aging body could have contributed to the one-sided findings of Glenn's experiment. Another example of remaining obstacles in the space community is the case of Sue Finley, a Jet Propulsion Laboratory employee who is the longest-serving woman at NASA. She was hired in 1958 as a computer and eventually

9 Josie A. Soper, interview by Rebecca Wright, April 19, 2006, pp. 71–72, Oral History Transcript, NASA Administrators Oral History Project, 19822, HRC.

10 *NOW pushing female Space Ex-Trainee*, October 29, 1998, 000336, HRC; *NOW pushing space mission for female Ex-Trainee*, October 30, 1998, 000336, HRC; Weitekamp, *Right Stuff, Wrong Sex. America's First Women in Space Program*, pp. 180–183.

promoted to the status of an engineer. However, in 2004 NASA degraded Finley's employment, stripping her of the title of engineer and cutting her salary because of a missing college degree.[11]

Although some hurdles remain for women in the space community, as in all areas of American society, the publicly expressed esteem for the 'lady engineers' and 'lady astronauts' has been immense in the last five years. For the first time, NASA selected a gender balanced astronaut class with four women and four men in 2013.[12] Also for the first time, space community women – women of color no less – took center stage in a space related movie. The Hollywood production of *Hidden Figures* portrays Katherine Johnson, Dorothy Vaughan und Mary Jackson as heroes of the space age on a level with the astronaut heroes.[13] Some of NASA's female space pioneers, namely Sally Ride, Mae Jemison, Nancy Roman, Margaret Hamilton, and Katherine Johnson, are available as Lego figures for children who dream of space and science to play with.[14] The picture of Margaret Hamilton standing next to the stacked-up code she and her MIT team had written for the Apollo project went viral on social media, proving the interest and respect people nowadays have for her work in general and as a woman in a male-dominated workspace in the 1960s in particular. Hamilton was honored for her important contributions to the American space effort with a Presidential Medal of Freedom in 2016. Katherine Johnson received hers in 2015, and Sally Ride was posthumously honored in 2013. In comparison, the Apollo 11 astronauts received their Presidential Medals of Freedom already in 1969 and the Apollo 13 astronauts and mission operation team were honored with the highest civil decoration awarded by the president in 1970. While other space community personalities, like John Glenn in 2012 and George Low in 1985 posthumously, also received their Presidential Medal of

---

11 Holt, 'How Sexism Held Back Space Exploration. America's Early Aerospace Engineers Ignored Computers Because They Considered Programming to Be Women's Work'.

12 'New Astronaut Candidates Take Center Stage at the Johnson Space Center', *NASA*, 2013 <https://www.nasa.gov/content/new-astronaut-candidates-take-center-stage-at-the-johnson-space-center> [accessed 23 March 2017]; 'NASA's Latest Class of Astronauts Is 50 Percent Female, and Could Be Heading to Mars', *New York Times*, 2016 <http://nytlive.nytimes.com/womenintheworld/2016/01/11/nasas-latest-class-of-astronauts-is-50-percent-female-and-could-be-heading-to-mars/> [accessed 23 March 2017].

13 Shetterly; Patrick Heidmann, 'Feelgood-Drama "Hidden Figures" Plötzlich Oscar-Kandidat', *Spiegel Online*, 2017 <http://www.spiegel.de/kultur/kino/hidden-figures-unerkannte-heldinnen-filmkritik-a-1132643.html> [accessed 23 March 2017].

14 Sarah Larimer, 'Everything Is Awesome for the Women of NASA Lego Set', *Washington Post*, 2017 <https://www.washingtonpost.com/news/speaking-of-science/wp/2017/03/01/everything-is-awesome-for-the-women-of-nasa-lego-set/> [accessed 23 March 2017].

Freedom after their active time in the space community, it is noticeable that women's contributions to the space program were only ever publicly honored years after their working life.[15] However, it was honored. The women's commendation not only reveals President Barack Obama's appreciation for the contribution of the women in the space community but yet another shift in the social and cultural structures of the United States in the twenty-first century. It remains to be seen if President Donald Trump will backtrack on this path to gender equality in the space community.

15 'John Glenn Given Medal of Freedom', *NASA*, 2013 <https://www.nasa.gov/multimedia/imagegallery/image_feature_2264.html> [accessed 28 January 2017]; 'Katherine Johnson Receives Presidential Medal of Freedom', *NASA*, 2015 <https://www.nasa.gov/image-feature/langley/katherine-johnson-receives-presidential-medal-of-freedom> [accessed 28 January 2017]; 'Margaret Hamilton, Apollo Software Engineer, Awarded Presidential Medal of Freedom', *NASA*, 2016 <https://www.nasa.gov/feature/margaret-hamilton-apollo-software-engineer-awarded-presidential-medal-of-freedom> [accessed 28 January 2017]; 'President Obama Awards Presidential Medal of Freedom to Sally Ride', *NASA*, 2013 <https://www.nasa.gov/content/president-obama-awards-presidential-medal-of-freedom-to-sally-ride/> [accessed 28 January 2017]; '13 ARE NAMED WINNERS OF MEDAL OF FREEDOM', *New York Times*, 1985 <http://www.nytimes.com/1985/04/09/us/13-are-named-winners-of-medal-of-freedom.html> [accessed 28 January 2017]; '121 – Remarks on Presenting the Presidential Medal of Freedom to Apollo 13 Mission Operations Team in Houston', *The American Presidency Project*, 1970 <http://www.presidency.ucsb.edu/ws/index.php?pid=2471> [accessed 28 January 2017]; Frances Romero, 'The Presidential Medal of Freedom', *Time*, 2009 <http://content.time.com/time/nation/article/0,8599,1915813,00.html> [accessed 28 January 2017].

Abbreviations

AAPIP	Asian American and Pacific Islander Program
ABMA	Army Ballistic Missile Agency
ACLU	American Civil Liberties Union
AFB	Air Force Base
AFGE	American Federation of Government Employees
AFPA	Armed Forces Procurement Act
AIAA	American Institute of Aeronautics and Astronautics
AIS	American Interplanetary Society (later American Rocket Society)
ALT	Approach and Landing Tests
ARC	Ames Research Center
ARPA	Advanced Research Project Agency
ARS	American Rocket Society
ASME	The American Society of Mechanical Engineers
AST	Aerospace Technology
BuAer	U.S. Navy Bureau of Aeronautics
CalTech	California Institute of Technology
CBS	Columbia Broadcasting System
CDR	Commander
CRS	Congressional Research Service
CSC	Civil Service Commission
CSM	Command/Service Module
CSRA	Civil Service Reform Act
CVT	Concept Verification Test
DFRC	Dryden Flight Research Center
DoD	Department of Defense
EEO	Equal Employment Opportunity
EEOAG	EEO Advisory Group
EEOC	Equal Employment Opportunity Commission
E.O.	Executive Order
EO	Equal Opportunity
ERA	Equal Rights Amendment
ESA	European Space Agency
EVA	Extra-vehicular Activity
FAA	Federal Aviation Administration
FAI	Federation Aeronautique International
FERA	Federal Emergency Relief Administration
FEW	Federally Employed Women
FFRDC	Federally funded research and development center (for example: JPL)
FWP	Federal Women's Program
FWPC	Federal Women's Program Committee
FY	Fiscal Year
GI	Term to describe the soldiers of all the U.S. military branches
GISS	Goddard Institute for Space Studies
GPL	General Purpose Laboratory

https://doi.org/10.1515/9783110629828-009

GS	General Schedule (U.S. government pay scale)
GSC	John H. Glenn Space Center
GSFC	Goddard Spaceflight Center
HEW	Department of Health, Education, and Welfare
HQ	NASA headquarters
HRC	NASA Historical Reference Collection
IDIC	Infinite Diversity and Infinite Combinations
IGY	International Geophysical Year
ILC	International Latex Corporation
IV&V	Independent Validation and Verification Facility
IWY	International Women's Year
JFK	John F. Kennedy
JFKPLM	John F. Kennedy Presidential Library and Museum
JPL	Jet Propulsion Laboratory
JSC	Lyndon B. Johnson Space Center
KSC	John F. Kennedy Space Center
LaRC	Langley Research Center
LBJ	Lyndon B. Johnson
LCRD	Lieutenant Commander
LeRC	Lewis Research Center (today GSC)
LM	Lunar Module
LOC	Launch Operations Center (today KSC)
LoC	Library of Congress
LULAC	League of United Latin American Citizens
MAF	Michoud Assembly Facility
MEAN	Minority Employees at NASA
MIT	Massachusetts Institute of Technology
MOCR	Mission Operations Control Room
MOL	Manned Orbiting Laboratory
MSC	Manned Spacecraft Center (today JSC)
MSFC	George C. Marshall Space Flight Center
MSRC	Moorland Spingarn Research Center, Howard University
NAACP	National Association for the Advancement of Colored People
NACA	National Advisory Committee for Aeronautics
NARA I	National Archives, College Park
NARA II	National Archives, Washington, D.C.
NASA	National Aeronautics and Space Administration
NASC	National Aeronautics and Space Council
NASM	National Air and Space Museum
NBL	Niels Bohr Library, College Park
NIH	National Institute of Health
NMI	NASA Management Instruction
NOW	National Organization for Women
NPD	NASA Policy Directive
NSDAP	Nazi Party
NSI	National Space Institute
NSPE	National Society of Professional Engineers

NSS	National Space Society
OFCC	Office of Federal Contract Compliance
OMB	Office of Management and Budget
P-1	Professional 1 (U.S. government pay scale)
PCSW	The Presidential Commission on the Status of Women
PI	Principal Investigator
PL	Public Law
R&D	Research and Development
RADM	Rear Admiral (Upper Half, 2 Star)
RAND	Project Research and Development
RBH	Ruth Bates Harris
RCA	Radio Corporation of America
RDML	Rear Admiral (Lower Half, 1 Star)
RIF	Reduction in Force
ROTC	Reserve Officer Training Corps
RPI	Rensselaer Polytechnic Institute
SAS	Small Astronomy Satellite
SCLC	Southern Christian Leadership Conference
SDS	Students for a Democratic Society
SNCC	Student Nonviolent Coordinating Committee
SS	Schutzstaffel (Protection Squadron)
SSC	John C. Stennis Space Center
SSTO	Single-Stage-to-Orbit (vehicle)
STEM	Science, Technology, Engineering, and Mathematics
STS	Space Transportation System
SWE	Society of Women Engineers
TOS	*Star Trek*: The Original Series
TWA	Trans World Airlines
USAAF	United States Army Air Forces
USAF	United States Air Force
USAFR	United States Air Force Reserve
USMC	United States Marine Corps
USN	United States Navy
VfR	Verein für Raumschiffahrt
VRC	Visitor Record Center
WFF	Wallops Flight Facility
WIM	Women in Motion Inc.
WISE	Women in Science and Engineering
WSTF	White Sands Test Facility

Archive Directory

John F. Kennedy Presidential Library and Museum, Boston, Mass.	JFKPLM
Johnson Space Center Archive, University of Clear Lake, Houston, TX	JSC
Library of Congress Archives, Washington, D.C.	LoC
Mills College Archive, Oakland, CA	Mills
Moorland-Spingarn Research Center Archives, Howard University, Washington, D.C.	MSRC
NASA Historical Reference Collection, Washington, D.C.	HRC
National Air and Space Museum Archives, Washington, D.C.	NASM
National Archives and Record Administration, Washington, D.C.	NARA I
National Archives and Record Administration, College Park, MD	NARA II
Niels Bohr Library and Archives, College Park, MD	NBL
Rensselaer Polytechnic Institute Archives, Troy, NY	RPI
Schlesinger Library, Radcliffe Institute, Harvard University, Boston, Mass.	Schlesinger

https://doi.org/10.1515/9783110629828-010

Bibliography

'121 – Remarks on Presenting the Presidential Medal of Freedom to Apollo 13 Mission Operations Team in Houston', *The American Presidency Project*, 1970 <http://www.presidency.ucsb.edu/ws/index.php?pid=2471> [accessed 28 January 2017]

'13 ARE NAMED WINNERS OF MEDAL OF FREEDOM', *New York Times*, 1985 <http://www.nytimes.com/1985/04/09/us/13-are-named-winners-of-medal-of-freedom.html> [accessed 28 January 2017]

Ackmann, Martha, *The Mercury 13. The Untold Story of Thirteen American Women and the Dream of Space Flight* (New York: Random House, 2003)

Agle, D.C., 'Playmates on the Moon', *Playboy*, 41 (1994), 138, 213

Altschuler, Glenn, and Stuart Blumin, *The GI Bill: The New Deal for Veterans* (Oxford: Oxford University Press, 2009)

Anderson, Benedict, *Imagined Communities: Reflections on the Origin and Spread of Nationalism* (London: Verso, 2006)

Anderson, Terry H., *The Sixties*, 2nd edn (New York: Pearson Longman, 2004)

Andrews, James T., and Asif A. Siddiqi, *Into the Cosmos: Space Exploration and Soviet Culture* (Pittsburgh: University of Pittsburgh Press, 2011)

Antrobus, Peggy, *The Global Women's Movement: Issues and Strategies for the New Century* (London, New York: Zed Books, 2004)

Asner, Glen, 'Space History from the Bottom Up: Using Social History to Interpret the Societal Impact of Spaceflight', in *Societal Impact of Spaceflight*, ed. by Steven J. Dick and Roger D. Launius (Washington, D.C.: National Aeronautics and Space Administration Office of External Relations History Division, 2007), pp. 387–406

Atkinson, Joseph D., and Jay M. Shafritz, *The Real Stuff. A History of NASA's Astronaut Recruitment Program* (New York: Praeger Publishers, 1985)

Attebery, Brian, *Decoding Gender in Science Fiction* (New York: Routledge, 2002)

Bacal, Vera, 'The Latest Fashion News from Moscow', *New York Times Magazine*, 20 November 1960, p. 125

Ball, Aimee Lee, 'When Mom Is an Astronaut', *Congressional Record Daily Edition*, 131 (1985)

Baumann, Zygmunt, *Gemeinschaften: Auf Der Suche Nach Sicherheit in Einer Bedrohlichen Welt*, (Frankfurt a. M.: Suhrkamp, 2009)

Baumann, Zygmunt, 'Utopia with No Topos', *History of the Human Science*, 16 (2003), 11–15

Baxandall, Rosalyn, and Linda Gordon, *Dear Sisters: Dispatches From The Women's Liberation Movement* (New York: Basic Books, 2000)

Begley, Sharon, 'Down to Earth', *Newsweek*, 1996, pp. 30–36

Belz, Herman, *Equality. A Quarter-Century of Affirmative Action* (New Brunswick: Transaction Publishers, 1992)

Benjamin, Marina, *Rocket Dreams* (London: Chatto & Windus, 2003)

Betson, Johnnie, and Robert R Secrest, 'Prospective Women Astronaut Selection Program. Rationale and Comments', *American Journal of Obstetrics & Gynecology*, 88 (1964), 421–23

Bibby, Cece, 'Friendship 7 with Cece Bibby' <http://freepages.genealogy.rootsweb.ancestry.com/~cecebibby/friendship-7/friendship_7.htm> [accessed 18 July 2016]

Bibby, Cece, 'Story Time: A Naked Lady' <http://freepages.genealogy.rootsweb.ancestry.com/~cecebibby/nasa-stories/naked-lady.htm> [accessed 18 July 2016]

https://doi.org/10.1515/9783110629828-011

Biddle, Wayne, *Dark Side of the Moon. Wernher von Braun, the Third Reich, and the Space Race* (New Yok: W. W. Norton & Company, 2012)

Billings, Linda, 'Cultural Functions of Space Exploration', in *Remembering the Space Age*, ed. by Steven J. Dick (Washington, D.C.: National Aeronautics and Space Administration, 2008), pp. 409–12

Billings, Linda, 'Eilene Galloway, the Woman Who Helped Create NASA, Dies at Age 102', 2009 <http://www.nasa.gov/topics/history/features/galloway_obit.html> [accessed 17 March 2016]

Billings, Linda, 'Overview: Ideology, Advocacy, and Saceflight: Evolution of a Cultural Narrative', in *Societal Impact of Spaceflight*, ed. by Steven J. Dick and Roger D. Launius (Washington, D.C.: National Aeronautics and Space Administration Office of External Relations History Division, 2007), pp. 483–500

Bilstein, Roger E., *Orders of Magnitude. A History of NACA and NASA, 1915–1990* (Washington, D.C.: Government Printng Office, 1989)

Bilstein, Roger E., *Stages to Saturn. A Technological History of the Apollo/Saturn Launch Vehicles* (Washington, D.C.: National Aeronautics and Space Administration, 1996)

Bilstein, Roger E., and Mary Amelia Taylor, 'Aerospace', in *The New Encyclopedia of Southern Culture: Volume 22: Science and Medicine*, ed. by James G. Thomas Jr. and Charles Reagan Wilson (Chapel Hill: University of North Carolina Press, 2012), pp. 29–33

Bix, Amy Sue, 'A History of Women's U.S. Engineering Education', *NWSA Journal*, 16 (2004), 27–49

Bix, Amy Sue, 'From "Engineeresses" to "Girl Engineers" to "Good Engineers": A History of Women's U.S. Engineering Education', in *Women in Engineering: Pioneers and Trailblazers*, ed. by Margaret E. Layne (Reston: ASCE Press, 2009), pp. 11–34

Blackhawk, Tony, *Key Concepts in Community Studies* (London: Sage, 2010)

Blackwell, Maylei, *¡Chicana Power!: Contested Histories of Feminism in the Chicano Movement* (Austin: University of Texas Press, 2011)

Blagonravov, Anatoli, 'Spaceship Designer Sergei Korolev', *Space World*, 1968, 9–11

Blanchard, Margaret A., ed., *History of the Mass Media in the United States: An Encyclopedia* (New York: Routledge, 1998)

Blotnick, Scully, *Otherwise Engaged* (Oxford: Facts on File Publications, 1985)

Borman, Frank, and Robert J. Serling, *Countdown* (New York: William Morrow, 1988)

Boyle, Alan, 'Why Sally Ride Waited until Her Death to Tell the World She Was Gay', *NBC News*, 2012 <http://www.nbcnews.com/science/science-news/why-sally-ride-waited-until-her-death-tell-world-she-f908942> [accessed 10 January 2017]

Breines, Winifred, *The Trouble Between Us: An Uneasy History of White and Black Women in the Feminist Movement* (Oxford: Oxford University Press, 2006)

Bridger, Sue, 'The Cold War and the Cosmos: Valentina Tereshkova and the First Woman's Space Flight', in *Women in the Khrushchev Era*, ed. by Melanie Ilic, Susan E. Reid, and Lynne Attwood (New York: Palgrave Macmillan, 2004), pp. 203–21

Bromberg, Joan Lisa, *NASA and the Space Industry* (Baltimore: Johns Hopkins University Press, 1999)

Brownmiller, Susan, *In Our Time: Memoir of a Revolution* (New York: Dell, 1999)

Brumberg, Johanna A., *Die Vermessung Einer Generation: Die Babyboomer Und Die Ordnung Der Gesellschaft Im US-Zensus Zwischen 1940 Und 1980* (Göttingen: Wallstein, 2015)

Brunner, Elgin Medea, *Foreign Security Policy, Gender, and US Military Identity* (New York: Palgrave Macmillan, 2013)

Burgess, Colin, *Liberty Bell 7: The Suborbital Mercury Flight of Virgil I. Grissom* (New York: Springer-Praxis Books, 2014)

Burgess, Colin, *Selecting the Mercury Seven: The Search for America's First Astronauts* (Chichester: Praxis Publishing, 2011)

Burgess, Colin, *Sigma 7: The Six Mercury Orbits of Walter M. Schirra, Jr.* (New York: Springer-Praxis Books, 2016)

Burgess, Colin, and Grace George Corrigan, *Teacher in Space: Christa McAuliffe and the Challenger Legacy* (Lincoln: University of Nebraska Press, 2000)

Burrelli, Joan S., and Jaquelina C. Falkenheim, 'Diversity in the Federal Science and Engineering Workforce', *National Science Foundation*, 2011 <https://www.nsf.gov/statistics/infbrief/nsf11303/> [accessed 6 December 2016]

Burrows, William E., *This New Ocean. The Story of the First Space Age* (New York: Random House, 1998)

Buss, Jared Scott, 'Willy Ley, the Science Writers, and the Popular Reenchantment of Science' (University of Oklahoma, 2014)

Butler, Judith, *Gender Trouble: Feminism and the Subversion of Identity* (New York: Routledge, 1999)

Butrica, Andrew J., 'The Right Stuff: The Reagan Revolution and the US Space Program', in *Remembering the Space Age*, ed. by Steven J Dick (Washington, D.C.: U.S. Government Printing Office, 2008), pp. 121–34

Butsch, Richard, *Media and Public Spheres* (Basingstoke: Palgrave Macmillan, 2007)

Byrnes, Mark E., *Politics and Space. Image Making by NASA* (Westport: Praeger Publishers, 1994)

Canan, James W., 'Conversations with Eile Galloway', *Aerospace America*, January (2008), 12–15

Canning, Kathleen, 'The Body as Method? Reflections on the Place of the Body in Gender History', *Gender and History*, 11 (1999), 499–513

Cargill-Hall, R., 'Civil-Military Relations in America's Early Space Program', in *The U.S. Air Force in Space, 1945 to the Twenty-First Century* (Washington, D.C.: USAF History and Museum Program, 1998), pp. 17–32

Carpenter, Scott M., Gordon L. Cooper, John H. Glenn, Virgil I. Grissom, Walter M. Schirra, Alan B. Shepard, and others, *We Seven: By the Astronauts Themselves* (New York: Simon and Schuster, 1962)

Casper, Monica J., and Lisa Jean Moore, 'Inscribing Bodies, Inscribing the Future: Gender, Sex, and Reproduction in Outer Space', *Sociological Perspectives*, 38 (1995), 311–33

Castillo, Greg, *Cold War on the Home Front. The Soft Power of Midcentury Design* (Minneapolis: University of Minnesota Press, 2010)

Catchpole, John, *Project Mercury. NASA's First Manned Space Programme* (Chichester: Praxis Publishing, 2001)

Chafe, William Henry, *The Unfinished Journey: America Since World War II* (New York: Oxford University Press, 1984)

Chambers, Joseph R., and Mark A. Chambers, *Emblems of Exploration. Logos of the NACA and NASA* (Washington, D.C.: National Aeronautics and Space Administration Office of Communication, 2015)

Cheng, John, *Astounding Wonder. Imagining Science and Science Fiction in Interwar America* (Philadelphia: University of Pennsylvania Press, 2012)

Chuang, Tamara, ‘Denver Artist Kacena Tapped by Sister to Design NASA Patch for Rodent Research IV Mission. The History of How Patches for Space Mission Are Created Is a Mystery’, *Denver Post*, 2017 <http://www.denverpost.com/2017/02/17/nasa-patches-cu-boulder-artist-kacena-spacex/> [accessed 20 February 2017]

Clarke, Arthur C., *Rendezvous with Rama* (London: Gollancz, 1973)

Clary, David A., *Rocket Man: Robert H. Goddard and the Birth of the Space Age* (New York: Hyperion, 2003)

Cobb, Jerrie, *Jerrie Cobb: Solo Pilot* (Sun City Center: Jerrie Cobb Foundation, 1997)

Cobb, Jerrie, and Jane Rieker, *Woman into Space* (Englewood Cliffs: Prentice-Hall, 1963)

Cobble, Dorothy Sue, *The Other Women's Movement: Workplace Justice and Social Rights in Modern America* (Princeton: Princeton University Press, 2004)

Cohen, Anthony P., *The Symbolic Construction of Community* (London: Routledge, 1985)

Collins, Gail, *When Everything Changed: The Amazing Journey of American Women from 1960 to the Present* (New York: Little Brown, 2009)

Collins, Michael, *Carrying the Fire: An Astronaut's Journeys* (New York: Cooper Square Press, 2001)

Connell, R.W., and James W. Messerschmidt, ‘Hegemonic Masculinity: Rethinking the Concept’, *Gender and Society*, 19 (2005), 829–59

Conner, Monroe, ‘NASA's SOFIA Observatory Flies Star Trek Icon and Educators on Science Mission’, *NASA*, 2015 <http://www.nasa.gov/centers/armstrong/features/sofia_flies_star_trek_icon.html> [accessed 8 August 2016]

Corrigan, Grace George, *A Journal for Christa: Christa McAuliffe, Teacher in Space* (Lincoln: University of Nebraska Press, 2000)

Cranny-Francis, Anne, Wendy Waring, Pam Stavropoulos, and Joan Kirkby, *Gender Studies. Terms and Debates* (London: Palgrave Macmillan, 2003)

Cristoforetti, Samantha, ‘There's Coffee in That Nebula’, 2015 <https://twitter.com/search?q=%23dragon%40astrosamantha&src=typd> [accessed 28 April 2015]

Critchlow, Donald, *Phyllis Schafly and Grassroots Conservatism: A Woman's Crusade* (Princeton: Princeton University Press, 2005)

Crouch, Margaret A., *Thinking About Sexual Harassment: A Guide for the Perplexed* (Oxford: Oxford University Press, 2001)

Crouch, Tom D., *Aiming for the Stars. The Dreamers and Doers of the Space Age* (Washington, D.C.: Smithsonian institution Press, 1999)

Cunningham, Walter, *The All-American Boys* (New York: Ibooks, 2003)

D'Amico, Francine, ‘Policing the U.S. Military's Race and Gender Lines’, in *Wives and Warriors. Women and the Military in the United States and Canada*, ed. by Laurie Weinstein and Christie C. White (New York: Praeger, 1997), pp. 199–234

Dark, Taylor E., ‘Reclaiming the Future: Space Advocacy and the Idea of Progress’, in *Societal Impact of Spaceflight*, ed. by Steven J. Dick and Roger D. Launius (Washington, D.C.: National Aeronautics and Space Administration Office of External Relations History Division, 2007), pp. 555–72

Davis, Flora, *Moving the Mountain: The Women's Movement in America since 1960* (New York: Simon&Schuster, 1999)

Dean, Jodi, *Aliens in America. Conspiracy Cultures from Outerspace to Cyperspace* (Ithaca: Cornell University Press, 1998)

Dean, Margaret Lazarus, ‘Warrior and Worriers: Risk, Masculinity, and the Anxiety of Individuality in the Literature of American Spaceflight’, in *Spacefarers. Images of*

Astronauts and Cosmonauts in the Heroic Era of Spaceflight, ed. by Michael J. Neufeld (Washington, D.C.: Smithonian Institution Scholarly Press, 2013), pp. 203–25

Deaux, Kay, and Mary E. Kite, 'Thinking about Gender', in *Analyzing Gender. A Handbook of Social Science Research*, ed. by Myra Marx Ferree and Beth B Hess (Newbury Park: Sage Publications, 1987), pp. 92–117

Decter, Midge, *The New Chastity: And Other Arguments against Women's Liberation* (New York: Coward, McCann & Geoghegan, 1972)

Deegan, Mary Jo, 'Sexism in Space: The Freudian Formula in Star Trek', in *Eros in the Minds's Eye. Sexuality and the Fantastic in Art and Film*, ed. by Donald Palumbo (New York: Greenwood Press, 1986), pp. 209–24

Delanty, Gerard, *Community* (London: Routledge, 2003)

Diamond, Edwin, *The Rise and Fall of the Space Age* (Garden City: Doubleday, 1964)

Dick, Steven J., 'Space, Time, and Aliens. The Role of Imagination in Outer Space', in *Imagining Outer Space. European Astroculture in the Twentieth Century*, ed. by Alexander C. T. Geppert (New York: Palgrave Macmillan, 2012), pp. 27–44

Dickens, Peter, and James S. Ormrod, *Cosmic Society: Towards a Sociology of the Universe* (London: Routledge, 2007)

Dickson, Paul, *A Dictionary of the Space Age* (Baltimore: The Johns Hopkins University Press, 2009)

Disler, Edith A, *Language and Gender in the Military. Honorifics, Narrative, and Ideology in Air Force Talk* (Amherst: Cambria Press, 2008)

Doel, Ronald E., 'Scientists, Secrecy, and Scientific Intelligence: The Challenges of International Science in Cold War America', in *Cold War Science and the Transatlantic Circulation of Knowledge*, ed. by Jeroen van Dongen (Leiden: Brill, 2015), pp. 11–35

Dongen, Jeroen van, ed., *Cold War Science and the Transatlantic Circulation of Knowledge* (Leiden: Brill Academic Publications, 2015)

Douglas, Susan G., *Where the Girls Are: Growing up Female with the Mass Media* (New York: Times Books, 1994)

Dow, Bonnie J., 'Feminism, Miss America, and Media Mythology', *Rhetoric & Public Affairs*, 6 (2003), 127–49

Dubbs, Chris, *Realizing Tomorrow: The Path to Private Spaceflight* (Lincoln: University of Nebraska Press, 2011)

Duke, Charlie and Dottie, *Moonwalker* (Rose Petal Press, 2011)

Dunn, Joe P., and Howard Lawrence Preston, *The Future South. A Historical Perspective for the Twenty-First Century* (Urbana: University of Illinois Press, 1991)

Eckel, Jan, and Samuel Moyn, *The Breakthrough: Human Rights in the 1970s* (Philadelphia: University of Pennsylvania Press, 2014)

Edmonston, Barry, and Charles Schultze, eds., *Modernizing the U.S. Census* (Washington, D.C.: National Academy Press, 1995)

Effrat, Marcia Pelly, 'Approaches to Community. Conflicts and Complementaries', in *The Community: Approaches and Applications*, ed.by Marcia Pelly Effrat (London: Collier Macmillan, 1974), pp. 1–32

Eisenhower, Dwight D., 'Are We Headed in the Wrong Direction?', *Saturday Evening Post*, 1962, pp. 19–25

Eisfeld, Rainer, *Mondsüchtig. Wernher von Braun Und Die Geburt Der Raumfahrt Aus Dem Geist Der Barbarei* (Fernwald: zuKlampen, 2012)

Ekbladh, David, *The Great American Mission: Modernization and the Construction of an American World Order* (Princeton: Princeton University Press, 2010)

Engerman, David C., and Corinna R. Unger, 'Towards a Global History of Modernization', *Diplomatic History*, 33 (2009), 375–85

Escolme-Schmidt, Libbie, *Glamour in the Skies: The Golden Age of the Air Stewardess* (Stroud: History Press, 2009)

Etzioni, Amitai, *The Moondoggle: Domestic and International Implications of the Space Race* (Garden City: Doubleday, 1964)

Evans, Ben, *Space Shuttle Challenger: Ten Journeys into the Unknown* (New York: Springer-Praxis Books, 2007)

Evans, Sara M., *Born for Liberty* (New York: Free Press Paperbacks, 1997)

Faludi, Susan, *Backlash: The Undeclared War Against American Women* (New York: Crown Publishing Group, 1991)

Faludi, Susan, *Stiffed. The Betrayal of the American Men* (New York: William Morrow, 1999)

Farrell, Amy Erdman, *Yours in Sisterhood: Ms. Magazine and the Promise of Popular Feminism* (Chapel Hill: University of North Carolina Press, 1998)

Fass, Paula S., *Outside In: Minorities and the Transformation of American Education* (Oxford: Oxford University Press, 1989)

'FEW History' <https://www.few.org/about-us/few-history/> [accessed 28 June 2016]

Filipink Jr., Richard M., *Dwight Eisenhower and American Foreign Policy during the 1960s: An American Lion in Winter* (Lanham: Lexington Books, 2015)

Foster, Amy, *Integrating Women into the Astronaut Corps. Politics and Logistics at NASA, 1972–2004* (Baltimore: The Johns Hopkins University Press, 2011)

Foster, Amy, 'The Gendered Anniversary: The Story of America's Women Astronauts', *The Florida Historical Quarterly*, 87 (2008), 150–73

Foucault, Michel, 'Andere Räume', in *Stadt-Räume*, ed. by Martin Wentz (Frankfurt a. M.: Campus, 1991), pp. 65–72

Freeman, Jo, *We Will Be Heard: Women's Struggles for Political Power in the United States* (Lanham: Rowman & Littlefield, 2004)

Frey, Marc, *Geschichte Des Vietnamkriegs. Die Tragödie in Asien Und Das Ende Des Amerikanischen Traums*, 10th edn (München: C.H.Beck, 2016)

Fries, Colin, 'Will It Go 'Round in Circles? A Colorful Mosaic of Art Flown in Space', *Quest*, 11 (2004), 12–21

Fries, Sylvia D., *NASA Engineers and the Age of Apollo* (Washington, D.C.: NASA Scientific and Technical Information Program, 1992)

Fuller, R. Buckminster, *Operating Manual for Spaceship Earth*, ed. by Jaime Snyder (Baden: Lars Müller Publishers, 1968)

Gainor, Chris, *To a Distant Day. The Rocket Pioneers* (Lincoln: University of Nebraska Press, 2008)

Galloway, Eilene, 'Sputnik and the Creation of NASA: A Personal Perspective', *NASA*, 2007 <http://www.nasa.gov/50th/50th_magazine/gallowayEsaay.html> [accessed 22 August 2016]

Galloway, Jonathan F., and Marcia S. Smith, 'Eilene M. Galloway (4.5.1906 – 2. 5.2009)', in *Pioneers of Space Law: A Publication of the International Institute of Space Law*, ed. by Stephan Hobe (Leiden, Boston: Brill, 2013), pp. 126–48

Gamson, William, and Andre Modigliani, 'The Changing Culture of Affirmative Action', in *Reseach in Political Sociology*, ed. by Richard Braungart (Greenwich: CT: JAI Press, 1987), pp. 137–77

Garber, Stephen J., and Roger D. Launius, 'A Brief History of the National Aeronautics and Space Administration', in *National Aeronautics and Space Administration (NASA): Background, Issues, Bibliography*, ed. by C.V. Anderson (New York: Nova Science Publishers, 2002), pp. 3–12

Garver, Lori, 'Women, Innovation and Aerospace. Speech at George Washington University', *NASA*, 2012 <https://www.nasa.gov/pdf/632664main_Deputy_Administrator_Lori_Garver_womens.pdf> [accessed 17 November 2015]

Gawdiak, Ihor, and Helen Fedor, *NASA Historical Databook, 1969–1978, Volume 4, NASA Resources* (Washington, D.C.: National Aeronautics and Space Administration, 1994)

Gelb, Joyce, 'Equal Employment Opportunity Policy in the United States', *PS Online*, 2004

Gerovitch, Slava, 'Creating Memories. Myth, Identity, and Culture in the Russian Space Age', in *Remembering the Space Age*, ed. by Steven J. Dick (Washington, D.C.: National Aeronautics and Space Administration External Relations, 2008), pp. 203–36

Gerovitch, Slava, *Soviet Space Mythologies: Public Images, Private Memories, and the Making of a Cultural Identity* (Pittsburgh: University of Pittsburgh Press, 2015)

Gibson, Karen, *Women in Space: 23 Stories of First Flights, Scientific Missions, and Gravity-Breaking Adventures* (Chicago: Chicago Review Press, 2014)

Gieseking, Jen Jack, William Mangold, Susan Saegert, Cindi Katz, and Setha Low, eds., *The People, Place & Space Reader* (New York: Routledge, 2014)

Giles, Cedric, 'The Rocket Societies', *Astronautics*, 60 (1944), 12–13

Gilman, Nils, *Mandarins of the Future: Modernization Theory in Cold War America* (Baltimore: Johns Hopkins University Press, 2003)

Golemba, Beverly E., 'Human Computers: The Women in Aeronautical Research', *Manuscript*, 1995 <http://crgis.ndc.nasa.gov/crgis/images/c/c7/Golemba.pdf> [accessed 8 September 2016]

Graham, Hugh Davis, *Civil Rights and the Presidency. Race and Gender in American Politics 1960–1972* (Oxford: Oxford University Press, 1992)

Graham, Patricia Albjerg, 'Expansion and Exclusion: A History of Women in American Higher Education', *Signs*, 3 (1978), 759–73

Greenfieldboyce, Nell, 'Pageant Protest Sparked Bra-Burning Myth', *Npr*, 2008 <http://www.npr.org/templates/story/story.php?storyId=94240375&from=mobile> [accessed 2 April 2016]

Griffith, Alison, *The National Aeronautics and Space Act: A Study of the Development of Public Policy* (Washington, D.C.: PublicAffairs Press, 1962)

Griswold, Robert L., '"Russian Blonde in Space": Soviet Women in the American Imagination, 1950–1965', *Journal of Social History*, 45 (2012), 881–907

Gutman, David, 'West Virginian of the Year: Katherine G. Johnson', *Charleston Gazette-Mail*, 2015 <http://www.wvgazettemail.com/news/20151226/west-virginian-of-the-year-katherine-g-johnson> [accessed 23 September 2016]

Habermas, Jürgen, *Strukturwandel Der Öffentlichkeit. Untersuchungen Zu Einer Kategorie Der Bürgerlichen Gesellschaft* (Berlin: Suhrkamp Taschenbuch Wissenschaft, 1990)

Hall, Rex D., Bert Vis, and David Shayler, *Russia's Cosmonauts* ([New York]: PraxisPublishing Ltd., 2005)

Hallion, Richard P., *On the Frontier: Flight Research at Dryden, 1946–1981* (Washington, D.C.: NASA, 1984)

Hamilton, Neil A., *The 1970s* (New York: Facts on File, 2006)

Handy, Bruce, 'Glamour with Altitude', *Vanity Fair*, 2014 <http://www.vanityfair.com/news/2002/10/stewardesses-golden-era> [accessed 28 July 2016]

Hannay, Alastair, *On the Public* (New York: Routledge, 2005)

Hardesty, Von, *Black Wings: Courageous Stories of African Americans in Aviation and Space History* (New York: HarperCollins, 2008)

Harford, James J., *Korolev: How One Man Masterminded the Soviet Drive to Beat America to the Moon* (New York: John Wiley and Sons, 1997)

Harlan, David M., *How NASA Learned to Fly in Space. An Exciting Account of the Gemini Missions* (Burlington: Apogee Books, 2004)

Harris, Ruth Bates, *Harlem Princess. The Story of Harry Delaney's Daughter* (New York: Vantage Press, 1991)

Harrison, Deborah, 'Violence in the Military Community', in *Military Masculinities: Identity and the State*, ed. by Paul Higate (Westport: Praeger, 2003), pp. 71–90

Harrison, Deborah, and Lucie Lalibeté, 'Gender, the Military, and Military Family Support', in *Wives and Warriors. Women and the Military in the United* States *and Canada*, ed. by Laurie Weinstein and Christie C. White (Santa Barbara: Praeger, 1997), pp. 35–53

Hartmann, Susan M., *From Margin to Mainstream: American Women and Politics since 1960* (New York: Knopf, 1989)

Hatch, Erin C., 'The National Aeronautics and Space Administration (NASA): History and Organization', in *National Aeronautics and Space Administration (NASA): Background, Issues, Bibliography*, ed. by C.V. Anderson (New York: Nova Science Publishers, 2002), pp. 13–42

Hauser, Gerard A., *Vernacular Voices: The Rhetoric of Publics and Public Spheres* (Columbia: University of South Carolina Press, 1999)

Heideking, Jürgen, Jörg Helbig, and Anke Ortlepp, *The Sixties Revisited: Culture, Society, Politics* (Heidelberg: Universitätsverlag Winter, 2001)

Heideking, Jürgen, and Christof Mauch, *Geschichte Der USA* (Tübingen: A. Francke Verlag, 2008)

Heidmann, Patrick, 'Feelgood-Drama "Hidden Figures" Plötzlich Oscar-Kandidat', *Spiegel Online*, 2017 <http://www.spiegel.de/kultur/kino/hidden-figures-unerkannte-heldinnen-filmkritik-a-1132643.html> [accessed 23 March 2017]

Heinemann, Isabel, 'Introduction', in *Inventing the Modern American Family. Family Values and Social Change in 20th Century United States*, ed. by Isabel Heinemann (Chicago: University of Chicago Press, 2012), pp. 7–30

Heinlein, Robert, 'Science Fiction: Its Nature, Faults, and Virtues', in *The Science Fiction Novel. Imagination and Social Criticism*, ed. by Basil Davenport (Chicago: Advent, 1969), pp. 14–48

Hersch, Matthew H., 'Checklist: The Secret Life of Apollo's "fourth Crewmember"', *The Sociological Review*, 57 (2009), 6–24

Hersch, Matthew H., *Inventing the American Astronaut* (New York: Palgrave Macmillan, 2012)

Hersch, Matthew H., 'Return of the Lost Spaceman: America's Astronauts in Popular Culture, 1959–2006', *Journal of Popular Culture*, 44 (2011), 73–92

Higate, Paul R., '"Soft Clerks" and "Hard Civvies". Pluralizing Military Masculinities', in *Military Masculinities: Identity and the State*, ed. by Paul R Higate (Westport: Praeger, 2003), pp. 27–42

Hilck, Karin, 'The Space Community and the Princess: Reworking the American Space Program's Public Image from "Miss NASA" to Princess Leia', in *A Galaxy Here and Now:*

Historical and Cultural Readings of Star Wars, ed. by Peter W. Lee (Jefferson: McFarland, 2016), pp. 33–61

Hillebrand, Jörg, Bernd Schneider, and Mike Okuda, 'NASA References in Star Trek', *Ex Astris Scientia*, 2015 <http://www.ex-astris-scientia.org/database/nasa_references.htm> [accessed 8 May 2015]

Hillery, George A., 'Definitions of Community, Areas of Agreement', *Rural Sociology*, 20 (1955), 111–23

Hitt, David, Owen K. Garriott, and Joe Kerwin, *Homesteading Space: The Skylab Story* (Lincoln: University of Nebraska Press, 2008)

Hockey, John, 'No More Heroes. Masculinity and the Infantry', in *Military Masculinities: Identity and the State*, ed. by Paul Higate (Westport: Praeger, 2003), pp. 15–26

Hodges, Jim, 'She Was a Computer When Computers Wore Skirts', *NASA*, 2008 <http://www.nasa.gov/centers/langley/news/researchernews/rn_kjohnson.html> [accessed 13 September 2016]

Hohler, Robert T., *I Touch the Future. The Story of Christa McAuliffe* (New York: Random House, 1986)

Holt, Nathalia, 'How Sexism Held Back Space Exploration. America's Early Aerospace Engineers Ignored Computers Because They Considered Programming to Be Women's Work', *The Atlantic*, 2016 <http://www.theatlantic.com/science/archive/2016/06/how-sexism-held-back-space-exploration/486644/> [accessed 20 October 2016]

Holt, Nathalia, *Rise of the Rocket Girls: The Women Who Propelled Us, from Missiles to the Moon to Mars* (New York: Little Brown, 2016)

Hooks, Gregory, 'Guns and Butter, North and South: The Federal Contribution to Manufacturing Growth, 1940–1990', in *The Second Wave: Southern Industrialization from the 1940s to the 1970s*, ed. by Philip Scranton (Athens: University of Georgia Press, 2001), pp. 255–85

Houchin III, Roy F., 'Air Force-Office of the Secretary of Defense Rivalry: The Pressure of Political Affairs in the Dyna-Soar (X–20) Program, 1957–1963', *Journal of the British Interplanetary Society*, 50 (1997), 162–268

Houchin III, Roy F, 'Interagency Rivalry: NASA, the Air Force, and MOL', *Quest: The History of Spaceflight Magazine*, 4 (1995), 40–45

Hoyt, Edwin P., *The Space Dealers. A Hard Look at the Role of American Business in Our Space Effort* (New York: The John Day Company, 1971)

Hoyt, Mary Finch, *American Women of the Space Age* (New York: Athenum, 1966)

HSA Space Exploration and Aviation Auction Catalog #6000 (Dallas: Heritage Air and Space, 2008)

Hubbard, Barbara Marx, *The Hunger of Eve. One Woman's Odyssey toward the Future* (Eastsound: Island Pacific NW)

Hughes, Richard L., 'National Organization for Women', in *Culture Wars: An Encyclopedia of Issues, Viewpoints and Voices*, ed. by Roger Chapman and James Ciment (New York: Routledge, 2014), p. 457

Hunley, J.D., 'The Enigma of Robert H. Goddard', *Technology and Culture*, 36 (1995), 327–50

Huzel, Dieter, *Peenemünde to Canaveral* (Englewood Cliffs: Prentice Hall, 1962)

Jackson, Kasi, 'Gender, Sex, and Sexuality in Speculative Fiction', in *Gender in Science Workbook*, ed. by Kasi Jackson (Dubuque: Kendall/Hunt Publishing, 2006), pp. 135–36

Jacobsen, Annie, *Operation Paperclip: The Secret Intelligence Program That Brought Nazi Scientists to America* (New York: Back Bay Books, 2015)

Jardins, Julie des, *The Madame Curie Complex: The Hidden History of Women in Science* (New York: Feminist Press, 2010)

Jemison, Mae, *Find Where the Wind Goes* (New York: Scholastic Press, 2001)

Jenkins, Dennis R., *Space Shuttle: The History of the National Space Transportation System – 3rd Edition: The History of the National Space Transportation System : The First 100 Missions*, 3rd edn (Forest Lake: Specialty Press, 2001)

Jewett, Andrew, *Science, Democracy, and the American University: From the Civil War to the Cold War* (Cambridge: Cambridge University Press, 2012)

'John Glenn Given Medal of Freedom', *NASA*, 2013 <https://www.nasa.gov/multimedia/imagegallery/image_feature_2264.html> [accessed 28 January 2017]

John, Mariel, 'NASA: Making Space for Women', *MIT Undergraduate Research Journal*, 11 (2004), 19–25

Jonas, Gerald, 'Judith Merril, 74, Science Fiction Editor and Writer', *The New York Times*, 1997 <http://www.nytimes.com/1997/09/17/arts/judith-merril-74-science-fiction-editor-and-writer.html>

Jones, Eric M., 'Apollo 12 CDR Cuff Checklist', 2013 <http://history.nasa.gov/alsj/a12/cuff12.html> [accessed 27 July 2016]

Kane, Paula, and Christopher Chzandler, *Sex Objects in the Sky: A Personal Account of the Stewardess Rebellion* (River Grove: Follett Pub. Co, 1974)

Káraman, Theodore von, and Lee Edson, *The Wind and Beyond* (Boston: Little Brown, 1967)

'Katherine Johnson Receives Presidential Medal of Freedom', *NASA*, 2015 <https://www.nasa.gov/image-feature/langley/katherine-johnson-receives-presidential-medal-of-freedom> [accessed 28 January 2017]

'Katherine Johnson: A Lifetime of STEM', *NASA*, 3013 <http://www.nasa.gov/audience/foreducators/a-lifetime-of-stem.html>

Kaufman, James L., *Selling Outer Space. Kennedy, the Media, and Funding for Project Apollo, 1961–1963* (Tuscaloosa: The University of Alabama Press, 1994)

Kay, W. D., *Defining NASA. The Historical Debate over the Agency's Mission* (Albany: State University of New York Press, 2005)

Kelkar, Kamala, 'Obama Signs Bill Eliminating "Negro," "Oriental" from Federal Laws', *PBS*, 2016 <http://www.pbs.org/newshour/rundown/obama-signs-bill-eliminating-negro-spanish-speaking-oriental-from-federal-laws/> [accessed 11 January 2016]

Kennedy, John F., 'Address before a Joint Session of Congress, May 25, 1961' <http://www.jfklibrary.org/Asset-Viewer/xzw1gaeeTES6khED14P1Iw.aspx> [accessed 1 August 2015]

Kessler-Harris, Alice, *Out of Work: A History of Wage-Earning Women in the United States* (New York: Oxford University Press, 1982)

Kevles, Bettyann Holtzmann, *Almost Heaven. The Story of Women in Space* (Cambridge: MIT Press, 2006)

Khazan, Olga, '"Pushy" Is Used to Describe Women Twice as Often as Men', *The Atlantic*, 2014 <https://www.theatlantic.com/business/archive/2014/05/pushy-is-used-to-describe-women-twice-as-often-as-men/371291/> [accessed 13 March 2016]

Kilgore, De Witt Douglas, *Astrofuturism. Science, Race, and Visions of Utopia in Space* (Philadelphia: University of Pennsylvania Press, 2003)

Kircher, Travis K., 'More Than Just a Merit Badge', *collectSpace*, 2000 <http://www.collectspace.com/resources/patches_astronauts.html> [accessed 20 February 2017]

Kitch, Carolyn, 'Women in Journalism', in *American Journalism: History, Principles, Practices*, ed. by W. David Sloan and Lisa Mullikin Parcell (Jefferson: McFarland, 2002), pp. 87–96
Knight, Damon, *A Century of Science Fiction* (New York: Simon & Schuster, 1962)
Kohonen, Ilina, 'The Space Race and Soviet Utopian Thinking', ed. by David Bell and Martin Parker (Malden: Blackwell Publishing/The Sociological Review, 2009), pp. 114–31
Kojevnikov, Alexei, 'The Cultural Spaces of the Soviet Cosmos', ed. by James T. Andrews and Asif A. Siddiqi (Pittsburgh: University of Pittsburgh Press, 2011), pp. 15–27
Koppel, Lily, *The Astronaut Wives Club* (New York: Gran Central Publishing, 2013)
Kraemer, Sven F., *Inside the Cold War From Marx to Reagan: An Unprecedented Guide to the Roots, History, Strategies, and Key Documents of the Cold War* (Lanham: UPA, 2015)
Kranz, Gene, *Failure Is Not an Option. Mission Control from Mercury to Apollo 13 and Beyond* (New York: Simon & Schuster, 2000)
Kulke, Ulli, *Weltraumstürmer. Wernher von Braun Und Der Wettlauf Zum Mond* (Berlin: Quadriga, 2012)
Kyle, Dave, 'The Legendary Hydra Club', 2000 <http://www.jophan.org/mimosa/m25/kyle.htm> [accessed 18 February 2016]
Ladewig, Nita, 'NITA and NASA: A Case History of Male Chauvinism in the Space Program', *NOW Acts*, 2 (1969), 16–19
Lamphier, Peg A., and Rosanne Welch, eds., *Women in American History: A Social, Political, and Cultural Encyclopedia and Document Collection. Volume 4: Cold War America to Today* (Santa Barbara: ABC-CLIO, 2017)
Laney, Monique, *German Rocketeers in the Heart of Dixie. Making Sense of the Nazi Past During the Civil Rights Era* (New Haven: Yale University Press, 2015)
Langley, Cory V., 'Library Analyst Helped Launch NASA', *Library of Congress Magazine*, May/June (2014), 20–21
Larimer, Sarah, 'Everything Is Awesome for the Women of NASA Lego Set', *Washington Post*, 2017 <https://www.washingtonpost.com/news/speaking-of-science/wp/2017/03/01/everything-is-awesome-for-the-women-of-nasa-lego-set/> [accessed 23 March 2017]
Lathers, Marie, *Space Oddities Women and Outer Space in Popular Film and Culture, 1960–2000* (London: Continuum International Publishing, 2010)
Launius, Roger D., 'Eisenhower, Sputnik, and the Creation of NASA: Technological Elites and the Public Policy Agenda', *Prologue: Quarterly of the National Archives and Records Administration*, 28 (1996), 127–43
Launius, Roger D., 'Heroes in a Vacuum: The Apollo Astronauts as Cultural Icon', *The Florida Historical Quarterly*, 87 (2008), 174–209
Launius, Roger D., 'History of Civil Space Activity and Spacepower', in *Toward a Theory of Spacepower: Selected Essays*, ed. by Peter L. Hays and Charles D. Lutes (Washington, D.C.: National Defense University Press, 2011)
Launius, Roger D., 'Public Opinion Polls and Perceptions of US Human Spaceflight', *Space Policy*, 19 (2003), 163–75
Launius, Roger D., and Dennis R. Jenkins, *Coming Home: Reentry and Recovery from Space* (Washington, D.C.: National Aeronautics and Space Administration, 2012)
Launius, Roger D., and Howard E. McCurdy, 'Epilogue: Beyond NASA Exceptionalism', in *Spaceflight and the Myth of Presidential Leadership*, ed. by Roger D.; McCurdy Howard E. Launius (Champaign: University of Illinois Press, 1997), pp. 221–50

Laville, Helen, *Cold War Women. The International Activities of American Women's Organisations* (Manchester: Manchester University Press, 2002)
Laville, Helen, 'Gender and Women's Rights in the Cold War', in *The Oxford Handbook of the Cold War*, ed. by Richard H. Immerman and Petra Goedde (Oxford: Oxford University Press, 2013), pp. 523–39
Layne, Margaret E., ed., *Women in Engineering: Pioneers and Trailblazers* (Reston: ASCE Press, 2009)
Lefebvre, Henri, *The Production of Space* (Oxford: Blackwell, 1991)
Leslie, Stuart W., *The Cold War and American Science: The Military-Industrial-Academic Complex at MIT and Stanford* (New York: Columbia University Press, 1994)
Levine Braun, Suzanne, and Mary Thom, *Bella Abzug: How One Tough Broad from the Bronx Fought Jim Crow and Joe McCarthy, Pissed Off Jimmy Carter, Battled for the Rights of Women and Workers, Rallied Against War and for the Planet, and Shook Up Politics Along the Way. An Oral History* (New York: Farrar, Straus and Giroux, 2007)
Levine, Arnold S., *Managing NASA in the Apollo Era* (Washington, D.C.: National Aeronautics and Space Administration, 1982)
Levine, Jay, 'Star Trek's Nichelle Nichols' Warp-Speed Visit to Dryden', *NASA*, 2012 <http://www.nasa.gov/centers/dryden/Features/nichols.html#.V6C1aKK2qE0> [accessed 4 August 2016]
Levine, Sol, *Your Future in NASA* (New York: Arco-Rosen Career Guidance Series, 1971)
Levy, Alan H., *The Political Life of Bella Abzug, 1920–1976: Political Passions, Women's Rights, and Congressional Battles* (Lanham: Lexington Books, 2013)
Levy, Lillian, *Space: Its Impact on Man and Society* (New York: Norton, 1965)
Lewis, Cathleen, 'From the Kitchen into Orbit. The Convergence of Human Spaceflight and Krushchev's Nascent Consumerism', in *Into the Cosmos Space Exploration and Soviet Culture*, ed. by James T. Andrews and Asif A. Siddiqi (Pittsburgh: University of Pittsburgh Press, 2011), pp. 213–39
Lewis, Cathleen, 'The Red Stuff: A History of the Public and Material Culture of Early Human Spaceflight in the U.S.S.R' (George Washington University, 2008)
Ley, Willy, 'Count von Braun', *Journal of the British Interplanetary Society*, 6 (1947), 154–56
Limerick, Patricia Nelson, 'Imagined Frontiers: Westward Expansion and the Future of the Space Program', in *Space Policy Alternatives*, ed. by Radford Byerly (Boulder: Westview Press, 1992), pp. 249–73
Linden-Ward, Blanche, and Carol Hurd Green, *American Women in the 1960s: Changing the Future* (New York: Twayne Publishers, 1993)
Llinares, Dario, *The Astronaut: Cultural Mythology and Idealised Masculinity* (Newcastle: Cambridge Scholars Publishing, 2011)
Lockney, Dan, 'NASA Spinoff', *NASA*, 2015 <https://spinoff.nasa.gov/> [accessed 10 July 2015]
Logsdon, John M., *Exploring the Unknown. Selected Documents in the History of the U.S. Civil Space Program. Vol. I: Organizing for Exploration* (Washington, D.C.: NASA History Office, 1995)
Lottmann, Herbert R., *Jules Verne: An Exploratory Biography* (New York: St. Martin's Press, 1996)
Lyon, Larry, *The Community in Urban Society* (Long Grove: Waveland Press, 1999)
MacLean, Nancy, *Freedom Is Not Enough. The Opening of the American Workplace* (Cambridge: Harvard University Press, 2008)
Mahler, Julianne G., and Maureen Hogan Casamayou, *Organizational Learning at NASA. The Challenger & Columbia Accidents* (Washington, D.C.: Georgetown University Press, 2009)

Maier, Mark, and James W. Messerschmidt, 'Commonalities, Conflicts and Contradictions in Organizational Masculinities: Exploring the Gendered Genesis of the Challenger Disaster', ed. by Mark Maier and James W. Messerschmidt, *Canadian Review of Sociology & Anthropology*, 35 (1998), 325–44

Makemson, Harlen, *Media, NASA, and America's Quest for the Moon* (New York: Peter Lang Publishing, 2009)

Mann, Adam, 'Strange Forgotten Space Station Concepts That Never Flew', 2012 <http://www.wired.com/2012/01/space-station-concepts/> [accessed 27 April 2016]

'Margaret Hamilton, Apollo Software Engineer, Awarded Presidential Medal of Freedom', *NASA*, 2016 <https://www.nasa.gov/feature/margaret-hamilton-apollo-software-engineer-awarded-presidential-medal-of-freedom> [accessed 28 January 2017]

Markoff, Helen S., 'The Federal Women's Program', *Public Administration Review*, 32 (1972), 144–51

Martin, Ralph G., *A Hero for Our Time. An Intimate Story of the Kennedy Years* (New York: Basic, 1985)

Martschukat, Jürgen, 'Gewalt in Den USA Der 1960er Und 1970er Jahre: Eine Einleitung', *Amerikastudien/American Studies*, 49 (2004), 321–28

May, Elaine Tyler, 'Explosive Issues: Sex, Women, and the Bomb', in *Recasting America: Culture and Politics in the Age of Cold War*, ed. by Lary May (Chicago: University of Chicago Press, 1989), pp. 154–70

May, Martha, *Women's Roles in Twentieth-Century America* (Westport: Greenwood Press, 2009)

Mazlish, Bruce, *The Railroad and the Space Program: An Exploration in Historical Analogy* (Cambridge: MIT Press, 1965)

McCray, W. Patrick, *The Visioneers: How a Group of Elite Scientists Pursued Space Colonies, Nanotechnologies, and a Limitless Future* (Princeton: Princeton University Press, 2013)

McCullough, Joan, 'The 13 Astronauts Who Were Left behind', *Ms.*, 3 (1973), 41–45

McCurdy, Howard E., 'Has Spaceflight Had an Impact on Society? An Interpretative Framework', in *Societal Impact of Spaceflight*, ed. by Steven J. Dick and Roger D. Launius (Washington, D.C.: National Aeronautics and Space Administration Office of External Relations History Division, 2007), pp. 3–17

McCurdy, Howard E., *Inside NASA. High Technology and Organizational Change in the U.S. Space Program* (Baltimore: Johns Hopkins University Press, 1993)

McCurdy, Howard E., *Space and the American Imagination* (Washington: Smithsonian Institution Press, 1997)

McDougall, Walter A., *The Heavens and the Earth. A Political History of the Space Age* (New York: Basic Books Publishers, 1985)

McGlynn, Lawrence, 'Breaking through the Glass Gantry', 2005 <http://www.collectspace.com/news/news-080705a.html> [accessed 18 July 2016]

McGlynn, Lawrence, 'Cece Bibby: The Mercury Artist', 2013 <http://www.spaceartarchive.com/2013/10/cece-bibby-the-mercury-artist-.html> [accessed 18 July 2016]

McIlwee, Judith, and Gregg Robinson, *Women in Engineering: Gender, Power, and Workplace Culture* (Albany: State University of New York Press, 1992)

McQuaid, Kim, 'Race, Gender, and Space Exploration: A Chapter in the Social History of the Space Age', *Journal of American Studies*, 41 (2007), pp. 405–34

McQuaid, Kim, '"Racism, Sexism, and Space Ventures": Civil Rights at NASA in the Nixon Era and Beyond', in *Societal Impact of Spaceflight*, ed. by Steven J. Dick and Roger D. Launius

(Washington, D.C.: National Aeronautics and Space Administration Office of External Relations History Division, 2007), pp. 421–50

MediaWorldWatch, *Hidden Figures: Katherine Johnson: One of the Black Woman Mathematicians Who Worked with NASA on the Space Race* (North Charleston: CreateSpace, 2017)

Meerman, David, and Richard Jurek, *Marketing the Moon. The Selling of the Apollo Lunar Program* (Cambridge: MIT Press, 2004)

Mergel, Thomas, 'Modernization', *European History Online*, 2012 <http://ieg-ego.eu/en/threads/models-and-stereotypes/modernization/thomas-mergel-modernization> [accessed 21 September 2016]

Merril, Judith, 'The Hydra Club', *Marvel Science Fiction*, November (1951), 128–29

Merril, Judith, and Emily Pohl-Weary, *Better to Have Loved: The Life of Judith Merril* (Between the Lines(CA), 2002)

Mersmann, Katy, 'Former Miss America Walks the Goddard Runway', *Goddard View*, 11 (2015), 8

Meyerowitz, Joanne, *Not June Cleaver: Women and Gender in Postwar America, 1945–1960* (Philadelphia: Temple University Press, 1994)

Michaud, Michael A. G., *Reaching for the High Frontier. The American Pro-Space Movement 1972–84* (New York: Praeger, 1986)

Miller, Brandon Marie, *Dressed for the Occasion: What Americans Wore 1620–1970* (Minneapolis: Lerner Publications Co., 1999)

Miller, Ron, 'Spaceflight and Popular Culture', in *Societal Impact of Spaceflight*, ed. by Steven J. Dick and Roger D. Launius (Washington, D.C.: National Aeronautics and Space Administration Office of External Relations History Division, 2007), pp. 501–12

'Miss America 2013 Mallory Hagan as a NASA Fitness Instructor', 2013 <https://www.youtube.com/watch?v=JGocgsZR8Dg> [accessed 27 June 2016]

'Miss NASA in Apollo 8', *Glenn Research Center*, 1971 <https://archive.org/details/C-1971-2942> [accessed 27 June 2016]

'Miss NASA Lewis Merri C. Fahnenbruck with Apollo Lunar Rock Display', *Glenn Research Center*, 1973 <https://archive.org/details/C-1973-1016> [accessed 27 June 2016]

'Miss NASA Lewis Research Center and Lewis Research Center Band on Agena Trailer', *Glenn Research Center*, 1970 <https://archive.org/details/C-1970-2578> [accessed 27 June 2016]

Morrow, Patrick, 'Those Sick Challenger Jokes', *Journal of Popular Culture*, 20 (1987), 175–84

Mullane, Mike, *Riding Rockets. The Outrageous Tales of a Space Shuttle Astronaut* (New York: Scribner, 2006)

Murphy, Thomas P., *Science, Geopolitics, and Federal Spending* (Heath Lexington: Lexington Books, 1971)

Myers, Polly Reed, *Capitalist Family Values: Gender, Work, and Corporate Culture at Boeing* (Lincoln: University of Nebraska Press, 2015)

NASA, 'Special Emphasis Programs', 2010 <https://www.nasa.gov/offices/oeod/special_programs/Programs.html> [accessed 7 October 2016]

'NASA – Final Frontier Astronauts Land on Star Trek', *NASA*, 2005 <https://www.nasa.gov/vision/space/features/Astros_on_StarTrek.html> [accessed 20 March 2015]

'NASA Space Shuttle Missions Get Hollywood-Style Movie Posters', *The Telegraph*, 2009 <http://www.telegraph.co.uk/news/science/picture-galleries/7511902/NASA-space-shuttle-missions-get-Hollywood-style-movie-posters.html> [accessed 28 July 2016]

'NASA's Latest Class of Astronauts Is 50 Percent Female, and Could Be Heading to Mars', *New York Times*, 2016 <http://nytlive.nytimes.com/womenintheworld/2016/01/11/nasas-latest-class-of-astronauts-is-50-percent-female-and-could-be-heading-to-mars/> [accessed 23 March 2017]

Neufeld, Michael J., *The Rocket and the Reich: Peenemunde and the Coming of the Ballistic Missile Era* (Washington, D.C.: Smithsonian institution Press, 1995)

Neufeld, Michael J., *Von Braun: Dreamer of Space, Engineer of War* (New York: Alfred A. Knopf, 2007)

'New Astronaut Candidates Take Center Stage at the Johnson Space Center', *NASA*, 2013 <https://www.nasa.gov/content/new-astronaut-candidates-take-center-stage-at-the-johnson-space-center> [accessed 23 March 2017]

Newell, Homer E., *Beyond the Atmosphere: Early Years of Space Science* (Washington, D.C.: National Aeronautics and Space Administration, 1980)

Nichols, Nichelle, 'Influence on Whoopi Goldberg', *Makers* <http://www.makers.com/moments/influence-whoopi-goldberg> [accessed 4 August 2016]

Nimmen, Janne van, Leonardo C. Bruno, and Robert L. Rosholt, *NASA Historical Databook, 1958–1968, Volume I, NASA Resources* (Washington, D.C.: National Aeronautics and Space Administration, 1976)

Notaker, Hallvard, David J. Snyder, and Giles Scott-Smith, *Reasserting America in the 1970s: U.S. Public Diplomacy and the Rebuilding of America's Image Abroad* (Manchester: Manchester University Press, 2016)

'NSS Statement of Philosophy', *NSS* <http://www.nss.org/about/philosophy.html> [accessed 17 February 2016]

O'Donnell, Franklin, *Explorer I* (Pasadena: California Institute of Technology, 2007) <https://pub-lib.jpl.nasa.gov/docushare/dsweb/Get/Document-2417

O'Leary, Brian, *The Making of an Ex-Astronaut* (London: Michael Joseph Ltd, 1970)

O'Mara, Margaret Pugh, *Cities of Knowledge: Cold War Science and the Search for the Next Silicon Valley* (Princeton: Princeton University Press, 2005)

Olcott, Jocelyn, *International Women's Year: The Greatest Consciousness-Raising Event in History* (Oxford: Oxford University Press, 2017)

Oreskes, Naomi, and John Krige, eds., *Science and Technology in the Global Cold War* (Cambridge: MIT Press, 2014)

Ortlepp, Anke, 'Flugreisen Und Männlichkeit Nach Dem Zweiten Weltkrieg', in *Väter, Soldaten, Liebhaber: Männer Und Männlichkeiten in Der Geschichte* Nordamerikas. *Ein Reader*, ed. by Jürgen Martschukat and Olaf Stieglitz (Bielefeld: transcript, 2007), pp. 293–312

Ortlepp, Anke, 'Friendly Skies? A Cultural History of Airtravel in Postwar America', *GHI Bulletin*, 115–125 (2007)

'P-489C, Miss Guided Missile, 1955', *Jet Propulsion Laboratory*, 1955 <https://pub-lib.jpl.nasa.gov/docushare/dsweb/Services/Document-808> [accessed 27 June 2016]

Parker, Patsy, 'The Historical Role of Women in Higher Education', *Administrative Issues Journal*, 5 (2015), 3–14

Parsons, Talcott, 'The American Family. It's Relations to Personality and the Social Structure', in *Family, Socialization and Interaction Process*, ed. by Talcott Parsons and Robert F. Bales (New York: Free Press, 1955), pp. 3–33

Paul, Richard, 'How NASA Joined the Civil Rights Revolution. Integration Came to the Nation's Space Agency in the Mid-1960s', *Air & Space Magazine*, 2014 <http://www.airspacemag.

com/history-of-flight/how-nasa-joined-civil-rights-revolution-180949497/> [accessed 20 October 2016]

Paul, Richard, and Steven Moss, *We Could Not Fail. The First African Americans in the Space Program* (Austin: University of Texas Press, 2015)

Peacock, Louisa, 'Being a Pan Am Air Stewardess Was Best Job for a Woman', *Telegraph*, 2013 <http://www.telegraph.co.uk/women/womens-life/9909671/Being-a-Pan-Am-air-stewardess-was-best-job-for-a-woman.html>

Pearlman, Robert, 'Cece Bibby, Artist behind NASA Space Capsule Logos, Dies', 2012 <http://www.foxnews.com/science/2012/11/16/cece-bibby-artist-behind-nasa-space-capsule-logos-dies.html> [accessed 19 July 2016]

Peebles, Curtis, *High Frontier. The U.S. Air Force and the Military Space Program* (Washington, D.C.: Air Force History and Museum Program, 1997)

Penley, Constance, *NASA/Trek. Popular Science and Sex in America* (New York: R.R. Donnelley & Sons, 1997)

Penley, Constance, 'Spaced Out: Remembering Christa McAuliffe', *Camera Obscura: Feminism, Culture, and Media Studies*, 29 (1992), 179–213

Peril, Lynn, *Swimming in the Steno Pool: A Retro Guide to Making It in the Office* (New York: W. W. Norton & Company, 2011)

Perlman, Allison, *Public Interests: Media Advocacy and Struggles Over U.S. Television* (New Brunswick: Rutgers University Press, 2016)

Phelps, J. Alfred, *They Had a Dream: The Story of African-American Astronauts* (Novato, CA: Presidio, 1994)

Philips, Susan U., 'Language Ideologies', in *The Handbook of Discourse Analysis*, ed. by Deborah Tannen, Heidi E. Hamilton, and Deborah Schiffrin (Chichester: Wiley-Blackwell, 2015), pp. 557–75

Pincus, Fred L., 'Discrimination Comes in Many Forms: Individual, Institutional, and Structural', in *Readings for Diversity and Social Justice*, ed. by Maurianne Adams and others (New York: Routledge, 2000), pp. 31–34

Pincus, Fred L., *Reverse Discrimination: Dismantling the Myth* (Boulder: Lynne Rienner, 2003)

'Playboy on the Moon?', *Playboy*, 2016 <https://www.playboy.com/articles/playboy-on-the-moon> [accessed 27 July 2016]

Potter, James M., and Mark D. Varien, 'The Social Production of Communities. Structure, Agency, and Identity', in *The Social Construction of Communities. Agency, Structure, and Identity in the Prehispanic Southwest* (Lanham: AltaMira Press, 2008), pp. 1–17

Powers, Sheryll Goecke, *Women in Flight Research at NASA Dryden Flight Research Center from 1946 to 1995* (Washington, D.C.: NASA History Office, 1997)

'President Obama Awards Presidential Medal of Freedom to Sally Ride', *NASA*, 2013 <https://www.nasa.gov/content/president-obama-awards-presidential-medal-of-freedom-to-sally-ride/> [accessed 28 January 2017]

Rayl, A.J.S., 'NASA Engineers and Scientists-Transforming Dreams Into Reality', *NASA*, 2008 <https://www.nasa.gov/50th/50th_magazine/scientists.html> [accessed 9 January 2017]

Reagan, Ronald, 'Proclamation 5775 of March 7, 1988. Women's History Month', 1988 <https://www.gpo.gov/fdsys/pkg/STATUTE-102/pdf/STATUTE-102-Pg4962.pdf> [accessed 16 November 2016]

'Resolution Adopted by the General Assembly. 3010 (XXVII). International Women's Year' <http://www.un-documents.net/a27r3010.htm> [accessed 16 November 2016]

Rhode, Deborah L., 'Media Images/Feminist Issues', in *Feminism, Media, and the Law*, ed. by Martha A. Fineman and Martha T. McCluskey (Oxford: Oxford University Press, 1997), pp. 8–21

Richers, Julia, *Die Erste Kosmonautin: Valentina Tereškova Und Der Transkontinentale Geschlechterkampf Im Kalten Krieg* (Zürich: Gender in Trans-it, 2009)

Romero, Frances, 'The Presidential Medal of Freedom', *Time*, 2009 <http://content.time.com/time/nation/article/0,8599,1915813,00.html> [accessed 28 January 2017]

Rose, Sonya, *What Is Gender History?* (Cambridge: Polity Press, 2010)

Rosen, Ruth, *The World Split Open. How the Modern Women's Movement Changed America* (New York: Penguin Books, 2000)

Rosenberg, Emily, 'Far Out: The Space Age in American Culture', in *Remembering the Space Age: Proceedings of the 50th Anniversary Conference*, ed. by Steven J. Dick (Washington, D.C.: National Aeronautics and Space Administration. Office of External Relations, 2008), pp. 157–84

Rosenberg, Rosalind, *Divided Lives: American Women in the Twentieth Century* (New York: Farrar, Straus and Giroux, 2008)

Rossiter, Margaret W., *Women Scientists in America. Vol. 2. Before Affirmative Action 1940–1972* (Baltimore: Johns Hopkins University Press, 1995)

Rossiter, Margaret W., *Women Scientists in America. Vol. 3. Forging a New World since 1972* (Baltimore: Johns Hopkins University Press, 2012)

Ross-Nazzal, Jennifer, 'Pioneering Women of Today: Space Travellers', *Sound Historian*, 2011, 23–32

Ross-Nazzal, Jennifer, 'You've Come a Long Way, Maybe. The First Six Women Astronauts and the Media', in *Spacefarers. Images of Astronauts and Cosmonauts in the Heroic Era of Spaceflight*, ed. by Michael J. Neufeld (Washington, D.C.: Smithonian Institution Scholarly Press, 2013), pp. 175–202

Ryan, Cornelius, 'Man Will Conquer Space Soon', *Collier's*, 1952, pp. 23–40

Ryan, Kevin, and James Michael Cooper, *Those Who Can, Teach* (Boston: Wadsworth Cengage Learning, 2010)

Safire, William, 'The Cold War's Hot Kitchen', *New York Times*, 2009 <http://www.nytimes.com/2009/07/24/opinion/24safire.html>

Sage, Daniel, 'Giant Leaps and Forgotten Steps: NASA and the Performance of Gender', *Sociological Review*, 2009, 146–63

Salamon, Milt, '"Goody-Two-Shoes" Glenn Got Eyeful', *Florida Today*, 17 November 1998, p. 12A

Sandler, Harold, and David L. Winter, *Physiological Responses of Women to Simulated Weightlessness: A Review of the Significant Findings of the First Female Bed-Rest Study* (Washington, D.C.: National Aeronautics and Space Administration, 1978)

Santy, Patricia A., *Choosing the Right Stuff. The Psychological Selection of Astronauts and Cosmonauts* (Westport: Praeger Publishers, 1994)

Sargeant, Winthrop, 'Through the Interstellar Looking Glass', *Life*, 1951, 127–40

Saviskaya, Svetlana, 'Gagarin Is Our Everything', *International Affairs: A Russian Journal of World Politics, Diplomacy & International Relations*, 57 (2011), 237–43

Schefter, James, *The Race: The Complete True Story of How America Beat Russia to the Moon* (New York: Anchor Books, 2000)

Schomp, Virginia, *American Voices from the Women's Movement* (New York: Marshall Cavendish Benchmark, 2007)

Schrag, Peter, *California: America's High-Stakes Experiment* (Berkeley: University of California Press, 2008)

Scott, Joan W., 'Gender: A Useful Category of Historical Analysis', *The American Historical Review*, 91 (1986), 1053–75

Shapiro, Fred R., 'Historical Notes on the Vocabulary of the Women's Movement', *American Speech*, 60 (1985), 3–16

Shavit, Uriya, *The New Imagined Community. Global Media and the Construction of National and Muslim Identities of Migrants* (Eastbourne: Sussex Academic Press, 2009)

Shayler, David J., *Gemini. Steps to the Moon* (London: Springer, 2001)

Shayler, David J., and Colin Burgess, *NASA's Scientist-Astronauts* (Chichester: Praxis Publishing, 2007)

Shayler, David J., and Ian A. Moule, *Women in Space. Following Valentina* (Chichester: Praxis Publishing, 2005)

Shchetinkov, Yevgeny S., 'Development of Winged Rockets in the USSR 1930–1939', in *First Steps toward Space*, ed. by Frederick C. Durant and George S. James (Washington, D.C.: Smithsonian Institution Press, 1974), pp. 247–57

Sheehan, Michael, *The International Politics of Space* (New York: Routledge, 2007)

Sheffield, Suzanne Le-May, 'Women's Advancement in Science since WWII', in *Gender in Science Workbook*, ed. by Jennifer Kasi Jackson (Dubuque: Kendall/Hunt Publishing, 2006), pp. 3–20

Sherrod, Robert, 'The Selling of the Astronauts', *Columbia Journalism Review*, 12 (1973), 16–25

Shetterly, Margot Lee, *Hidden Figures: The Untold Story of the African American Women Who Helped Win the Space Race* (London: Harper Collins, 2016)

Shirley, Donna, *Managing Martians* (New York: Broadway Books, 1998)

Siddiqi, Asif A., *Challenge to Apollo: The Soviet Union and the Space Race, 1945–1974* (Washington, D.C.: National Aeronautics and Space Administration NASA History Division, 2000)

Siddiqi, Asif A., 'Cosmic Contradictions. Popular Enthusiasm and Secrecy in the Soviet Space Program', ed. by James T. Andrews and Asif A. Siddiqi (Pittsburgh: University of Pittsburgh Press, 2011), pp. 47–76

Siddiqi, Asif A., 'Spaceflight in the National Imagination', in *Rembering the Space Age*, ed. by Steven J. Dick (Washington, D.C.: National Aeronautics and Space Administration, 2008), pp. 17–36

Slayton, Donald K., and Michael Cassutt, *Deke! U.S. Manned Space: From Mercury to the Shuttle* (New York: Tom Doherty Associates Book, 1994)

Smith, Michael L., 'Selling the Moon: The U.S. Manned Space Program and the Triumph of Commodity Scientism', in *The Culture of Consumption: Critical Essays in American History, 1880–1980*, ed. by T.J. Jackson Lears and Richard Wightman Fox (New York: Pantheon, 1983), pp. 175–209

Smith, Wilson, and Thomas Bender, eds., *American Higher Education Transformed, 1940–2005: Documenting the National Discourse* (Baltimore: The Johns Hopkins University Press, 2008)

Smith, Yvette, 'Katherine Johnson: The Girl Who Loved to Count', *NASA*, 2015 <http://www.nasa.gov/feature/katherine-johnson-the-girl-who-loved-to-count> [accessed 13 September 2016]

Soeters, Joseph L., Donna J. Winslow, and Alise Weibull, 'Military Culture', in *Handbook of the Sociology of the Military*, ed. by Giuseppe Caforio (New York: Springer, 2003), pp. 237–54

'Solid Propellant Chemistry Gains Fitzgerald Transducer', *Lab-Oratory*, January/Fe (1956) <https://beacon.jpl.nasa.gov/mystery-photo/page/6?mopt=1>

'Space' <http://www.oxforddictionaries.com/de/definition/englisch/space> [accessed 31 August 2015]

'Space and Exploration Are Focus of Let's Read! Let's Move! Event in Washington, D.C.', 2013 <http://www.nasa.gov/content/space-and-exploration-are-focus-of-lets-read-lets-move-event-in-washington-dc/#.V3EErKK2qE0> [accessed 27 June 2016]

Spector, Dina, 'Meet The Beautiful Women Who Send Rockets Into Space', *Business Insider*, 2012 <http://www.businessinsider.com/the-women-of-nasa-2012-8?IR=T> [accessed 6 December 2016]

Spiegel, Lynn, 'White Flight', in *The Revolution Wasn't Televised: Sixties Television and Social Conflict*, ed. by Lynn Spigel and Michael Curtin (New York: Routledge, 1997)

Spiller, James, 'Nostalgia for the Right Stuff: Astronauts and Public Anxiety about a Changing Nation', in *Spacefarers. Images of Astronauts and Cosmonauts in the Heroic Era of Spaceflight*, ed. by Michael J. Neufeld (Washington, D.C.: Smithonian Institution Scholarly Press, 2013), pp. 57–80

Spires, David N., *Beyond Horizons: A Half Century of Air Force Space Leadership* (Peterson Air Force Base: Air Force Space Command, 1997)

Spires, David N., and Rick W. Sturdevant, ' '... to the Very Limit of Our Ability... ': Reflections on Forty Years of Civil-Military Partnership in Space Launch', in *To Reach the High Frontier: A History of U.S. Launch Vehicles*, ed. by Roger D. Launius and Dennis R. Jenkins (Lexington: University Press of Kentucky, 2002), pp. 470–501

Stainback, Kevin, and Donald Tomaskovic-Devey, *Documenting Desegregation: Racial and Gender Segregation in Private Sector Employment Since the Civil Rights Act* (New York: Russel Sage Foundation, 2012)

Stanton, Neville A., *Trust in Military Teams* (Boca Raton: CRC Press, 2011)

Starr, Kevin, *California: A History* (New York: Modern Library, 2005)

Stein, Marc, *Rethinking the Gay and Lesbian Movement* (New York: Routledge, 2012)

Stienstra, Deborah, *Women's Movements and International Organizations* (Houndmills, Basingstoke, Hampshire: Macmillan, 1994)

Stoiko, Michael, *Soviet Rocketry. Past, Present, and Future* (New York: Holt, Rineheart and Winston, 1970)

Subtirelu, Nic, 'Beyond Bossy: More on Our Gendered Characterizations of Leadership and Authority', *Linguistic Pulse*, 2014 <https://linguisticpulse.com/2014/05/19/beyond-bossy-more-on-our-gendered-characterizations-of-leadership-and-authority/> [accessed 13 March 2016]

Swenson Jr., Loyd S., James M. Grimwood, and Charles C. Alexander, *This New Ocean. A History of Project Mercury* (Washington, D.C.: National Aeronautics and Space Administration, 1966)

Sylvester, Roshanna P., '"Let's Find Out Where the Cosmonaut School Is". Soviet Girls and Cosmic Visions in the Aftermath of Tereshkova', in *Soviet Space Culture: Cosmic Enthusiasm in Socialist Societies*, ed. by Eva Maurer and others (Basingstoke: Palgrave Macmillan, 2011), pp. 121–38

Sylvester, Roshanna P., 'She Orbits over the Sex Barrier. Soviet Girls and the Tereshkova Moment', ed. by James T. Andrews and Asif A. Siddiqi (Pittsburgh: University of Pittsburgh Press, 2011), pp. 195–213

Talbot, Mary M., *Language and Gender* (Cambridge: Polity, 2010)

'Tennessee State's Lady Engineer. Professor Splits Time between Teaching, Space Project Work', *Ebony*, 1964, pp. 75–82

Tereshkova, Valentina, 'Women in Space', *Impact of Science on Society*, 20 (1970), 5–12

Terkildsen, Nayda, and Frauke Schnell, 'How Media Frames Move Public Opinion: An Analysis of the Women's Movement', *Political Research Quarterly*, 50 (1997), 879–900

Thom, Mary, *Inside Ms: 25 Years of the Magazine and the Feminist Movement* (New York: Holt, 1997)

Tokaty, G.A., 'Foundations of Soviet Cosmonautics', *Spaceflight*, 10 (1968), 335–46

Turner, Frederick Jackson, 'The Significance of the Frontier in American History', in *Major Problems in the History of the American West*, ed. by Clyde A. Milner II (Lexington: D.C. Heath and Company), pp. 2–21

Unger, Corinna R., 'Cold War Science: Wissenschaft, Politik Und Ideologie Im Kalten Krieg', *Neue Politische Literatur*, 51 (2006), 49–68

Vaughan, Diane, 'Autonomy, Interdependence, and Social Control: NASA and the Space Shuttle Challenger', *Administrative Science Quarterly*, 35 (1990), 225–57

Vis, Bert, 'Soviet Women Cosmonaut Flight Assignments 1963–1989', *Spaceflight*, 41 (1999), 474–80

Waechter, Matthias, *Die Erfindung Des Amerikanischen Westens. Die Geschichte Der Frontier-Debatte* (Freiburg: Rombach, 1996)

Wagner, Jens-Christian, ed., *Konzentrationslager Mittelbau-Dora: 1943–1945. Begleitband Zur Ständigen Ausstellung in Der KZ-Gedenkstätte Mittelbau-Dora* (Göttingen: Wallstein Verlag, 2007)

Wainwright, Loudon, *The Great American Magazine: An Inside History of Life* (New York: Knopf, 1986)

Waldock, Michael, and Johanna Omelia, *Come Fly With Us: A Global History of the Airline Hostess* (Portland: Collectors Press, 2003)

Walsh, David M., *The Military Balance in the Cold War: US Perceptions and Policy, 1976–85* (New York: Routledge, 2008)

Walsh, Margaret, 'The Frontier and the West: Realities, Myths and the Historians', in *The American West. Visions and Revisions* (Cambridge: Cambridge University Press, 2005), pp. 1–18

Wang, Jessica, 'Scientists and the Problem of the Public in Cold War America, 1945–1960', in *Science and Civil Society*, ed. by Lynn K. Nyhart and Thomas H. Broman (Chicago: The University of Chicago Press, 2002), pp. 323–47

Ward, Barbara, *Spaceship Earth* (New York: Columbia University Press, 1966)

Ward, Bob, *Dr. Space: The Life of Wernher von Braun* (Annapolis: Naval Institute Press, 2005)

Watkins, Elizabeth Siegel, *On the Pill: A Social History of Oral Contraceptives, 1950–1970* (Baltimore: Johns Hopkins University Press, 1998)

Wegner, Phillip, *Imaginary Communities. Utopia, the Nation, and the Spatial Histories of Modernity* (Berkeley: University of California Press, 2002)

Weinstock, Maia, 'Scene at MIT: Margaret Hamilton's Apollo Code', *MIT News*, 2016 <http://news.mit.edu/2016/scene-at-mit-margaret-hamilton-apollo-code-0817> [accessed 9 January 2017]

Weitekamp, Margaret A., *Right Stuff, Wrong Sex. America's First Women in Space Program* (Baltimore: The Johns Hopkins University Press, 2004)

Weitekamp, Margaret A., 'Selling Education in the Shape of a Shuttle', *The Florida Historical Quarterly*, 87 (2008), 210–34

Weitekamp, Margaret A., 'The "Astronautrix" and the "Magnificent Male": Jerrie Cobb's Quest to Be the First Woman in America's Manned Space Program', in *Impossible to Hold. Women and Culture in the 1960s*, ed. by Avital H. Bloch and Lauri Umansky (New York University Press, 2005), pp. 9–28

Werth, Karsten, *Ersatzkrieg Im Weltraum. Das US-Raumfahrtprogramm in Der Öffentlichkeit Der 1960er Jahre* (Frankfurt: Campus Verlag, 2006)

Wetterer, Angelika, 'Konstruktion von Geschlecht: Reproduktionsweisen Der Zweigeschlechtlichkeit', in *Handbuch Frauen- Und Geschlechterforschung*, ed. by Ruth Becker and Beate Kortendiek (Wiesbaden: VS Verlag für Sozialwissenschaften, 2008), ii, 126–36

'What Census Calls Us: A Historical Timeline', *PewResearchCenter*, 2015 <http://www.pewsocialtrends.org/interactives/multiracial-timeline/> [accessed 11 November 2016]

'What Is FEW' <https://www.few.org/about-us/what-is-few/> [accessed 28 June 2016]

White, Steve, 'Star Trek Creator Gene Roddenberry's Ashes to Be Blasted into Deep Space', *Mirror*, 2013 <http://www.mirror.co.uk/news/world-news/star-trek-creator-gene-roddenberrys-1975563>

Whitelegg, Drew, *Working the Skies: The Fast-Paced, Disorienting World of the Flight Attendant* (New York: New York University Press, 2007)

Williams, Miriam F., 'Reimagining NASA: A Cultural and Visual Analysis of the U.S. Space Program', *Journal of Business & Technical Communication*, 26 (2012), 368–89

Winter, Frank H., *Prelude to the Space Age. The Rocket Societies 1924–1940* (Washington, D.C.: Smithsonian Institution Press, 1983)

Winter, Frank H., *Rockets into Space* (Cambridge: Harvard University Press, 1990)

Wolfe, Audra J., *Competing with the Soviets: Science, Technology, and the State in Cold War America* (Baltimore: Johns Hopkins University Press, 2013)

Wolfe, Tom, *The Right Stuff* (New York: Farrar-Straus-Giroux, 1983)

Wong, Wilson W.S., and James Fergusson, *Military Space Power. A Guide to the Issues* (Santa Barbara: Praeger, 2010)

'World Conferences on Women' <http://www.unwomen.org/en/how-we-work/intergovernmental-support/world-conferences-on-women> [accessed 16 November 2016]

Wright, Rebecca, Sandra Johnson, and Steven J. Dick, eds., 'Interview with Michael D. Griffin, Administrator', in *NASA at 50. Interviews with NASA's Senior Leadership* (Washington, D.C.: Office of Communications Public Outreach Division NASA History Program Office, 2008), pp. 1–23

Yano, Christine Reiko, *Airborne Dreams: 'Nisei' Stewardesses and Pan American World Airways* (Durham: Duke University Press, 2011)

Zarefsky, David, *President Johnson's War on Poverty: Rhetoric and History* (Tuscaloosa: University of Alabama Press, 1986)

Zaretsky, Natasha, *No Direction Home: The American Family and the Fear of National Decline, 1968–1980* (Chapel Hill: University of North Carolina Press, 2007)

Zirkel, Perry A., Sharon Nalbone Richardson, and Steven S. Goldberg, *A Digest of Supreme Court Decisions Affecting Education*, 4th edn (Bloomington: Phi Delta Kappa Educational Foundation, 2001)

Acknowledgements

I owe a lot of thanks to a lot of people, both for encouragement and academic support in writing this book. First I would like to thank my parents, Martin and Edith Hagen, for raising me to believe I could achieve anything I set my mind to and supporting me in the quest to do just that in every way possible. Thank you for always being there for me.

I would also like to thank my husband, Felix Hilck, for all the love and support in general, but especially during the time I was working on this book. I could not have done it without you.

I also want to express my gratitude to Prof. Dr. Corinna Unger for helping me bring order to the chaos of ideas that was this project at the very beginning. I would also like to thank Prof. Dr. Marc Frey, Prof. Dr. Anke Ortlepp and Prof. Dr. Isabel Heinemann.

A special thanks goes to Chris Shaw and to my partner in crime Craig Koifman-Taylor and all the archivists and librarians who helped me track down information and sources. Thank you to all my friends, favorite neighbors, the drumming people and the chocolate people.

Furthermore, I would like to thank the FAZIT Foundation, the German Academic Exchange Service, the Walther Blohm Foundation and the German Historical Institute for supporting my research in the U.S. as well as the printing of this book.

Last but not least, I would like to express my special appreciation and thanks to my high school history teacher Jürgen Pach, without whom I might have never known how exciting history can be.

https://doi.org/10.1515/9783110629828-012

Index

https://doi.org/10.1515/9783110629828-013